# Industrial Control
# Systems Design

# Industrial Control Systems Design

**Michael J Grimble**

*Industrial Control Centre, University of Strathclyde,
Glasgow, UK*

**JOHN WILEY & SONS, LTD**
Chichester • New York • Weinheim • Brisbane • Singapore • Toronto

Copyright © 2001 by John Wiley & Sons Ltd,
Baffins Lane, Chichester,
West Sussex PO19 1UD, UK

National 01243 779777
International (+44) 1243 779777
e-mail (for orders and customer service enquiries): cs-books@wiley.co.uk

Visit our Home Page on http://www.wiley.co.uk or http://www.wiley.com

*Other Wiley Editorial Offices*

John Wiley & Sons, Inc., 605 Third Avenue,
New York, NY 10158-0012, USA

Wiley-VCH Verlag GmbH, Pappelallee 3,
D-69469 Weinheim, Germany

Jacaranda Wiley Ltd, 33 Park Road, Milton,
Queensland 4064, Australia

John Wiley & Sons (Canada) Ltd, 22 Worcester Road,
Rexdale, Ontario M9W 1L1, Canada

John Wiley & Sons (Asia) Pte Ltd, Clementi Loop #02-01,
Jin Xing Distripark, Singapore 129809

*Library of Congress Cataloguing-in-Publication Data*

Grimble, Michael J.
    Industrial control systems design / Michael J. Grimble.
      p. cm.
    Includes bibliographical references and index.
    ISBN 0-471-49225-6
    1. Production control.   2. Automatic control.   I. Title.
    TS155.8.G75 2000
    670.42′75 – dc21                          00-059433

*British Library Cataloguing in Publication Data*
A catalogue record for this book is available from the British Library

ISBN 0 47149225 6

Produced from camera-ready copy supplied by the author
Printed and bound in Great Britain by Bookcraft (Bath) Ltd, Midsomer Norton
This book is printed on acid-free paper responsibly manufactured from sustainable forestry,
in which at least two trees are planted for each one used for paper production.

*I am pleased to dedicate this text to my wife Wendy and children Claire and Andrew. It has been a privilege to live in Scotland for almost the last two decades and to work with my colleagues at the University of Strathclyde.*

# Contents

## II State Space and Frequency Response Descriptions 281

# 9   QFT and Frequency Domain Design                                      379

# III   Industrial Applications                                            419

# 10   Power Generation and Transmission                                   421

# 11   Design of Controllers for Metal Processing                              **459**

# Preface

The text provides an overview of different advanced control design methods for industrial systems. In many application areas there are limits on the types of advanced design technique that can be applied, but it is believed that even when classical methods continue to be used, there are lessons to be learned from the analysis, design and synthesis techniques presented. The book is separated into three parts and in the early chapters (2 to 6) the system is assumed to be represented in polynomial systems or frequency domain form. Since the text is mainly concerned with discrete-time systems, the models used are often ARMA models, similar to those employed in the self-tuning control or plant identification literature. The second part of the book considers systems represented in discrete state equation model form, which is very suitable for systems which are defined by sets of physical difference equations. The final part of the text is concerned with different application areas.

The modern control engineer is required to be a systems scientist as much as a regulating loop designer. This is reflected in the broad overview presented in Chapter 1 which discusses intelligent as well as model based robust control methods. The design of fault tolerant control systems and the use of fault-monitoring and detection systems is explored in this chapter and in later chapters the so-called reliable control method is also discussed, where systems can have sensor or actuator failures and still maintain a reasonable performance. An example of reliable control is presented in Chapter 11, based on a hot rolling problem, where the sensor can fail and yet good tension and thickness control may still be maintained.

A good example of the insights $H_2$ optimal methods can provide is in the design of feedforward control systems discussed in Chapter 2. It is surprising that, from a classical design viewpoint feedforward control has not received much attention and in fact most books simply deal with the structure rather than the design of the controller. The result is that, in practice, feedforward controllers are often simplistic and do not provide the benefits which are possible. The optimal control design methods to be presented provide a clear understanding of the way in which feedforward controllers operate and given this understanding, engineers may wish to use classical or other approaches. The benefits of the theoretical techniques introduced are often in the intuitive understanding gained and the measure provided of what could be achievable, given the necessary computing and skills resources.

Predictive control methods have perhaps become the most successful modern control design technique, at least as regards the number of real applications on large multivariable systems. The $H_2$ approach to predictive control is introduced in Chapter 3. The *Generalized Predictive Control* algorithm is the best known and is introduced first. However, there are some well known difficulties in using this algorithm on non-minimum phase and open-loop unstable systems. The so-called Linear Quadratic Gaussian Predictive Control algorithm is therefore also presented, which has much improved stability and robustness properties.

The $H_2$ or LQG design methods are natural multivariable control techniques and this is illustrated in Chapter 4. The solution of multivariable problems follows the same route as in scalar problems of Chapter 2. To make the solution of the multivariable problem more interesting, two particular situations are considered. The first involves an extension of the usual $H_2$ optimal control problem, where fault estimates are also required. The so-called *combined fault estimation and control problem* is considered.

The second situation considered in Chapter 4 is the polynomial solution of multivariable problems using a *separation principle* type of approach. This involves generating an LQG controller, but separated into filter and control gain blocks, using polynomial descriptions. Possible advantages for implementation are discussed.

The $H_\infty$ optimal synthesis and design methods are now becoming very popular and their potential in the aerospace industry has been researched extensively. These techniques are introduced in Chapter 5. The $H_\infty$ predictive control methods are not so well known and they are presented in a form where constrained optimisation algorithms can be used. In fact, an important reason for using multi-step cost functions is to be able to write the equations in a form where quadratic dynamic programming may be applied.

The polynomial approach to the solution of estimation problems is considered in Chapter 6. An unusual feature is to use probabilistic models to represent the system uncertainty. This can then be combined with unstructured uncertainty models and solved using $H_\infty$ methods. The standard system model for solving general filtering problems is considered in this chapter and this enables deconvolution and inferential estimation problems to be solved.

The second part of the text concerns state-space system modelling and the relationship to transfer-function models.

Many specific problems which occur in real industrial systems are discussed at length. For example, in rolling processes, the effect of transport delays is significant. This also applies to process control systems and hence the design of $H_2$ optimal controllers for systems with transport delays is considered in Chapter 7 of Part II. To put the material into perspective, the more traditional transport delay compensation technique, involving Smith Predictors, is discussed. The Smith Predictor only provides a structure for a solution and does not indicate how the controller should actually be designed. Once more, the optimal synthesis methods provide insights and even design guidance for classically designed Smith predictor systems. An understanding of the different blocks in the Smith predictor is gained and the optimal solution, which involves a Kalman filter, has many advantages for implementation and design. This chapter also summarizes some of the main results in the solution of state-space $H_\infty$ optimal control and estimation problems.

The importance of the predictive control design methods for industrial applications has already been noted. For large scale systems, the state-space methods are often considered preferable to frequency-domain modelling methods. The state-space approach to Predictive control is therefore introduced in Chapter 8. State-space versions of both GPC and LQGPC algorithms are presented.

Chapter 9 on Quantitative Feedback Theory Design involves techniques which are very different to the optimal control synthesis methods in other chapters. The QFT methods provide insight into how to obtain real robustness in control systems. The main contribution of this chapter is therefore the perspective it provided on robust control design. This is not a natural multivariable control design method, although the technique may be used for solving multivariable problems.

The third and final part of the book includes a number of application chapters. Problems in electrical power generation and transmission are discussed in Chapter 10. The use of advanced control methods in metal rolling processes, including tandem hot strip mills, is considered in Chapter 11. The design of sway, yaw and roll motion control systems for ships is discussed in Chapter 12. In the final Chapter, 13, the design of aero-engine and flight control systems is considered. Most of the design examples are based on the results of industrial projects and they demonstrate the significant performance improvements that may be possible.

In some industrial sectors there is a ready acceptance of new technology and in this case the design methods presented provide realistic and practical solutions. This is indicated in the applications chapters. All of the techniques described have shown promise in different applications. A strongly held belief is that a wide range of different tools are necessary to cope with varying industrial requirements and demands.

# Acknowledgements

I am grateful for the continuing help and support of my colleagues at the Industrial Control Centre at the University of Strathclyde. I am particularly indebted to my friend and colleague Professor Michael Johnson for his help in building the Centre and our joint research projects.

The many contributions on joint Centre research projects from Dr. Reza Katebi, Dr. Jacqueline Wilkie and Dr. Andrzej Ordys are also much appreciated. I am pleased to note that all of the academic staff of the centre including Dr. Bill Leithead, Dr. Akis Petropoulakis, Dr. Joe McGhee and Dr. Ian Henderson have helped in various ways.

I am particularly grateful for the technical typing and organizational skills of Mrs. Ann Frood who kindly managed the production of this manuscript.

I should also like to acknowledge the contributions from my colleagues on the hot mill project, *Dr. Gerrit van der Molen, Dr. Gerald Hearns, Dr. Gordon McNeilly, Mr. Barish Bulut and Mr. David Greenwood.* In fact Dr. Hearns kindly provided many of the metal processing simulation results.

The help with simulation results and assistance of the following research staff of the Industrial Control Centre are acknowledged: *Dr. Steven de la Salle, Dr. Stephen Breslin, Dr. Ilyas Eker, Dr. Stephen Forrest, Dr. Innes McLaren, Dr. Ender St. John Olcayto, Mr. Evert van de Waal, Professor Kenneth Hunt, Dr. Demos Fragopoulus, Dr. Nigel Hickey and Mr. Peter Martin.*

I should like to acknowledge the kind support and assistance of Mr. Andrew Buchanan and Mr. Jim Hamilton of Industrial Systems and Control Limited, Glasgow, who provided a continuous flow of challenging technical problems and who enthusiastically promoted the benefits of advanced control. We are also grateful for the support provided by the manager of the Advanced Control Technology Club, Mr. Andrew Clegg.

I would like to thank the many companies and staff that contributed to the results presented and supported the industrial projects. These include: *Mr. Colin Cloughly,* John Brown Engineering, Clydebank, *Mr. Roger Farnham and Mr. John Davis,* Scottish Power, Glasgow, *Mr. Brian Gee, Dr. Chris Fielding, Dr. Jonathan Irving, Mr. Brian Caldwell, Dr. Steven Ravenscroft,* British Aerospace, Warton, *Dr. Peter Dootson, Dr. Mike North, Mr. Arthur Sutton,* Lucas Aerospace, Hall Green, Birmingham, *Dr. Mark Brewer, Mr. Ron Cowan, Mr. John Warren and Mr. Philip Fendenczuk,*

B.P. Oil, Grangemouth, *Dr. Vittorio Arcidiacono, Dr. Sandro Corsi and Mr. Claudia Brasica,* ENEL, Milan, *Mr. Dilip Dholiwar, Mr. Graham Dadd and Mr. Maurice Porter,* Defence Research Agency, *Dr. John Jamieson, Mr. Bill McDiarmid, Mr. Rob Melville and Dr. David Wood,* Brown Brothers, Edinburgh, *Mr. Mike Dean, Mr. Paul Corney and Dr. Mel Hague,* British Steel plc, Teesside and South Wales, *Dr. Chris Davenport,* Alcan International, Banbury, *Mr. Robert Gronbech, Dr. Mick Clarke and Mr. Russel Mayor,* Kvaerner Metals, Sheffield, *Mr. Alan Kidd, Dr. Peter Reeve* and *Dr. Richard Bond,* Alstrom Drives and Controls, Rugby, Mr. Nuncio Bonavita, ABB, Genoa, Italy, *Mr. Ian M. Allan,* Smith Kline Beecham, Irvine, *Mr. Terry Madden and Dr. Barry Scott,* Vosper Thornycroft Controls, Portsmouth, *Mr. Kevin Wright and Mr. Jim Crowe,* Roche Products, Dalry.

International collaboration with various visitors to the centre was appreciated and in particular: *Professor Joseph Bentsman,* University of Illinois, USA, Professor *Paul Kalata*, Drexel University, USA, *Dr. Steen Toffner-Clausen,* Department of Electrical Engineering, Aalborg University, Denmark, *Professor Jacob Stroustrop,* Department of Control Engineering, Aalborg University, *Professor Mogens Blanke,* Institute of Automation, Technical University of Denmark, *Dr. Mads Hanystrap,* Department of Control Engineering, Aalborg University.

The cooperation over many years with Professor Peter Thompson (System Technology Inc), who developed the very user friendly package PROGRAM CC, is much appreciated.

I am indebted to Professor Constantine Houpis and Captain Steven Rasmussen of the US Airforce Institute of Technology and Wright Labs (Wright-Patterson Airforce Base, Dayton, Ohio) for their cooperation and help in a cooperative project on Quantitative Feedback Theory. The support of the Principal Engineer Duane Rubertus and of the British Aerospace project leader Chris Fielding were also much appreciated.

We are grateful for the cooperation with Peter Kock of the Technical University of Hamburg, and Jan-F Hansen of the Thondheim University for their help with the predictive control results.

Finally, I would like to record the inspiration George Zames provided to our control engineering community and our sadness at his very premature demise.

**Michael J. Grimble**

# 1

# Introduction to Advanced Industrial Control

## 1.1    Introduction

The aim of this chapter is to introduce some of the wider issues involved in industrial control systems engineering, before turning to the more specific synthesis and design problems of later chapters. The overview is slightly wider than the main themes of the text and it should provide a useful background against which the specific techniques described can be judged.

The area of linear control systems design has advanced rapidly over the last 50 years. In 1942 Norbert Wiener introduced the term *cybernetics* meaning a steersman in Greek. His early work on optimal control and filtering provided the first rigorous synthesis theory and the results were first applied in military research. However, the report he produced was considered to be mathematically difficult for engineers at the time and it became known as the yellow peril because of its yellow covers (see Wiener, 1949).

Most of the work up to the 1960's was concerned with the design of controllers in the frequency domain using Nyquist, Bode and Root-Loci plots (Nyquist 1932, Bode 1945, Evans 1954). Some of the measures of performance and robustness were unit-step response overshoot, and the gain and phase margins. These graphically based design methods were, however, difficult to extend to the multivariable case.

During the 1960's the state-space based multivariable control and filtering methods were developed, where the system model was assumed to be completely known. Interest in these methods was stimulated by the needs of the aerospace and defence industries. The Kalman filter and Linear Quadratic Gaussian (LQG) control design methods were particularly successful (Kalman, 1960). The linear quadratic design procedures were the first methods to tackle the multivariable design problem directly, rather than by a sequence of single-loop design steps. In fact, it may still be argued that optimal methods provide the only real multivariable design approaches.

In the 1980's it was recognised that LQG controllers for output feedback systems could exhibit poor robustness properties, although with hindsight it is now known that

1

much improved performance and robustness can be obtained by following the correct design and modelling procedures. However, the recognition of these potential problems led to the development of methods to improve LQG design (Doyle and Stein, 1981), such as *multivariable gain and phase margins* and *loop transfer recovery techniques* (Safanov and Athans, 1977) and more importantly stimulated the development of a different optimal control design approach.

George Zames (1981) recognised that minimising a new measure of system performance, namely the $H_\infty$ norm, could provide significant advantages, particularly for systems that were uncertain. The $H_\infty$ space is one member of a family of spaces (Hardy, 1915), introduced by the mathematician G.H. Hardy[1]. This is the space of functions on the complex plane, that are analytic and bounded in the right-half plane (Duren, 1970). Although sensitivity and robustness problems had been considered by other researchers such as Horowitz (1963), it was the work of Zames and Francis (1981) which led to an explosion of interest in this particular problem. The $H_\infty$ norm is well suited to the design of uncertain systems, where the uncertainty can be frequency response bounded. Moreover, although initially it was difficult to develop good numerical algorithms, the theoretical basis of $H_\infty$ design provided an analytic solution which enabled its robustness properties to be established. This compared well with previous approaches at improving robustness which were often ad hoc and empirical.

The link between $H_\infty$ optimization and classical frequency domain design approaches was also important. The intuition engineers gained on classical frequency domain methods can be employed in the $H_\infty$ design approach. The actual algorithms for $H_\infty$ design can utilise a polynomial or a state space setting which is often numerically convenient. However it is possible to approach the design process completely using frequency domain concepts and ideas.

In parallel with the development of the state space based $H_\infty$ algorithms (Doyle et al., 1989), a polynomial systems approach was followed by researchers such as Kwakernaak (1983, 1987) and Grimble (1986, 1987). There are of course advantages to using both types of approach, depending upon the applications area considered. For example, the polynomial systems approach is particularly valuable for adaptive or self-tuning control systems.

Initially $H_\infty$ controllers were calculated for systems in the usual classical feedback control loop configuration. However, more recently emphasis has switched to the solution of the so called *standard $H_\infty$ control problem*. This enables a wide class of different $H_\infty$ optimal control problems to be considered which is particularly valuable in the development of standard software tools. Unfortunately, the standard system model approach does not always give the same insights which can be obtained by considering particular system configurations directly. The structural simplifications to equations which are possible from considering say the feedforward/feedback optimal problem, are not so apparent when using standard system models. However, the standard system model is very convenient for obtaining the solutions to new optimal control problems, when system models include additional features.

In the 1990's intelligent control, covering areas like expert systems, neural net-

---

[1] Hardy's convexity theorem is in his 1915 paper which is now regarded as the historical starting point of the theory of $H_p$ spaces. The $H$ in $H_\infty$ does of course relate to Hardy.

works, fuzzy control, neuro-fuzzy control became the focus of attention. Although these methods have been employed over the last decade or so, for use in consumer products, it is over the last few years that there has been growing industrial interest.

It is very likely that significant advances will soon be made in nonlinear control systems design, since this is one of the few remaining areas where scientifically based practical solutions are often not available. There is therefore an industrial need to provide a sound theoretical basis for the subject, which also enables practical designs to be produced. It is unlikely that one overall solution will be developed covering all nonlinear system possibilities. It is more likely that nonlinear control problem will be solved in stages by finding design approaches which are particularly suitable for certain classes of nonlinear systems.

### 1.1.1 Hierarchical Modelling and Control

Most of the text will be concerned with the lower regulating loop levels of the controller hierarchy. Predictive control is considered in several chapters and has a significant role at the supervisory level. It will be useful to define the various levels of the control hierarchy at this point (see Fig. 1.1). There is no agreed notation for the various levels and different texts use different numbering systems, but the following seems logical.

**Level 1** : *Lower level regulating loop control*

- PID Classical lead/lag design (90% of all loops)

- Loop level fault monitoring and detection

**Level 2** : *Multivariable regulating loop control*

- Classical, GPC, LQG, $H_\infty$, LTR, QFT, INA, CL, SRD

- Multivariable fault monitoring and detection

**Level 3** : *Dynamic upper level control*

- Multivariable control of total systems

- Constrained optimisation of inputs and outputs using MBPC

- Supervisory level fault detection

**Level 4** : *Optimisation*

- Static and dynamic calculation of optimal conditions and setpoints

- Model adaptation

**Level 5** : *Scheduling*

- Optimal use of resources

- Determination of specifications for Levels 2 and 3

- Operational research and supply chains

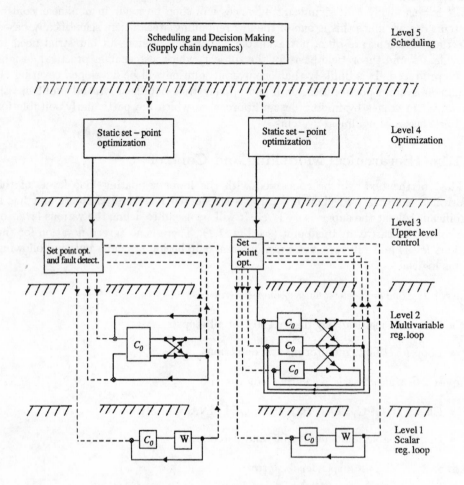

**Fig. 1.1** : Multilevel Industrial Control Structure

## 1.1.2   Benefits of Advanced Process Control

A typical process plant has between 2000 and 4000 loops and a typical control loop is a £20,000 asset which has to be utilised effectively to maximise economic returns. The mission for most advanced process control systems is to increase the economic yield of the plant. Typical benefits include :

- Increased throughput

- Increased yield of more valuable products

- Decrease in cost of utilities/unit of feedstock

- Improvements in product quality.

The economic benefits can also accrue from indirect causes, such as the reduction in thermal stresses that better control can achieve. The main advantage of tighter control is that the variance of the output can be reduced, so that the setpoint can be moved closer to the operational boundaries. The best economic performance can often be achieved by operating closer to such boundaries and hence the improved financial return can be linked directly to the incremental changes in setpoint level.

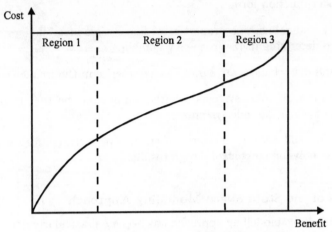

**Fig. 1.2** : Cost Versus Benefit Characteristic
Region 1 : *Regulator control region using DCS or PLC systems*
Region 2 : *Multivariable and advanced control region*
Region 3 : *Optimisation region*

Figure 1.2 illustrates the type of improvements that can be obtained by using different types of control solution. Note that the greatest benefit, relative to cost, is due to the use of multivariable or advanced control algorithms. The following quotation is out of date but is indicative of the benefits achievable with advanced process control:

> *...Industry realizes that excellent payoffs can be achieved from process control projects. As an illustration, Du Pont's process control technology panel recently reported that by improving process control throughout the corporation, it could save as much as half a billion dollars per year* ...[2]

### 1.1.3 Frequency Domain and State Space Methods

Both frequency domain and state space modelling methods are considered in the following, since both have their own merits in particular industrial applications. The polynomial systems approach to frequency-domain modelling and design is preferred, since this is more appropriate for numerical algorithm development than the transfer-function based equations.

---

[2] Source : National Reseach Council, Frontiers in Chemical Engineering : Research needs and opportunities, National Academy Press, Washington DC (1988)

**Advantages of the Frequency Domain Modelling Approach**

The polynomial or frequency domain approach to optimal control has many advantages:

- Engineers often have a better feel for the transfer function models of systems than for state-space representations.

- If the plant is identified from system measurements, it will normally be obtained in transfer-function form.

- Stability and robustness properties of linear time-invariant systems are much easier to determine in the frequency domain.

- Noise and disturbances are easier to characterise in the frequency domain.

- Manipulation of total systems is often simpler in the frequency domain (for example forming cascade systems).

- Self-tuning and adaptive systems often involve a straightforward extension of the basic polynomial control design results.

**Advantages of the State Space Modelling Approach**

The state space modelling approach has become particularly dominant in North America and in particular industries, such as the aerospace industry. Some of its advantages are as follows:

- There is a common belief that state space methods have significant advantages for calculations involving large systems.

- Physical models of systems are normally obtained in terms of nonlinear partial or ordinary differential equations, and linear state space models can therefore be linked to the underlying physical quantities involved.

- The ability to estimate state variables using a state space based Kalman filter is often valuable for control, monitoring and diagnostic purposes.

- State space models often provide a convenient mechanism for scheduling the controller, as the nonlinear system operating points change.

- Some control design methods, like eigenstructure assignment, naturally involve a state-equation structure.

- Numerical algorithms for time-invariant state space models involve manipulation and calculation with constant coefficient matrices, which are the simplest objects with which to perform numerical calculations and are therefore likely to be the most robust numerically.

- There are more commerically available software packages for control synthesis and design, using state-equation models then for frequency-domain modelling methods.

## 1.2  Robust Control Design

The design of controllers for systems that are uncertain and where guaranteed performance or stability requirements are to be met, has been one of the main topics for research over the last decade (Doyle et al., 1989 and Glover and McFarlane, 1989). The design of controllers which are robust to such variations is of great importance industrially, since consistency and reliability have a very high premium in both process and manufacturing plants.   In fact there still remains a debate as to whether the mathematical framework for representing uncertainty is totally representative of real systems.

### 1.2.1  Models for Plant Uncertainty

The mismatch between a linear plant model and the actual system can be represented by two types of uncertainty, either structured, that requires detailed knowledge about the variation in plant parameters and structure (Keel and Bhattacharyya, 1994), or unstructured uncertainty (non-parametric), where only information about the gain variations in the plant (as a function of frequency) is known. The modelling uncertainty can of course be a mixture of both structured and unstructured uncertainties. In a typical design structured uncertainty models are useful for representing uncertainty in the low to medium frequency range. Unstructured uncertainty models which are normally defined using $H_\infty$ norms, are valuable for modelling high frequency uncertainty.   This is one of the main justifications for the use of $H_\infty$ control design (see the introduction in Grimble and Johnson, 1991) that attempts to maximise robustness margins in terms of the $H_\infty$ norm (Chiang and Safonov, 1988).   An alternative Quantitative Feedback Theory (QFT) approach to robust control is discussed in later chapters (Doyle, 1986).

**Robustness and Sensitivity Functions**

In the early 1980's the connection between robustness to unmodelled dynamics and certain closed loop transfer functions, called the *sensitivity functions*, was emphasised by several researchers in the robust control community. For example, for a scalar unity-feedback system, with discrete-time plant transfer-function $W(z^{-1})$ and controller $C_0(z^{-1})$, the following sensitivity functions may be defined:

$$S(z^{-1}) = 1/(1 + W(z^{-1})C_0(z^{-1})) \tag{1.1}$$

$$M(z^{-1}) = C_0(z^{-1})/(1 + W(z^{-1})C_0(z^{-1})) \tag{1.2}$$

$$T(z^{-1}) = W(z^{-1})C_0(z^{-1})/(1 + W(z^{-1})C_0(z^{-1})) \tag{1.3}$$

where $S(z^{-1}), M(z^{-1})$ and $T(z^{-1})$, are the so called *sensitivity*, the *control sensitivity* and the *complementary sensitivity* functions, respectively. These functions also determine other properties of the system, such as disturbance rejection, measurement noise attenuation and output regulation but these will be discussed in more detail later. The following perturbation structures may also be introduced:

- **Additive uncertainty :**

$$\tilde{W}(z^{-1}) = W(z^{-1}) + \Delta(z^{-1}) \tag{1.4}$$

- **Multiplicative uncertainty :**

$$\tilde{W}(z^{-1}) = W(z^{-1})(1 + \Delta(z^{-1})) \tag{1.5}$$

- **Inverse multiplicative uncertainty :**

$$\tilde{W}(z^{-1}) = W(z^{-1})/(1 + \Delta(z^{-1})) \tag{1.6}$$

Now suppose a multiplicative uncertainty description has been found to be adequate and that the uncertainty $\Delta(z^{-1})$ is bounded by some frequency dependent scalar $\ell(\omega)$ :

$$|\Delta(e^{-j\omega})| < \ell(\omega), \quad \forall \omega \geq 0 \tag{1.7}$$

Furthermore, assume that the true plant $\tilde{W}(z^{-1})$ has the same number of unstable poles as the nominal plant model $W(z^{-1})$ and that a controller $C_0(z^{-1})$ has been designed that stabilises $W(z^{-1})$ . Then it may be shown that $C_0(z^{-1})$ stabilises the entire plant family $\tilde{W}(z^{-1})$ , if and only if:

$$|T(e^{-j\omega})| \leq \ell^{-1}(\omega), \quad \forall \omega \geq 0 \tag{1.8}$$

Thus, in the frequency range where the multiplicative uncertainty gain is large, the complementary sensitivity of the closed-loop system must be small. Applying the same arguments to additive and inverse multiplicative uncertainty and assuming a bound of (1.7), the results given in Table 1.1 may be derived. It follows that, robustness to norm bounded perturbations may be assessed through inspection of the closed-loop sensitivity functions (Skogestad and Postlethwaite, 1996 and Zhou et al,. 1995, 1996). In the next section it will be shown how performance requirements also impose limits on the sensitivity functions.

**Table 1.1 :** Uncertainty and Influence of the Sensitivity Functions

| Perturbation | Stability requirement | Sensitivity bound |
|---|---|---|
| Additive uncertainty | Control sensitivity small | $|M(e^{-j\omega})| \leq \ell^{-1}(\omega)$ |
| Multiplicative uncertainty | Comp sensitivity small | $|T(e^{-j\omega})| \leq \ell^{-1}(\omega)$ |
| Inverse multi uncertainty | Sensitivity small | $|S(e^{-j\omega})| \leq \ell^{-1}(\omega)$ |

### 1.2.2   Uncertainties in Multivariable Systems

The perturbation structures introduced above apply to multivariable systems, with the added complication that is necessary to distinguish between uncertainties acting at the plant inputs and the plant outputs. Furthermore, the size of a perturbation is normally measured by using its maximum singular value. Thus, introduce the following models:

- **Additive uncertainty :**

$$\tilde{W}(z^{-1}) = W(z^{-1}) + \Delta(z^{-1}) \tag{1.9}$$

- **Input multiplicative uncertainty :**

$$\tilde{W}(z^{-1}) = W(z^{-1})(1 + \Delta(z^{-1})) \tag{1.10}$$

- **Output multiplicative uncertainty :**

$$\tilde{W}(z^{-1}) = (1 + \Delta(z^{-1}))W(z^{-1}) \tag{1.11}$$

- **Inverse input multiplicative uncertainty:**

$$\tilde{W}(z^{-1}) = W(z^{-1})(1 + \Delta(z^{-1}))^{-1} \tag{1.12}$$

- **Inverse output multiplicative uncertainty :**

$$\tilde{W}(z^{-1}) = (1 + \Delta(z^{-1}))^{-1}W(z^{-1}) \tag{1.13}$$

where the uncertainty $\Delta(z^{-1})$ satisfies:

$$\sigma_{\max}(\Delta(e^{-j\omega})) < \ell(\omega), \quad \forall \omega \geq 0 \tag{1.14}$$

Consider, for example, the block diagram in Fig. 1.3, where the nominal plant is assumed to be perturbed by an output multiplicative dynamic uncertainty. The true plant is thus assumed to belong to the set:

$$\tilde{W}(z^{-1}) = (1 + \Delta(z^{-1}))W(z^{-1})$$

where $W(z^{-1})$ is the nominal model and the perturbation $\Delta$ is bounded as in (1.14). Assume that the true plant has the same number of unstable poles as $W(z^{-1})$ (that is, $\Delta(z^{-1})$ is assumed to be stable), and assume that a nominal stabilising controller $C_0(z^{-1})$ has been found. Then $C_0(z^{-1})$ will stabilise the entire plant family if and only if no characteristic loci of $\tilde{W}(z^{-1})C_0(z^{-1})$ pass through the Nyquist point, thereby changing the number of encirclements.

If the loci pass through the Nyquist point the characteristic equation $\det(I + \tilde{W}(z^{-1})C_0(z^{-1}))$ will have a root on the imaginary axis. Consequently for robust stability it is required that for $z = e^{j\omega}$ and all frequencies $\omega$, the $\det(I + \tilde{W}(z^{-1})C_0(z^{-1})) \neq 0$. A necessary requirement for this condition to be fulfilled is:

$$\sigma_{\max}(T_0(e^{-j\omega})) \leq \ell^{-1}(\omega), \quad \forall \omega \geq 0$$

where $T_o = WC_0S_o$ and $S_o = (I + WC_0)^{-1}$ are referred to as the *output complementary sensitivity* and output sensitivity functions, respectively. It may be noted that the related input complementary and sensitivity functions are defined as: $T_i = C_0WS_i$ and $S_i = (I + C_0W)^{-1}$. Applying similar arguments to the other types of uncertainty, the sensitivity requirements stated in Table 1.2 may be derived.

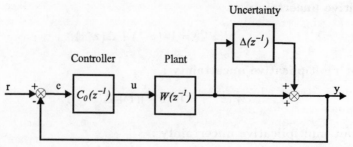

**Fig. 1.3** : Plant Perturbed by Output Multiplicative Uncertainty

Generally both the input and output sensitivities should be checked, since for ill-conditioned plants, stability margins may be satisfactory at one point, but can be very poor at the other loop breaking point.

**Table 1.2** : Uncertainty Descriptions and their Influence on the Sensitivity Functions for the Multivariable Case

| Perturbation | Stability demand | |
|---|---|---|
| Additive uncertainty | Control sensitivity small | $\sigma_{\max}(M(e^{-j\omega})) \leq \ell^{-1}(\omega)$ |
| Input multi. uncertainty | Comp. sens. (input) small | $\sigma_{\max}(T_i(e^{-j\omega})) \leq \ell^{-1}(\omega)$ |
| Output multi. uncertainty | Comp. sens. (output) small | $\sigma_{\max}(T_0(e^{-j\omega})) \leq \ell^{-1}(\omega)$ |
| Inverse input multi. uncertainty | Input sensitivity small | $\sigma_{\max}(S_i(e^{-j\omega})) \leq \ell^{-1}(\omega)$ |
| Inverse output multi. uncertainty | Output sensitivity small | $\sigma_{\max}(S_0(e^{-j\omega})) \leq \ell^{-1}(\omega)$ |

**Table 1.3** : Performance Objectives for Multivariable Systems

| Desired performance objectives | Constraints on sensitivity functions |
|---|---|
| Good reference tracking | $\sigma_{\max}(S_0(z^{-1}))$ *small* |
| Good disturbance rejection | $\sigma_{\max}(S_0(z^{-1}))$ *small* |
| Good sensor noise rejection | $\sigma_{\max}(T_0(z^{-1}))$ *small* |
| Control magnitude limitations | $\sigma_{\max}(M_0(z^{-1}))$ *small* |

## 1.2.3  Robust Stability and Performance

Control systems must have both *robust stability* and *robust performance*. A system is *robustly stable* when the closed-loop is stable for any chosen plant within the specified uncertainty set. The system has *robust performance* if the closed-loop system satisfies performance specifications for any plant model within the specified uncertainty description. In classical design the *stability margins* are normally used as the main indicators of the degree of stability. The *gain* and *phase margins* are the most common and these denote the amount of gain and phase variations in the open-loop transfer function that can be tolerated, before the closed-loop system becomes unstable. Note that when calculating these margins, either gain or phase variations are assumed; they do not guarantee stability if simultaneous gain and phase uncertainty are present.

### Robust Decoupling Methods

Early work on decoupling multivariable systems utilised open loop compensation methods. More recently methods have been developed of parameterising the controller so that a decoupled closed-loop system can be designed. This approach enables a controller to be produced which simultaneously achieves internal stability and closed-loop decoupling, without going through the open-loop decoupling stage (Kiong, 1996). In most cases plants cannot be completely decoupled and hence approximate decoupling methods are needed. In this situation the closed-loop transfer function has to be close to a diagonal transfer-function, in the sense that some error norm is satisfied within a given frequency range. Frequency dependent weighting functions can be employed and an $H_\infty$ optimisation problem can be constructed to provide a solution to this type of decoupling problem. The notion of *robust decoupling* refers to the ability of a compensator to provide a guaranteed maximum level of interaction, in the presence of given uncertainty levels and for the specified nominal system description.

## 1.2.4   Robust Controllers: Limitations and Advantages

Some of the problems, limitations and advantages of the $H_\infty$ robust control design approach are summarised below.

### Disadvantages

- Companies understand the need for robust control but are not very aware of recent design procedures and an educational gap exists.

- Design procedures for multivariable systems, to ensure desired industrial specifications are met, still involve too much trial and error and require expert understanding. Simpler design procedures for the multivariable case are essential if they are to be widely adopted.

- Most advanced control algorithms provide high order controllers and although these can be model reduced, this is a disadvantage.

- Multivariable $H_\infty$ robust control laws which enable constraints to be satisfied, particularly on outputs, can be very complicated and cannot be analysed easily.

### Advantages

- For particular industrial problems, like the development of roll stabilisers for ships, straightforward guidelines can be produced for robust design and relatively poor ship models can be used whilst still achieving certain guaranteed performance characteristics.

- Safety critical systems require the greatest reliability and robustness, and robust design techniques can provide formal methods to achieve desired stability margins.

- Many existing multivariable industrial systems have been designed using single-input single-output techniques and substantial improvements can be obtained by using truly multivariable designs.

- The process industries can often put a cash value on the accuracy at which they satisfy output requirements in a multivariable system. Robust $H_\infty$ predictive control methods can provide a mechanism for satisfying constraints.

- It is very difficult to de-skill the design engineering task using classical control methods. Advanced control design methods, like $H_\infty$, offer the basis for the development of formalised design procedures.

### 1.2.5   Structured Singular Values

For multivariable systems the size of the uncertainty may be specified using *singular values*. The singular value is a measure of the size of the matrix when structural information is not available. However, in many systems the structure of the system is constrained so that variations cannot occur in certain elements. For example, consider a 3 x 3 transfer function where there is no interaction so that the matrix is diagonal. For such a system the off-diagonal elements cannot possibly include the uncertainty, even though the diagonal terms may include large modelling errors. Unfortunately, the singular value measure does not distinguish between a matrix with completely general elements and one where its structure is restricted in this way. Designs based upon singular value uncertainties therefore tend to be very conservative, since they allow for variations which may not be possible in the real system.

The *structured singular-value* (Doyle et al., 1990) enables the structure of the uncertainty to be specified corresponding to the physical systems of interest. The measure of the size of the uncertainty is similar in principle to the singular-value measure but it only accounts for the variations which are possible in the system of restricted structure. Thus, in this case designs are less conservative, since all possible perturbations are those which might apply in the real system. The properties of the structured singular value are very similar to those for the singular value. The actual definition of the structured singular value can be a little confusing since the size of the matrix so measured must be related to the structure involved. It also leads to high order controllers, which are rather impractical. However, it does have significant potential in special applications, and the technique is reasonably straightforward to use (see Balas et al., 1993).

### 1.2.6   Linear Matrix Inequalities

A number of advances over the past two decades have resulted in numerical solution methods becoming more relevant for problems arising in systems and control. There has also been the growth in computing power, recent breakthroughs in optimisation theory, algorithms and linear algebra. As a consequence, numerical methods can be used to solve problems in systems and control for which no analytical solutions exist. This is especially true with regard to convex optimisation methods. Problems that reduce to finite-dimensional convex optimisation problems are in principle no harder to solve than a system of simultaneous linear equations. Optimisation problems involving *linear matrix inequalities* constitute a special class of convex optimisation problems that have provided practical and valuable results.

A Linear Matrix Inequality or LMI is a matrix inequality of the form:

$$F(\zeta) \;=\; F_0 + \sum_{i=1}^{m} \zeta_i F_i > 0 \tag{1.15}$$

where $\zeta \in R^m$ is the variable, and the symmetric matrices $F_i = F_i^T \in R^{n \times n}$ , $i = 0,...,m$ are given. The inequality symbol in (1.15) means that $F(\zeta)$ is positive-definite, i.e. $u^T F(\zeta) u > 0$ for all non-zero $u \in R^n$ . The set $\{\zeta | F(\zeta) > 0\}$ is convex. Several software packages for solving LMI optimisation problems are currently available. The relevance of LMI's to control theory stems from the fact that several important problems from control theory can be reformulated as LMI optimisation problems (Boyd and Barratt, 1991).

## 1.3 Fault-tolerant Control Systems

The increasing use of automation has generated interest in more sophisticated and intelligent systems. The main requirement for any manufacturing or process control system is to ensure reliable and continuous operation. Faults or failures can cause unacceptable danger or undesirable economic consequences. There is therefore great interest in the development of fault-tolerant control systems which undergo graceful degradation when faults arise (Siljak, 1980; Patton, 1997; Joshi, 1986; Jacobson and Nett, 1991; Stoustrup et al., 1997). This enables human or automatic systems to put in place corrective measures before the system fails totally.

High levels of reliability, maintainability and performance are now needed to ensure safe operation in hazardous human or environmental situations. The consequences of faults and failures in flight controls, chemical plants, nuclear plants, vehicle systems etc. are well known. The fault diagnosis function is one of the critical elements in a fault-tolerant control system.

*Reliable control* involves generating controllers that can cope with sensor failures (or actuator failures), since the possibility of failure is allowed for by treating the failure as an uncertainty (Viellette, 1995 and Viellette et al., 1992). The resulting control solution is stabilising, even when the sensor fails (assuming a clean failure with no spurious signal output). *Combined fault monitoring and control* utilises a similar model but represents the fault conditions by signals which can be estimated. The general subject of fault-tolerant control is explored further in Chapter 4.

The development of high reliability and fault-tolerant control systems is one of the most important innovations in advanced control systems design theory. There are many possible approaches but there is significant interest in $H_2/H_\infty$ model based methods. The early detection of faults is a further important component in the total solution and this is referred to as the fault detection problem. Some of the terms commonly utilised in *Fault-tolerant Control* and in *Fault Monitoring* are summarised in the following two sections.

## 1.3.1  Summary of Terms in Fault-tolerant Control

There follows a list of some of the main terms used in safety critical control systems design.

*Malfunction* : Intermittent irregular behaviour that often results when measurement or actuator systems do not perform to specification.

*Fault condition* : Unacceptable deviation of at least one characteristic property or variable of the system from regular or usual behaviour.

*Failure* : Major interruption of a systems ability to perform a required function, under specified operating conditions, that is often permanent.

*Fault-tolerant control* : A generic term representing a control system which can tolerate certain fault conditions.

*Reliable control* : A control law which allows for possible sensor or actuator failure by treating this as a possible uncertainty in the design models. Such a control law is fixed (non-adaptive) but is designed so that if a failure arises the system remains stable and provides adequate performance

*High integrity* : A control system is referred to as having high integrity if when different loops are broken the remaining closed-loop system remains stable.

*Reconfigurable control* : A control system whose structure may be adapted when faults arise (for example, utilising a different combination of actuators when one fails).

*Combined fault estimation and control* : A control law which provides estimates of the signals representing the fault condition, in addition to generating control action for the closed-loop that may or may not include a fault.

## 1.3.2  Summary of Terms in Fault Monitoring

A very effective robust, or reliable, control system can hide gradually developing fault conditions, which can be a disadvantage in certain industrial situations. A parallel but equally important subject that has emerged is that of fault monitoring (Frank, 1994; Chen and Patton, 1996; Patton and Chen, 1997). By combining an effective fault-tolerant control scheme, with a fault monitoring and detection system, this problem of hiding fault conditions can be avoided.

The terms that are often used in this condition/fault monitoring area are listed below:

*Symptom* : Change of an observable quantity indicating abnormal behaviour.

*Perturbation* : An input or change in a system which results in a temporary departure from steady state.

*Residual* : Fault indicator, based on deviation between measurements and model-based estimates.

*Fault detection* : Determination of faults occurring in a system and the time of detection.

*Fault isolation* : Determination of the kind, location and time of detection of the fault that follows the fault detection stage.

*Fault identification* : Determination of the size and time-varying behaviour of a fault that follows the fault isolation stage.

*Fault diagnosis* : Determination of the kind, location, size and time of detection of the fault. Follows the fault detection stage and includes fault isolation and identification.

*Monitoring* : Continuous real-time task of determining the condition of a physical system, by recording information, recognising and then indicating any anomalies of behaviour.

*Supervision* : Monitoring a physical system and resulting decisions and actions to maintain satisfactory operation in the case of faults.

*Protection* : Means by which potentially damaging behaviour of the system is suppressed, and the mechanism by which the consequences of very poor behaviour is avoided.

*Quantitative model* : The use of static and dynamic relations among system variables and parameters, in order to describe system behaviour in quantitative mathematical terms.

*Qualitative model* : The use of static and dynamic relations among system variables and parameters, in order to describe system behaviour in qualitative terms, such as causalities or if-then rules.

*Diagnostic model* : A set of static or dynamic relations which link specific input variables (the symptoms), to specific output variables (the faults).

*Analytical redundancy* : The use of two or more (not necessarily identical ways) to determine a variable, where at least one uses a mathematical process model in analytical form.

*Reliability* : The ability of a system to perform a required function under stated conditions, during a given period of time. Let

$$MTBF = 1/\lambda$$

where MTBF = Mean Time Between Failure and $\lambda$ = rate of failures (e.g. failures per year)

### 1.3.3  Fault Monitoring and Diagnosis

There are ever increasing demands for higher reliability, availability and security of industrial processes. Many different tools are now being employed for more reliable control, fault monitoring and diagnosis systems (Dailly, 1990). This includes techniques such as fuzzy logic, neural networks, neuro-fuzzy systems and model based systems, (Watanabe and Himmelblau 1983a,b; Watanabe and Hou Liya 1993). Failure and detection algorithms are needed, as illustrated in Fig. 1.4:

(a) *To detect the occurrence of a failure.*

(b) *To isolate the failed component or sub-system.*

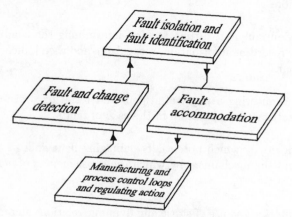

**Fig. 1.4** : Levels of Fault Diagnosis, Isolation, Identification
and Accommodation Functions

The detection, isolation and diagnosis of fault conditions in process or manufacturing systems (Himmelblau, 1978) can be considered from two perspectives. The first approach is to use control engineering theory and quantitative modelling. The second method is to employ qualitative modelling and reasoning based techniques developed by the artificial intelligence community.

There are several benefits to be gained from effective fault diagnosis systems. In process control systems these benefits include improved plant safety and efficiency, reduced down time and the safety of plant operation in the presence of faults. The techniques of fault diagnosis include intelligent systems, statistical methods, quantitative and qualitative models, inferential estimation schemes and robust estimation methods.

**Faults and Failures**

The terms *fault* and *failure* are now understood to have rather different technical meanings. A fault may present an unexpected and undesirable deviation in the system characteristics so that the desired purpose is not fulfilled by the system. There may not be a physical failure or breakdown but a fault or malfunction changes the normal operation of a system, resulting in a degradation of performance, and a dangerous situation may possibly arise. The term fault is normally used to denote a malfunction rather than the more serious catastrophic situation a failure entails. When a failure occurs a complete breakdown of the system is assumed to arise whilst a fault may only indicate that a tolerable malfunction has arisen. The aim of fault diagnosis and isolation systems is to detect a fault before any serious consequences occur and to isolate the source of the fault (Ding et al., 1991).

The main components in a fault monitoring, diagnosis and isolation system are as follows:

**Fault Detection Systems :** The aim of this component in the system is to detect the possible occurrence of faults and to make a decision as to when a real fault occurs. The fault detection subsystem utilises residuals calculated from mathematical models and measurements. The residual signals carry information on the operational state of the system.

**Fault Isolation :** If a fault is determined based upon the information contained in the residuals then it can be isolated. The isolation function is concerned with the location of faults and their time of occurrence in the particular sub-systems or components, including actuators, process plant or sensors.

**Fault Identification :** The fault identification stage involves the determination of the nature of the fault and its importance. The type, size, magnitude and significance of the fault can be identified.

**Fault Accommodation :** After a fault has been isolated and identified, fast action must be taken to ensure the fault is accommodated and the disturbance to normal operation is mitigated.

The problems of fault identification and accommodation are particularly important when the system is to be reconfigured to avoid the consequences of the fault condition. The first two functions are, however, common requirements in most systems and fault detection and isolation are given the abbreviations FDI. A failure detection test is normally used to determine when a fault exists in the plant. To identify the source of the failure, a failure isolation test is required. A possible approach is to perform a failure detection test and if a failure is found, then a *failure isolation* hypothesis test can be employed.

**Previous Work on Failure Detection**

Mangoubi (1998), working at the Draper Laboratory of MIT, presented an overview of state-space based robust failure detection and isolation algorithms and proposed a particular solution which was shown to be very effective in applications. This exploited the latest developments in $H_\infty$ robust filtering theory, both for discrete and continuous time systems. His failure detection test involved a hypotheses test between a set of unfailed plants and a set of failed plants, allowing for the possible presence of uncertainty. This is an example of *Model Based* fault detection methods. Authors like Watanabe (1989), Hoskins et al. (1991), Venkatasubramanian and Chan (1993), Tzafestas and Dalianis (1994) and Kwang et al. (1995) describe various examples of the alternative neural network fault diagnosis techniques.

**Component Based Fault Analysis**

Jorgensen (1995) has described graphical methods of enabling a designer to determine the fault propagation through a system. This enables the severity of each fault to be investigated and the formal reliability analysis to be undertaken. A fault tree diagram can be employed, consisting of logic symbols to show the inter relationships between final events and faults which cause these events. The fault tree is constructed by working backwards from the final undesirable event. There are a number of techniques that utilise physical models to identify possible component faults and *Failure Mode and Effect Analysis* (FMEA) is one of the most popular in industry (Bogh, 1997).

**Ideal Fault Detection Filter Response**

The transfer function between an input failure and the failure estimate should be close to unity over a wide frequency range in the ideal situation. The filter should be as fast as possible but the probability of false alarms should be minimised. Disturbances should not of course result in failures because of a mis-interpretation of results. Mangoubi (1998) used quadratic forms of the failure estimate to determine the detection and isolation functions. These quadratic forms provide estimates of the energy in the underlying signals which is an intuitively reasonable way of detecting system failures. Model uncertainties can seriously degrade the performance of fault detection algorithms. There is therefore a natural role for $H_\infty$ design methods in this type of problem. The modelling of uncertainties in robust fault diagnosis problems was considered by Patton et al. (1992). In fact the determination of suitable models is often the most difficult problem and the computation of suitable estimators is relatively straightforward.

## 1.3.4   Redundancy in Fault and Failure Diagnosis Methods

Failure detection algorithms normally use hardware techniques or *analytical redundancy* to determine anomalies in the system behaviour. A simple method to undertake fault diagnosis is to monitor the level of signals and take action when they reach a certain threshold level. However, false alarms can occur in the presence of noise and a single fault can result in the appearance of multiple faults, so that fault isolation is difficult to achieve. An alternative approach to fault diagnosis is to use *hardware redundancy*. A voting method is often used for hardware redundancy checking but this involves the duplication of physical devices, which is expensive. Multiple sensors can be used with the voting method and the outputs of these sensors can be compared to check for discrepancies between the measured signals. This is illustrated in Fig. 1.5, which shows a Fault Detection and Isolation (FDI) scheme based on analytical redundancy.

**Hardware Redundancy**

Fly-by-wire flight control systems employ *hardware redundancy* techniques but there are problems due to the additional complexity of the solution, the cost, maintenance and space requirements. A possible failure in a sensor can be detected and isolated by comparing a number of outputs that should give similar results. In many systems the outputs which are to be estimated, or controlled, cannot be measured directly. In such cases the models of the process or system might be used to provide an indirect measure of signals. Inconsistencies in behaviour can then be detected to determine failure conditions. This is of course an inferential estimation problem, discussed in more detail later.

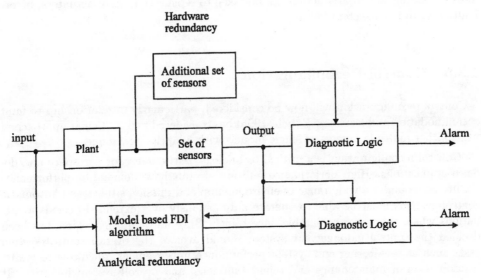

**Fig. 1.5** : FDI Hardware and Analytical Redundancy Scheme

### Analytic Redundancy

A model based approach is to employ *analytic redundancy* which involves the use of the functional relationships between measured variables to provide a cross-checking capability. Additional equipment may not be needed in such a circumstance, since existing measurements are simply used to provide estimates of other variables. A residual signal is defined based upon the difference generated from consistency checks. The residual will be of value zero during normal operation and will diverge from zero in the presence of fault conditions. This type of approach relies upon the use of a model and therefore falls under the category of *model-based fault diagnosis methods*. Analytical redundancy is potentially more reliable than hardware redundancy, since it does not require additional hardware.

The *residual* signal is used as a fault indicator and it requires an analysis method to determine when the fault has occurred. The residual should ideally be zero when the system is operating normally. The main advantage of model based FDI algorithms is that additional sensors are often not required.

Modelling uncertainties cause difficulties in model based fault diagnosis systems. It is difficult to determine when a fault has occurred if the changes may be due to the natural variations in system parameters. False alarms can arise or alarms may not be raised when actual faults occur. It is therefore essential that *robust* FDI algorithms are used. These must be insensitive to modelling errors and disturbances.

*Incipient faults* are those which arise from small and slowly developing fault conditions which are often referred to as *soft* faults. In fact, the diagnosis of substantial abrupt faults is not so difficult in the presence of modelling errors, since thresholds on the *residual* may be employed. The detection of incipient faults which have a small effect on the *residuals* and are difficult to separate from modelling uncertainty is much more difficult. Robustness of the estimation algorithm is essential in such a case. Although the determination of soft faults is a more difficult case, it is also the most rewarding, since early action can be taken to replace sensors or actuators, before faults lead to total system failure.

## 1.3.5   Control Reconfiguration

Control reconfiguration will now be considered, particularly in relationship to fault accommodation and learning. Rauch (1995) has considered two particular approaches, using either multiple models or single models with adaptive techniques.

Control reconfiguration is a critical technology, particularly for aerospace and defence applications. High performance military aircraft have demanding performance requirements under a wide range of environmental and mission situations. Automatic control systems must be able to operate autonomously over extended periods in situations where there is considerable uncertainty. Once an operating regime has been decided the control management system should oversee the overall control system tasks such as monitoring and system performance improvement, diagnosis of faults, co-ordination of maintenance and repair functions, and overall system integrity. It is desirable that such systems should be autonomous but human operators will often require some override capabilities to enable the system to enter regions which are forbidden but where mission demands create a priority.

### Reconfigurable Control in Aircraft

The subject of *control reconfiguration* has been stimulated by the requirements of high performance aircraft which have a rapid response to control surface faults. A *Fault Diagnosis, Isolation and Reconfiguration* (FDIR) system for an aircraft is illustrated in Fig.1.6. The fault detection system continuously monitors sensors and compares the measured system response to a model representing a healthy system. Control reconfiguration is normally based on stored control laws which are tailored to each anticipated fault condition. Control laws can of course be calculated from on-line algorithms based upon particular estimated fault conditions. Such faults can occur in elevators, ailerons and the rudder. The elevator and aileron controls operate in pairs and hence there are a total of five control surfaces. The probability of occurence of particular fault models (such as the fraction of a control surface being destroyed) can be calculated using well defined algorithms.

Control design approaches which have been used for reconfiguration include eigenstructure and pole assignment methods, model reference adaptive control, implicit model following control, feedback linearisation methods and pseudo-inverse controllers.

The aircraft stabilisers and engines can sometimes be used to counteract disturbances that arise through damaged control surfaces. These actuator effects are usually too slow to be used together with the fast control surfaces but they can produce large forces and moments to help recover an aircraft from certain failures.

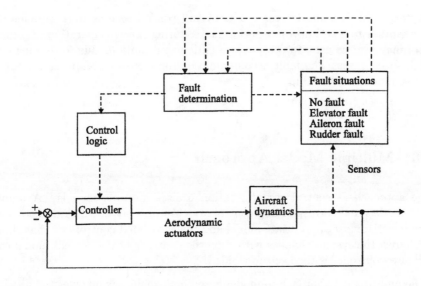

**Fig. 1.6** : Fault Detection, Isolation and Reconfiguration

**Reconfigurable Control in SWATH Vessels**

There are significant control problems in the motion control of Small Water Plane Area Twin Hull vessels (SWATH). High speed SWATH vessels can include four submerged cylindrical lower hulls located below the water surface. These provide buoyancy. The deck is supported above the water level by struts. The contention is that this arrangement of hulls and struts makes such vessels less sensitive to wave action than a monohull or catamaran. A SWATH vessel will provide improved performance in rough or high sea conditions. It can also provide high speed in waves and good manoeuvring and course keeping at low speeds.

A SWATH vessel can be powered by propellers in the submerged hulls. The directional and motion control functions can be performed by a number of control surfaces. Typically there are two forward control surfaces (canards) and two aft stabilisers. These control surfaces enable heading, pitch and roll motions to be controlled. Some vessels have two submerged hulls but the four hulls provide improved operation at higher speeds. Adaptive and reconfigurable controls can perform the following functions:

1. Self defining control (automatic adjustment of controller parameters during sea trials)

2. Self tuning (continuous adaptation of control parameters during normal operation)

3. Graceful degradation or reconfiguration (automatic controller adaption to assumed fault conditions).

The actuators for such systems are very nonlinear and require their own nonlinear compensation in the actuator channels. The ship dynamics are also very nonlinear to

both static and dynamic forces.    The overall control scheme must provide continuous adaption to give improved performance during normal operation. Feedforward
control may also be used to compensate anticipated motions due to dominant waves
or high speed turns. The fault accommodation must ensure compensation for major
propulsion or actuator failures.

## 1.3.6   Multiple Model Approach

One approach at control reconfiguration utilises multiple models.  A number of
system models and corresponding control laws are first obtained. A decision element
determines the most appropriate model and the associated control law that should be
used. When the system changes with time the estimate of the correct model changes
and a new controller is then recommended.

The multi-model approach to model based fault diagnosis involves the use of models to represent either particular fault conditions or normal no-fault conditions.  A
multiple-model estimator is illustrated in Fig. 1.7, where the outputs from the individual filters are pooled, using the weightings determined by the hypothesis decision
making system.  This diagram is based on the ideas of Maybeck (1999) but utilises
robust $H_\infty$ filters rather than Kalman filters.

**Multiple Model Adaptive Control**

The use of Multiple Model Adaptive Control (MMAC) is a technique to allow for
continuous parameter variations in systems using real time adaptation. The method
has been developed particularly for flight control systems, for military aircraft, where
compensation for changes due to damage, degradation or sensor/actuator areas is
needed. Major advances were made by Stepaniak (Second Lieutenant working with
Maybeck at the US Airforce Institute of Technology, Dayton) who applied this approach to the VISTA F-16 Fighter Aircraft. The thesis (December 1995) considered
novel MMAC algorithms for this problem but these have much wider applications.

Combining the multiple-model robust estimator in Fig. 1.7, with a robust control
law calculation gives the adaptive system shown in Fig. 1.8. An alternative strategy,
where the control law is applied directly to the individual estimates, is illustrated
in Fig. 1.9.   The algorithms can be extended, as shown in Fig. 1.10, to generate
parameter estimates and provide a true adaptive capability.  In this case:

- Filters utilise different possible fault condition models, parameters $a_1, ..., a_k$

- The filters allow for modelling errors via $H_\infty$ design

- Probability weights selection follows Maybeck's method

- $H_\infty$ feedback control computations via standard algorithms.

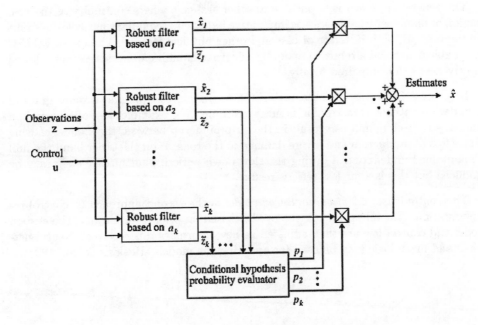

**Fig. 1.7** : Multiple Model Adaptive Robust Estimator

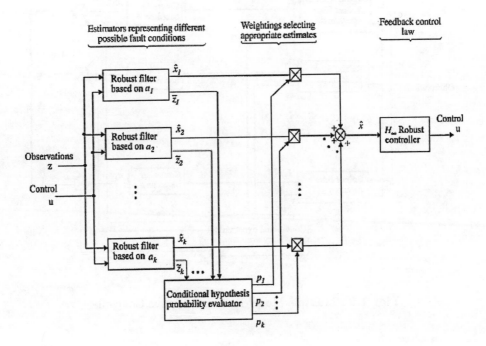

**Fig. 1.8** : Multiple Model Robust Adaptive Estimation and Control Using State Estimation

The general approach is applicable to other systems, where continuity of the production or manufacturing process is imperative but where redundant hardware systems are impractical, both in terms of cost and other physical factors. The use of MMAC can enable a crippled aircraft to maintain adequate handling qualities so that mission effectiveness is not impaired totally.

In process control or manufacturing the technique can provide a more graceful degradation without catastrophic failure conditions. For example, loss of sensors in a tandem strip rolling mill can result in the strip piling up between the stands, causing a days loss of production and severe damage to the rolls. If on the other hand the mill is maintained under control during the slow down period, poor quality strip may be produced but this is a far less serious result.

The *continuously adaptive nonlinear model* is the second approach to control reconfiguration. The initial model employed is based on prior information. The system model and control law are then adjusted as new information is received. Neural networks and fuzzy logic may be used for updating the models (Hoskins et. al., 1992).

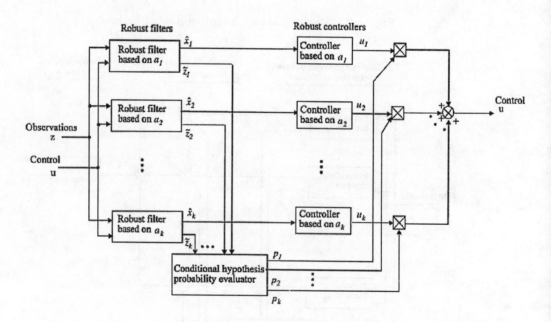

Fig. 1.9 : Multiple Model Robust Adaptive Controller

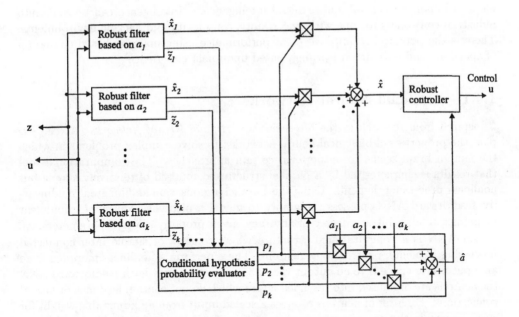

**Fig. 1.10** : Multiple Model Adaptive Robust Estimation and Control Including
Parameter Estimation

### 1.3.7  Safety Critical Control System Design

One of the most important questions with new control designs for safety critical
systems is whether they can be applied on the actual plant safely.   Both the commissioning and the normal operational phases need to be considered.   Safety can
be improved by validating designs as far as possible off-line.   The question which
arises is *what criteria must be satisfied before a design is ready for implementation?*
Although there are many aspects to the validation and implementation of a design,
the safety aspect focuses attention on questions of stability, robustness and transient
performance.

It is straightforward to define desired stability margins, such as the smallest gain
and phase margins which can be accepted.   However, for some systems at given
operating points it may be impossible for any controller to provide such margins.   An
alternative strategy is therefore to require that the best robustness be provided for
the given plant.   This then becomes a robustness optimization problem.   There is of
course also a need to consider performance requirements.

## 1.4   Intelligent Control and Artificial Intelligence

The text is mostly concerned with model based robust and predictive control design
methods. However, large industrial systems involve both intelligent and model based
control techniques. The so-called intelligent control methods are often used at the supervisory levels of the hierarchy. It will therefore be helpful to review some of the basic

ideas in intelligent control and artificial intelligence. Intelligence can be built into industrial controllers to cope with uncertainty, complexity and varying environments. There is the potential to improve plant performance using machine based emulation of operators and automated learning, based upon past experience[3].

## 1.4.1   Artificial Neural Networks

Inspired from research in the life sciences, *Artificial Neural Networks* (ANNs) exploit the properties of biological neural networks, to solve complex problems in which the human brain seems to outperform certain approaches. The computational tool that results is characterised by a parallel structure composed of relatively simple but nonlinear processing elements. Using this parallel organisation and inherent nonlinearity, feedforward ANNs possess the ability to model a piecewise-continuous nonlinear mappings, to an arbitrary degree of accuracy, given properly selected parameters.

An ANN is a computational structure which consists of simple inter-connected processing elements, referred to as neurons. These perform a nonlinear mapping from an input vector space U to an output vector space Y. In a multilayer feedforward ANN the neurons are organised into cascaded layers which do not contain feedback or lateral connections. Each set of neurons receives a scaled input from an adjustable weight for every neuron in the preceding layer. A multilayer feedforward ANN includes an input layer, a number of hidden layers, and an output layer generating the output signal.

Although the popular method of back propagation has been successful in a range of ANN applications, it can suffer from the inherent limitations of gradient search techniques. That is, slow rates of convergence and the inability to distinguish between local and global minimum points.

Neural networks are playing an increasing role in intelligent control. A class of networks called associative memory neural networks have many desirable properties including learning, convergence, local generalisation and real time adaptation etc. This class of neural network includes a certain class of fuzzy systems with a common framework of presentation and implementation.

Neural networks are used for building models from data. These networks are analogous to the human brain in that they consist of interconnected neurons. An input to a neuron results in a signal being sent to the next neuron and so on. The way in which neurons respond to signals is changed during the learning process until the neural network model accurately fits the data. Neural networks can deal with noisy and uncertain data and handle complex, nonlinear relationships. The main disadvantage of neural networks over model based methods is that the intuitive understanding of the model is lacking and the workings of the model are largely hidden.

The brain can be viewed as a complex but highly structured dynamical system in which feedback plays a major role. If the quality of feedback deteriorates certain human neurological dysfunctions can occur. These observations have been used to inspire novel artificial neural systems with interesting dynamic behaviour, information storage and processing capabilities. In analogy to biological structures, artificial

---

[3] Further background material may be found in the Department of the Environment Good Practice Guides and in particular Guide 215 *Reducing Energy Costs in Industry with Advanced Computing and Control.*

neural networks take advantage of distributed information processing capabilities and they therefore provide the potential for parallel computations to be performed.

The conventional wisdom is that neural networks are very valuable for use in plant identification (Win et al. 1992), stochastic forecasting, data analysis and modelling. However, many would question the use of neural networks in closed-loop control, although successful solutions have been proposed (Psalties et al. 1988, Narendra et al. 1990 ; Iiguni et al. 1991).

## 1.4.2  Fuzzy Control

Fuzzy set theory allows linguistic inexact data to be manipulated and logical decisions taken. Fuzzy logic has attracted considerable interest worldwide in various areas. The original proposal to use fuzzy control was by Zadeh in the sixties and it stimulated considerable interest, particularly in Japan, for consumer goods applications.

Fuzzy logic might be considered to be an extension of expert or rule based systems since it allows rules to be expressed in an imprecise manner. For example, the concepts of hot and warm are imprecise definitions. It is possible to define *very nearly*, *almost* and so on using truth values.  Very complex relationships can be represented by a rule that is simple to express and easy to understand. The combination of appropriate rules can constitute a control system which can be developed without the need for advanced mathematical results (Zadeh, 1978).

The benefits of fuzzy systems are as follows:

- Applications knowledge can be built into the rules.

- No programming is needed.

- They can deal with noisy and uncertain data.

- The operation can be understood intuitively and be reasonably fast.

- Development times for the algorithms can be very short.

- Implementation in hardware is possible for a greater speed of operation and cost effectiveness.

Fuzzy rules can be used in control systems, or for decision support applications. In this latter case they are similar to rule based expert systems but with the added advantage of being able to represent and reason using fuzzy concepts.

A fuzzy control system incorporates experience or expert knowledge from a human operator in order to derive an improved strategy for the control of a process. The main advantage of a fuzzy logic controller is its ability to implement rule of thumb experience and heuristic rules. It is also claimed that good performance can be achieved without the construction of models for the process. The fuzzy control technique applies to both linear and nonlinear systems.

### 1.4.3   Expert Systems and Knowledge Based Systems

An expert system is a computer program which can assimilate human expertise and then allow this expertise to be applied consistently, quickly and accurately. Early expert systems contained expertise in the form of rules that were separated from the algorithms which were used to interpret these rules. This interpretation was conducted in the inference engine. This type of approach is now referred to as rule based systems. Some alternatives to rule based reasoning include:

- Model based reasoning

- Constraint based reasoning.

An expert system is normally considered to be a hybrid system in the sense that expertise is included in the form of rules, procedures, models and past examples etc. Such a system will include rule based reasoning, rule induction, neural networks, genetic algorithms, case based reasoning, constraint solving, fuzzy logic and a range of other possible learning techniques.   These systems are used for:  Rule based control, Planning and scheduling, Intelligent monitoring, Fault diagnosis, Advisory systems, Configuration of equipment.

The major benefits of using rule based and expert systems include:

- Transferring expertise from the skilled operator to the novice.

- Improving knowledge, through the development process.

- Preserving corporate knowledge.

- Making knowledge more widely and readily available.

### 1.4.4   Genetic Algorithms

The term genetic algorithm stemmed from its mode of operation which is to mimic the evolution of plants and animals. They can be used for a range of optimisation problems and they enable optimum solutions to be obtained. The key steps are as follows:

(i)  A population of random solutions for the problem of interest is selected.

(ii)  A criterion is then used to assess the fitness of the solutions.

(iii)  The next generation of solutions is then produced by combining the better solutions and discarding the rest. This involves the survival of the fittest and the hope is that the new solutions are fitter than their parents.

(iv)  This process is repeated until a suitably fit solution is obtained.

The approach has some advantages:

- There are no assumptions about the nature of the problem, like the linearity or nonlinearity of the solution.

- Continuous and discrete variables can be combined.

- Expert mathematical knowledge is not required to use the approach.

## 1.5   Total Control and Systems Integration

Modern industrial systems use digital computers for several levels of the hierarchy including measurement, control, monitoring, supervision as well as data communications, networking and co-ordination, as illustrated in Fig. 1.11. The computer systems use a wide range of disparate technologies to establish links with the system hardware, sensors, drives and actuators. To achieve the system goals, these technologies should be integrated in a unified framework with some desired qualities such as conformance, modularity, interoperability, extensibility and reusability for individual components. Traditionally, the term *integration* refers to the interoperability of disparate, heterogeneous computer systems and the ability to communicate digital data and information. With the growing complexity of modern industrial and social systems, system integration is now concerned with the systematic development of complex systems by means of knowledge across several disciplines. Such systems are composed of a multitude of technologies and methodologies for both continuous and discrete industrial systems.

The traditional strategies, tools and techniques for complex systems design and evolution are not sufficient to solve the new problems introduced by the ever increasing complexity of systems. The development of effective tools to combine technological, organisational, ergonomic, environmental, and societal aspects can help to achieve economical, safe, and efficient systems. Such tools are needed in many industries including defence, aerospace, marine industries, power systems, transportation, air traffic, chemical plants and discrete manufacturing.

The theory of large-scale systems has been developed over several decades. Hierarchical and decentralised structures with different degrees of complexity have been formalised. The problems of aggregation and decomposition have been investigated and systems with several parallel time scales have been analysed. The control of large-scale systems has also been under investigation. Methods and algorithms such as decoupled controllers, decentralised controllers, singular perturbation methods, predictive control algorithms have all been utilised. In recent years, the design of intelligent controllers using expert systems, fuzzy logic and neural network methods have also been considered. For the design and development of complex systems, a number of techniques exist, such as optimisation and simulation, cost-benefit analysis and sensitivity analysis.

Some of the areas of system integration of current research include:

- Standardisation, optimisation, implementation and assessment procedures for novel algorithms and architectures in complex control systems.

- Production of standardised tools for production design.

- Development of safe procedures to allow new control systems to be implemented and optimised with reliable guarantees.

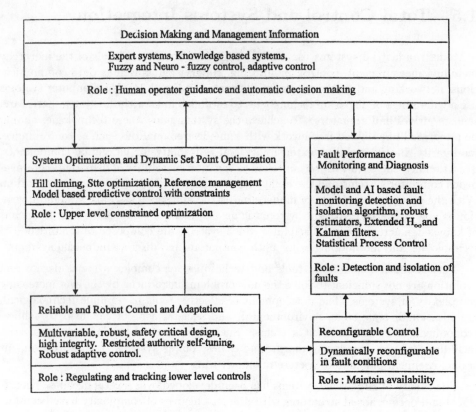

**Fig. 1.11** : Control Problems in Systems Integration

## 1.5.1   Systems Integration and Fault Conditions

Most large scale industrial systems include significant levels of redundancy since downtime due to fault conditions is expensive and can be dangerous. In general this is achieved using both on-line and off-line standby mechanisms. Since wholesale duplication of plant is expensive, significant returns can be generated if the system can be made fault tolerant using alternative mechanisms. At the same time the complexity of modern real time systems which incorporate a number of individual sensor and processing systems is increasing.   The delays and potential loss of data introduced through reconfiguration using off-line standby mechanisms for fault tolerant control can be unacceptable and on-line dynamic reconfiguration may be preferred. There are many problems in achieving full dynamic reconfiguration. However, the use of common standards and open systems architectures now make this a practical possibility. The potential benefits of integrated real time multi sensor/multi tasking/dynamically reconfigurable processing systems is substantial.

Verification and validation are essential steps in project implementation and final certification of a complex system. These terms refer to the following.

- *Verification* : the process of determining whether or not the products of a given phase in the system design cycle fulfil the requirements established during the definition phase.

- *Validation* : the process of evaluating the system design at the end of the development process to ensure compliance with system requirements.

**General Requirements of a System**

In addition to the specific requirements of a control system, including a degree of fault tolerance, the system should normally exhibit a number of other properties:

1. *Correctness* : Reflects the ability of the system to perform given tasks as defined in the requirements and specifications.

2. *Robustness* : This is the ability of a system to function even under abnormal conditions with incomplete information.

3. *Extensibility* : This is the ease by which system components may be adapted to changes of system specifications.

4. *Reusability* : This corresponds to the ability of system components to be reused in whole or in part in different applications.

5. Compatibility: This is the ease by which system components may be combined.

## 1.6  Concluding Remarks

Most of the following text is concerned with regulating loop and supervisory level model based control design. The present chapter has tried to draw attention to some of the wider issues of systems engineering. The chapter has also reflected the growing importance of condition monitoring and fault tolerant control design procedures. The control engineer of the future will be faced with total system design responsibilities and will be required to achieve higher levels of fault tolerance to provide more reliable operation.

## 1.7  References

Balas, G.J., Doyle, J.C., Glover, K., Packard, A. and Smith, R., 1993, *User's Guide for $\mu$−Analysis and Synthesis* Toolbox in MATLAB, The MathWorks, Inc.

Bode, H.W., 1945, *Network analysis and feedback amplifier design,* Van Nostrand : Princeton, NJ.

Bogh, S.A., 1997, *Fault tolerant control systems - a development method and real-life case study,* PhD Thesis, Aalborg University, Dept of Control Eng., Denmark.

Boyd, S.P. and Barratt, C.H., 1991, *Linear controller design, limits of performance,* Prentice Hall, Englewood Cliffs.

Chen, J.A. and Patton, R.J., 1996, Robust fault detection and isolation systems, *Control and Dynamic Systems*, Vol. **74**, 171-2,23.

Chiang, R.Y. and Safonov, M.G., 1988, *User's Guide for Robust Control Toolbox in MATLAB*, The MathWorks, Inc.

Dailly, C., 1990, Fault monitoring and diagnosis, *Computing and Control Eng. Journal*.

Ding, X., and Frank, P.M., 1991, Frequency domain approach and theshold selector for robust model-based fault detection and isolation, *IFAC Fault Detection, Supervision and Safety for Technical Processes*, Baden Baden, Germany, pp. 271-276.

Doyle, J.C., 1986, QFT and robust control, *Proceedings ACC*,(1691-8), Seattle.

Doyle, J.C., Glover, K., Khargonekar, P.P. and Francis, B.A., 1989, State-space solutions to standard $H_2$ and $H_\infty$ control problems, IEEE Trans. Automatic Control, Vol. 34, No.4, pp. 831-846.

Doyle, J.C., and Stein, G., 1981, Multivariable feedback Design : concepts for a classical/modern synthesis, *IEEE Transactions on Automatic Control*, Vol. AC-**26,**

Doyle, J.C., Stein, G., Banda, S.S. and Yeh, H.H., 1990, Lecture Notes for the American Control Conference workshop on $H_\infty$ and $\mu$ method for robust control, San Diego, California.

Duren, P.L., *Theory of $H^p$ spaces,* Academic Press, 1970.

Evans, W.R., 1954, *Control system dynamics*, McGraw Hill, New York.

Frank, P.M., 1994, Enhancement of robustness in observer-based fault detection, *Int. J. Control*, Vol. **59**, No. 4, pp. 955-981.

Glover, K. and McFarlane, D., 1989, Robust stabilization of normalised coprime factor plant descriptions with $H_\infty$ bounded uncertainty, *IEEE Trans. Automatic Control*, Vol. **34**, No. 8, (821-830).

Grimble, M.J. and Johnson, M.A., 1991, $H_\infty$ robust control design - a tutorial review, *IEE Computing and Control Engrg. Journal*, Vol. **2**, No. 6, (275-281).

Grimble, M.J., 1986, Optimal $H_\infty$ robustness and the relationship to LQG design problems, *Int. J. Control*, Vol. **43**, No.2, 351-372.

Grimble, M.J, 1987, $H_\infty$ robust controller for self-tuning applications, part 1 : Controller design, *Int. J. Control*, Vol. **46**, No. 4, 1429-1444.

Hardy, G.H., 1915, The mean value of the modulus of an analytic function, *Proc. London Math. Soc,* **14**, pp. 269-277.

Himmelblau, D.M., 1978, *Fault detection and diagnosis in chemical and petrochemical processes*, Elsevier Scientific Publishing Company, New York.

Horowitz, I.M., 1963, Synthesis of feedback systems, Academic Press, New York.

Hoskins, J.C., Kaliyur, K.M. and Himmelblau, D.M., 1991, Fault diagnosis in complex chemical plants using artificial neural networks, AIChE J., 37-1 pp. 137-141.

Hoskins, D.A., Hwang, J.N. and Vagners, J., 1992, Iterative inversion of neural networks an its application to adaptive control, *IEEE Trans. on Neural Networks*, Vol. **3**, No. 2, pp. 292-301.

Iiguni, Y., Sakai, H. and Tokumaru, H.,1991, A nonlinear regulator design in the presence of systems uncertainties using multilayered neural networks, *IEEE Trans. On Neural Networks*, Vol. **2**, No. 4, pp. 410-417.

Jacobson, C.A. and Nett, C.N., 1991, An integrated appraoch to controls and diagnostics using the four parameter control, *IEEE Control Systems Magazine*, 22-29.

Jorgensen, 1995, *Development and test methods for fault detection and isolation*, PhD thesis, Dept. of Control Eng., Aalborg University, Denmark.

Joshi, S.M., 1986, Failure-accommodating control of large flexible space craft, *American Control Conference*, Seattle.

Kalman, R.E., 1960, A new approach to linear filtering and prediction probelms, *J. Basic Eng.*, Vol. **82**, pp. 35-45.

Keel., L.H. and Bhattacharyya, S.P., 1994, Control system design for parametric uncertainty, Int. J. Robust and Nonlinear Control.

Kiong, Q.C., 1996, Minimum interaction designs in robust multivariable control systems, PhD thesis, Department of Electrical Engineering, National University of Singapore.

Kwakernaak, H., 1983, Robustness optimization of linear feedback systems, 22nd CDC, San Antonio, Texas.

Kwakernaak, H., 1987, A polynomial approach to $H_\infty$ optimization of control systems, NATO ASI Series : Modelling, Robustness and Sensitivity Reduction, Springer Verlag.

Kwang, B, C., Saif, M. and Jamshidi, M., 1995, Fault detection and diagnosis of a nuclear power plant using artificial neural networks, Journal of Intelligent and Fuzzy Systems, Vol. 3, John Wiley & Sons, pp. 197-213.

Mangoubi, R.S., 1998, *Robust estimation and failure detection : a concise treatment*, Springer Verlag, Berlin.

Maybeck, P.S., 1999, Multiple model adaptive algorithms for detecting and compensating sensor and actuator/surface failures in aircraft flight control systems, *Int. J. of Robust and Nonlinear Control*, Vol. **9**, pp. 1051-1070.

Narendra, K.S. and Parthasarathy, K. 1990, Identification and control of dynamical systems using neural networks, *IEEE Trans.* NN-1, No. **1**, pp. 4-27.

Nyquist, H., 1932, Regeneration theory, *Bell Syst. Tech J.*

Patton, R.J., 1997, Fault-tolerant control : the 1997 situation (survey), *preprints of IFAC Symposium on Fault Detection, Supervison and Safety for Technical Processes : Safeprocess 97*, Vol. **2**, 1033-1055, University of Hull,.

Patton, R.J. and Chen, J., 1997, Observer-based fault detection and ioslation robustness and applications, *IFAC Journal : Control Engineering Practice*, Vol. **5**, No. 5, pp. 671-682.

Patton, R.J., Zhang, H.Y. and Chen, J., 1992, Modelling of uncertainties of robust fault diagnosis, *Proc of 31st CDC Conf.* Tucson, Arizona, pp. 921-926.

Psalties, D., Sideris, A. and Yamamura, A.A., 1988, A multilayered neural network controller, *IEEE Control Systems Magazine*, pp. 17-21.

Rauch, H.E., 1995, Autonomous control reconfiguration, *IEEE Control Systems Magazine, pp. 37-48.*

Safonov, M.G. and Athans, M., 1977, gain and pahse margin for multiloop LQG regulators, *IEEE Trans. Autom. Control.* Vol. AC-**22**, No. 2.

Stepaniak, M.J. 1995, *Multiple model adaptive control of VISTA F-16*, Airforce Institute of Technology, Wright Patterson Airforce Base, Dayton, Ohio, Report AFIT, GE/ENG/95D-26.

Siljak, D.D., 1980, Reliable control using multiple control systems, *Int. J. Control.* Vol. **31**, No. 2, pp. 303-329..

Skogestad, S., and Postlethwaite, I., 1996, *Multivariable feedback control,* John Wiley, Chichester, UK.

Stoustrup, J., Grimble, M.J. and Niemann, H., 1997, Design of integrated systems for the control and detection of actuator/sensors faults, *Sensor Review*, Vol. **17**, No. 2, pp. 138-149.

Tzafestas, S.G., and Dalianis, P.J., 1994, Fault diagnosis in complex systems using artificial neural networks, 3rd IEEE CCA Conference, Glasgow, Scotland.

Veillette, R.J., 1995, Reliable linear quadratic state feedback control, *Automatica,* Vol. **131**, No. 1, 137-143.

Veillette, R.J., Medanic J.V. and Perkins, W.R., 1992, Design of reliable control systems, *IEEE Trans. Automatic Control.* Vol. 37, No. 3, pp. 290-304.

Venkatasubramanian, V. and Chan, K., 1993, A neural network methodology for process fault detection, AIChE J., 35 pp.

Watanabe, K, 1989 Incipient fault diagnosis of chemical process via artificial neural networks, AIChE J., 35-11 pp. 1803-1812.

Watanabe, K. and Himmelblau, D.M., 1983a, Fault diagnosis in nonlinear chemical process, Part-1 theory, AIChE J., 29-2, pp. 2243-2249.

Watanabe, K. and Himmelblau, D.M., 1983b, Fault diagnosis in nonlinear chemical process Part-11 theory, AIChE J., 29-2, pp. 2250-2261.

Watanabe, K., and Hou Liya, 1993, Application to macro-structured neural network technique to incipient multiple fault diagnosis in processes, Trans. SICE 29-11, pp. 1369-1378.

Wiener, N., 1949, Extrapolation, interpolation and smoothing of stationary time series, with engineering application, *New York Technology Press and Wiley* (issued in Feb. 1942 as a classified Nat. Defense Council Report).

Win, S.Z., Su, H. and McAvoy, T.J., 1992, Comparison of four neural net learning methods of dynamic systems identification, *IEEE Trans. on Neural Networks*, Vol. **3**, No. 1, pp. 122-130

Zadeh, L.A., 1978, Fuzzy sets as a basis for a theory of probability, Fuzzy sets and systems, Vol. I, pp. 3-28

Zames, G., 1981, Feedback and optimal sensitivity : Model reference transformations, multiplicative seminorms and approximate inverses, IEEE Trans Automatic Control. Vol. 26, No.2,

Zames, G. and Francis, B.A., 1981, A new approach to classical frequency methods : Feedbackand minimax sensitivity, CDC, San Diego, California.

Zhou, K., Doyle, J.C. and Glover, K., 1996, *Robust and optimal control,* Prentice Hall, New York.

Zhou, K., Khargonekar, P.P., Stoustrup, J. and Niemann, H.H., 1995, Robust performance of systems with structured uncertainties in state space, *Automatica,* **3** **1(2)** : 249-255.

# Part I

# Polynomial System
# Descriptions

# 2

# H$_2$ Optimal and Feedforward Control

## 2.1   Introduction

The Linear Quadratic Gaussian (LQG) optimal control laws remain some of the most popular of the class of advanced control design methods. Although all optimal control techniques should be referred to as synthesis methods, they are in fact normally utilised as design techniques. The optimal control cost-functions normally do not have a real physical importance and the optimal control problem is only used to provide a framework for design. For scalar systems and output feedback control problems, one scalar tuning variable is all that is necessary in the cost-function. In this case the controller is easier to tune than a classical PID controller. A more general cost-function is introduced in the following where the cost function weightings include frequency dependent dynamic components. Dynamic cost function weightings provide greater generality but more general tuning variables then arise.

In current terminology this class of LQG optimal control problem is often referred to as $H_2$ optimisation. This is particularly true when very general control system structures are introduced. Thus, although $H_2$ optimisation has a precise mathematical definition it often refers to the class of very general least squares LQ or LQG optimisation problems. The solution procedure to be followed in this chapter was motivated by the early work of *Kučera* (1979). The polynomial solution obtained has particular advantages when analysing the frequency response properties of the system. The pole zero behaviour of the system, noise and disturbance rejection properties and even the stability characteristics, are easier to establish through the polynomial systems solution. In later chapters attention will turn to the state space methods which are not so transparent, but provide reliable numerical algorithms for large systems, in well proven commercial toolboxes. The industrial case studies established in later chapters therefore concentrate more on the state space results but the engineering insights are provided by these early frequency domain polynomial system chapters.

A theoretical introduction to polynomial techniques and the tools of Youla parameterisation were presented in (Grimble and Johnson, 1988) and industrial applications of polynomial systems were described in (Grimble, 1994). There are in fact two basic approaches to frequency domain optimal control, following either the *Kučera* school (1979) or the Youla et al. (1976a,b) parameterisation approaches. The following solution will mainly utilise the *Kučera* (1980) philosophy but reference will also be made to the Youla parameterisation results, that provide very useful insights into the design process.

The chapter begins by considering the standard unity feedback loop optimal control problem but added interest is introduced by including feedforward action. The feedforward control problem is in fact of independent interest, since feedforward action is often used in industrial processes and yet the feedforward design process is not so well understood. The optimal feedforward/feedback solution discussed provides a valuable structure for design and gives many insights into how such systems should be designed, even when using a classical control framework.

Feedforward control (Seraji, 1987) can provide substantial improvements in disturbance rejection properties but the results are also very sensitive to knowledge of the models utilised for design. The LQG solution of feedforward control problems has been considered by Grimble (1988) and by Hunt (1988) but in these papers no independent tuning facility was available between the feedforward and feedback controllers. Feedforward control design has also been considered by Sternad and Ahlen (1987) who also discussed interesting dual estimation problems.

The second part of the chapter considers a more general feedforward optimal control problem, where the cost function includes terms that allow separate tuning of the feedforward, tracking and feedback controllers. This is of particular value when disturbance models are poorly known and hence caution must be exercised in the amount of feedforward control action introduced. The usual optimal control problem, that is considered at the start of the chapter, does not provide any freedom in tuning these different feedback and feedforward controllers in an independent manner.

The inferential control problem, where the quantities to be controlled are not directly measurable, may also be considered using a similar approach (Grimble, 1998). There are in fact many industrial applications, where it is physically impossible to measure a quantity that must be controlled. In some cases, it is unsafe to make such a measurement or the cost is prohibitive.

The computational procedures and the types of results obtained are illustrated by process control and marine applications. The examples include computational results and frequency and time responses are also presented.

## 2.1.1   Industrial Controller Structures

There are five controller structures often employed in industrial systems. Namely, one, two, three, two and a half, and three and a half degrees of freedom (DOF) structures. The results derived in the following can be specialised to each of these special cases. The $2\frac{1}{2}$ and $3\frac{1}{2}$ DOF structures have received little attention in the literature and these apply to 2, or 3, DOF structures that have useful additional properties. The $\frac{1}{2}$ DOF is not of course to be taken literally, it is simply helpful notation to draw attention to the different controller structures employed.    It is normally the case that the number of degrees of freedom in the control action is taken to be the number of independent transfer-functions that can be chosen to design the controller.

Most control system designs use the one DOF control structure shown in Fig. 2.1(a). However, this structure has the disadvantage that the feedback loop properties cannot be designed independently to the reference tracking transfer-function. This implies that a compromise must always be made when choosing between robustness properties and set-point following.

These limitations of the one DOF control structure are overcome in the two DOF structure (Grimble, 1988 ) that is shown in Fig. 2.1(b). In this case the feedback-loop controller $C_0(z^{-1})$ can be chosen independently to the reference following controller. However, there is still a problem with this structure which does not apply to the classical one DOF structure. If, for example, the classical feedback controller includes integral action, then in the steady-state, a constant reference input will lead to a constant output of the same magnitude. The two DOF control structure does not provide such a solution, unless the zero-frequency gain of the reference controller $C_1(z^{-1})$ is chosen appropriately.

The $2\frac{1}{2}$ DOF control structure is shown in Fig. 2.1(c). The feedback controller $C_0(z^{-1})$ can be chosen to shape robustness properties and to include integral action. The controller therefore has both the zero-frequency tracking properties of one DOF designs and the freedom to shape robustness and tracking properties independently, as in two DOF solutions. This is therefore referred to as a $2\frac{1}{2}$ DOF control structure (Grimble 1992, 1994). Adding feedforward control action gives the so-called $3\frac{1}{2}$ DOF control solution.

Note that there is a class of problems where having an additional degree of freedom does not provide much improvement over one degree of freedom solutions. Consider for example, a stable open-loop system. If the requirement to reject disturbances demands a wide bandwidth in the closed-loop system, it is not possible to use a low gain within the feedback loop to achieve greater stability robustness. The two DOF designs are not therefore so valuable in this case and a single DOF solution may provide satisfactory tracking and disturbance rejection properties.

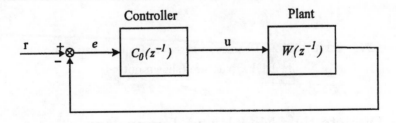

**Fig. 2.1(a)** : One Degree of Freedom Control Structure

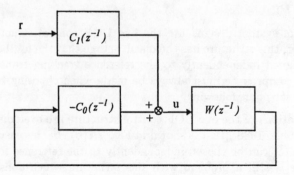

**Fig. 2.1 (b):** Two Degrees of Freedom Control Structure

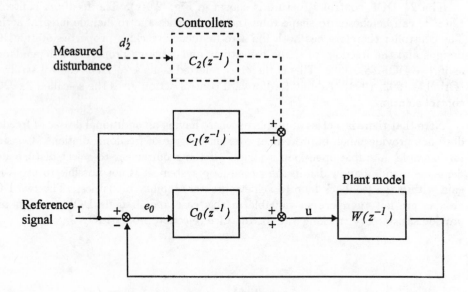

**Fig. 2.1(c) :** A $2\frac{1}{2}$ Degrees of Freedom Control Structure and
$3\frac{1}{2}$ DOF Structure (shown dotted)

## 2.1.2   Discrete-time Models and Terminology

Discrete-time systems, represented by difference equations, can be described by polynomial-functions in the unit delay operator $z^{-1}$. The resulting models have a number of well known forms, such as the **A**uto-**R**egressive **M**oving **A**verage (ARMA) models. The following table summarises some of the most common terms.

**Table 2.1** : List of Common System Abbreviations

| Discrete Delay Polynomial Models Notation | |
|---|---|
| ARX | A-AUTO |
| ARMA | R-REGRESSIVE |
| ARIX | M-MOVING |
| ARMAX | A-AVERAGE |
| ARIMAX | I-INTEGRATED |
| AREMAX | X-eXogenous or eXternal input |
| CARIMA | E-EXTENDED |
| | C-CONTROLLED |

## 2.2  Wiener Feedforward/Feedback Solution

The solution of the LQG optimal control problem for a scalar system represented in polynomial form is described below. The type of solution presented is very similar to that used for the more complicated multi-step cost or multivariable control problems in later chapters. The three-degrees-of-freedom (3 DOF) solution involves the synthesis of feedback, tracking and feedforward controllers.

Most fixed and self-tuning control systems are based upon a classical feedback controller configuration. Feedforward action is normally added later in a non-optimal non-adaptive manner. The cascade controller employed has the difference of the reference and observations signals as an input. This type of controller is known as a single-degree-of-freedom controller but two-degrees-of-freedom controllers provide several additional advantages (Grimble, 1988). Such a controller has separate inputs for the reference and observations signals and may enable lower optimal cost values to be achieved. An LQG two-degrees-of-freedom controller, which also includes feedforward action is referred to as a 3 DOF controller.

It is clearly desirable that the feedforward controller should be an integral part of the total LQG design. The polynomial solution given below requires that a diophantine equation be solved for the feedback controller and this is no more complicated than when no feedforward control is used. The feedforward controller requires the solution of one additional diophantine equation which depends upon the equations for the previously calculated feedback controller.

Feedforward control action is very desirable and effective when a fast disturbance rejection response is required. However, there are several problems:

(i) The disturbance entering the system will not be the same as the measured disturbance signal (this is accounted for here by modelling the disturbance as two components, only one of which may be measured, and assuming that the spectrum for this component is corrupted by the feedforward path and feedforward measurement system dynamics).

(ii) Good disturbance models for feedforward design are usually not available (the identification stage in a self-tuning algorithm can sometimes provide the missing data).

(iii) There is no simple traditional method of obtaining a good feedforward controller design, allowing for the presence and influence of the feedback controller (the integrated philosophy proposed overcomes this difficulty).

The design of two-degrees-of-freedom controllers for the multivariable case has been considered by Sebek (1983a,b) but no disturbance feedforward control action was included. The polynomial approach follows the general philosophy developed by *Kučera* (1980) and used in solutions of both the deterministic and stochastic tracking problems (*Kučera* and Sebek, 1984a,b). A further generalisation of the problem is to allow for the presence of coloured measurement noise and dynamic cost-function weighting terms. Youla et al. (1976a,b), Shaked (1979) and Whitbeck (1981) also considered a general system description of this type but from the Wiener-Hopf transfer-function point of view (Wiener, 1949). The LQG optimal control approach can also be used for multivariable system decoupling, as described by Bongiorno and Lee (1991) and Lee et al. (1993).

## 2.2.1   System Model and Signals

The system is assumed to be linear, discrete-time and single-input single-output. The total system model is shown in Fig. 2.2. The system transfer functions are assumed to be functions of the delay operator $z^{-1}$ in the time domain, or the z-transform complex number in the complex frequency domain. The latter interpretation should be applied when complex integral expressions are considered. For notional simplicity the $z^{-1}$ arguments are often omitted.

The external white noise sources drive colouring filters representing the reference $(W_r(z^{-1}))$, disturbance $(W_{d_1}(z^{-1}),\ W_{d_2}(z^{-1}))$ and measurement noise (or output disturbance) subsystems $(W_n(z^{-1}))$. The reference or tracking controller is denoted by $C_1(z^{-1})$, the feedback loop controller by $C_0(z^{-1})$ and the feedforward controller by $C_2(z^{-1})$. The system equations, by reference to Fig. 2.2, become:

**Observations :**

$$z(t) = v(t) + H_f(y(t) + n(t)) \tag{2.1}$$

**Output :**

$$y(t) = d(t) + Wu(t) \tag{2.2}$$

**Input disturbance :**

$$d(t) = d_1(t) + d_2(t) = W_{d_1}\xi_1(t) + W_{d_2}\xi_2(t) \tag{2.3}$$

**Measurement noise :**

$$n(t) = W_n\omega(t) \tag{2.4}$$

**Reference signal:**

$$r(t) = W_r\zeta(t) \tag{2.5}$$

**Control:**

$$u(t) = C_1 r(t) - C_0 z(t) - C_2 W_s d_2(t) \tag{2.6}$$

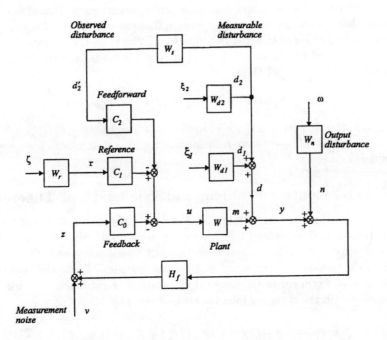

**Fig. 2.2** : Feedback and Feedforward Controller Structure
and Plant Model

## Assumptions

(i) The white noise sources $\zeta$, $\xi_1$, $\xi_2$, $\omega$ and $v$ are zero mean and mutually statistically independent. The covariances for these signals are without loss of
generality taken to be unity, with the exception of the measurement noise $v$
whose covariance is denoted by $R > 0$.

(ii) The plant $W$ and feedback transfer-function $H_f$, are free of unstable hidden
modes and the reference ($W_r$), input disturbance ($W_{d_1}$ and $W_{d_2}$) and measurement noise ($W_n$) subsystems are asymptotically stable.

(iii) The reference and disturbance models can be taken to be minimum phase. It is
also assumed that the reference ($W_r$) and disturbance models ($W_s$ and $W_{d_2}$) do
not include pure transport delay terms.

(iv) The feedback transfer-function $H_f$ is minimum phase and asymptotically stable
but may include a $k_f$-steps measurement system delay.

Such a system structure is to be found in marine autopilot design (Grimble et al.
1984, Katebi et al. 1985). In this case the measurable disturbance ($W_{d_2}$) represents the
wind speed and direction and the unmeasurable disturbance ($W_{d_1}$) models the current
and tidal forces acting on the vessel. The output disturbance represents the first order
waveforces on the vessel. In this problem the input disturbances must be counteracted

by use of the rudder actions whereas the output disturbances (waveforces) must be ignored, so that the rudder does not respond to these relatively high frequency motions.

In the minimisation of the cost-function by a Wiener method a causal transform is introduced, signified by the term $\{.\}_+$ . That is, if $f(z^{-1})$ is the two-sided z-transform of a signal $f(t)$, $t \, \epsilon(-\infty, \infty)$ then

$$\{f(z^{-1})\}_+ \;=\; \mathcal{Z}_2\{f(t)U(t)\}$$

where $U(t)$ is the Heaviside unit step function ($U(t) = 1$ for $t \geq 0, U(t) = 0$ for $t < 0$). In other words this represents the transform of a signal that is zero for $t < 0$ and may be considered causal.

### 2.2.2  LQG Control Problem and Wiener-Hopf Theorem

The LQG cost-function to be minimised can be expressed in terms of the expected value of the squared error and control signals measured, over all time.  The noise and disturbance signals are assumed to be zero mean and hence the cost-function represents the variance of the error and control signals.  Since the optimal control solution is to be obtained in the frequency domain it is appropriate to use Parseval's theorem to obtain the complex integral form of the LQG criterion:

$$J \;=\; \frac{1}{2\pi j} \oint_{|z|=1} \{Q_c(z^{-1}) \, \Phi_{ee}(z^{-1}) + R_c(z^{-1}) \, \Phi_{uu}(z^{-1})\} \, \frac{dz}{z} \qquad (2.7)$$

where $\Phi_{ee}$ and $\Phi_{uu}$ represent the error and control power spectral densities and $Q_c$ and $R_c$ represent weighting terms which are positive semi-definite and positive-definite respectively on $|z| = 1$. The optimal control problem is to minimise (2.7) for the system shown in Fig. 2.2.

A solution to the problem, based upon polynomial system models, is required but on route to this solution it is straightforward to derive the Wiener or transfer-function based controller (Borgiorno, 1969). The Wiener-Hopf result is summarised below.

**Theorem 2.1** : *Wiener-Hopf Feedback/Feedforward Controllers*

Consider the system described in Section 2.2.1 and the cost-function (2.7). The LQG optimal controller to minimise (2.7) has three component parts which may be calculated from the spectral factor and partial fraction expressions. Let the generalised spectral factors $Y_c$ and $Y_f$ be defined so that $Y_c^{-1}$ and $Y_f^{-1}$ are asymptotically stable transfer functions which satisfy:

$$Y_c^* Y_c = W^* Q_c W + R_c \qquad (2.8)$$

$$Y_f Y_f^* = H_f(\Phi_{nn} + \Phi_{d_1 d_1})H_f^* + \Phi_{vv} \qquad (2.9)$$

The optimal closed-loop *control sensitivity* transfer-function $M$ may be computed as:

$$M = Y_c^{-1}\{Y_c^{*-1}Q_c\Phi_{d_1 d_1}W^* H_f^* Y_f^{*-1}\}_+ Y_f^{-1} \qquad (2.10)$$

The optimal feedback loop controller $C_0$ follows as:

$$C_0 = (1 - MH_fW)^{-1}M \tag{2.11}$$

The reference-input controller becomes:

$$C_1 = Y_c^{-1}\{Y_c^{*-1}Q_cW_rW^*\}_+S^{-1}W_r^{-1} \tag{2.12}$$

The feedforward controller is given as:

$$C_2 = (Y_c^{-1}\{Y_c^{*-1}Q_cW_{d_2}W^*\}_+W_{d_2}^{-1} - MH_f)S^{-1}W_s^{-1} \tag{2.13}$$

where the sensitivity function : $S = 1 - MH_fW$.

The solution of this Wiener-Hopf control problem is presented in the following section.

### 2.2.3   Wiener-Hopf Solution of 3 DOF Control Problem

The three important sensitivity functions were defined in Chapter 1. Two of these appear in the above theorem and are therefore summarised below. Define the closed-loop transfer function from the noise input to the control signal as:

$$M = C_0(1 + H_fWC_0)^{-1} \tag{2.14}$$

Thus, the sensitivity function:

$$S = 1 - H_fWM = (1 + H_fWC_0)^{-1} \tag{2.15}$$

The control, plant output and tracking error signals may easily be derived from Fig. 2.2 as:

$$u = -M(v + H_f(n + d)) + S(C_1r - C_2W_sd_2)$$

$$= -M(v + H_f(n + d_1)) + SC_1r - (MH_f + SC_2W_s)d_2 \tag{2.16}$$

$$y = -WM(v + H_f(n + d_1)) + WS(C_1r - C_2W_sd_2) + d - WMH_fd_2 \tag{2.17}$$

$$e = r - y = (1 - WSC_1)r - (1 - WMH_f)d + WM(v + H_fn) + WSC_2W_sd_2$$

$$= (1 - WSC_1)r + WM(v + H_fn) - (1 - WMH_f)d_1$$

$$- (1 - WS(C_0H_f + C_2W_s))d_2 \tag{2.18}$$

The LQG cost-function to be minimised was defined in equation (2.7) as:

$$J = \frac{1}{2\pi j}\oint I_c(z^{-1})\frac{dz}{z} = \frac{1}{2\pi j}\oint_{|z|=1}\{Q_c\Phi_{ee} + R_c\Phi_{uu}\}\frac{dz}{z} \tag{2.19}$$

Substituting from (2.16) and (2.18), and noting the independence of the noise sources, the integrand of (2.19) becomes:

$$I_c = Q_c[(1 - WSC_1)\,\Phi_{rr}(1 - WSC_1)^* + (1 - WMH_f)\,\Phi_{d_1 d_1}(1 - WMH_f)^*$$

$$+WM(\Phi_{vv} + H_f \Phi_{nn} H_f^*)M^* W^* + (1 - WSC_{02})\Phi_{d_2 d_2}(1 - WSC_{02})^*]$$

$$+R_c[M(\Phi_{vv} + H_f(\Phi_{nn} + \Phi_{d_1 d_1})H_f^*)M^* + SC_1 \Phi_{rr} C_1^* S^* + SC_{02}\Phi_{d_2 d_2} C_{02}^* S^*]$$

where to simplify the expression define the function : $C_{02} = C_0 H_f + C_2 W_s$

There follows some straightforward algebraic manipulation to simplify the cost-function (Grimble and Johnson, 1988). Let the adjoint operator $W^*(z^{-1}) = W(z)$, then the criterion becomes:

$$J = \frac{1}{2\pi j} \oint_{|z|=1} \{(W^* Q_c W + R_c)SS^*[C_{02}\Phi_{d_2 d_2} C_{02}^* + C_1 \Phi_{rr} C_1^* + C_0(H_f \Phi_{d_2 d_2} H_f^*$$

$$+\Phi_{vv} + H_f \Phi_{nn} H_f^*)C_0^*] + Q_c(\Phi_{dd} + \Phi_{rr}) - Q_c \Phi_{d_2 d_2}(C_{02}^* S^* W^* + WSC_{02})$$

$$-Q_c \Phi_{rr}(C_1^* S^* W^* + WSC_1) - Q_c \Phi_{d_1 d_1}(WMH_f + H_f^* M^* W^*)\}\frac{dz}{z} \qquad (2.20)$$

A procedure will soon be followed, referred to as *completing the squares*, which is essentially the same as that employed in simple algebraic least squares optimisation problems. In the present case a squared term does not result but a self-adjoint term of the form $AA^* + B$, where $A$ depends upon the control law selection and $B$ is invariant, whatever the choice of control. However, before getting to this stage two spectral factors $Y_c$ and $Y_f$ must be introduced. The first spectral factor is referred to as the control spectral factor $Y_c$ and is dependent on the cost-function weightings. The second is the filter spectral factor $Y_f$ that depends upon the noise and disturbance models.

**Spectral Factors**

Introduce the *Generalised Spectral Factors* (Shaked, 1976) $Y_c$ and $Y_f$ :

$$Y_c^* Y_c = W^* Q_c W + R_c$$

$$Y_f Y_f^* = H_f(\Phi_{nn} + \Phi_{d_1 d_1})H_f^* + \Phi_{vv}$$

There are standard algorithms and commercial packages that enable the spectral factors $Y_c$ and $Y_f$ to be computed. However, it is easier to compute the polynomial spectral factors, defined later, rather than these transfer function forms.

## Completing the Squares

The completing the squares stage will now be undertaken and this solution technique will be utilised repeatedly in the following chapters. First substitute in (2.20) for the control $Y_c$ and filter $Y_f$ spectral factors. Substituting for these terms, the cost-function integrand may be simplified as:

$$I_c = [(Y_c^* Y_c S^* S [C_{02} \Phi_{d_2 d_2} C_{02}^* + C_1 \Phi_{rr} C_1^* + C_0 Y_f Y_f^* C_0^*]$$

$$+ \Phi_0 - \Phi_{h2} C_{02}^* S^* - \Phi_{h2}^* S C_{02} - \Phi_{h1} C_1^* S^* - \Phi_{h1}^* S C_1 - \Phi_{h0} M^* - \Phi_{h0}^* M \qquad (2.21)$$

where the following spectral terms may be introduced:

$$\Phi_0 = Q_c (\Phi_{dd} + \Phi_{rr}), \quad \Phi_{h0} = Q_c \Phi_{d_1 d_1} W^* H_f^*, \quad \Phi_{h1} = Q_c \Phi_{rr} W^*$$

$$and \quad \Phi_{h2} = Q_c \Phi_{d_2 d_2} W^*$$

The cost function includes independent compensator terms and the completion of squares procedure may be applied to each of these terms. Consider for example the feedback controller related terms involving the control sensitivity $M = SC_0$. Now (2.21) includes terms in $MM^*, M$ and $M^*$. The completion of squares procedure involves rewriting the cost function in a form where the coefficients of all of these functions remain the same. Completing the squares in (2.21) and assuming $W_r, W_{d_2} \neq 0$ the cost-function becomes:

$$J = \frac{1}{2\pi j} \oint_{|z|=1} \{ (Y_c M Y_f - \frac{\Phi_{h0}}{Y_c^* Y_f^*})(Y_c M Y_f - \frac{\Phi_{h0}}{Y_c^* Y_f^*})^*$$

$$+ (Y_c S C_1 W_r - \frac{\Phi_{h1}}{Y_c^* Y_r^*})(Y_c S C_1 W_r - \frac{\Phi_{h1}}{Y_c^* Y_r^*})^*$$

$$+ (Y_c S C_{02} W_{d_2} - \frac{\Phi_{h2}}{Y_c^* W_{d_2}^*})(Y_c S C_{02} W_{d_2} - \frac{\Phi_{h2}}{Y_c^* W_{d_2}^*})^* + \Phi_{01} \} \frac{dz}{z} \qquad (2.22)$$

where the following term is independent of the choice of control action:

$$\Phi_{01} = \Phi_0 - \frac{1}{Y_c^* Y_c} \left( \frac{\Phi_{h0} \Phi_{h0}^*}{Y_f^* Y_f} + \frac{\Phi_{h1} \Phi_{h1}^*}{W_r W_r^*} + \frac{\Phi_{h2} \Phi_{h2}^*}{W_{d2} W_{d2}^*} \right) \qquad (2.23)$$

To gain confidence in the *completing the squares* procedure, referred to above, the integrand of equation (2.22) can be expanded and may easily be shown to be equal to (2.21). The solution of the optimal control problem is required in polynomial system form. However, a brief review of the Wiener-Hopf transfer-function solution to the problem will be presented below to obtain an insight into the form of the solution from this perspective.

### 2.2.4   Wiener-Hopf Cost-function Minimisation

Note that the final term in (2.22), denoted $\Phi_{01}$, does not depend upon the controllers and does not therefore enter into the minimisation argument. The first three terms in (2.22) cannot simply be set to zero to minimise $J$, otherwise non-causal controllers will result. The best that can be achieved is to choose $M, C_1$ and $C_{02}$ so that the causal components of the cost terms are set to zero. The resulting expressions will then only involve causal components. Thus, let $\{f(z^{-1})\}_+$ denote the transform of the causal or positive-time component of the signal.    Then the cost-function (2.22) is minimised by causal controllers satisfying (see Grimble and Johnson, 1988):

$$M = Y_c^{-1}\{Y_c^{*-1}\Phi_{h0}Y_f^{*-1}\}_+ Y_f^{-1} \tag{2.24}$$

$$C_1 = Y_c^{-1}\{Y_c^{*-1}\Phi_{h1}W_r^{*-1}\}_+ S^{-1}W_r^{-1} \tag{2.25}$$

$$C_{02} = Y_c^{-1}\{Y_c^{*-1}\Phi_{h2}W_{d2}^{*-1}\}_+ S^{-1}W_{d2}^{-1} \tag{2.26}$$

Simplifying these equations, by substituting for the signal spectra, obtain:

$$M = Y_c^{-1}\{Y_c^{*-1}Q_c\Phi_{d_1 d_1}W^*H_f^*Y_f^{*-1}\}_+ Y_f^{-1} \tag{2.27}$$

$$C_1 = Y_c^{-1}\{Y_c^{*-1}Q_cW_rW^*\}_+ S^{-1}W_r^{-1} \tag{2.28}$$

$$C_2 = (Y_c^{-1}\{Y_c^{*-1}Q_cW_{d2}W^*\}_+ S^{-1}W_{d2}^{-1} - C_0H_f)W_s^{-1} \tag{2.29}$$

Having calculated the optimal control sensitivity function $M$ the controller $C_0$ follows from (2.14) as : $C_0 = (1 - MH_fW)^{-1}M$, and the tracking and feedforward controllers $C_1, C_2$ then follow directly from the last two equations.

### 2.2.5   Structure of the Wiener-Hopf Optimal Solution

It is interesting that because of the problem construction the reference input $(C_1)$ and feedforward controller $(C_2)$ calculations depend upon the feedback controller calculation $(C_0)$ but not vice versa. The reference input and feedback controllers (see (2.28) and (2.27)) do not depend upon the type and character of the feedforward controller (2.29). These are therefore unchanged by the presence (or absence) of the feedforward action and the measurable disturbance $d_2$.

To further explore the form of the control signal let the control signal be written as : $u^o(t) = (u_r(t) + u_b(t) + u_f(t))$. That is, the optimal control:

$$u^o(t) = C_1 r(t) - C_0 z(t) - C_2 W_s d_2(t)$$

$$= S^{-1}\left(M_2 r(t) - M\left(z(t) - H_f d_2(t)\right) - M_3 d_2(t)\right) \tag{2.30}$$

where the following closed-loop control operators are defined:

$$M = C_0 S = Y_c^{-1} \{ Y_c^{*-1} Q_c \Phi_{d_1 d_1} W^* H_f^* Y_f^{*-1} \}_+ Y_f^{-1} \tag{2.31}$$

$$M_2 = Y_c^{-1} \{ Y_c^{*-1} Q_c W_r W^* \}_+ W_r^{-1} \tag{2.32}$$

$$M_3 = Y_c^{-1} \{ Y_c^{*-1} Q_c W_{d_2} W^* \}_+ W_{d_2}^{-1} \tag{2.33}$$

The attenuating action of the two controllers $C_0$ and $C_2$ on the disturbance $d_2$ may now be determined. Note that $C_2$ in (2.29) includes the term $-C_0 H_f W_s^{-1}$ which leads to the signal $-M(z(t) - H_f d_2(t))$ in (2.30). After substituting for $z(t)$ this term is completely independent of $d_2$. The effective feedforward compensating action of the two controllers is therefore due to the term $-S^{-1} M_3 d_2(t)$ in (2.30).

Thus, the control signal may be split into the following components:

$$u^o = u_r + u_b + u_f \tag{2.34}$$

where the feedback signal $u_b = -C_0 z = u_{b1} + u_{b2}$ and $u_{b1} = -S^{-1} M (z - H_f d_2)$ is independent of $d_2$ and $u_{b2} = -C_0 H_f d_2$. The feedforward signal $u_f = -C_2 W_s d_2 = u_{f1} + u_{f2}$ but $u_{f2} = C_0 H_f d_2 = -u_{b2}$ and $u_{f1}$ represents the optimal control signal for an equivalent open-loop control problem. That is, $u_{f1}$ is the optimal open-loop feedforward control, given the measured disturbance $d_2$. The total control signal follows as:

$$u^o = u_r + u_{b1} + u_{f1}$$

where the signals $u_r, u_{b1}$ and $u_{f1}$ are mutually statistically independent.

Another way of interpreting the above results is that the optimal control signal for a system with feedforward and feedback action may be computed from a two step procedure:

**Step 1**: Compute the optimal feedback controller for a system with measurable disturbance $d_2 \equiv 0$ to obtain the optimal control signal $u_r + u_{b1}$.

**Step 2** : Compute the optimal feedforward control signal, assuming the feedback loop is open to obtain $u_{f1}$. The total optimal control signal is then given as : $u^o = (u_r + u_{b1}) + u_{f1}$.

## 2.3  Polynomial Feedforward/Feedback Solution

A polynomial approach is now taken to the solution of the 3 DOF LQG control problem. The same system and cost index are considered but the solution will be obtained in a more convenient form for numerical calculations.

## 2.3.1  Polynomial System Description and Solution

Let the system shown in Fig. 2.2 be represented in polynomial equation form (Kailath, 1980; Callier and Desoer, 1982), where the system transfer-functions become:
Plant :

$$W = A^{-1}B \tag{2.35}$$

Measurement noise:

$$W_n = A_n^{-1}C_n \tag{2.36}$$

Feedback dynamics :

$$H_f = A_h^{-1}B_h z^{-k_f} \tag{2.37}$$

Reference generator :

$$W_r = A_e^{-1}E_r \tag{2.38}$$

Unmeasurable disturbance :

$$W_{d_1} = A_{d_1}^{-1}C_{d_1} \tag{2.39}$$

Disturbance measurement :

$$W_s = A_s^{-1}B_s \tag{2.40}$$

Measurable disturbance:

$$W_{d_2} = A_{d_2}^{-1}C_{d_2} \tag{2.41}$$

The various polynomials $A, B, A_n, C_n$ etc. are not necessarily coprime but all the system elements are assumed to be free of unstable hidden modes. In state-space terms this is the same as assuming the system includes no unstable modes which are not controllable, or not observable. If there were of course such modes it would be impossible to find a stabilising control law.

The cost-function weighting elements will be represented in polynomial form as:

$$Q_c = A_q^{*-1}Q_n A_q^{-1} \quad \text{and} \quad R_c = A_r^{*-1}R_n A_r^{-1} \tag{2.42}$$

**Polynomial Spectral Factors**

The spectral factors $Y_c$ and $Y_f$ may now be represented in the polynomial form : $Y_c = A_c^{-1}D_c$ and $Y_f = A_f^{-1}D_f$ . Thus, by substituting from the above polynomial results into (2.8) and (2.9), the *control polynomial spectral factor* satisfies:

$$A_c^{*-1}D_c^* D_c A_c^{-1} = (AA_q)^{*-1}B^* Q_n B(AA_q)^{-1} + A_r^{*-1}R_n A_r^{-1}$$

$$= (AA_q A_r)^{*-1}(A_r^* B^* Q_n BA_r + A^* A_q^* R_n A_q A)(AA_q A_r)^{-1} \tag{2.43}$$

and the *filter polynomial spectral* factor satisfies:

$$A_{f1}^{-1} D_{f1} D_{f1}^* A_{f1}^{*-1} = A_h^{-1} B_h (A_n^{-1} C_n C_n^* A_n^{*-1} + A_{d_1}^{-1} C_{d_1} C_{d_1}^* A_{d_1}^{*-1}) B_h^* A_h^{*-1} + R$$

$$= (A_h A_n A_{d_1})^{-1} (B_h (A_{d_1} C_n C_n^* A_{d_1}^* + A_n C_{d_1} C_{d_1}^* A_n^*) B_h^*$$

$$+ A_h A_n A_{d_1} R A_{d_1}^* A_n^* A_h^*)(A_h^* A_n^* A_{d_1}^*)^{-1} \tag{2.44}$$

where $D_c$ and $D_{f1}$ are strictly Schur and $D_c$, $A_c$ and $D_{f1}$, $A_{f1}$ are coprime pairs. Now let $A_f = A A_{f1}$ , where $A$ is the plant pole polynomial, and let $D_f$ be a Schur polynomial satisfying $D_f D_f^* = A A^* D_{f_1} D_{f_1}^*$. Also define the following polynomials: $A_c = A A_q A_r$ and $A_f = A_h A_n A_{d_1} A$ .

**Theorem 2.2** : *Three DOF Linear Quadratic Gaussian Controller*

The LQG controller, for the system represented in polynomial system form, can be calculated as follows:

(a) Compute the strictly Schur spectral factors $D_c$ and $D_f$ using:

$$D_c^* D_c = A_r^* B^* Q_n B A_r + A^* A_q^* R_n A_q A \tag{2.45}$$

$$D_{f_1} D_{f_1}^* = B_h (A_{d_1} C_n C_n^* A_{d_1}^* + A_n C_{d_1} C_{d_1}^* A_n^*) B_h^* + A_h A_n A_{d_1} R A_{d_1}^* A_n^* A_h^* \tag{2.46}$$

where $D_f = A_{s_1} D_{f_1}$ and $A_{s_1}$ is strictly Schur and satisfies $A_{s_1} A_{s_1}^* = A A^*$.

(b) Calculate the solution $(G_0, H_0, F_0)$, with $F_0$ of smallest degree:

$$D_c^* G_o z^{-g} + F_o A_2 = Q_n B^* A_r^* D_f z^{k_f - g} \tag{2.47}$$

$$D_c^* H_o z^{-g} - F_o B_2 = R_n A^* A_q^* D_f z^{-g} \tag{2.48}$$

where $A_2 = A A_{d_1} A_n A_q B_h$ and $B_2 = B A_{d_1} A_n A_r B_h z^{-k_f}$. $\tag{2.49}$

(c) Obtain the solution $(L_0, P_0)$, with $P_0$ of smallest degree:

$$D_f^* D_c^* L_o z^{-g_o} + P_o A_3 = (B_h C_n C_n^* B_h^* + A_n A_h R A_h^* A_n^*) Q_n B^* A_r^* A^* A_{d_1}^* z^{k_f - g_o} \tag{2.50}$$

where $A_3 = A_n A_q B_h$ .

(d) Calculate the solution $(M_0, N_0)$, with $N_0$ of smallest degree:

$$D_c^* z^{-g_1} M_o + N_o A_q A_e = Q_n E_r B^* A_r^* z^{-g_1} \tag{2.51}$$

(e) Obtain the solution $(X_0, Z_0)$, with respect to $Z_0$ of smallest degree:

$$D_c^* z^{-g_2} X_o + Z_o A_q A_{d_2} = Q_n C_{d_2} B^* A_r^* z^{-g_2} \tag{2.52}$$

The scalars $g, g_0, g_1,$ and $g_2$ are the smallest positive integers to ensure that the above equations (2.47) to (2.52) are polynomial equations in $z^{-1}$. The feedback, tracking and feedforward controllers now follow as:

**Feedback :**

$$C_0 = T_f^{-1} N_f A_r A_h B_h^{-1} \tag{2.53}$$

**Reference :**

$$C_1 = T_f^{-1} M_o A_r D_f E^{-1} \tag{2.54}$$

**Feedforward :**

$$C_2 = T_f^{-1} (X_o D_f - C_{d2} z^{-k_f} N_f) A_r A_s B_s^{-1} C_{d2}^{-1} \tag{2.55}$$

where $T_f = A_q H_o + L_o A_{d_1} A_r z^{-k_f} B$ and $N_f = G_o - L_o A_{d_1} A$

**Sensitivity function:**

$$S = (D_c D_f)^{-1} A T_f \tag{2.56}$$

**Complementary sensitivity function:**

$$T = (D_c D_f)^{-1} A_r B z^{-k_f} N_f \tag{2.57}$$

**Optimal Solution**

Before presenting the proof of the theorem, the nature of the optimal solution will be discussed. To compute the optimal feedback, feedforward and reference input controllers two equations must be spectrally factored and five diophantine equations must be solved. The number of equations to be solved does of course decrease in less general problems. In fact one of the advantages of the current problem formulation is that a diophantine equation is associated with each of the control actions. Thus, in the following theorem, which summarises the computational procedure, the diophantine equations represent:

(a) The regulating feedback action (equations (2.47) and (2.48)).

(b) The presence of coloured measurement noise or output disturbance (see equation (2.50)).

(c) The presence of a reference signal and reference input subsystem (see equation (2.51)).

(d) The use of feedforward control signal and controller (see equation (2.52)).

These results can be simplified considerably in special cases. If, for example, the setpoint or reference input is zero then step (c) may be omitted in the calculation of the controller.

## 2.3.2   Three DOF Control Law Proof

The optimal cost-function was expanded in Section 2.2 in terms of the system trans-fer functions and a *completing the squares* argument was used to derive (2.22). Each of the terms in this equation may be simplified by using the polynomial equations (2.47) to (2.52). The existence and uniqueness of the solutions to these diophantine equations will first be established.

The terms $D_f^* D_c^*$ and $A_3$ in (2.50), $D_c^*$ and $A_q A_e$ in (2.51) and $D_c^*$ and $A_q A_{d_2}$ in (2.52) are each coprime pairs. It follows that the minimal degree solutions exist and are unique (see Jezek, 1982). Consider now the diophantine equations (2.47) and (2.48) :

$$\begin{bmatrix} D_c^* z^{-g} & A_2 & 0 & G_o \\ 0 & -B_2 & D_c^* z^{-g} & F_o \end{bmatrix} = \begin{bmatrix} Q_n B^* A_r^* D_f z^{k_f - g} \\ R_n A^* A_q^* D_f z^{-g} \end{bmatrix}$$

Note that $\det\left\{\begin{bmatrix} A_2 & Q_n B^* A_r^* D_f z^{k_f - g} \\ -B_2 & R_n A^* A_q^* D_f z^{-g} \end{bmatrix}\right\} = D_c^* D_c D_f z^{-g} A_{d_1} A_n B_h$ and write the above coupled equations as $TX = Y$. These equations have a solution if and only if $rank[T] = rank[T, Y]$. The greatest common divisors of all 1x1 and 2x2 minors of these matrices are 1 and $D_c^* z^{-g}$ , respectively. The rank of both matrices is at least unity. The only values of $z$ for which $T$ is rank 1 coincide with the values of $z$ for which $[T, Y]$ is rank 1. It follows that the bilateral diophantine equations have a solution (see Kucera, 1979).

To show that the solution $(G_0, F_0, H_0)$ is unique, note that $[-A_2 \; D_c^* z^{-g} \; B_2]$ is a basis for the kernel of $T$. Let $(G_0, F_0, H_0)$ denote a particular solution, so that

$$\begin{bmatrix} G \\ F \\ H \end{bmatrix} = \begin{bmatrix} G_o \\ F_o \\ H_o \end{bmatrix} + \begin{bmatrix} -A_2 \\ D_c^* z^{-g} \\ B_2 \end{bmatrix} p(z^{-1})$$

where $p(z^{-1})$ is an arbitrary polynomial. The particular solution satisfying $n_{fo} < g$ is unique since $F_0$ is the least degree representative of $F$, modulo $D_c^* z^{-g}$.

### Simplification of the Cost-function Integrand Terms

The feedback control ($M$ dependent) term in the cost-function (2.22) may now be simplified. From the polynomial models, (2.9) and (2.21) obtain:

$$\frac{\Phi_{h_0}}{Y_c^* Y_f^*} = \frac{Y_f Q_c W^*}{H_f Y_c^*} - \frac{(H_f \Phi_{nn} H_f^* + \Phi_{vv}) Q_c W^*}{H_f Y_c^* Y_f^*}$$

$$= \frac{Q_n B^* A_r^* D_f z^{k_f}}{A_{d_1} A_n A A_q B_h D_c^*} - \frac{(B_h C_n C_n^* B_h^* + A_n A_h R A_h^* A_n^*) Q_n B^* A_r^* A^* A_{d_1}^* z^{k_f}}{A_n A_q B_h D_c^* D_f^*}$$

$$= \left(\frac{G_o}{A_2} + \frac{F_o z^g}{D_c^*}\right) - \left(\frac{L_o}{A_3} + \frac{P_o z^{g_o}}{D_f^* D_c^*}\right)$$

Now $M = C_0(1 + H_f W C_0)^{-1}$ and hence if the feedback controller is written as $C_0 = C_{0d}^{-1} C_{0n}$ then,

$$Y_c M Y_f = \frac{D_c D_f C_{0n}}{A_q A_n A_{d_1} A_r A (C_{0d} A_h A + B_h z^{-k_f} B C_{0n})}$$

The first (feedback related) squared term in (2.22) therefore becomes:

$$Y_c M Y_f - \frac{\Phi_{h0}}{Y_c^* Y_f^*} = \frac{D_c D_f B_h C_{0n} - (C_{0d} A_h A + B_h z^{-k_f} B C_{0n}) A_r (G_o - L_o A_{d_1} A)}{A_r A_2 (C_{0d} A_h A + B_h z^{-k_f} B C_{0n})}$$

$$- \left( \frac{F_o z^g}{D_c^*} - \frac{P_o z^{g_o}}{D_f^* D_c^*} \right)$$

This may be simplified by noting that if (2.47) and (2.48) are added (after appropriate multiplication and simplification) the following equation is obtained:

$$D_c D_f = G_o A_r B z^{-k_f} + H_o A_q A$$

This is referred to as the *implied equation*, since it is determined by the pair of equations (2.47) and (2.48). Substituting and simplifying, using this equation, gives:

$$Y_c M Y_f - \frac{\Phi_{h0}}{Y_c^* Y_f^*}$$

$$= \left[ \frac{C_{0n}(A_q B_h H_o + L_o A_{d_1} B_h z^{-k_f} A_r B) - C_{0d}(G_o - L_o A_{d_1} A) A_h A_r}{A_3 A_{d_1} A_r (C_{0d} A_h A + B_h z^{-k_f} B C_{0n})} \right]$$

$$- \left( \frac{F_o z^g}{D_c^*} - \frac{P_o z^{g_o}}{D_f^* D_c^*} \right) \qquad (2.58)$$

This equation may clearly be written in the form:

$$Y_c M Y_f - \frac{\Phi_{h0}}{Y_c^* Y_f^*} = T_0^+ + T_0^- \qquad (2.59)$$

where the term within the square brackets is denoted by $T_0^+$. This term is asymptotically stable since $A_3 A_{d_1} A_r$ is strictly Schur and the closed-loop characteristic polynomial that determines stability: $\rho_c = (C_{0d} A_h A + B_h z^{-k_f} B C_{0n})$ is required to be strictly Schur for $J_{\min} < \infty$. The final term in (2.59) is strictly unstable, since $D_c^*$ and $D_f^*$ are strictly non-Schur.

**Tracking Reference Controller ($C_1$ Dependent Term )**

From (2.21) and the definitions of $A_c$ and $A_f$ obtain:

$$\frac{\Phi_{h1}}{Y_c^* W_r^*} = \frac{Q_c \Phi_{rr} W^*}{Y_c^* W_r^*} = \frac{Q_n E_r B^* A_r^*}{(D_c^* A_q A_e)} = \left( \frac{M_o}{A_q A_e} + \frac{N_o z^{g_1}}{D_c^*} \right)$$

The second (tracking related) term in the cost expression (2.22) now becomes:

$$Y_c S C_1 W_r - \frac{\Phi_{h1}}{Y_c^* W_r^*} = \frac{D_c C_1 E_r}{A_c (1 + \frac{B_h}{A_h} z^{-k_f} \frac{BC_0}{A}) A_e} - \left( \frac{M_o}{A_q A_e} + \frac{N_o z^{g_1}}{D_c^*} \right)$$

and if the tracking controller is written as $C_1 = C_{1d}^{-1} C_{1n}$ then,

$$Y_c S C_1 W_r - \frac{\Phi_{h1}}{Y_c^* W_r^*} = \left[ \frac{C_{1n} D_c E_r A_h C_{0d} - C_{1d} M_o A_r (A_h A C_{0d} + B_h z^{-k_f} B C_{0n})}{C_{1d} A_q A_e A_r (A_h A C_{0d} + B_h z^{-k_f} B C_{0n})} \right]$$

$$- \frac{N_o z^{g_1}}{D_c^*} \tag{2.60}$$

The tracking controller $C_1$ is outside the loop and must therefore be asymptotically stable for the cost $J_{\min} < \infty$. The terms $A_q, A_e$ and $A_r$ and the characteristic polynomial $\rho_c$ are strictly Schur polynomials. Thus, the tracking cost term (2.60) may be written in the form:

$$Y_c S C_1 W_r - \frac{\Phi_1}{Y_c^* W_r^*} = T_1^+ + T_1^- \tag{2.61}$$

where $T_1^+$ again denotes the term within the square brackets in (2.60).

**Feedforward Controller ($C_2$ Dependent Term )**
From (2.21) and the definitions of $A_c$ and $A_f$ obtain:

$$\frac{\Phi_{h2}}{Y_c^* W_r^*} = \frac{Q_c \Phi_{rr} W^*}{Y_c^* W_r^*} = \frac{Q_n E B^* A_c^*}{(D_c^* A_q A^*)} = \left( \frac{X_o}{A_q A_{d_2}} + \frac{Z_o z^{g_2}}{D_c^*} \right) \tag{2.62}$$

The third (feedforward related) term in (2.22) follows as:

$$Y_c S C_{02} W_{d_2} - \frac{\Phi_{h2}}{Y_c^* W_{d_2}^*} = \frac{D_c C_{02} C_{d_2}}{A_c (1 + \frac{B_h}{A_h} z^{-k_f} \frac{BC_0}{A}) A_{d_2}} - \left( \frac{X_o}{A_q A_{d_2}} + \frac{Z_o z^{g_2}}{D_c^*} \right)$$

and if the feedforward controller is written as $C_{02} = C_{02d}^{-1} C_{02n}$ then,

$$Y_c S C_{02} W_{d_2} - \frac{\Phi_{h2}}{Y_c^* W_{d_2}^*}$$

$$= \left[ \frac{C_{02n} D_c C_{d2} A_h C_{0d} - C_{02d} X_o A_r (A_h A C_{0d} + B_h z^{-k_f} B C_{0n})}{A_q A_{d_2} A_r (A_h A C_{0d} + B_h z^{-k_f} B C_{0n}) C_{02d}} \right] - \frac{Z_o z^{g_2}}{D_c^*} \tag{2.63}$$

The term within the square brackets must be asymptotically stable for the cost $J_{\min} < \infty$. This may be confirmed by first deriving the transfer-function between the measured disturbance and the control signal. Setting other noise sources to zero:

$$u = -C_2 W_s d_2 - C_0 H_f d_2 - C_0 H_f W u$$

giving

$$u = -(1 + C_0 H_f W)^{-1}(C_2 W_s + C_0 H_f)d_2 = -SC_{02}d_2$$

$$= \frac{-C_{0d}A_h AC_{02n}d_2}{(A_h AC_{0d} + B_h z^{-k_f} BC_{0n})C_{02d}}$$

This transfer-function must clearly be asymptotically stable if the variance of the control signal is to be finite. Thus, equation (2.63) may be written in the form:

$$Y_c SC_{02}W_{d_2} - \frac{\Phi_{h2}}{Y_c^* W_{d_2}^*} = T_2^+ + T_2^- \tag{2.64}$$

where $T_2^+$ denotes the term within the square brackets in (2.63).

## 2.3.3   Cost-function Minimisation

Given the simplification of terms in the cost-function presented above, the cost minimisation procedure may be followed. Note that the cost-function (2.22) may be written using (2.59), (2.61) and (2.64) as:

$$J = \frac{1}{2\pi j} \oint_{|z|=1} \{(T_0^+ + T_0^-)(T_0^+ + T_0^-)^* + (T_1^+ + T_1^-)(T_1^+ + T_1^-)^*$$

$$+ (T_2^+ + T_2^-)(T_2^+ + T_2^-)^* + \Phi_{o1}\} \frac{dz}{z} \tag{2.65}$$

From the Residue theorem of complex analysis (Churchill, 1948) the integrals of the cross-terms $T_j^+ T_j^{-*}$, $T_j^- T_j^{+*}$, for $j = \{0, 1, 2\}$, are zero. This result follows because $\oint T_j^+ T_j^{-*} dz/z = -\oint T_j^- T_j^{+*} dz/z$ but the term $T_j^- T_j^{+*}$ is analytic for $|z| < 1$, so that the sum of the residues obtained in calculating $\oint T_j^- T_j^{+*} dz/z$ is zero. The cost-function therefore simplifies as:

$$J = \frac{1}{2\pi j} \oint_{|z|=1} \sum_{j=0}^{2} \{(T_j^+ T_j^{+*} + T_j^- T_j^{-*}) + \Phi_{o1}\} \frac{dz}{z} \tag{2.66}$$

Since the terms $T_j^-$ and $\Phi_{o1}$ are independent of the choice of the controller elements the criterion $J$ is minimised when the terms $T_j^+ = 0$, for each $j = \{0, 1, 2\}$. Thus, from (2.58), (2.60) and (2.63) obtain:

**Feedback controller :**

$$C_0 = \left((A_q H_o + L_o A_{d_1} A_r z^{-k_f} B)B_h\right)^{-1} (G_o - L_o A_{d_1} A)A_h A_r \tag{2.67}$$

**Reference controller :**

$$C_1 = (D_c E_r A_h C_{0d})^{-1} M_o A_r (A_h AC_{0d} + B_h z^{-k_f} BC_{0n}) \tag{2.68}$$

**Feedforward controller :**

$$C_2 = \left((D_c C_{d_2} A_h C_{0d})^{-1} X_o A_r (A_h A C_{0d} + B_h z^{-k_f} B C_{0n}) - C_0 H_f\right) W_s^{-1} \qquad (2.69)$$

The expressions for the reference and feedforward controllers may be further simplified by substituting for the feedback controller polynomials $C_0 = C_{0d}^{-1} C_{0n}$ and by then cancelling common terms. The resulting feedforward controller structure is illustrated in Fig. 2.3. The minimum value for the cost-function:

$$J_{\min} = \frac{1}{2\pi j} \oint_{|z|=1} \sum_{j=0}^{2} \{T_j^- T_j^{-*} + \Phi_{o1}\} \frac{dz}{z} \qquad (2.70)$$

**Sensitivity function:**

$$S = (1 + H_f W C_0)^{-1} \qquad (2.71)$$

and substituting from (2.53) and using the implied equation obtain:

$$S = T_f (T_f + z^{-k_f} \frac{B}{A} N_f A_r)^{-1} = (D_c D_f)^{-1} A T_f \qquad (2.72)$$

**Complementary sensitivity function :**

$$T = 1 - S \qquad (2.73)$$

and substituting from the *implied equation* obtain:

$$T = (D_c D_f)^{-1}(D_c D_f - A T_f) = (D_c D_f)^{-1} N_f A_r B z^{-k_f} \qquad (2.74)$$

This completes the proof of the results summarised in Theorem 2.2. The stability properties of the solution may now be considered.

**Lemma 2.1** : The degree of stability of the closed-loop system may be determined from the implied diophantine equation:

$$G_o A_r B z^{-k_f} + H_o A_q A = D_c D_f \qquad (2.75)$$

where $D_c$ and $D_f$ are strictly Schur polynomials.

**Proof** : Adding (2.47) $\times A_r B z^{-k_f}$ to (2.48) $\times A A_q$ , using (2.45) and dividing, gives the implied equation that defines the characteristic polynomial for the closed-loop system:

$$G_o A_r A_h (B B_h z^{-k_f}) + H_o A_q B_h (A A_h) = D_f D_c A_h B_h$$

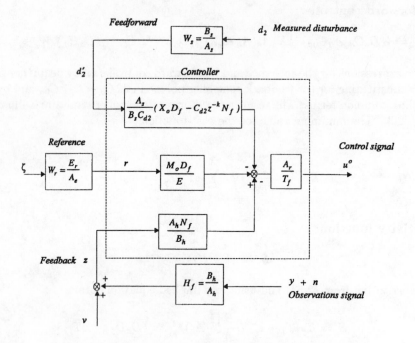

Fig. 2.3 : **Structure of the Two Degrees of Freedom and Feedforward LQG Controller (3 DOF Design)**

## 2.3.4   Optimal Solution: Properties and Structure

The optimal solution has several interesting properties which may be noted here:

(a) With the given assumptions on $W_s$ and $H_f$ the feedforward or feedback controllers simply cancel these transfers, respectively.

(b) Each of the controllers include zeros due to the $A_r$ weighting term.

(c) If the output disturbance $n \equiv 0$ , $H_f = 1$ and the error weighting $Q_c =$ *constant* then $L_0 = 0$ (see (2.50)) and the feedback controller simplifies as $C_0 = (A_q H_o)^{-1}(G_o A_r)$ , showing that desired pole and zero terms may easily be introduced into the controller via the choice of $A_q$ and $A_r$ .

(d) The reference or tracking controller depends upon the reference model $A_e^{-1}E$ but the spectral factors $D_c$ and $D_f$ , and the feedback/feedforward controllers are independent of the reference model.

(e) Both the reference input and feedback controllers are optimal even if the measurable disturbance $d_2$ is set to zero.

(f) Letting $C_{d2} = k C'_{d2}$ and substituting in (2.55), as $k$ is varied the feedforward controller $C_2$ remains unchanged. This feature will clearly be valuable in a self-tuning system (where $W_s$ is known and $C_{d2}$ is estimated) if errors in the $C_{d2}$

polynomial are mainly of this form, since the feedforward controller calculation is unaffected by errors in the gain $k$.

The system structure used above may be generalised slightly by passing the measureable disturbance $d_2$ through a colouring filter $W_{d3} = A_{d3}^{-1} C_{d3}$ , before it is added to the plant output. The disturbance being measured is then different to that acting on the system. However, $W_{d3}$ can be assumed to be stable and minimum phase and hence the previous results also provided the solution to this problem, by redefining:

$$A_{d2} \rightarrow A_{d2} A_{d3}, \quad C_{d2} \rightarrow C_{d2} C_{d3}, \quad A_s \rightarrow A_s A_{d3}, \quad B_s \rightarrow B_s A_{d3}$$

**Example 2.1** : *Feedback and Feedforward Controller Calculations*
Let the system shown in Fig. 2.2 have unity variance white noise sources and let the following transfer-functions be defined:

$$H_f = W_s = 1, \quad W_n = 0, \quad R = r$$

$$W_r = W_{d2} = \frac{1}{1 - \alpha_2 z^{-1}}, \quad W_{d1} = \frac{1}{1 - \alpha_1 z^{-1}}$$

$$W = \frac{z^{-1}}{1 - az^{-1}}, \quad |\alpha_1| < 1, \quad |\alpha_2| < 1$$

The optimal control cost function weights are defined as:

$$Q_c = 1 / \left( (1 - \alpha_q z^{-1})(1 - \alpha_q z) \right) \quad and \quad R_c = r_c$$

The control spectral factor $D_c$ follows from equation (2.45):

$$D_c^* D_c = 1 + (1 - az^{-1})(1 - \alpha_q z^{-1}) r_c (1 - \alpha_q z)(1 - az)$$

giving $D_c = d_{co} + d_{c1} z^{-1} + d_{c2} z^{-2}$, where $A_c = (1 - az^{-1})(1 - \alpha_q z^{-1})$.  The filter spectral factor follows from (2.46):

$$D_{f_1} D_{f_1}^* = 1 + (1 - \alpha_1 z^{-1}) r (1 - \alpha_1 z)$$

giving $D_{f1} = d_{flo} + d_{f11} z^{-1}$ and $D_f = A_{s1} D_{f1} = d_{fo} + d_{f1} z^{-1} + d_{f2} z^{-2}$.
The diophantine equations follow from (2.47), (2.48) and (2.50), noting $g = 2$,

$$(d_{c2} + d_{c1} z^{-1} + d_{co} z^{-2})(g_o + g_1 z^{-1} + g_2 z^{-2})$$

$$+ (f_o + f_1 z^{-1})(1 - az^{-1})(1 - \alpha_1 z^{-1})(1 - \alpha_q z^{-1})$$

$$= z(d_{fo} + d_{f1} z^{-1} + d_{f2} z^{-2}) z^{-2}$$

$$(d_{c2} + d_{c1}z^{-1} + d_{co}z^{-2})(h_o + h_1z^{-1} + h_2z^{-1}) \ - (f_o + f_1z^{-1})z^{-1}(1 - \alpha_1z^{-1})$$

$$= r_c(1 - az)(1 - \alpha_qz)(d_{fo} + d_{f1}z^{-1} + d_{f2}z^{-2})z^{-2}$$

Since $C_n$ and $R$ are null, from (2.50), $L_o = P_o = 0$:

$$(d_{c2} + d_{c1}z^{-1} + d_{co}z^{-2})(m_o + m_1z^{-1}) \ + (n_o + n_1z^{-1})(1 - \alpha_qz^{-1})(1 - \alpha_2z^{-1}) = z^{-1}$$

From equation (2.52):

$$(d_{c2} + d_{c1}z^{-1} + d_{co}z^{-2})(x_o + x_1z^{-1}) \ + (z_o + z_1z^{-1})(1 - \alpha_qz^{-1})(1 - \alpha_2z^{-2}) = z^{-1}$$

Clearly in this problem $M_0 = X_0$ and $N_0 = Z_0$ . Thence, the optimal controllers may be expressed in the form:

**Feedback controller:**

$$C_0 \ = \ \frac{(g_o + g_1z^{-1} + g_2z^{-2})}{(1 - \alpha_qz^{-1})(h_o + h_1z^{-1} + h_2z^{-2})}$$

**Reference controller:**

$$C_1 \ = \ \frac{(x_o + x_1z^{-1})(d_{fo} + d_{f1}z^{-1} + d_{f2}z^{-2})}{(1 - \alpha_qz^{-1})(h_o + h_1z^{-1} + h_2z^{-2})}$$

**Feedforward controller:**

$$C_2 \ = \ \frac{(x_o + x_1z^{-1})(d_{fo} + d_{f1}z^{-1} + d_{f2}z^{-2}) - (g_o + g_1z^{-1} + g_2z^{-2})}{(1 - \alpha_qz^{-1})(h_o + h_1z^{-1} + h_2z^{-2})}$$

An explicit self-tuner (Grimble, 1984, 1985) can be constructed for such a system by estimating the unknown coefficients $a$, $\alpha_1$ and $\alpha_2$ , and the polynomial $D_f$ , and by solving the diophantine equations for the polynomials $G_0, H_0$ and $X_0$.

The LQG controller has many potential applications in process control problems (Warren, 1992). The process control example which follows is for the simplest case of a one degree-of-freedom structure with no feedforward action.

**Example 2.2** : *Hydrogen Reformer LQG Control Problem*

The design of the controllers for a Hydrogen Reformer in a petrochemical plant is now considered. The objective of the hydrogen reformer process is to produce hydrogen from desulphurised hydrocarbons by using a catalysis. To generate hydrogen the hydrocarbons are mixed with superheated steam before entering the reformer tubes, where a nickel catalyst is heated at high temperature (about 750°C) to provide the hydrogen. The high temperature is necessary to speed up the reaction and is produced by burning fuel in the reformer. The fuel flow must be controlled to maintain the desired temperature of the catalyst. A schematic diagram of the system is shown in Fig. 2.4.

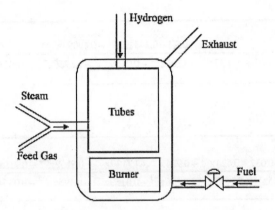

**Fig. 2.4** : Hydrogen Reformer System

The control system should maintain the desired catalyst temperature by modifying the amount of fuel which is fed into the reformer. The catalyst temperature and the fuel flow are measured to implement a cascaded control, as shown in Fig. 2.5.

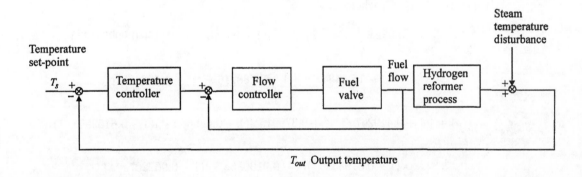

**Fig. 2.5** : Block Diagram of Hydrogen Reformer Process

In the actual process there are several disturbances, such as varying feed flow, fuel gas quality, steam temperature, etc. The most important disturbance is the steam temperature variation, which modifies the catalyst temperature directly. Hence, the control system should attenuate this disturbance as much as possible by acting on the fuel flow set-point. The flow control loop can be considered adequate for present purposes, so only the design of the controller for the outer loop must be performed, with the main objective of providing tighter control of the output temperature, without deteriorating the response to command changes.

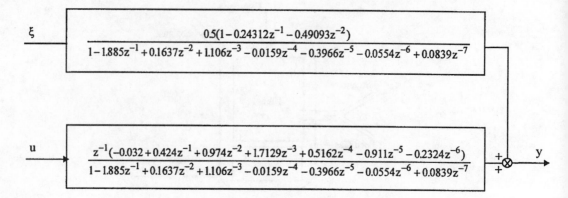

**Fig. 2.6** : Identified Process and Disturbance Model

Consider the process control system shown in Fig. 2.5, which represents a cascaded system with an outer temperature control loop and an inner flow control loop. From identification experiments on the process (Shakoor, 1996) the total process and inner-loop control system has the discrete ARMAX form shown in Fig. 2.6.

**Definition of System Polynomials**

The plant transfer function $W = A^{-1}B$ and (unmeasurable) disturbance model $W_{d1} = W_d = A^{-1}C_d$ where

$$B = -0.032z^{-1}(1 + 0.2453z^{-1})(1 - 0.623257z^{-1})(1 + 0.99994z^{-1})$$

$$\times (1 - 15.4484z^{-1})[(1 + 0.788216z^{-1})^2 + 1.566429^2]$$

$$A = [(1 + 0.419267z^{-1})^2 + 0.379915^2](1 + 0.58958z^{-1})(1 - 0.615995z^{-1})$$

$$\times (1 - 0.81983z^{-1})(1 - 0.910085z^{-1})(1 - 0.967207z^{-1})$$

$$C_d = 0.0594(1 - 0.24312z^{-1} - 0.49093z^{-2})$$

The plant is clearly stable and non-minimum phase, and has approximate integral action. It contains a unit-delay with a sample interval of $T = 30s$.

Instead of increasing the order of the system by introducing a new pole for the reference model $W_r$, let

$$W_r = E_r/A_e = 0.1/(1 - 0.967207z^{-1})$$

Since this is a pole due to the $A$ polynomial it will not inflate the order of the controller and at the same time it represents the required approximate integrator. The measurement noise model $W_n = 0$ in this case.

**Weighting Definitions**

Writing the error and control weightings $Q_c = H_{qc}^* H_{qc}$ and $R_c = H_{rc}^* H_{rc}$ define these functions:

$$H_{qc} = \frac{(1 - 0.9z^{-1})^2}{(1 - 0.999z^{-1})} \quad \text{and} \quad H_{rc} = 300(1 - 0.92z^{-1})^2$$

The former weighting function involves an approximate integrator and lead terms in the mid-frequency region. This will ensure approximate integral action at low frequencies and that some error costing will still be included in the mid-frequency region. The control weighting involves a lead term to ensure the controller rolls off in the high frequency region. This is important since a measurement noise model was not included and hence to ensure reasonable high frequency behaviour the control weighting must penalise the high frequency control signal components.

The relative gains between the control and error weightings was selected to ensure the crossover frequency between the frequency response plots of $H_{rc}$ and $H_{qc}W$ was appropriate. A rule of thumb is that this frequency will often correspond with the unity-gain crossover-frequency for the open-loop system which is close to the closed-loop bandwidth frequency.

**Computed LQG Feedback Controller**

$$C_0 = \frac{\begin{array}{c} 2.2944 \times 10^{-3}[(1 + 0.4192667z^{-1})^2 + 0.3799^2 z^{-2}] \\ \times (1 + 0.5895834z^{-1})(1 - 0.615057z^{-1})(1 - 0.820574z^{-1}) \\ \times (1 - 0.91091z^{-1})(1 - 0.934995z^{-1}) \end{array}}{\begin{array}{c} [(1 + 0.4013995z^{-1})^2 + 0.366497^2 z^{-2}][(1 - 0.51521z^{-1})^2 + 0.1874618^2 z^{-2}] \\ \times (1 + 0.589384z^{-1})(1 - 0.6496282z^{-1}) \\ \times [(1 - 0.805577z^{-1})^2 + 0.340925^2 z^{-2}](1 - 0.999z^{-1}) \end{array}}$$

This controller is of 9th order, and may be model reduced with little loss of accuracy. An appropriate reduced order controller becomes:

$$\tilde{C}_0 = \frac{10^{-3}(2.593706 - 6.91605z^{-1} + 6.137723z^{-2} - 1.81268z^{-3})}{\begin{array}{c} (1 - 3.64057z^{-1} + 5.364865z^{-2} - 3.995954z^{-3} \\ + 1.501476z^{-4} - 0.2297715z^{-5}) \end{array}}$$

The open loop gain is found to be $W(1)C_0(1) = 226.846$. Note that the plant gain includes a scaling factor of 1000 because of the physical units used.

**Results**

The plant frequency response is shown in Fig. 2.7. The disturbance and reference models are shown in Fig. 2.8. Notice that the reference model frequency response has a lower gain at low frequency, since disturbance rejection is considered more important than tracking properties. This suggests that the gain of the disturbance model should be chosen to be dominant through the most important frequency range .

The frequency responses of the dynamic weighting functions are shown in Fig. 2.9. Using the above rule of thumb enables the relative gains of the two weighting functions to be selected. The lead term in the control weighting function will ensure the controller rolls off at high frequency and the integral term in the error weighting function will ensure high gain in the controller at low frequencies. The resulting feedback controller frequency response is shown in Fig. 2.10.

The open loop frequency response for the system is shown in Fig. 2.11. There is some overshoot on the sensitivity and complementary sensitivity functions, as shown in Fig. 2.12. This suggests that there will be corresponding peaks in the disturbance and tracking time-domain step responses.

The peak in the complementary sensitivity function will translate into a peak overshoot on the time response but this is not of major importance in this problem. The control sensitivity function frequency response is shown in Fig. 2.13. This rolls of at high frequency since the controller frequency response also rolls off.

### Time Responses

The unit step response of the closed loop system is shown in Fig. 2.14. The overshoot is due to the lack of importance attached to the step response model, as indicated in the relative sizes of the disturbance and reference models. The output of the system when only the disturbance is present is also indicated and the disturbance rejection is adequate. The effectiveness of the disturbance rejection is illustrated in Fig. 2.15. This shows both the disturbance and the output due to the disturbance under closed-loop control. Clearly the closed-loop system should significantly attenuate the disturbances.

The control signal response due to the reference change and due to the disturbance inputs is shown in Fig. 2.16. The control signal variations also have physically realistic rates of change and magnitude variations.

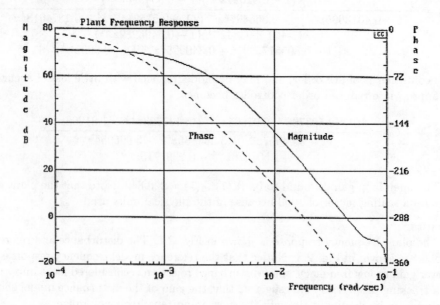

Fig. 2.7 : Plant Frequency Response Gain and Phase Plots

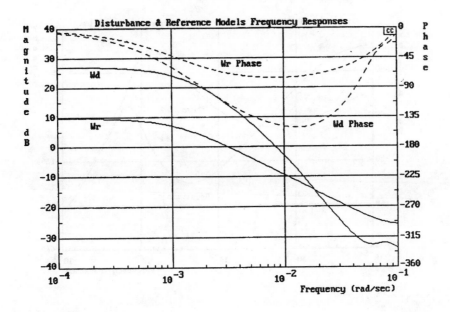

Fig. 2.8 : Frequency Response of Disturbance and Reference Models

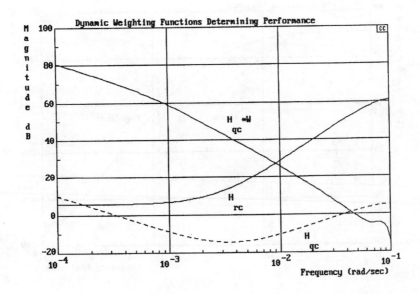

Fig. 2.9 : Frequency Response of Dynamic Weighting Functions

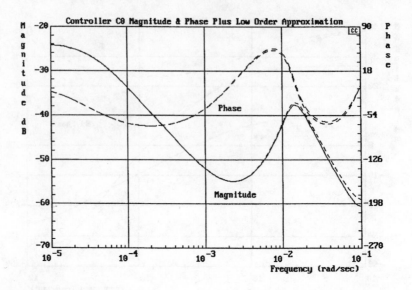

Fig. 2.10 : Feedback Controller Frequency Response and Low Order Approximation

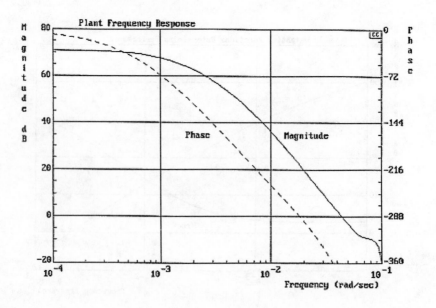

Fig. 2.11 : Open Loop Transfer Function Frequency Responses

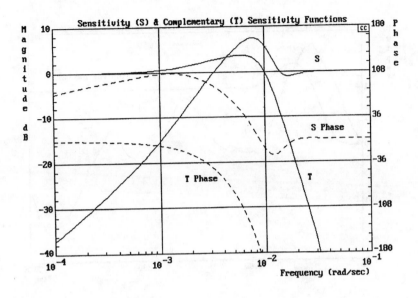

**Fig. 2.12** : Sensitivity and Complementary Sensitivity Function
Frequency Responses

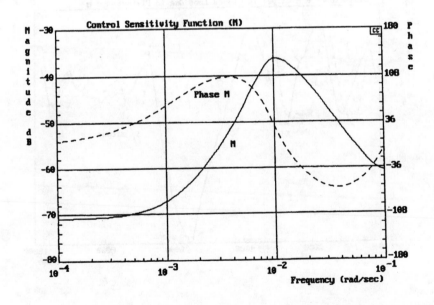

**Fig. 2.13** : Frequency Response of Control Sensitivity Function

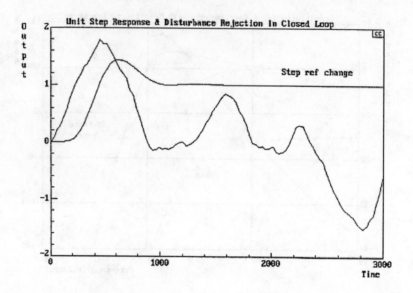

Fig. 2.14 : Unit Step Response and Output due to Disturbance Inputs

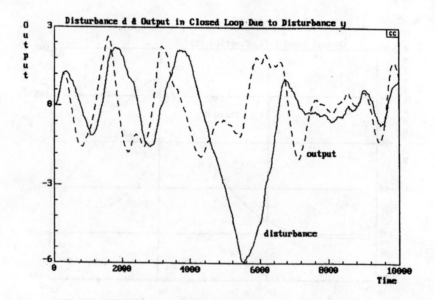

Fig. 2.15 : Disturbance Signal and the Output of the Closed-loop System
due to the Disturbance

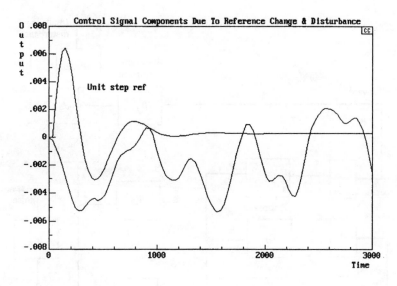

**Fig. 2.16** : Components of the Control Signal due to the Reference Change

and due to the Disturbance

# 2.4  Multiple DOF LQG Control Problem

The solution of a more general $H_2$ control problem will now be considered, where the controller can utilise the $2\frac{1}{2}$ or $3\frac{1}{2}$ DOF structures, often utilised in the aerospace industry. The cost-function will also provide greater generality by enabling the feedback, feedforward and tracking signals to be costed independently. This provides valuable additional tuning capabilities. The analysis will again begin by considering a stochastic setting for the system description.

The linear, discrete-time, single-input, single-output, feedback system of interest is shown in Fig. 2.17. The white noise sources $\{\zeta(t)\}, \{\xi_1(t)\}, \{\xi_2(t)\}$ and $\{\omega(t)\}$ are assumed to be zero mean and mutually statistically independent. The covariances for $\{\zeta(t)\}, \{\xi_1(t)\}, \{\xi_2(t)\}$ and $\{\omega(t)\}$ are, without loss of generality, taken to be unity. The plant $W(z^{-1})$ is assumed to be free of unstable hidden modes and the reference $W_r(z^{-1})$, and the output disturbance $W_n(z^{-1})$, subsystems are assumed to be asymptotically stable.

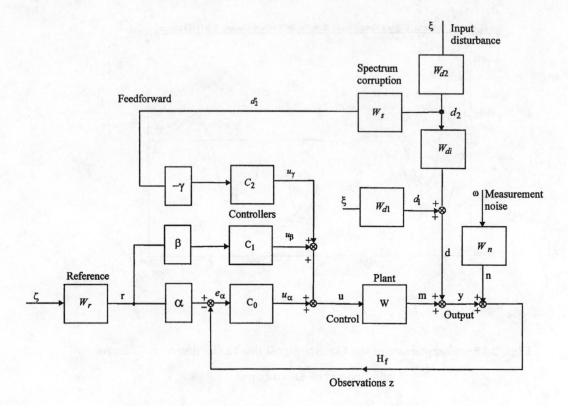

**Fig. 2.17** : Canonical $3\frac{1}{2}$DOF feedback System with Input Disturbance
and Measurement Noise
($\alpha = \beta = \gamma = 1$ for $3\frac{1}{2}$ DOF and $\alpha = 0, \beta = \gamma = 1$ for 3 DOF)

The plant output $y(t) = m(t) + d(t)$ is formed from the controlled output $m(t) = W(z^{-1})u(t)$, together with the input disturbance signal: $d(t) = d_0(t) + d_1(t)$ where $d_1(t) = W_{d1}(z^{-1})\xi_1(t), d_0(t) = W_{di}(z^{-1})d_2(t)$ and $d_2(t) = W_{d2}(z^{-1})\xi_2(t)$. The subsystem $W_{d1}$ represents the unmeasured disturbance and $W_{d2}$ represents the generation of that component of the disturbances which can be measured. The output $\{y(t)\}$ is required to follow the reference signal $\{r(t)\}$ to minimise the model following tracking error:

$$e(t) = W_i r(t) - y(t) \qquad (2.76)$$

where $W_i$ denotes the transfer of an ideal response model. Model following techniques are often utilised in flight controls and the addition of $W_i$ adds a little complexity but provides greater flexibility. The output disturbance or measurement noise model which includes a white noise component is represented by the signal $n(t) = W_n(z^{-1})\omega(t)$. The total observations signal output from the system is given as:

$$z(t) = y(t) + n(t) \qquad (2.77)$$

and the controller input from the error channel is denoted by $e_\alpha(t)$:

$$e_\alpha(t) = \alpha r(t) - z(t) \tag{2.78}$$

The value of $\alpha$ depends upon the controller structure employed. The various controller structures (Grimble, 1994) are obtained by appropriately defining the scalars $\alpha, \beta$, and $\gamma$ in Fig. 2.17. This is clarified in Table 2.2, where $\beta_1 = 1 - \beta$ and $\gamma_1 = 1 - \gamma$.

**Table 2.2 : Controller Structure Path Switch Settings**

| Controller structure | $\alpha$ | $\beta$ | $\gamma$ | $\beta_1$ | $\gamma_1$ |
|---|---|---|---|---|---|
| 1 | 1 | 0 | 0 | 1 | 1 |
| 2 | 0 | 1 | 0 | 0 | 1 |
| 3 | 0 | 1 | 1 | 0 | 0 |
| $2\frac{1}{2}$ | 1 | 1 | 0 | 0 | 1 |
| $3\frac{1}{2}$ | 1 | 1 | 1 | 0 | 0 |

## 2.4.1 Polynomial System Description

The various subsystems have the following polynomial system descriptions (Kucera, 1979):

**Plant:**

$$W = A^{-1}B \tag{2.79}$$

**Unmeasurable disturbance:**

$$W_{d1} = A^{-1}C_{d1}$$

**Measurable disturbance:**

$$W_{d2} = A_{d2}^{-1}C_{d2} \quad , \quad W_s = A_s^{-1}B_s z^q \quad , \quad W_{di} = A_{di}^{-1}C_{di} \tag{2.80}$$

where $q \geq 0$, ($q > 0$ if future disturbance signal knowledge available).

**Measurement noise:**

$$W_n = A_n^{-1}C_n \tag{2.81}$$

**Reference:**

$$W_r = A^{-1}E$$

These subsystems are assumed to be causal ($A(0) = A_n(0) = A_s(0) = 1$). The plant may also be represented in the coprime polynomial form $W = A_0^{-1}B_0$ where the greatest common divisor of $A$ and $B$ is denoted by $U_0$ and $A = A_0U_0$, $B = B_0U_0$. The common denominator employed above can also be used for the measurable disturbance and tracking models. That is,

$$W_{di}W_{d2} = (A_{di}A_{d2})^{-1}C_{di}C_{d2} = A^{-1}C \quad \text{and} \quad W_r = A_e^{-1}E_r = A^{-1}E \tag{2.82}$$

The use of a common denominator involves no lack of generality and is valid even if the plant is unstable ($C$ and $E$ would then include cancelling common factors). The models $W_s$ and $W_{di}$ in Fig. 2.17 describe the different dynamics through which the disturbance $d_2$ is measured and through which the plant output is affected, respectively.

**Sensitivity functions**

The following sensitivity functions are again required:

$$S = (1 + WC_0)^{-1}, \quad M = C_0(1 + WC_0)^{-1}, \quad T = WC_0(1 + WC_0)^{-1}$$

**Ideal response model**

The model following capability can be provided by introducing the ideal response model $W_i$. That is, the system output can be required to follow the output of an ideal response model, rather than following the reference signal. By this means abrupt changes in the reference signal can be filtered to give smoother more realistic variations in the desired response signal. The ideal response model is driven by the reference signal, to obtain the desired output response signal: $y_d(t) = W_i(z^{-1})r(t)$ where in polynomial system model form: $W_i = A_i^{-1}B_i$.

**Assumptions**

1(i) The plant can be stable or unstable but any common factor $U_0$ must be strictly Schur.

1(ii) With little loss of generality $E_r$ and $C_{d2}$ can be assumed to be strictly Schur polynomials.

1(iii) The measureable disturbance signal path model $W_{di}$ will be assumed to be strictly stable and $W_s$ is assumed to be minimum-phase and strictly stable.

1(iv) The ideal response model $W_i$ for the system is necessarily assumed to be strictly stable

## 2.4.2    Feedback/Feedforward System Equations

From inspection of the feedforward/feedback system in Fig. 2.17 the equations, after substituting for the sensitivities, become:

**Output:**

$$y = Wu + d = W(C_0(\alpha r - n - y) + C_1\beta r - C_2\gamma d_2') + d$$

$$= S(W[C_0\alpha r - C_0 n - C_2\gamma d_2'] + d + WC_1\beta r) \qquad (2.83)$$

**Observations:**

$$z = y + n = S(W[C_0\alpha r - C_2\gamma d_2'] + d + n + WC_1\beta r) \qquad (2.84)$$

**Controller input:**

$$e_\alpha = \alpha r - z = S((\alpha r - d - n) + WC_2\gamma d_2' - WC_1\beta r)$$

**Tracking error :**

$$e = W_i r - y = W_i r - d - SW(C_0(\alpha r - d - n) - C_2 \gamma d_2' + C_1 \beta r)$$

$$= (W_i - WSC_{01})r - (1 - WM)d_1 - (W_{di} - WSC_{02})d_2 + WMn \qquad (2.85)$$

where $C_{01} = C_0 \alpha + C_1 \beta$ and $C_{02} = C_0 W_{di} + C_2 \gamma W_s$ .

**Control signal:**

$$u = u_\alpha + u_\beta + u_\gamma = C_0(\alpha r - n - d - Wu) + C_1 \beta r - C_2 \gamma W_s d_2$$

$$= SC_{01}r - Md_1 - SC_{02}d_2 - Mn \qquad (2.86)$$

## 2.4.3 Error and Control Signal Costing Terms

The cost-function to be defined must penalise both the error and control signal terms. Inspection of (2.85) and (2.86) reveals that the various terms are statistically independent. This simplifies later calculations, since the variance of the different signals involves a sum of (statistically independent) terms. It is explained below that this property is maintained if these expressions ((2.85) and (2.86)) are written as:

$$e = (W_i - WSC_{01})\beta r + (W_i - WM)\beta_1 r - (1 - WM)d_1$$

$$-(W_{di} - WSC_{02})\gamma d_2 - (1 - WM)\gamma_1 d_0 + WMn \qquad (2.87)$$

and

$$u = SC_{01}\beta r + M\beta_1 r - Md_1 - SC_{02}\gamma d_2 - M\gamma_1 d_0 - Mn \qquad (2.88)$$

where $\beta_1 = 1 - \beta$ and $\gamma_1 = 1 - \gamma$. From Table 2.2 it is clear that only one of the first two terms in (2.87) and (2.88) is included depending on the choice of ($\beta_1 = 1$ when $\beta = 0$ and $\beta_1 = 0$ when $\beta = 1$ ). Similarly only one of the last two terms in (2.87) and (2.88) is included ($\gamma_1 = 1$ when $\gamma = 0$ and $\gamma_1 = 0$ when $\gamma = 1$). This simple device enables each of the one and two, or higher degrees of freedom solutions, to be obtained from the single optimal control problem.

Let the error and control signal terms in (2.87) and (2.88) be separated into the following statistically independent components:

$$e = e_0 + e_1 + e_2 \text{ and } u = u_0 + u_1 + u_2$$

where the error and control signal components may be defined as follows:

$$e_0 = W_i \beta_1 r - \gamma_1 d_0 - d_1 - WM(\beta_1 r - \gamma_1 d_0 - d_1 - n) \qquad (2.89)$$

$$e_1 = (W_i - WSC_{01})\beta r \qquad (2.90)$$

$$e_2 = -(W_{di} - WSC_{02})\gamma d_2 \qquad (2.91)$$

and

$$u_0 = M(\beta_1 r - \gamma_1 d_0 - d_1 - n) \qquad (2.92)$$

$$u_1 = SC_{01}\beta r \qquad (2.93)$$

$$u_2 = -SC_{02}\gamma d_2 \qquad (2.94)$$

These signals have the following physical significance:

$(e_0, u_0)$ : Component of the error and control signals due to the feedback controller action including that for the one DOF tracking problem.

$(e_1, u_1)$ : Tracking error and control signals resulting from the reference signal input for the two or higher DOF designs.

$(e_2, u_2)$ : Component of the error and control signals due to the measurable component of the disturbances.

The above observations suggest that each of the terms might be costed separately, providing the significant advantage of independent tuning knobs for the feedback, tracking and feedforward control actions. Two other signals can also be defined $\{e_s(t)\}$ and $\{u_s(t)\}$ where,

$$e_s = Sf \quad and \quad u_s = Mf \qquad (2.95)$$

and where the signal $f$ is defined as:

$$f = \beta_1 r - \gamma_1 d_0 - d_1 - n \qquad (2.96)$$

These signals depend directly upon the *sensitivity* and *control sensitivity* functions, respectively and are introduced into the cost-function, to enable the $S(z^{-1})$ and $M(z^{-1})$ sensitivity terms to be costed.

### 2.4.4   Cost-function and Optimal Control Solution

The so called *Dual-criterion* (Grimble, 1986a) involves the minimisation of the usual error and control signal variances and also the sensitivity and control sensitivity costing terms which affect the robustness properties. The LQG dual criterion is defined, in terms of the power and cross-power spectra, for the error and control signal components, as:

$$J = \frac{1}{2\pi j} \oint_{|z|=1} \{[Q_{c0}\Phi_{e0} + R_{c0}\Phi_{u0} + G_{c0}\Phi_{u_0 e_0} + G_{c0}^*\Phi_{e_0 u_0} + M_{c0}\Phi_{e0}$$

$$+N_{c0}\Phi_{u0}]\Sigma_0 + [Q_{c1}\Phi_{e1} + R_{c1}\Phi_{u1} + G_{c1}\Phi_{u_1 e_1} + G_{c1}^*\Phi_{e_1 u_1}]\Sigma_1$$

$$+[Q_{c2}\Phi_{e2} + R_{c2}\Phi_{u2} + G_{c2}\Phi_{u_2 e_2} + G_{c2}^*\Phi_{e_2 u_2}]\Sigma_2\} \; dz/z \qquad (2.97)$$

where $\Phi_{e0}$ denotes the power spectrum of the signal $e_0(t)$. Let the index $i = 0, 1, 2$ correspond to the regulating, tracking and feedforward controller signal components. The weightings in the above criterion have the following significance:

$Q_{ci}, R_{ci}, G_{ci}$ : Error, control and cross-weighting terms.

$\Sigma_i$ : Common so called *robustness* weighting elements.

$M_{c0}, N_{c0}$ : Sensitivity and control sensitivity weightings for feedback loop.

The weighting functions can be dynamical and the sensitivity function weighting role of the last two elements follows since:

$$M_{c0}\Phi_{e0} + N_{c0}\Phi_{u0} = (S^* M_{c0} S + M^* N_{c0} M)\Phi_{ff}$$

where $\Phi_{ff}$ denotes the power spectrum of the signal $f = \beta_1 r - d_1 - n$. The cost-function is very general but it may be simplified in practice, since only a small subset of the above weighting elements is needed.

From an obvious identification of terms, the criterion (2.97) can clearly be written as the sum of three components:

$$J = J_0 + J_1 + J_2 = \frac{1}{2\pi j} \oint_{|z|=1} \{I_0 \Sigma_0 + I_1 \Sigma_1 + I_2 \Sigma_2\} \; dz/z \qquad (2.98)$$

where $I_0, I_1$ and $I_2$ correspond to the respective square bracketed terms in (2.97). The polynomial forms of the weightings in the cost index may be defined, for $i = \{0, 1, 2\}$, as:

$$Q_{ci} = \frac{Q_{ni}}{A_{wi}^* A_{wi}} \; , \qquad R_{ci} = \frac{R_{ni}}{A_{wi}^* A_{wi}} \; , \qquad G_{ci} = \frac{G_{ni}}{A_{wi}^* A_{wi}}$$

$$\Sigma_i = \frac{B_{\sigma i}^* B_{\sigma i}}{A_{\sigma i}^* A_{\sigma i}}, \qquad M_{c0} = \frac{M_{n0}}{A_{w0}^* A_{w0}} \; , \qquad N_{ci} = \frac{N_{n0}}{A_{w0}^* A_{w0}} \qquad (2.99)$$

There is little loss in generality in assuming that the denominators of the weightings $A_{wi}$ and $A_{\sigma i}$ are strictly Schur.

**Theorem 2.3** : $3\frac{1}{2} DOF$ *Robust Feedback/Feedforward* $H_2$ *Controller*

The $H_2$ optimal controller to minimise the dual-criterion (2.98), may be found by first computing the strictly-Schur spectral factors $D_{c0}, D_{c1}, D_{c2}$ and $D_f$ using:

$$D_{c0}^* D_{c0} = B_0^*(Q_{n0} + M_{n0})B_0 + A_0^*(R_{n0} + N_{n0})A_0 - B_0^* G_{n0} A_0 - A_0^* G_{n0} B_0 \qquad (2.100)$$

$$D_{c1}^* D_{c1} = B_0^* Q_{n1} B_0 + A_0^* R_{n1} A_0 - B_0^* G_{n1} A_0 - A_0^* G_{n1}^* B_0 \qquad (2.101)$$

$$D_{c2}^* D_{c2} = B_0^* Q_{n2} B_0 + A_0^* R_{n2} A_0 - B_0^* G_{n2} A_0 - A_0^* G_{n2}^* B_0 \qquad (2.102)$$

$$D_f D_f^* = A_n(\beta_1^2 EE^* + \gamma_1^2 CC^* + C_{d1}C_{d1}^*)A_n^* + AC_n C_n^* A^* \qquad (2.103)$$

and compute the Schur spectral-factor $A_{ds}$ using $A_{ds}A_{ds}^* = A_{d2}A_{d2}^*$.

**Regulating loop equations** : Compute $(G, H, F)$, with $F$ of smallest degree:

$$D_{c0}^* G z^{-g} + F(A A_{w0} A_n) = (B_0^*(Q_{n0} + M_{n0}) - A_0^* G_{n0}^*)D_f z^{-g} \qquad (2.104)$$

$$D_{c0}^* H z^{-g} - F(B A_{w0} A_n) = (A_0^*(R_{n0} + N_{n0}) - B_0^* G_{n0})D_f z^{-g} \qquad (2.105)$$

**Measurement noise equation** : Compute $(L, P)$, with $P$ of smallest degree:

$$D_{c0}^* D_f^* L z^{-g_0} + P(A_e A_{w0} A_n A_i) = (B_0^* Q_{n0} - A_0^* G_{n0}^*)(\beta_1^2 E_r E^* A_n A_n^*(A_i - B_i)$$

$$+ A_e C_n C_n^* A^* A_i)z^{-g_0} \qquad (2.106)$$

**Reference equation** : Compute $(X_1, Y_1)$, with $Y_1$ of smallest degree:

$$D_{c1}^* X_1 z^{-g_1} + Y_1(A_e A_w A_i) = (B_0^* Q_{n1} - A_0^* G_{n1}^*)E_r B_i z^{-g_1} \qquad (2.107)$$

**Feedforward equation** : Compute $(X_2, Y_2)$, with $Y_2$ of smallest degree:

$$D_{c2}^* X_2 z^{-g_2} + Y_2(A_{ds} A_{w2} A_{di}) = (B_0^* Q_{n2} - A_0^* G_{n2}^*)C_{d2}C_{di} z^{-g_2} \qquad (2.108)$$

**Robustness feedback equation** : Compute $(N_0, F_0)$, with $F_0$ of smallest degree:

$$D_f^* D_{c0}^* N_0 z^{-g_0} + F_0 A_{\sigma 0} = (P - F D_f^* z^{-g_0 + g})B_{\sigma 0} \qquad (2.109)$$

where $P$ is obtained from the solution of (2.106).

**Robustness tracking equation:** Compute $(N_1, F_1)$, with $F_1$ of smallest degree:

$$D_{c1}^* N_1 z^{-g_1} + F_1 A_{\sigma 1} = Y_1 B_{\sigma 1} \qquad (2.110)$$

**Robustness feedforward equation** : Compute $(N_2, F_2)$, with $F_2$ of smallest degree:

$$D_{c2}^* N_2 z^{-g_2} + F_2 A_{\sigma 2} = Y_2 B_{\sigma 2} \qquad (2.111)$$

The scalars, $g, g_o, g_1, g_2$ are chosen to be the smallest positive integers which ensure the above equations are polynomial in the indeterminate $z^{-1}$, that is :

$$g \geq \deg(D_{c0}), \quad g_1 \geq \deg(D_{c1}), \quad g_0 \geq \deg(D_f D_{c0}), \quad g_2 \geq \deg(D_{c2})$$

**Feedback Controller:**

$$C_0 = C_{0d}^{-1} C_{0n} = (H + (K_1 + K_2)B)^{-1}(G - (K_1 + K_2)A) \qquad (2.112)$$

where the Youla gains:

$$K_1 = (A_e A_i)^{-1}L \quad \text{and} \quad K_2 = B_{\sigma 0}^{-1} A_{w0} A_n N_0$$

**Tracking Controller :**

$$C_1\beta = C_{01} - C_0\alpha \tag{2.113}$$

where

$$C_{01} = C_{0d}^{-1}(X_1 B_{\sigma 1} + N_1 A_{w1} A_e A_i) D_f \left( \frac{D_{c0} A_e B_{\sigma 0}}{D_{c1} E_r B_{\sigma 1}} \right) \tag{2.114}$$

and

$$C_{0d} = (HA_e A_i B_{\sigma 0} + (LB_{\sigma 0} + A_{w0} A_n A_e A_i N_0)B)$$

**Feedforward Controller :**

$$C_2\gamma = (C_{02} - C_0 W_{di})/W_s \tag{2.115}$$

where

$$C_{02} = C_{0d}^{-1}(X_2 B_{\sigma 2} + N_2 A_{w2} A_{ds} A_{di}) D_f \left( \frac{D_{c0} A_e A_i B_{\sigma 0}}{D_{c2} C_{d2} A_{di} B_{\sigma 2}} \right) \tag{2.116}$$

**Proof :** The proof also employs a Kučera (1979) completing the squares type of solution procedure and is presented below.

**Lemma 2.2 :** *Robust Weighted $3\frac{1}{2}$ DOF $H_2$ Controller Properties*
   The *characteristic polynomial* and the related implied equation which determine closed-loop stability, are given respectively as:

$$\rho_c = AC_{0d} + BC_{0n} = D_{c0} D_f A_e A_i U_0 B_{\sigma 0} \tag{2.117}$$

$$A_0 H + B_0 G = D_{c0} D_f \tag{2.118}$$

**Sensitivity:**

$$S = A_0[HA_e A_i B_{\sigma 0} + (LB_{\sigma 0} + A_w A_n A_e A_i N_1)B]/(D_{c0} D_f A_e A_i B_{\sigma 0}) \tag{2.119}$$

**Complementary sensitivity:**

$$T = B_0[GA_e A_i B_{\sigma 0} - (LB_{\sigma 0} + A_w A_n A_e A_i N_1)A]/(D_{c0} D_f A_e A_i B_{\sigma 0}) \tag{2.120}$$

The *minimum value* for the cost-function can be computed using:

$$J_{\min} = \frac{1}{2\pi j} \oint_{|z|=1} X_{\min}(z^{-1}) \frac{dz}{z} \tag{2.121}$$

where the value of the integrand at the minimum:

$$X_{\min} = \left( \frac{F_0 F_0^*}{D_f D_f^* D_{c0} D_{c0}^*} + \Sigma_0 T_0 \right) + \beta^2 \left( \frac{F_1 F_1^*}{D_{c1} D_{c1}^*} + \Sigma_1 T_1 \right)$$

$$+\gamma^2 \left( \frac{F_2 F_2^*}{D_{c2} D_{c2}^*} + \Sigma_2 T_2 \right) \tag{2.122}$$

$$T_0 = Q_{c0}(\beta_1^2 W_i \Phi_{rr} W_i^* + \gamma_1^2 \Phi_{d_0 d_0} + \Phi_{d_1 d_1}) + M_{c0} \Phi_{ff}$$

$$-Y_{c0}^{*-1} \Phi_{h0} Y_f^{*-1} Y_f^{-1} \Phi_{h0}^* Y_{c0}^{-1} \tag{2.123}$$

$$T_1 = \left( Q_{c1} W_i W_i^* - Y_{c1}^{*-1} \Phi_{h1} \Phi_{h1}^* Y_{c1}^{-1} \right) \Phi_{rr} \tag{2.124}$$

$$T_2 = \left( Q_{c2} W_{di} W_{di}^* - Y_{c2}^{*-1} \Phi_{h2} \Phi_{h2}^* Y_{c2}^{-1} \right) \Phi_{d_2 d_2} \tag{2.125}$$

$$\Phi_{h0} = W^*[Q_{c0}(\beta_1^2 W_i \Phi_{rr} + \gamma_1^2 \Phi_{d_0 d_0} + \Phi_{d_1 d_1}) + M_{c0} \Phi_{ff}]$$

$$-G_{c0}^*(\beta_1^2 W_i \Phi_{rr} + \gamma_1^2 \Phi_{d_0 d_0} + \Phi_{d_1 d_1}) \tag{2.126}$$

$$\Phi_{h1} = (W^* Q_{c1} - G_{c1}^*)W_i \quad and \quad \Phi_{h2} = (W^* Q_{c2} - G_{c2}^*)W_{di} \tag{2.127}$$

$$Y_{c0} = (A_0 A_{w0})^{-1} D_{c0}, \quad Y_{c1} = (A_0 A_{w1})^{-1} D_{c1}, \quad Y_{c2} = (A_0 A_{w2})^{-1} D_{c2}$$

$$Y_f = (A A_n)^{-1} D_f, \quad Y_r = A_e^{-1} E_r, \quad Y_{d2} = A_{ds}^{-1} C_{d2}$$

**Proof** : These results are obtained from the same type of analysis as in §2.3, and the proof is presented in full in Grimble (1995).

**Remarks**

(i) The characteristic polynomial for the closed-loop system:

$$\rho_c = A C_{0d} + B C_{0n} = D_{c0} D_f A_e A_i U_0 B_{\sigma 0}$$

includes the pole polynomial for the ideal response model $W_i = A_i^{-1} B_i$

(ii) To realise the feedback, tracking and feedforward controllers any common elements in each controller should be implemented in a common element within the feedback loop. This particularly applies to any integral terms.

(iii) The only terms in the tracking and feedforward controllers, defined by (2.113) and (2.115), respectively which may be unstable are included within the feedback controller denominator term $C_{0d}$ and can be implemented in a common element within the feedback loop. This is necessary if stability of the tracking and feedforward signals is to be guaranteed.

(iv) Let the measurement noise be zero at zero-frequency and the gain $W_i(1) = 1, (A_i(1) = B_i(1) = 1)$. Also assume that the weighting $\Sigma_0$ is a constant and $Q_{c0}$ includes integral action $(A_{w_0}(z^{-1}) = 1 - z^{-1})$. Then, from (2.105) and (2.106) $L(1) = 0$ and $H(1) = 0$. It follows from (2.112) that the controller $C_0(z^{-1})$ includes integral action $(C_{0d}(1) = 0)$. That is, to introduce integral-action the ideal response gain, to zero-frequency signals, should be defined as unity and the integral of error should be penalised in the cost-function.

(v) Only a small set of the above diophantine equations must be solved in any particular case. A null solution to (2.106) is obtained if measurement noise is null and the solutions to (2.109) to (2.111) are null if the weightings $\Sigma_0$, $\Sigma_1$ and $\Sigma_2$ are constants.

(vi) Equation (2.106) clearly simplifies considerably, in the usual case, where the ideal response model $W_i = 1$, since then $(A_i - B_i) = 0$.

(vii) The optimal control problem solution involves $C_{01}$ and $C_{02}$ and the tracking $(C_1)$ and feedforward $(C_2)$ controllers ((2.114) and (2.116)) depend upon the difference between these functions and the feedback controller $C_0$. This explains why the controller frequency responses for the tracking and feedforward designs, in the $3\frac{1}{2}$ DOF solutions, are of an unusual form.

(viii) Although the presence of measurement noise is allowed for in the feedback loops, it is not included on the feedforward signal. It is therefore important to include the high frequency weighting of the feedforward component of the control signal. Measurement noise can of course easily be included in the feedforward signal model, but it is simpler to include a lead term on the feedforward control weighting, to account for noise.

## 2.4.5 Invariance Property

There is an interesting invariance property for the magnitude of the variance of certain signals or correspondingly the forward path gains in certain channels. That is, the feedforward controller will not change if the magnitude of the driving white noise signal for the measurable disturbance block is multiplied by any non-zero scalar gain. The feedback and tracking controllers are also unchanged in this situation.

The same applies if the driving noise for the reference generation block is multiplied by any non-zero scalar gain. None of that transfer function for the controllers will be changed. This property is a consequence of the uncorrelated nature of the different noise sources which enables the different terms in the criterion to be separated out. If one of the aforementioned noise sources is multiplied by a factor this only changes the absolute level of the corresponding cost term but not its optimum controller. It is a valuable feature of this particular problem construction that the tracking and feedforward controllers are invariant to errors in the variances of these noise signals.

There is a slightly different and more well known result for the feedback controller which depends upon a different set of the noise source inputs. In this case, the controller will be unchanged if each of the noise sources which affect the filter spectral factor $Y_f$ are multiplied by the same non-zero scalar. This result is also true of Kalman filtering optimal control schemes (Grimble and Johnson, 1988), where if both the disturbance and noise models are multiplied by the same factor the actual Kalman filter gains are unchanged.

## 2.5 Design of Feedforward Controllers

The equations presented in Section 2.4.3 provide a clue as to how to design a good feedforward controller. The feedforward controller can clearly be seen to add a parallel path (equation (2.91)) which can be used to negate the main disturbance component. If the feedback loop does not have high gain, a constant disturbance input would result in a steady-state error. However, if a measurement of the disturbance is available it is a relatively straightforward matter to determine the feedforward controller gain which would ensure zero steady-state error. In practice, such a solution is probably unrealistic since errors occur in the measurement of the disturbance and in most cases the feedback loop has high gain at zero frequency making such a procedure unnecessary.

It is therefore likely that feedforward control is needed more in the mid frequency range where high gain in the feedback loop cannot be achieved, since it corresponds with the unity gain crossover frequency region. In fact, in most cases feedforward control is cited as a method of providing a fast response to sudden disturbance effects. It is well known that if the control signal is only determined by a feedback controller some delay in responding to a disturbance input is inevitable since the feedback error must build up. However, if a feedforward signal is used control action is taken immediately. Thus, feedforward control is very useful to reject sudden disturbance inputs of a significant magnitude.

This realisation of the main objective that feedforward control should fulfil, brings with it a recognition of the major control difficulty. That is, if the performance of the feedforward controller is to improve mid frequency characteristics, the design problem is not straightforward and the design will be model dependent. Moreover, the controllers obtained are likely to have a high order if they are to change the basic response characteristics in the mid frequency range of the total disturbance signal path.

Lessons can be learned from the optimal solution for the classical design of feedforward controllers. The effective open loop compensator $C_{02}$ must be determined and as in the previous analysis the actual feedforward controller can then be obtained. From equation (2.91) it might be thought that simply choosing $C_{02}$ to make this expression zero would be adequate. However, there are two problems. The first is that a non-physically realisable compensator is likely to be obtained. The second is that the controller obtained may produce excessive control signal variations. Thus in practice, just as in the optimal control case, the component of error due to the disturbance cannot be made exactly zero if realistic control signal changes are to be obtained. Nevertheless, this equation does provide motivation for what is needed. In the frequency range where the disturbance rejection is required, the compensator $C_{02}$ should be chosen so that the gain of the transfer function in equation (2.91) is small. This is provided automatically via the optimal control approach, which also limits the feedforward control signal variations through the control weighting term $R_{c2}$.

**Cost-function Weighting Selection**

The design of feedforward controllers has received relatively little attention. In the optimal case this reduces to the choice of the feedforward controller cost weighting terms and the elements of the system associated with the feedforward action. Note that there is often freedom in the choice of the disturbance models, since these are not known precisely, and they can therefore be used as design variables.

The following notes help to summarise the design process to be used:

1. In most cases the feedback loop design will ensure high gain at low frequency and hence a sensitivity function which is small at low frequencies. Disturbance rejection in the very low frequency should therefore be adequate and the feedforward control action is mainly needed in the mid-frequency region.

2. Good feedforward action often requires significant lead terms in the feedforward controller $C_2$. This results in high gain in the mid and higher frequencies which may become worse if $W_s$ has several lag terms.

3. The disturbance sub-system $W_{di}$ acts rather like the ideal response model in the tracking problem. For example, see equations (2.90) and (2.91), and note that low frequency disturbance errors require the second feedforward term to approximate the first signal component. In most applications $W_{di}$ will be a low pass filter. This follows because disturbances acting on the system are more likely to encounter dynamic terms with low pass characteristics.

4. The feedforward and feedback controllers enter the optimal control equations through the term $C_{02}$. The frequency response of this term is usually much more logical and easy to explain than that for the feedforward controller (Grimble, 1986b), which is formed from the difference of the responses (2.115).

5. The error weighting term $Q_{c2}$ directly affects the feedforward error term. The controller weighting term $R_{c2}$ directly affects the compensator term $C_{02}$ and not the controller $C_2$ itself (see equation (2.94)).

The design steps will be illustrated in a ship positioning example, which now follows.

## 2.5.1   Feedforward in Dynamic Ship Positioning Systems

In the design of conventional Dynamic Positioning (DP) systems, although the feedback controller is designed by optimal methods the feedforward controller is usually designed by empirical methods. However, it is of course possible to design feedback, feedforward and tracking controllers using one unified $H_2$ approach.

The LQG state-space design approach involving a Kalman filtering solution has been applied extensively in ship positioning systems by Balchen et al. (1976) and by Grimble et al. (1979). However, feedforward control is normally introduced using classical control ideas and it is not normally treated as part of the optimal solution. The following design will illustrate the benefits of including feedforward in the total optimal control solution.

The ship positioning problem will be similar to that outlined in Grimble et al. (1980) and Fotakis et al. (1982). However, wind feedforward measurements will be incorporated in this case. The ship models follow the approach in Wise and English (1975) and the wave modelling involves a linear approximation to standard sea spectra (Fossen, 1994). The $H_\infty$ version of this problem is considered in Chapter 12.

**Ship model :**

$$W = \frac{B_0}{A_0} = \frac{0.16}{s(s + 0.0546)(s + 1.55)}$$

This represents the transfer-function between actuator input and position output and $1/1.55$ denotes the thruster time-constant

**Wave model :**

$$W_h = \frac{C_h}{A_h} = \frac{3.52s^2}{(s + 0.086385)((s + 0.09043)^2 + 0.4487074^2)(s + 8.28776)}$$

**Wave and measurement noise model :**

$$W_n = \frac{C_n}{A_n} = \frac{0.282843\left((s + 0.110228)^2 + 0.033757^2\right)(s + 1.38548)(s + 8.14656)}{A_h}$$

where $C_n C_n^* = C_h C_h^* + A_h R_f A_h^*$ and the noise variance $R_f = 1/8$ . Realistic control designs require some white-noise component to be included but this is relatively small in the mid-frequency range where the wave model must dominate for effective filtering action to be introduced.

**Input disturbance model :**

$$W_{d1} = \frac{0.01}{s^2(s + 0.0546)}$$

This represents an integrator disturbance model feeding the ship dynamics.  In fact the disturbance forces enter the ships equations at the same point as the thruster force model. The magnitude of this disturbance is chosen so that it dominates at low frequencies but the noise model $W_n$ dominates at high frequencies.

**Measureable disturbance model :**

$$W_{d2} = \frac{C_{d2}}{A_{d2}} = \frac{0.25}{s^2} \quad , \quad W_{di} = 1$$

The measurable disturbance model corresponds to the wind force in this problem. This model was chosen to illustrate the benefits of feedforward control and should normally be fitted to a Davenport wind spectrum.

**Reference model :**

$$W_r = \frac{E_r}{A_e} = \frac{0.1}{s}$$

The reference model is of comparable magnitude in the mid-frequency range to the disturbance model.

**Spectrum corruption :**

$$W_s = \frac{B_s}{A_s} = \frac{1}{s+1}$$

**Ideal response model:**

$$W_i = \frac{B_i}{A_i} = \frac{1}{1+5s}$$

### System Description

The system and disturbance models are sometimes treated as known quantities which leaves the design problem concerned with the choice of cost-function weightings. However, in practice there is considerable uncertainty in the noise and disturbance models and hence these can also be chosen to meet design objectives.

The ship sway motion frequency response is shown in Fig. 2.18. Note that the ship is essentially a low pass filter which includes integral action. The wave model frequency response is shown in Fig. 2.19, together with the wave model response augmented by the measurement noise model ($W_n$). Some measurement noise component must be included, otherwise the feedback controller gain will be unrealistic at high frequencies. Note that the wave model frequency response includes a peak at the dominant wave frequency and is an approximation to the Pierson-Moscowitz spectrum for the sea-state 9.

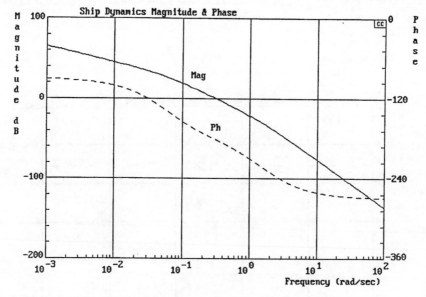

**Fig. 2.18**: Ship Dynamics Magnitude and Phase

### Feedback Loop Design

The feedback loop design depends upon the measurement noise model $W_n$ and the input disturbance model $W_{d1}$ shown in Fig. 2.20. These models were chosen so that the wave model dominates in the mid-frequency region, where good wave filtering

action is required.  Also note that at low frequencies, the input disturbance model
looks like a double integrator formed from the combined effect of the disturbance
and the integrator which is present due to the calculation of position given velocity.
The measurable input disturbance model, $W_{d2}$ is also shown in Fig. 2.20 and this
represents a double integrator action.  Note that the magnitude of this gain, relative
to the disturbances which affect the feedback loop, is not important since (as explained
earlier), the feedforward controller calculation is essentially decoupled from that for
the feedback controller.

**Feedback Weightings :**

$$Q_c = 100 \frac{(-s + 0.1)(s + 0.1)}{(-s^2)} \text{ and } R_c = 0.01$$

The second design stage involves the selection of the optimal cost-function which
amounts to the specification of the cost-function weightings.  The frequency response
of the feedback controller weightings, defined above, are shown in Fig. 2.21.  The
error weighting $Q_c$ has integral action to ensure integral action is introduced into the
controller.  To ensure that some error weighting is applied in the mid frequency range
the integral action is removed by a lead term that enters before the desired unity-gain
crossover frequency ($\omega_c$) for the feedback loop.

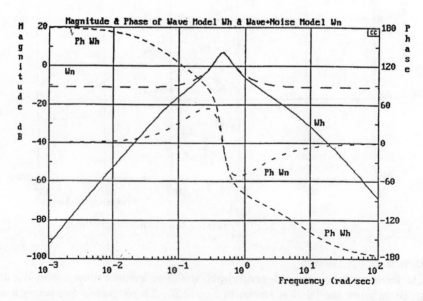

Fig. 2.19:  Wave Model $W_h$ and Wave Plus Noise Model $W_n$

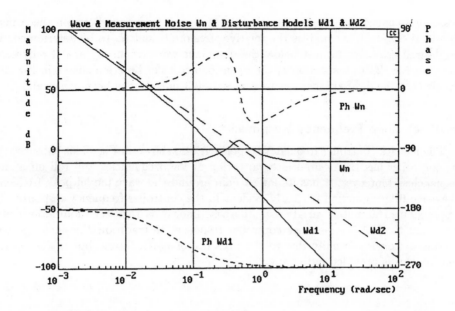

**Fig. 2.20** : Wave and Measurement Noise model $W_n$, Disturbance Models $W_{d1}$ and $W_{d2}$.

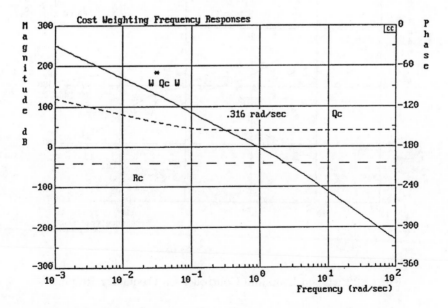

**Fig. 2.21** : Cost Function Weightings for Feedback Controller Frequency Responses

Note from Fig. 2.19 that the dominant wave frequency is at 0.46 rad/sec and hence the objective must be to ensure the controller gain is low around this frequency range. This will occur in part because of the choice of the output disturbance model $W_n$ and partially because of the choice of weightings. A rule of thumb is that the value of the crossover frequency $\omega_c$ is often close to the crossover frequency between the gains of $R_c$ and $W^*Q_cW$. This enables the relative sizes of $Q_c$ and $R_c$ to be determined. Let the desired crossover be just below the dominant wave frequency at 0.3 rad/sec, as shown in Fig. 2.21. Inspection of the open-loop transfer function shown in Fig. 2.22 reveals that the desired crossover frequency of 0.3 rad/sec has been achieved.

### Feedback Loop Frequency Response

The feedack controller frequency response is also shown in Fig. 2.22. Observe that the controller has the desired integral action at low frequencies and roll off at high frequencies. Moreover, it has its lowest gain around the wave model peak frequency. The wave filtering action is clearly evident in the controller frequency response. The suppression of actuator variations (otherwise known as thruster modulation) is an important facet of the feedback controller response. In traditional designs, the wave filtering action is normally due to the presence of notch filters, but in the optimal solution this is provided automatically.

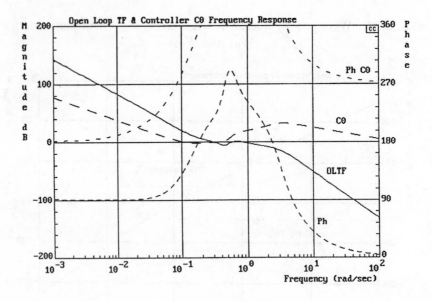

Fig. 2.22 : Open Loop and Controller $C_0$ Frequency Responses

The sensitivity and complementary sensitivity function magnitude and phase plots for the feedback loop are shown in Fig. 2.23. The wave filtering action in the controller accounts for the dual peaks in the sensitivity functions in the mid frequency. The control sensitivity function is shown in Fig. 2.24. The peak of this function is well above the dominant wave frequency.

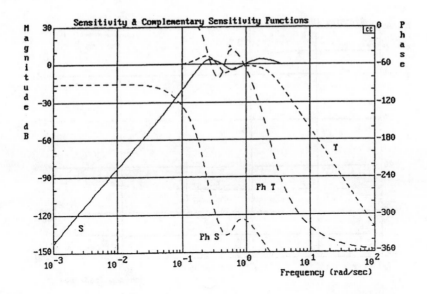

**Fig. 2.23** : Sensitivity and Complementary Sensitivity Functions

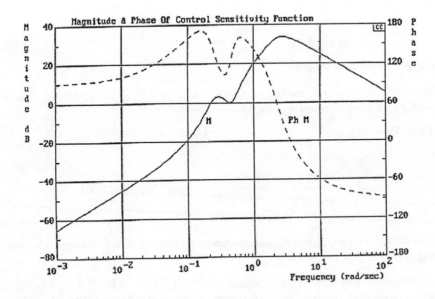

**Fig. 2.24** : Frequency Response of Control Sensitivity Function

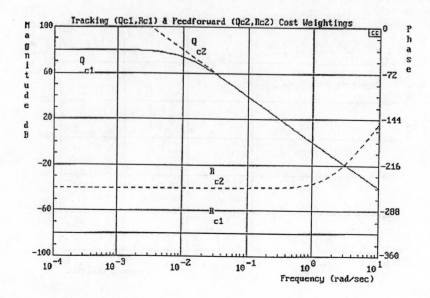

**Fig. 2.25** : Tracking $(Q_{c1}, R_{c1})$,  and Feedforward $(Q_{c2}, R_{c2})$ Cost Function
Weightings

**Tracking Controller Design**

The weightings for the tracking controller, $Q_{c1}$ and $R_{c1}$, are shown in Fig. 2.25. The weightings were selected as follows:

**Tracking Weightings:**

$$Q_{c1} = \frac{(1 + 0.02s)(1 - 0.02s)}{(0.01 + s)(0.01 - s)} \quad \text{and} \quad R_{c1} = 0.0001$$

The gain of the error weighting $Q_{c1}$ is chosen to be large at low frequencies to penalise tracking errors at low frequencies.

The frequency response of the tracking controller for the 3 DOF design is shown in Fig. 2.26. Note that the frequency response of this controller provides the difference between the ideal response controller for a 2 DOF and that which is already present due to the feedback controller. In fact the frequency response of the controller $C_{01}$ essentially represents that for a 2 DOF design. Note that the integral action at low frequency must be implemented in a common integral block.

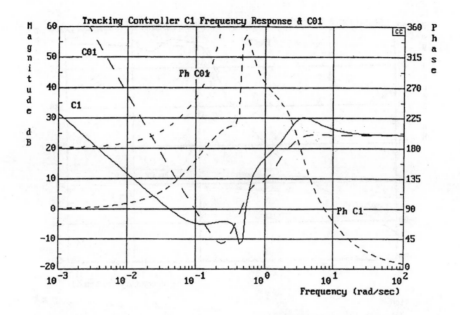

Fig. 2.26 : Frequency Responses of tracking Controller $C_1$,and $C_{01}$

## Feedforward Controller Design

The weightings for the feedforward controller, $Q_{c2}$ and $R_{c2}$, were also shown in Fig. 2.25 and were chosen as follows:

**Feedforward weights:**

$$Q_{c2} = \frac{(1 + 0.001s)(1 - 0.001s)}{(-s^2)} \quad \text{and} \quad R_{c2} = 0.01(1 + 0.5s)^2(1 - 0.5s)^2$$

The integral error costing is included to ensure good low frequency disturbance rejection. The feedforward controller frequency response is shown in Fig. 2.27. The response of the compensator term $C_{02}$ is also shown. This is of course the equivalent of an open loop feedforward controller and is not to be used directly but is needed in the calculation of the feedforward controller $C_2$.

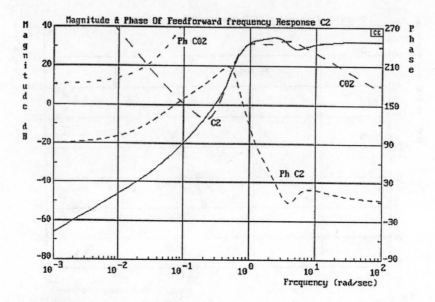

Fig. 2.27 : Frequency Responses of Feedforward Controller
$C_2$ and Compensator Term $C_{02}$

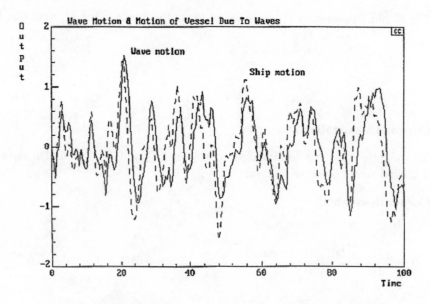

Fig. 2.28 : Wave Motion and Motion of Vessel to Waves

**Time Responses**

    The wave motion and the resulting position of the vessel due to the waves is shown in Fig. 2.28. Note that the ships motion is very similar to the basic wave motion. This suggests that the thrusters are not being used to offset wave motions, which was a major objective of the design. It demonstrates that the notch filtering action which is present in the feedback controller is working effectively. These filters are of course an inherent characteristic of the controller.

The feedback loop unit-step response is shown in Fig. 2.29. Note that since a $3\frac{1}{2}$DOF is employed, there is no necessity for the feedback loop step response to be good for tracking purposes. However, the tracking response which is essentially due to a 2 or $2\frac{1}{2}$DOF design must be suitable and this is also shown. The response clearly approximates the form of the ideal response model (see equation (2.90)).

The effectiveness of feedforward action is illustrated in Fig. 2.30. This shows the motions of the vessel due to the measurable disturbance, with and without the aid of feedforward action. Clearly the feedforward control is very effective at reducing motions due to the measurable disturbance. The measurable and unmeasurable disturbances are both shown in Fig. 2.31. The beneficial effect of feedforward action is again demonstrated in this figure where the measurable disturbance is attenuated greatly.

If the measurement of the disturbance is thought to include significant errors the amount of feedforward action should be reduced. If for example $R_{c2}$ is increased by a factor of 100, then the disturbance rejection is much less impressive, as shown in Fig. 2.32. The tuning freedom which is provided in the above solution is clearly very useful. The beneficial effects of feedforward control are also illustrated in Fig. 2.33. This reveals the impulse-response from the measurable disturbance input, with and without the feedforward controller.

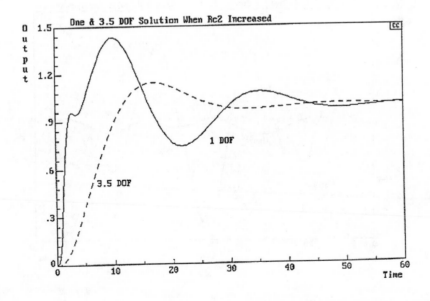

**Fig. 2.29** :Feedback Loop Unit Step Responses and $3\frac{1}{2}$DOF
Tracking Responses

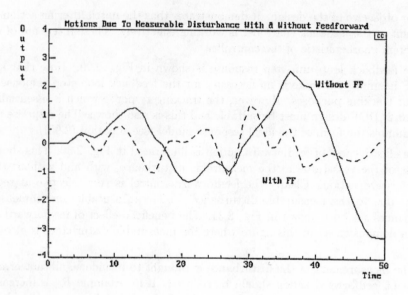

**Fig. 2.30** : Vessel Motions due to Measurable Disturbance
With and Without Feedforward Action

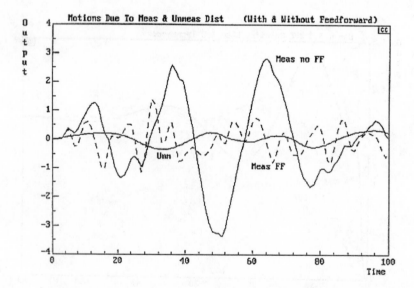

**Fig. 2.31** : Vessel Motions due to Unmeasurable and Measurable
Disturbances With and Without Feedforward Action

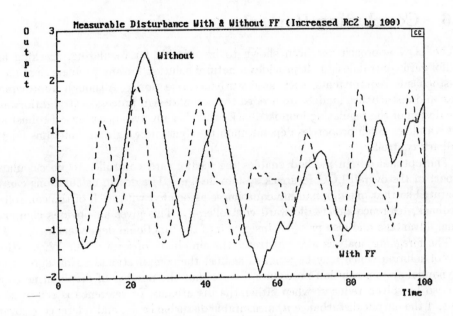

Fig. 2.32 : Measurable Disturbance with and without FF
(Control Weighting $C_1$ increased by Factor of 100)

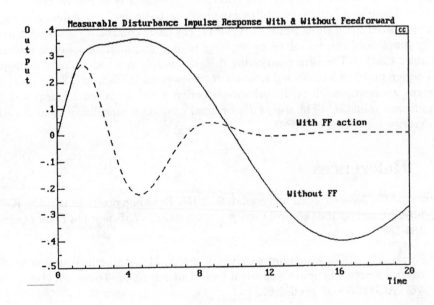

Fig. 2.33 : Measurable Disturbance Impulse Response With and Without
Feedforward

## 2.6   Concluding Remarks

The LQG approach has been shown to be suitable for regulating, tracking and feedforward control design. It provides a natural solution for any problem dominated by stochastic requirements, such as disturbance rejection.  Although approximate noise and disturbance models are needed the actual design process is straightforward. The design of the regulating loop feedback controller involves questions of robustness and stability.  Such properties depend upon the dynamic weighting functions for the feedback controller.

The optimal design method enables the feedforward controller to be calculated as part of the overall LQG feedback system design.  The design calculations can be separated so that one diophantine equation is associated with each of the controllers (feedback, reference and feedforward controllers).  This gives an obvious computational advantage and also provides insights into the functional dependencies.

The foregoing analysis also illustrates the simplicity of form of the Wiener-Hopf type of solution which may be weighed against the computational advantages of the polynomial system solution procedure.  Note that the number of diophantine equations to be solved reduces, when either the disturbance or reference is set to zero. Thus, if the output disturbance $n$, measurable disturbance $d_2$, and reference $r$ covariances are set to zero, the related diophantine equations have zero solution and the resulting regulating problem involves the solution of only the two coupled regulating loop diophantine equations.

The combined feedforward and two degrees of freedom feedback controller is important in a range of real applications.  Moreover, there is a growing need in self-tuning systems for the added flexibility and speed of response a feedforward controller can provide.  There are of course identification problems to be resolved to derive an explicit self-tuner based upon these results.  However, the basic controller expressions are relatively simple and can be solved by existing on-line diophantine solution techniques (Grimble, 1984).   The ship positioning design example is a particularly appropriate LQG design problem (considered from an $H_\infty$ viewpoint in Chapter 12).  In fact there are many applications where disturbance rejection is the main requirement and LQG solutions are valuable.  The use of $H_2$ optimal control in machinery control systems was described in Grimble (1984).

## 2.7   References

Balchen, J.G., Jenssen, N.A. and Saelid, S., 1976, Dynamic positioning using Kalman filtering and optimal control theory, *Automation in Offshore Oil Field Operation*, 183-188.

Bongiorno J.J., 1969, Minimum sensitivity design of linear multivariable feedback control systems by matrix spectral factorization, *IEEE Trans. on Auto. Contr.* Vol. AC-14, No. 6, pp 665-673.

Bongiorno, J.J., Jr and Lee Hai-Ping, 1991, On the design of optimal decoupled multivariable feedback control systems, *ACC Boston*, Massachusetts, pp. 3040-3041.

Callier, F.M. and Desoer, C.A., 1982, *Multivariable feedback systems*, Springer Verlag.

Callier, F.M. and Nahum, C.D., 1975, Necessary and sufficient conditions for the complete controllability and observability of systems in series using the coprime factorization of a rational matrix, *IEEE Trans. on Circuits and Systems*, Vol. **22**, pp. 90-95.

Churchill, R.V., 1948, *Complex variables and applications*, McGraw Hill, New York.

Fossen, T.I., 1994, *Guidance and control of ocean vehicles*, John Wiley Chichester,

Fotakis, J., Grimble, M.J. and Kouvartakis, B., 1982, A comparison of charactertistic locus and optimal designs for dynamical ship positioning systems, *IEEE Trans on Automatic Control*, Vol. **27**, No. 6, pp. 1143-1157.

Ganesh, C. and Pearson, J.B., 1989, $H_2$ optimization with stable controllers, *Automatica*, **25(4)**:629-634.

Grimble, M.J., 1984, Implicit and explicit LQG self-tuning controllers, *Automatica*, Vol. **20**, 5, 661-669.

Grimble, M.J., 1985, Controllers for LQG self-tuning applications with coloured measurement noise and dynamic costing, *IEE Proceedings*, Part **D**, pp 18-29.

Grimble, M.J., 1986a, Dual criterion stochastic optimal control problem or robustness improvement, *IEEE Trans.* Vol. AC-**31**, No. 2, pp. 181-185.

Grimble, M.J., 1986b, Feedback and feedforward LQG controller design, *American Control Conference*, Seattle.

Grimble, M.J., 1988, Two-degrees of freedom feedback and feedforward optimal control of multivariable stochastic systems, *Automatica*, **24**, 6, pp. 809-817.

Grimble, M.J. 1992, Youla parameterized $2\frac{1}{2}$ degrees of freedom LQG controller and robustness improvement cost weighting, *IEE Proceedings*, Pt. D., Vol. **139**, No. 2, pp. 147-160.

Grimble, M.J., 1994, *Robust Industrial Control*, Prentice Hall, Hemel Hempstead.

Grimble, M.J., 1994, Two and a half degrees of freedom LQG controller and application to wind turbines, *IEEE Trans.* AC, Vol. **39**, No. 1, pp. 122-127

Grimble, M.J., 1995, $3\frac{1}{2}$ DOF polynomial solution of the feedforward $H_2/H_\infty$ control problem, *IEEE CDC Conference*, New Orleans.

Grimble, M.J., 1988, $3\frac{1}{2}$ DOF polynomial solution of the inferential $H_2/H_\infty$ control problem with application to metal rolling, *Journal of Dynamic Systems, Measurement and Control*, Vol. **120**, pp. 445-455.

Grimble, M.J., Patton, R.J. and Wise, D.A., 1979, The design of dynamic ship positioning control systems using extended Kalman filtering technique, *Oceans 79 Conf.* (IEEE), San Diego, California, pp. 488-497.

Grimble, M.J., Patton, R.J. and Wise, D.A., 1980, Use of Kalman filtering techniques in dynamic ship positioning systems, *IEE Proc.* Vol. **127**, Pt. D. No.3, pp. 93-102.

Grimble, M.J., Katebi, M.R. and Wilkie, J., 1984, Ship steering control systems modelling and control design, *Proc. 7th Ship Control Systems Symp*, Vol. **2**, 105-116, Bath.

Grimble, M.J. and Johnson, M.A., 1988, *Optimal multivariable control and estimation : theory and applications : Vol. I and II*, John Wiley.

Hunt, K.J., 1988, General polynomial solution to the optimal feedback/feedforward stochastic tracking problem, *Int. J. Control*, Vol. **48**, No. 3, pp. 1057-1073.

Jezek, J., 1982, New algorithm for minimal solution of linear polynomial equations, *Kybernetika*, Vol. **18**, No. 6, 505-516.

Kailath, T., 1980, *Linear Systems*, Prentice Hall, Englewood Cliffs N.J.

Katebi, M.R., Grimble, M.J. and Byrne, J., 1985, LQG adaptive autopilot design, *IFAC Conference on Identification*, University of York.

Kučera, V., 1979, *Discrete linear control*, Cambridge, John Wiley & Sons.

Kučera, V. and Sebek, M., 1984a, A polynomial solution to regulation and tracking : Part 1, Deterministic problem, *Kybernetika*, Vol. **20**, No. 3, 177-188.

Kucera, V. and Sebek, M., 1984b, A polynomial solution to regulation and tracking : Part 2, Stochastic problem, *Kybernetika*, Vol. **20**, No. 4, 257-282.

Kučera, V., 1980, Discrete stochastic regulation and tracking, *Kybernetika*, Vol. **16**, No. 3, 263-272.

Lee, Hai-Ping and Bongiorno, J.J. Jr., 1993, Wiener-Hopf design of optimal decoupled multivariable feedback control systems, *IEEE Trans. on Auto. Control*. Vol. **38**, No. 12,

Sebek, M., 1983a, Stochastic multivariable tracking : A polynomial equation approach, *Kybernetika*, Vol. **19**, No. 6, 453-459.

Sebek, M., 1983b, Direct polynomial approach to discrete-time stochastic tracking, *Problems of Contr. And Inform. Theory*, 12(4), 293-302.

Seraji, H., 1987, Design of feedforward controllers for multivariable plants, *Int. J. Control*, Vol. **46**, No. 5, pp. 1633-1651.

Shaked, U., 1979, A transfer function approach to the linear discrete stationary filtering and the steady-state discrete optimal control problems, *Int. J. Control*, **29**, 2, pp. 279-291.

Shakoor, B.S., 1996, *Industrial uses of predictive control*, MPhil Thesis, Industrial Control Centre, Unviersity of Strathclyde.

Sternard, M. and Ahlen, A., 1987, A correspondence between input estimation and feedforward control, Research Report UPTEC 87 24 R Teknikum Institute of Technology, Uppsala University, Sweden.

Warren, J., 1992, Model based-control of catalytic cracking, *C & I Magazine*, pp. 57-58.

Whitbeck, R.F., 1981, Direct Wiener-Hopf solution of filter/observer and optimal coupler problems, *J. Guidance and Control*, Vol 4, 3, pp. 329-336.

Wiener, N., 1949, Extrapolation, Interpolation and smoothing of stationary time series, with engineering applications, New York: Technology Press and Wiley (issued in Feb. 1942 as a classified National Defence Research Council Report).

Wise, D.E., and English, J.W., 1975, Tank and wind tunnel tests for a drill ship with dynamic position control, *Offshore Technology Conference*, Dallas, OTC2345.

Youla, D.C., Bonjiorno, J.J. and Jabr., H.B., 1976a, Modern Wiener-Hopf design of optimal controller, Part I : the single-input-output case, *IEEE Trans. on Automatic Control*, Vol. AC-21, No. 1, p3-13.

Youla, D.C., Jabr, H.A. and Bongiorno, J.J., 1976b, Modern Wiener-Hopf design of optimal controllers - Part II : the multivariable case, *IEEE Trans. on Auto. Contr.*, Vol. AC-21, No. 3, pp. 319-338.

# 3

# H₂ Predictive Optimal Control

## 3.1 Introduction

In the late 1970's, two predictive control strategies were developed by industrial groups in France (Richalet, et al. 1978) and in the United States (Cutler and Ramaker, 1980). Both methods were developed to control large multivariable systems which could include inequality constraints on the input and output variables. Predictive control methods have been applied successfully in the petrochemicals industry for about two decades.

*Long Range Predictive Control* (LRPC) methods were introduced by Richalet and coworkers in the so called IDentification and COMmand (IDCOM) algorithm. The model employed was, however, rather restrictive, containing only moving-average terms (only zeros in the z-transfer-function). This was superseded by the forerunner of many of the predictive control laws which was introduced by Cutler and Ramaker (1980), and referred to as the Dynamic Matrix Control (DMC) design approach. Garcia, Prett and Morari (1989) have shown that many model based predictive control schemes have a similar structure.

Clarke et al. (1987a,b, 1989) called this class of algorithms *Generalized Predictive Control* (GPC) laws, which is a terminology adopted here to cover non-LQG predictive optimal controllers. The Clarke GPC approach probably represents the most popular LRPC algorithm, and is based on a modified Auto Regressive Moving Average eXogenous (ARMAX) type of system description. The Clarke, Mohtadi and Tuffs (1987a,b) formulation of the problem and solution is simple to understand, and may be applied in adaptive systems (Peterka, 1984).

Because of the computational simplicity of this type of algorithm, it soon found application in self-tuning systems. A novel approach, referred to as the MUlti-Step Multivariable Adaptive Regulator (MUSMAR), was developed by Mosca (1982). Comparisons and introductions to other predictive control methods have been described by De Keyser et al. (1988) and Grimble (1992). The first part of the chapter provides a simple introduction to GPC control laws.

### 3.1.1   The LQGPC Solution Strategy

The second approach to predictive control is introduced later in §3.4 and uses a similar framework to the LRPC model based algorithms, but solves the problem in the same manner as for LQG optimal control problems. It is referred to as the *Linear Quadratic Gaussian Predictive Control* (LQGPC) design method (Grimble, 1995). A crucial feature of the solution presented is the fact that the assumptions lead to an LQG type of solution which has the same matrix structure as in the usual GPC control problem. The stability properties that result are identical to the usual single step LQG cost-function problem (Grimble, 1984a,b, 1986b).

### 3.1.2   Applications of Predictive Control

Many companies have reported successful applications of predictive optimal control laws, such as BP, Shell, Bayer, Enichem and ICI. Some of the processes include: distillation columns, catalytic cracking units, batch polymerisation and furnace temperature controls, (Lewis et al. 1991). The market place is very unpredictable in the oil industry, because of the fluctuations in the oil price. The process plants are very suitable for the application of predictive methods, being large scale and multivariable. There is also a need to produce different grades of oil from varying feedstock. Richalet (1993) reported that at the time of the paper there were around 300-400 commissioned applications of predictive control. The applications included chemical, petrochemical, steel, glass manufacturing, robot manipulators, aviation and tracking systems. There are currently closer to 10,000 applications and the number is growing.

The DMC controllers have proved to be particularly effective in large multivariable petrochemical computer control problems. The system model employed in DMC was very relevant to the process industries, since it could be found from open loop step-response testing which is often used in chemical plants. It was not, of course, appropriate for open-loop unstable systems, although the cost-function did include both output and control costing terms, so that non-minimum phase processes could be controlled. Typical payback periods from the use of LRPC schemes are less than a year. An example is the BP oil refinery in Grangemouth, Scotland, from which savings of around $1M per year have been reported, which has easily repaid the investment in advanced control (Warren, 1992). The major benefits are obtained by introducing optimisation to limit energy use or to ensure higher quality product or throughput. The LRPC schemes are able to include such inequality constraints in a relatively simple algorithm, (Shakoor, 1995).

### 3.1.3   Predictive Control Terminology

Some acronyms used in predictive control follow in Table 3.1:

**Table 3.1** : Acronyms Used in Predictive Control

| | |
|---|---|
| MV | Minimum Variance |
| GMV | Generalised Minimum Variance |
| LQGPC | Linear Quadratic Guassian Predictive control |
| IDCOM | IDentification and COMmand |
| MUSMAR | MUltiStep Multivariable Adaptive Regulator |
| PCA | Predictive Control Algorithm |
| DMC | Dynamic Matrix Control |
| IMC | Internal Model Control |
| LRPC | Long Range Preditive Control |
| EHAC | Extended Horizon Adaptive Control |
| GPC | Generalised Predictive Control |
| EPSAC | Extended Predictive Self Adaptive Control |
| LDMC | Linear Dynamic Matrix Control |
| QDMC | Quadratic Dynamic Matrix Control |
| GPP | Generalised Pole Placement |
| LRPI | Long Range Predictive Identification |
| PFC | Predictive Functional Control |
| GPCC | Generalised Predictive Cascade Control |
| CRHPC | Constrained Receding Horizon Predictive Control |
| CGPC | Constrained Generalised Predictive Control |
| WPC | Weighted Predictive Control |
| MBPC | Model Based Predictive Control |
| EHAC | Extended Horizon Adaptive Control |
| MAC | Model Adaptive Control |

## 3.2 GPC System Description and Prediction

The most common predictive control law is the very well known *Generalised Predictive Control* (GPC) algorithm (Grimble, 1992). The chapter will therefore begin with an analysis of this problem. The system of interest in GPC control is assumed to be linear and time invariant and to be represented by a discrete delay operator model. The system description is similar to an ARMAX model, except that the disturbance model is assumed to include an integrator.

The **C**ontrolled **A**uto **R**egressive **I**ntegrated **M**oving **A**verage (CARIMA) model, representing the plant model in GPC design, is defined as:

$$A(z^{-1})y(t) = B_k(z^{-1})u(t-k) + \frac{C(z^{-1})}{\Delta}\xi(t) \tag{3.1}$$

where $\Delta = 1 - z^{-1}$ and $\xi(t)$ denotes a white noise sequence. The disturbance model: $C(z^{-1})$ is assumed to be strictly minimum phase. The transport-delay $k \geq 1$, and $B(z^{-1}) = B_k(z^{-1})z^{-k}$ . Without loss of generality $A(0)$ can be defined as: $A(0) = 1$.

### 3.2.1 Optimal Linear Predictor

The solution of the optimal control problem requires the introduction of a least squares predictor. This enables the output at times $t + k + 1$, $t + k + 2$,... to be calculated (assuming that the disturbance at future times is null). The cost-function to be minimised, which defines the optimal least-squares predictor, is defined as:

$$J = E\{\tilde{y}(t+j\,|t)^2\}$$

where the estimation error is defined as:

$$\tilde{y}(t+j\,|t) = y(t+j) - \hat{y}(t+j\,|t) \tag{3.2}$$

and $\hat{y}(t+j\,|t)$ defines the predicted value of $y(t)$ for $j$ steps ahead.

**Diophantine Equation for Optimal Linear Predictor**

Let $j$ denote the prediction interval, where the output is predicted at the time $t+j$. The following diophantine equation must be solved for the minimal-degree solution $(H_j, E_j)$, with respect to $E_j$, of the equation:

$$E_j(z^{-1})A(z^{-1})\Delta(z^{-1}) + z^{-j-k}H_j(z^{-1}) = C(z^{-1}) \tag{3.3}$$

Multiplying the plant equation by $E_j z^{j+k}$, obtain:

$$E_j A\Delta y(t+j+k) = E_j B_k \Delta u(t+j) + E_j C\xi(t+j+k).$$

Substituting from the above diophantine equation for $E_j A\Delta$, obtain:

$$Cy(t+j+k) = E_j B_k \Delta u(t+j) + E_j C\xi(t+j+k) + H_j y(t) \tag{3.4}$$

The degree of the polynomial $E_j$ is $j+k-1$, and hence the white noise components in $E_j\xi(t+j+k)$ include $\xi(t+j+k)$ ,..., $\xi(t+1)$, which are all at future times (greater than $t$). Define, recalling that $C(z^{-1})^{-1}$ is assumed to be stable,

$$y_f(t) = \frac{1}{C(z^{-1})}\,y(t) \qquad \text{and} \qquad u_f(t) = \frac{1}{C(z^{-1})}\,u(t) \tag{3.5}$$

Then (3.4) may be written as:

$$y(t+j+k) = [E_j B_k \Delta u_f(t+j) + H_j y_f(t)] + (E_j\xi(t+j+k)) \tag{3.6}$$

where $E_j\xi(t+j+k) = e_o\xi(t+j+k) + ... + e_{j+k-1}\xi(t+1)$ .

### 3.2.2 Derivation of the Predictor

The optimal predictor of the output at time $(t+j+k)$, given observations up to time $t$, can now be derived. Assume that the measured output data up to this time, and the values of $u(t),..,u(t+j)$, are known. Since the future control action is found by feedback, which depends upon the random disturbance input, this latter assumption is not strictly valid, although the future control action in most LRPC schemes is assumed to be known and constant. An alternative assumption which is sometimes made is that the future control input is independent of the future noise sequence. In either case, the expected value of the square [.] and round (.) bracketed terms in (3.6) is zero.

The optimal predictor to minimise the variance of the estimation error (3.2), given that the cross terms in the cost are null, then follows from (3.2) and (3.6) as:

$$\hat{y}(t+j+k\,|t) \;=\; [E_j B_k \Delta u_f(t+j) \;+\; H_j y_f(t)] \tag{3.7}$$

and the prediction error:

$$\tilde{y}(t+j+k\,|t) = y(t+j+k\,|t) - \hat{y}(t+j+k\,|t) \;=\; (E_j \xi(t+j+k)) \tag{3.8}$$

Note from equation (3.3) that the term: $E_j \;=\; (C - z^{-j-k} H_j) A^{-1}/\Delta$ and hence it may be computed by considering the z-transform of the first $j+k$ terms in the disturbance model step response.

### Alternative Form of the Predictor

A second diophantine equation may now be introduced to break up the term $E_j B_k$ into a part with a $(j+1)$ step delay and a part depending on $C(z^{-1})$. Thus, for $j \geq 0$ introduce the following diophantine equation with a minimal degree solution, denoted by $(G_j, S_j)$ :

$$C(z^{-1})G_j(z^{-1}) \;+\; z^{-j-1} S_j(z^{-1}) \;=\; E_j(z^{-1}) B_k(z^{-1}) \tag{3.9}$$

Thus, $deg(G_j) = j$, and hence for $j \geq 0$, obtain:

$$\hat{y}(t+j+k\,|t) \;=\; CG_j \Delta u_f(t+j) \;+\; S_j \Delta u_f(t-1) \;+\; H_j y_f(t) \tag{3.10}$$

The degree of $G_j(z^{-1})$ is $j$, and hence the first term in (3.10) involves $u(t+j), ..., u(t)$. Define the signal $f_j(t)$, in terms of past outputs and controls, as:

$$f_j(t) \;=\; S_j \Delta u_f(t-1) \;+\; H_j y_f(t) \tag{3.11}$$

Thus, (3.10) gives for $j \geq 0$,

$$\hat{y}(t+j+k\,|t) \;=\; CG_j \Delta u_f(t+j) \;+\; f_j(t) \;=\; G_j \Delta u(t+j) \;+\; f_j(t) \tag{3.12}$$

It is clear from this equation that the signal $f_j(t)$ represents the free response prediction of $y(t+j+k)$, assuming that the control increments for $u(t+i), i \geq 0$ are all zero. That is, the signal $f_j(t)$ represents the response of the plant assuming that all future controls equal the previous control $u(t+j-1)$, so that the control increments are null.

### Significance of the $G_j(z^{-1})$ Polynomial

From the diophantine equations (3.3) and (3.9):

$$CG_j \;+\; z^{-j-1} S_j \;=\; E_j B_k \;=\; (C - z^{-j-k} H_j) \frac{B_k}{A\Delta}$$

Hence obtain:

$$G_j \;=\; \frac{B_k}{A\Delta} \;-\; z^{-j}\left( z^{-k} \frac{H_j B_k}{A\Delta} \;+\; z^{-1} S_j \right)/C \tag{3.13}$$

Recall that the degree of $G_j$ is $j$ and $k \geq 1$. It follows that $G_j$ may be evaluated by finding the first $j$ points in the plants sampled step-response. It is interesting that the $G_j$ polynomials have a similar form:

$$
\begin{aligned}
G_0 &= g_0 \\
G_1 &= g_0 + g_1 z^{-1} \\
&\;\;\vdots \\
G_j &= g_0 + g_1 z^{-1} + ... + g_j z^{-j}
\end{aligned}
$$

Equation (3.13) may alternatively be written as:

$$
\frac{B_k}{A\Delta} = \frac{E_j B_k}{C} + z^{-j-k} H_j B_k / (A\Delta C)
$$

$$
= G_j + z^{-j-1} S_j / C + z^{-j-k} H_j B_k / (A\Delta C) \tag{3.14}
$$

The polynomial $G_j$ therefore includes the first $(j + 1)$ Markov parameters $g_i$ of the transfer-function $B_k / (A\Delta)$ . This confirms that these parameters may be found from pulse-response tests on the model $B_k / (A\Delta)$, or step-response tests on the plant model $A^{-1} B_k$ .

**Matrix Representation of the Predictor Equations**

The equation (3.12) may be used to obtain the following vector equation for the predicted output at $n_1 + 1$ successive values of time:

$$
\begin{bmatrix} \hat{y}(t+k\,|t) \\ \hat{y}(t+1+k\,|t) \\ \vdots \\ \hat{y}(t+n_1+k\,|t) \end{bmatrix} = \begin{bmatrix} g_0 & 0 & \cdots & 0 \\ g_1 & g_0 & & 0 \\ \vdots & \vdots & & \vdots \\ g_{n_1} & g_{n_1}-1 & \cdots & g_0 \end{bmatrix} \begin{bmatrix} \Delta u(t) \\ \Delta u(t+1) \\ \vdots \\ \Delta u(t+n_1) \end{bmatrix} + \begin{bmatrix} f_0(t) \\ f_1(t) \\ \vdots \\ f_{n_1}(t) \end{bmatrix} \tag{3.15}
$$

If the initial predicted time is changed to $t + k + n_0$ the equation becomes:

$$
\begin{bmatrix} \hat{y}(t+n_0+k\,|t) \\ \hat{y}(t+n_0+1+k\,|t) \\ \vdots \\ \hat{y}(t+n_1+k\,|t) \end{bmatrix} = \begin{bmatrix} g_{n_0} & g_{n_g-1} & \cdots & g_0 & & 0 \\ g_{n_0+1} & g_{n_0} & \cdots & g_1 & g_0 & 0 \\ \vdots & & & & & \\ g_{n1} & g_{n_1-1} & \cdots & & \cdots & g_0 \end{bmatrix} \begin{bmatrix} \Delta u(t) \\ \Delta u(t+1) \\ \vdots \\ \Delta u(t+n_1) \end{bmatrix}
$$

$$
+ \begin{bmatrix} f_{n_0}(t) \\ f_{n_0+1}(t) \\ \vdots \\ f_{n_1}(t) \end{bmatrix}
$$

If the control horizon $n_2$ is smaller than the prediction horizon $n_1 - n_0$ and $\Delta u(t+i) = 0$ for $i \geq n_2 + 1$, then this equation becomes:

$$
\begin{bmatrix}
\hat{y}(t + n_0 + k \,|t) \\
\hat{y}(t + n_0 + 1 + k \,|t) \\
\vdots \\
\hat{y}(t + n_1 + k \,|t)
\end{bmatrix}
=
\begin{bmatrix}
g_{n_0} & g_{n_g-1} & \cdots & g_0 & 0 \\
g_{n_0+1} & g_{n_0} & \cdots & g_1 & g_0 & 0 \\
\vdots & & & & & g_0 \\
g_{n1} & g_{n_1-1} & \cdots & & \cdots & g_{n_1-n_2}
\end{bmatrix}
\begin{bmatrix}
\Delta u(t) \\
\Delta u(t+1) \\
\vdots \\
\Delta u(t+n_2)
\end{bmatrix}
$$

$$
+
\begin{bmatrix}
f_{n_0}(t) \\
f_{n_0+1}(t) \\
\vdots \\
f_{n_1}(t)
\end{bmatrix}
$$

The vector form of the prediction error : $E_j\xi(t + j + k)$ may be written, corresponding to (3.15), as:

$$
\tilde{y}(t) =
\begin{bmatrix}
e_0\xi(t+k) + \dots + e_{k-1}\xi(t+1) \\
e_0\xi(t+1+k) + \dots + e_k\xi(t+1) \\
\vdots \\
e_0\xi(t+n_1+k) + \dots + e_{n_1+k-1}\xi(t+1)
\end{bmatrix}
$$

## 3.3  Generalised Predictive Control Law

The GPC control law will now be derived, but the cost-function to be minimised will first be introduced (Mohtadi and Clarke, 1986). Let the GPC cost-function be defined as:

$$
J = E\left\{ \sum_{j=n_o}^{n_1} (y(t+j+k) - r(t+j+k))^2 + \sum_{j=0}^{n_2} \lambda_j^2 (\Delta u(t+j))^2 \,|t \right\} \tag{3.16}
$$

where $\lambda_j$ is a weighting function that may be time-varying, and $E\{.|t\}$ denotes the expectation conditioned on measurements up to and including time $t$. Notice that the criterion involves future values of the reference signal $\{r(t)\}$. It is often reasonable to assume that the future variations of the reference signal $\{r(t)\}$ are predetermined, at least over a fixed future horizon. For example, batch operations in the petrochemicals industry often involve known future desired setpoint trajectories, and this applies to spray painting robots, and thermal processes.

### 3.3.1  Solution for the GPC Optimal Control Law

The optimal control solution is obtained below using relatively simple matrix manipulations. The prediction equation (3.15) must first be written in a vector-matrix form as :

$$
\hat{Y} = G\Delta U + F \tag{3.17}
$$

where

$$\hat{Y}^T \ = \ [\hat{y}(t + n_o + k\,|t),\ \hat{y}(t + n_o + 1 + k\,|t), ..., \hat{y}(t + n_1 + k\,|t)]$$

$$\Delta U^T \ = \ [\Delta u(t),\ \Delta u(t + 1), ..., \Delta u(t + n_2)]$$

$$F^T \ = \ [f_{n_0}(t),\ f_{n_0+1}(t), ..., f_{n_1}(t)]$$

The optimal control cost-function may now be written in vector matrix form as:

$$J \ = \ E\left\{(Y - R)^T Q_c (Y - R) \ + \ \Delta U^T R_c \Delta U | t\right\} \qquad (3.18)$$

where $Q_c \geq 0$ and $R_c = diag\{\lambda_o^2,\ \lambda_1^2, ...,\ \lambda_{n_2}^2\} > 0$ are diagonal matrices. Thence, obtain:

$$J \ = \ E\left\{(\hat{Y} - R + \tilde{Y})^T Q_c (\hat{Y} - R + \tilde{Y}) \ + \ \Delta U^T R_c \Delta U | t\right\}$$

The prediction errors $\{\tilde{y}(t+j|t)\}$ consists of future values of $\{\xi(t)\}$ which were assumed to be independent of future controls. The reference and disturbance signals are also assumed to be statistically independent, and hence the cost-function may be written as:

$$J \ = \ E\left\{(\hat{Y} - R)^T Q_c (\hat{Y} - R) \ + \ \Delta U^T R_c \Delta U \ + \ \tilde{Y}^T Q_c \tilde{Y} | t\right\} \qquad (3.19)$$

Let the first two terms be denoted by $J_0$, and note that the last term does not enter the minimisation process, since it does not depend on the control signal.

Substituting (3.17) into the cost function (3.19) obtain:

$$J_o \ = \ E\left\{(G\Delta U + F - R)^T Q_c (G\Delta U + F - R) \ + \ \Delta U^T R_c \Delta U | t\right\}$$

$$= \ E\{\Delta U^T (G^T Q_c G + R_c)\Delta U \ + \ 2\Delta U^T (G^T Q_c (F - R))$$

$$+ \ (F - R)^T Q_c (F - R) | t\}$$

This expression involves the conditional expectation and from (3.11) the vector of signals is known at time $t$. The computation of the optimal control signal therefore reduces to a deterministic optimisation problem. Setting the gradient to zero gives the optimal control as:

$$\Delta U \ = \ -(G^T Q_c G + R_c)^{-1} G^T Q_c (F - R) \qquad (3.20)$$

Although the vector of future control increments is given by the vector $\Delta U$, only the element at time $t$ is actually implemented. That is, only the top row of (3.20) need be evaluated. This is equivalent to the use of a deterministic receding horizon optimal control philosophy, where the control and output cost intervals recede (Kwon and Pearson, 1977). There are many variations to this basic algorithm that provide improved stability characteristics, such as *Weighted Predictive Control* (Duan 1993, Duan et al. 1997).

### 3.3.2 Algorithm for the GPC Optimal Control Law

The main steps in the calculation of the GPC controller may now be listed.

(1) Calculate the plant step-response coefficients $\{g_i \; : \; i = 1 \; to \; n_1\}$ , for either step-response tests on the plant, or from the plant model $A^{-1}B_k$:

$$g_t \; = \; (A^{-1}B_k)(\frac{1}{\Delta}\delta(t)) \; = \; A^{-1}B_kU(t)$$

where $U(t)$ here denotes the Heaviside Unit Step Function.

(2) Compute the free response of the system for $j = 1$ to $n_1$ using:

$$A\Delta y(t+j+k) \; = \; B_k\Delta u(t+j)$$

with $\Delta u(t) \; = \; \Delta u(t+1) \; = \; \Delta u(t+2) \; = \; ... \; = \; \Delta u(t+j) \; = \; 0$ , using the known measured values of the output $\{y'(\tau)\}$ for time $\tau \; < \; t+j+k$ . Let $f_j(t) \; = \; y'(t+j+k)$ . Alternatively compute $f_j(t)$ from the diophantine equations (3.3) and (3.9) and equation (3.11).

(3) Calculate the inverse $(G^TQ_cG+R_c)^{-1}$ using a UD factorisation and find $\Delta u(t) = [1, \; 0 \; , .., \; 0]\Delta U(t)$, where the vector $\Delta U$ is given as :

$$\Delta U(t) \; = \; (G^TQ_cG + R_c)^{-1}G^TQ_c(R - F).$$

**Output and Control Horizons**

The minimum output horizon can be set to zero, that is $n_0 = 0$. The value of $n_1$ should be chosen so that it is at least as large as the dominant plant time constant. The control horizon $n_2$ should be taken equal to $n_1$ to be consistent since there is no physical reason for costing the two signals over different time intervals. However, in practice $n_2 \; < \; n_1$ might be desirable since this can ensure the $(n_2 + 1)$ square matrix $(G^TQ_cG + R_c)$ is invertible even when $R_c$ tends to the zero matrix. The projected control increments at times greater than $n_2$ in such a case are assumed to be null. This can be thought of as placing infinite control weightings on the control variations at times greater than $n_2$. If the plant is open-loop unstable, the assumption is unrealistic and the results are unpredictable. However, the use of $n_2 \; < \; n_1$ does significantly reduce the computational burden.

**Example 3.1** : *First Principles Optimisation Calculation*

Although $n_0$, $n_1$ and $n_2$ are treated as design variables in GPC algorithms, the most practical choices of these quantities are $n_0 \; = \; 0$, $n_1 \; = \; n_2$. It is reasonable to cost the control signal over the same number of points as the output signal. That is, the $n_2 + 1$ control signal values which influence the $n_2 + 1$ outputs can be costed. To illustrate the solution which is obtained in such a case consider the situation where $n_0 \; = \; 0$, $n_1 \; = \; 2$, $n_2 \; = \; 2$ . Then the cost-function:

$$J_0 \; = \; E\left\{\sum_{j=0}^{2}(\hat{y}(t+j+k\,|t) - r(t+j+k))^2 \; + \; \lambda_j^2(\Delta u(t+j))^2|t\right\}$$

$$= E\left\{(g_0\Delta u(t) + f_0(t) - r(t+k))^2 + \lambda_0^2\Delta u(t)^2\right.$$

$$+ (g_0\Delta u(t+1) + g_1\Delta u(t) + f_1(t) - r(t+1+k))^2 + \lambda_1^2\Delta u(t+1)^2$$

$$+ (g_0\Delta u(t+2) + g_1\Delta u(t+1) + g_2\Delta u(t) + f_2(t)$$

$$\left. -r(t+2+k))^2 + \lambda_2^2\Delta u(t+2)^2|t\right\}$$

The optimal control can be found by standard variational arguments. Three equations are obtained by differentiating with respect to $\Delta u(t)$, $\Delta u(t+1)$ and $\Delta u(t+2)$, respectively, and equating these expressions to zero. Thus obtain.:

$$g_0\left(g_0\Delta u(t) + f_0(t) - r(t+k)\right) + \lambda_0^2\Delta u(t)$$

$$+ g_1(g_0\Delta u(t+1) + g_1\Delta u(t) + f_1(t) - r(t+1+k))$$

$$+ g_2(g_0\Delta u(t+2) + g_1\Delta u(t+1) + g_2\Delta u(t) + f_2(t) - r(t+2+k)) = 0$$

$$g_0(g_0\Delta u(t+1) + g_1\Delta u(t) + f_1(t) - r(t+1+k)) + \lambda_1^2\Delta u(t+1)$$

$$+ g_1(g_0\Delta u(t+2) + g_1\Delta u(t+1) + g_2\Delta u(t) + f_2(t) - r(t+2+k)) = 0$$

$$g_0(g_0\Delta u(t+2) + g_1\Delta u(t+1) + g_2\Delta u(t) + f_2(t) - r(t+2+k))$$

$$+ \lambda_2^2\Delta u(t+2) = 0$$

The optimal control signal increments therefore satisfy:

$$(g_0^2 + g_1^2 + g_2^2 + \lambda_0^2)\Delta u(t) + (g_0g_1 + g_1g_2)\Delta u(t+1) + g_0g_2\Delta u(t+2)$$

$$= g_0(r(t+k) - f_0(t)) + g_1(r(t+1+k) - f_1(t)) + g_2(r(t+2+k) - f_2(t)) \quad (3.21)$$

$$(g_0g_1 + g_1g_2)\Delta u(t) + (g_0^2 + g_1^2 + \lambda_1^2)\Delta u(t+1) + g_0g_1\Delta u(t+2)$$

$$= g_0(r(t+1+k) - f_1(t)) + g_1(r(t+2+k) - f_2(t)) \quad (3.22)$$

$$g_0g_2\Delta u(t) + g_0g_1\Delta u(t+1) + (g_0^2 + \lambda_2^2)\Delta u(t+2)$$

$$= g_0(r(t+2+k) - f_2(t)) \quad (3.23)$$

These equations may be written in the vector-matrix form (3.20), and can be solved for the optimal control signal $\Delta u(t)$, $\Delta u(t+1)$, $\Delta u(t+2)$ .

**Optimal Control for Zero Control Weighting**

If the control weighting is null, the cost is minimised by setting each of the error terms in the cost-function to zero. That is, if $(\hat{y}(t+j+k\,|t) - r(t+j+k)) = 0$ for j =1, 2, 3, the equations which determine the optimal control become:

$$g_0\Delta u(t) + f_0(t) - r(t+k) = 0 \qquad (3.24)$$

$$g_0\Delta u(t+1) + g_1\Delta u(t) + f_1(t) - r(t+1+k) = 0 \qquad (3.25)$$

$$g_0\Delta u(t+2) + g_1\Delta u(t+1) + g_2\Delta u(t) + f_2(t) - r(t+2+k) = 0 \qquad (3.26)$$

Clearly (3.24) may be solved for $\Delta u(t)$ , equation (3.25) can be solved for $\Delta u(t+1)$ given $\Delta u(t)$, and finally (3.26) can be solved for $\Delta u(t+2)$ given $\Delta u(t)$ and $\Delta u(t+1)$

.

The above technique applies to any predictive problem where the values of the control weighting are set to zero, and the cost-function has the same number of control and output terms. The control increment u(t) will be identical to the Generalised Minimum Variance (GMV) control signal (Grimble, 1988a) in such a case. This is of course the only control action which is actually implemented, since in the spirit of *receding-horizon optimal control*, at time $t+1$ the $\Delta u(t+1)$ that is implemented is not that found above, but that which satisfies (3.24). That is,

$$g_0\Delta u(t+1) + f_0(t+1) - r(t+k+1) = 0$$

To demonstrate this result in the general case, note that when the control weighting is null the problem corresponds to minimum-variance control. From (3.10):

$$\hat{y}(t+j+k\,|t) - r(t+j+k) = G_j\Delta u(t+j)$$

$$+(S_j/C)\Delta u(t-1) + (H_j/C)y(t) - r(t+j+k) = 0$$

or

$$(CG_j + z^{-j-1}S_j)z^j\Delta u(t) = Cr(t+j+k) - H_jy(t)$$

Substituting from equation (3.9) the control signal, which sets the predicted error to zero, is given as:

$$\Delta u(t+j) = (Cr(t+j+k) - H_jy(t))/(E_jB_k) \qquad (3.27)$$

Recall that when the control weighting tends to zero, the LQG and minimum variance control laws become identical. This provides an important clue as to the development of LQG based predictive control laws. That is, one fixed feedback controller cannot be used to control both the feedback system, and to predict forward in time optimally. As in finite-time LQ optimal control, the optimal predicted control should be found from a $t$-dependent control law. This is discussed later in the chapter. The relationship between the GPC law and the *minimum variance*, or *generalised minimum-variance* control laws, is explored further in the following sections.

**Optimal Control for Single Step-ahead Cost-function**

The optimal control, if only the output is weighted and a single cost term ($j = 0$) is used, can again be chosen to set the predicted error to zero, to obtain:

$$\hat{y}(t+k\,|t) - r(t+k) = 0$$

or

$$G_0\Delta u(t+1) = r(t+k) - f_0(t) = r(t+k) - S_0\Delta u_f(t-1) - H_0 y_f(t)$$

$$(CG_0 + z^{-1}S_0)\Delta u(t) = Cr(t+k) - H_0 y(t)$$

$$\Delta u(t) = (Cr(t+k) - H_0 y(t))/(E_0 B_k)$$

This corresponds with the minimum-variance optimal control law for minimum phase systems. If the system is non-minimum phase, the $1/B_k$ term leads to an unstable pole zero cancellation, and the closed-loop system is unstable. It follows that the GPC control law results in an unstable system under low control weighting when only one step is used in the criterion. The implication is that the system could be unstable, particularly when a small number of terms are used in the cost-index summation. The controller is identical to the Generalised Minimum Variance (GMV) control law, when the control weighting is small (Grimble, 1988a, 1992).

### 3.3.3   Relationship Between GMV and GPC Control

The objective in this section is to show that, even in the *multi-step* cost function case, the control law reduces to the GMV law when the control weighting tends to zero. First note that a physically realistic cost-function has $n_0 = 0$ and $n_2 = n_1$. If in addition $\lambda_i = 0$ for $i = 0, ..., n_1$, then the cost-function reduces to the sum of a number of prediction errors. The matrix G in this case has the form shown in (3.15). That is, $G$ is square and lower triangular. The cost term to be minimised in this case becomes:

$$J_0 = E\left\{(G\Delta U - (R-F))^T Q_c(G\Delta U - (R-F))|t\right\} \tag{3.28}$$

and if the delay is correctly estimated $G$ is invertible so that the optimal control $\Delta U = G^{-1}(R-F)$ and $J_{0\,min} = 0$.

In practice, as in the discussion of the previous example, $G^{-1}$ need not be evaluated, since $\Delta u(t)$ can be found from the first equation, followed by $\Delta u(t+1)$ (using $\Delta u(t)$) and so on. The expression for $\Delta u(t)$ which follows from the first equation, and is implemented in the spirit of the *receding horizon* control philosophy (Bitmead et al. 1989a,b), therefore satisfies:

$$(\hat{y}(t+k\,|t) - r(t+k)) = 0$$

or

$$(g_0 \Delta u(t) + f_0(t) - r(t+k)) = 0$$

giving

$$\Delta u(t) = (r(t+k) - f_0(t))/g_0$$

This is just the GMV controller for a system that is assumed minimum phase, and it has the stability problems referred to above on non-minimum phase systems (Grimble, 1988a).

### 3.3.4 Predictive Control Laws Properties and Problems

As described in the previous sections, the GPC algorithms are rather inconsistent in the theoretical basis employed, and the suboptimality that results may often be a disadvantage in comparison with LQG designs. However, there are many interesting features and techniques introduced via the GPC approach which are valuable in real applications. Despite problems with the theoretical basis of GPC design, the technique has proven to be very suitable for adaptive control (Clarke and Mohtadi, 1989). There are also claims that it is robust in practice, despite the fact that there are the stability problems mentioned.

Although GPC design considers only a finite number of control increments, it is really a steady-state control law based on the receding-horizon principle (Kwon and Pearson, 1977), where the cost is measured from time $t$ up to time $t+p$. The use of the receding time frame ensures that, in this case, the control law is not time-varying.

In different versions of GPC algorithms there are alternative assumptions about the form of the control signal after the control horizon $n_2$ is exceeded. The future control is often assumed to become constant (Clarke and Mohtadi, 1989) which may be appropriate in some systems, but would not be valid for marine, aerospace, wind turbine, and other applications where stochastic disturbances dominate.

The GPC algorithm does not account for the different variances of the disturbance signals entering the plant, since in effect it ignores the stochastic nature of the problem. This may be appropriate when only initial condition responses need to be regulated, but it seems inappropriate when the system includes substantial stochastic terms. In effect, GPC is more like a deterministic control law, since stochastic properties of the noise and disturbances entering the system are neglected. This may be confirmed from the optimisation argument in the previous section.

The stability properties of GPC algorithms are difficult to establish. Unlike LQG algorithms, which have a direct relationship between the cost weights and the closed loop pole positions (see the return-difference relationships in Chapter 7). There is no such relationship for GPC designs. Indeed, the link between the cost weightings and the pole positions is rather complex for the GPC algorithm.

As noted, the GPC control law is not always stabilising. If the control and output horizons are the same, which is physically realistic, and the control weighting is small (as recommended by Clarke and Mohtadi 1987a,b), then the GPC controller is identical to a single-step GMV controller. This is unstable on non-minimum phase processes when the control weighting is small, and was the main reason that other alternatives were considered by Clarke and his coworkers. Despite this difficulty, GPC designs can be stabilised by careful choice of weightings and cost horizons, and there are also several alternative methods of ensuring stability but with some increase in computational complexity (Kouvaritakis et al. 1992, and Rossiter et al. 1998).

The main justification for the use of this class of control law is the good results many researchers have reported in both simulated and real applications. A better understanding of the theoretical basis of the method has been achieved through the seminal work of Bitmead, Gevers and Wertz (1990).

## 3.4  Linear Quadratic Gaussian Predictive Control

### 3.4.1  Introduction

Although many of the model based LRPC algorithms appear to be stochastic in nature, once the predictor is introduced into the equations they almost become equivalent to deterministic optimal control problems. The approach to be followed below uses a similar framework to the LRPC model based algorithms, but the problem is solved in the same manner as for LQG stochastic optimal control problems (Grimble, 1986a,b). The approach is sufficiently similar to enable insights to be gained into the GPC algorithms, and yet at the same time the good stability properties of LQG solutions are obtained (Grimble, 1994). The so called Linear Quadratic Gaussian Predictive Control (LQGPC) law is derived below.

A crucial feature of the solution presented is the assumptions, that must be made (Grimble, 1994). These assumptions, which were motivated by the GPC literature, lead to an LQG like solution which has the same matrix structure, as in the usual GPC control problem. The following analysis is conducted using a single degree of freedom controller structure. Note that the *receding horizon control law* (Kwon and Pearson, 1977) is not employed in this situation, and hence the controller does not have the same stability problems that arise as in GPC design. In fact, the resulting optimal system has the same stability results as in traditional LQG solutions (Grimble, 1988b, 1995). Bitmead et al. (1989a,b; 1990) motivated interest in this class of problems using a state-space based approach.

The multistep cost function is constructed so that the expected value of the error at time $t$ is minimised, and the sum of the future predicted error and control signals are also minimised for future times. A physically realistic problem is therefore constructed which does not have the same contradictions as appear in the receding horizon cost problem. The control law obtained is different to those which attempt to minimise an expected cost at time $t$, and at the same time minimise future predicted control signal variations, using the same time-invariant controller (Grimble, 1997).

The optimal control solution obtained below involves a nested set of controllers having different coefficients for different prediction intervals, and has parallels in the solution of finite-time optimal control problems, where the state-feedback gain matrix is time-varying. Note, however, that these optimal controllers do not have to be computed if the vector-matrix form of the future control signals algorithm is employed. In this case, an expression which is similar to that for the GPC controller is obtained, with the exception that the lower triangular structure of the matrix involved ensures that the control at time t is not affected by the future control computations. It follows that the controller that is implemented within the feedback loop of the actual system is the same as in an LQG design. The stability properties (if constraints are not applied) are therefore identical to the usual single step cost term LQG control problem, where the cost-function involves only the first term in the multistep criterion.

## 3.4.2 Plant and Disturbance Description

The single-input single-output one degree-of-freedom discrete-time system shown in Fig. 3.1 is assumed to be linear and time-invariant. The external white noise sources drive colouring filters which represent the reference $W_r(z^{-1})$ and disturbance $W_d(z^{-1})$ subsystems. The system equations become:

**Plant output :**

$$m(t) = Wu(t)$$

**System output :**

$$y(t) = d(t) + Wu(t)$$

**Disturbance :**

$$d(t) = W_d\xi(t)$$

**Reference :**

$$r(t) = W_r\zeta(t)$$

**Error :**

$$e(t) = r(t) - y(t)$$

**Control signal :**

$$u(t) = C_0(r(t) - y(t))$$

**System Assumptions**

(i) The white noise sources $\{\zeta(t)\}$ and $\{\xi(t)\}$ are zero-mean and mutually statistically independent. The variances for these signals are without loss of generality taken to be unity.

(ii) The plant $W$ is free of unstable hidden modes, and the reference $W_r$ and disturbance $W_d$ subsystems are asymptotically stable. The plant $W$ includes a $k$-steps delay.

(iii) The disturbance model $W_d$ will be assumed to be strictly minimum phase.

**Signals and Nominal Sensitivity Functions**

The following sensitivity, complementary sensitivity, and control sensitivity functions may be defined:

$$S = (1 + WC_0)^{-1}, \quad T = 1 - S = WC_0S, \quad M = C_0S = C_0(1 + WC_0)^{-1}$$

Expressions may now be derived for the output, error and control signals as:

$$y(t) = WC_0Sr(t) + Sd(t) \tag{3.29}$$

$$e(t) = r(t) - y(t) = (1 - WC_0S)r(t) - (1 - WC_0S)d(t) \tag{3.30}$$

$$u(t) = SC_0(r(t) - d(t)) \tag{3.31}$$

**Polynomial Form of Plant Model**

The controlled auto regressive integrated moving average plant model used in the GPC control law, was defined as:

$$A_o(z^{-1})y(t) = B_0(z^{-1})u(t - k) + \frac{C(z^{-1})}{\Delta}\xi(t)$$

where $\Delta = 1 - z^{-1}$, and $\{\xi(t)\}$ denotes a white noise sequence. The transport-delay $k \geq 1$ and without loss of generality let $A_o(0) = 1$. This is equivalent to the following more general ARMAX model for the system:

$$A(z^{-1})y(t) = B(z^{-1})u(t) + C(z^{-1})\xi(t) \tag{3.32}$$

This is the model to be employed below but may easily be related to the CARIMA case by defining:

$$A = A_o\Delta \quad \text{and} \quad B = B_o\Delta z^{-k}.$$

The reference or set-point signal can be represented by the ARMA model:

$$A(z^{-1})r(t) = E_r(z^{-1})\zeta(t)$$

where $\{\zeta(t)\}$ denotes a white noise sequence, uncorrelated with the sequence $\{\xi(t)\}$. Without loss of generality, the white noise signals are assumed to be zero-mean and of unity-variance, and the system models are assumed to have the common denominator polynomial $A(z^{-1})$.

**Polynomial Form of Plant Transfer Functions**

The system shown in Fig. 3.1 may therefore be represented in polynomial form, where the plant, disturbance, and tracking transfer-functions may be written as:

$$W = A^{-1}B = A^{-1}B_0\Delta z^{-k}, \quad W_d = A^{-1}C, \quad W_r = A^{-1}E_r$$

The system models are all assumed to be free of unstable hidden modes.

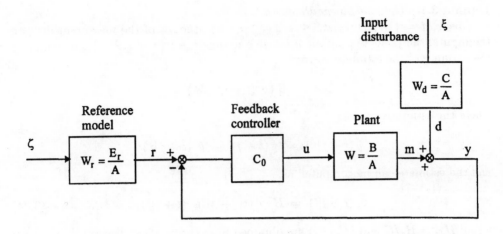

**Fig. 3.1** : One Degree of Freedom Controller and System Models

### 3.4.3 Optimal Linear Predictor

An LQ cost-function involves minimising the variance of the error signal $e(t+k) = r(t+k) - y(t+k)$ and the variance of the corresponding control signal $u(t)$. In GPC control problems, prediction error and control terms may be added into the cost-function, and at the next sample instant these will involve the prediction error : $\hat{e}_j$ $(t+k+1|t+k)$ and control signal $\hat{u}_j(t+1|t)$. The prediction error for $j$ further steps ahead is defined as:

$$\hat{e}_j(t+k+j|t+k) = r(t+k+j) - \hat{y}_j(t+k+j|t+k)$$

where $\hat{y}_j$ is the predicted system output.

**Assumptions**

The future values of the control signal are unknown at time $t$ and hence there is an assumption which must made when deriving the expression for the predicted system output:

**Ass(1)** : The predicted values of $u(t)$ can be substituted to obtain:

$$\hat{y}_j(t+j|t) \; = \; H_{f_j}d(t) \; + \; W\hat{u}_j(t+j|t)$$

A second assumption is also required which enables the close relationship to the GPC equations to be developed.

**Ass(2)** : Assume that the future values of the control signal (unknown at time $t$) are equal to their predicted values, so that $u(t+j) = \hat{u}_j(t+j|t)$.

**Lemma 3.1** : *Optimal Linear Predictor*

Given $M = \{y(t_1), u(t_2), t_1 \leq t, \ t_2 \leq t - k\}$, the set of the measurements and the inputs, the predicted output $\hat{y}(t + j|t)$ at time $t + j \ \{j = 1, 2, ...\}$, to minimise the variance of the estimation error:

$$J = E\{\tilde{y}^2(t + j|t)|M\} \tag{3.33}$$

where the estimation error:

$$\tilde{y}(t + j|t) = y(t + j) - \hat{y}_j(t + j|t)$$

and the estimate can be computed as:

$$\hat{y}_j(t + j|t) = H_{fj}d(t) + W\hat{u}_j(t + j|t) \tag{3.34}$$

where $H_{fj} = H_j/C$ and $(H_j, E_j)$ are obtained from the smallest-degree solution with respect to $E_j$ of the diophantine equation:

$$E_j A + z^{-j} H_j = C$$

**Proof** : Similar to that in §3.2.1 and given in Grimble (1994, 1995).

**Prediction of Output, Error and Control Signals**

The predicted error signal may now be defined as:

$$\hat{e}_j(t + jt) = r(t + j) - \hat{y}_j(t + j|t)$$

and the predicted control signal can be based upon the predicted output as:

$$\hat{u}_j(t + j|t) = C_j \hat{e}_j(t + j|t) = C_j\left(r(t + j) - \hat{y}_j(t + j|t)\right)$$

After substituting from (3.34):

$$\hat{u}_j(t + j|t) = C_j r(t + j) - C_j\left(H_{fj}d(t) + W\hat{u}_j(t + j|t)\right)$$

$$= S_j C_j(r(t + j) - H_{fj}d(t)) = M_j(r(t + j) - H_{fj}d(t)) \tag{3.35}$$

where the *jth sensitivity* and *control sensitivity* functions may be defined as:

$$S_j = (1 + WC_j)^{-1} \text{ and } M_j = C_j S_j$$

The estimated plant output, given by (3.34), can now be written as:

$$\hat{y}_j(t + jt) = H_{fj}d(t) + W\hat{u}_j(t + j|t)$$

$$= (1 - WM_j)H_{fj}d(t) + WM_j r(t + j) = S_j H_{fj}d(t) + WM_j r(t + j) \tag{3.36}$$

The *predicted error* signal now follows from (3.36) as:

$$\hat{e}_j(t + j|t) = r(t + j) - \hat{y}_j(t + j|t) = (1 - WM_j)r(t + j) - (1 - WM_j)H_{fj}d(t)$$

$$= S_j(r(t+j) - H_{fj}d(t)) \tag{3.37}$$

Note that the future predicted control signals are given by controllers $C_j(z^{-1})$ which are $j$ dependent. This is equivalent in state equation form to allowing the feedback gain matrix over the future time interval $[0, n_1]$ to be time-varying. The optimal control solution will determine whether a $j$ dependent solution is required. The prediction model, under the above assumption, which determines the time-response of the future control and output signals, is shown in Fig. 3.2.

Also note that the above equations reduce to those for the actual feedback loop if $j = 0$, $H_j = H_0 = C$ and $H_{fj} = H_{fo} = 1$.

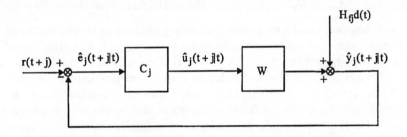

**Fig. 3.2** : Model for the Prediction of the jth Step Time Responses

### 3.4.4 Stochastic Generalised Predictive Optimal Control

The Predictive Control law is derived in the following to minimise a cost-function which is related to both LQG and GPC cost terms. The weighting functions in the criterion can be dynamic, and may be set to zero to represent different error and control prediction intervals. The cost-function is defined as:

$$J = E\left\{(H_{q0}(r(t+k) - y(t+k)))^2 + (H_{ro}u(t))^2\right\}$$

$$+E\left\{\sum_{j=1}^{n_1}\{(H_{qj}(r(t+k+j) - \widehat{y}_j(t+k+j|t+k)))^2 + (H_{rj}\widehat{u}_j(t+j|t))^2\}\right\} \tag{3.38}$$

where $E\{.\}$ denotes the unconditional expectation operator, $H_{qo}$ and $H_{ro}$ denote the usual LQG error and control signal weighting terms, and $H_{qj}$ and $H_{rj}$ $\{j = 1, 2, .., n_1\}$ denote the weightings on the prediction error and predicted control signals, respectively. The assumption will now be made that $H_{ro}^* H_{r0} > 0$ on $|z| = 1$. Noting $\widehat{y}(t+k|t+k) = y(t+k)$, the criterion becomes:

$$J = E\left\{\sum_{j=0}^{n_1}\{(H_{qj}\widehat{e}_j(t+k+j|t+k))^2 + (H_{rj}\widehat{u}_j(t+j|t))^2\}\right\}$$

For greater generality, cross-product terms may also be introduced into the cost-function, which may then be written in the complex-frequency domain as:

$$J = \frac{1}{2\pi j} \oint_{|z|=1} \left\{ \sum_{j=o}^{n_1} \left\{ Q_{cj}\Phi_{\hat{e}\hat{e}} + R_{cj}\Phi_{\hat{u}\hat{u}} + G_{cj}\Phi_{\hat{u}\hat{e}} + \Phi_{\hat{e}\hat{u}}G_{cj}^* \right\} \right\} \frac{dz}{z} \qquad (3.39)$$

where the dynamic weightings:

$$Q_{cj} = H_{qj}^* H_{qj} \quad , \quad R_{cj} = H_{rj}^* H_{rj} \quad , \quad G_{cj} = H_{pj}^* H_{pj}$$

and these have the following polynomial descriptions:

$$H_{qj} = B_{qj}/A_{wj} \ , \ H_{rj} = B_{rj}/A_{wj}, \ H_{pj} = B_{pj}/A_{wj} \text{ and } H_{fj} = B_{fj}/A_{wj}$$

The cost function employed is not a receding horizon criterion, although it has similar terms. The cost index certainly involves calculating a control signal which minimises a cost function running from time $t$. However, since the problem is stochastic, the additional terms in the cost index involve future predicted error and control signals. Thus, at any given time $t$, the optimal control is calculated for the actual feedback loop, and the future predicted controls can also be found to minimise the additional terms in the criterion. In deterministic receding horizon control laws, it is of course the future values of the error and the control signals which are to be minimised on a sliding cost horizon. There is therefore an inconsistency in receding horizon control, since the future control calculated at time $t$ will never be implemented. That is, at time $t+1$ a new optimal control signal is found, and the value computed previously is ignored.

The same logical inconsistency does not arise in the cost function above, since future control signals are not evaluated at time t, but only their predicted values. In practice this is all that is required, since it is necessary to know whether future predicted control and error signals will lie outside allowable limits. A causal controller could not, of course, compute future control signals in a system with random input disturbances. The starting point for the present analysis therefore appears to be theoretically sound, since the cost function to be minimised can be justified physically.

### 3.4.5  Solution of the LQGPC Problem

The proof which follows is related to the approach first established by Kucera (1980). The main step in the analysis is to again write the cost-function in a squared form, using a completing the squares type of argument. Equations (3.35) and (3.37), for $\hat{e}_j(t+j|t)$ and $\hat{u}_j(t+j|t)$ reveal that these can be written for all $j = \{0, 1, ..., n_1\}$ as:

$$\hat{u}_j(t+j|t) = M_j \left( r(t+j) - H_{f_j}d(t) \right) \qquad (3.40)$$

$$\hat{e}_j(t+jt) = (1 - WM_j)(r(t+j) - H_{f_j}d(t)) \qquad (3.41)$$

where if $j = 0, H_{fj} = 1$, which implies : $H_j = C, M_0 = M, S_0 = S$, and $\hat{e}_0(t|t) = e(t)$. To simplify these expressions let,

$$f_j(t) = r(t+j) - H_{fj}d(t) \qquad (3.42)$$

**Cost-function with Future Predicted Error and Control Terms**

The cost function (3.39) to be minimised was defined as:

$$J = \frac{1}{2\pi j} \oint_{|z|=1} \sum_{j=0}^{n_1} \{Q_{cj}\Phi_{\hat{e}\hat{e}} + R_{cj}\Phi_{\hat{u}\hat{u}} + G_{cj}\Phi_{\hat{u}\hat{e}} + \Phi_{\hat{e}\hat{u}}G_{cj}^*\}\} \frac{dz}{z}$$

Substituting from (3.40), and (3.41) and recalling the independence of the noise sources, obtain:

$$J = \frac{1}{2\pi j} \oint_{|z|=1} \{\sum_{j=0}^{n_1} \{I_j(z^{-1})\}\} \frac{dz}{z}$$

where the $j$th term in the integrand $I_j$ is defined as:

$$I_j = Q_{cj}(1 - WM_j)\Phi_{f_j f_j}(1 - WM_j)^*$$

$$+R_{cj}M_j\Phi_{f_j f_j}M_j^* + G_{cj}M_j\Phi_{f_j f_j}(1 - WM_j)^* + (1 - WM_j)\Phi_{f_j f_j}M_j^*G_{cj}^*$$

$$= (WQ_{cj}W^* - G_{cj}W^* - WG_{cj}^* + R_{cj})(M_j\Phi_{f_j f_j}M_j^*)$$

$$-Q_{cj}(WM_j\Phi_{f_j f_j} + \Phi_{f_j f_j}M_j^*W^*) + G_{ij}M_j\Phi_{f_j f_j} + \Phi_{f_j f_j}M_j^*G_{cj}^* + Q_{cj}\Phi_{f_j f_j} \quad (3.43)$$

**Definition of the Generalised Spectral Factors**

Introduce the following *generalised spectral factors* :

$$Y_{cj}^*Y_{cj} = W^*Q_{cj}W + R_{cj} - G_{cj}W^* - WG_{cj}^*$$

$$Y_{fj}Y_{fj}^* = \Phi_{f_j f_j}$$

for all $j = 0, 1, 2, ..., n_1$.   The $j$th term in the cost integrand (3.43) may now be written as:

$$I_j = Y_{cj}^*Y_{cj}(M_j Y_{fj}Y_{fj}^* M_j^*) - (\Phi_h^* M_j + \Phi_h M_j^*) + Q_{cj}\Phi_{f_j f_j}$$

where $\Phi_h = (W^*Q_{cj} - G_{cj}^*)\Phi_{f_j f_j}$. *Completing the squares,* now obtain:

$$I_j = (Y_{cj}M_j Y_{fj} - Y_{cj}^{*-1}\Phi_h Y_{fj}^{*-1})(Y_{fj}^* M_j^* Y_{cj}^* - Y_{fj}^{-1}\Phi_h^* Y_{cj}^{-1}) + I_{j0} \quad (3.44)$$

where the control independent term:

$$I_{j0} = Q_{cj}\Phi_{f_j f_j} - Y_{cj}^{*-1}(\Phi_h Y_{fj}^{*-1}Y_{fj}^{-1}\Phi_h^*)Y_{cj}^{-1}$$

**Equivalent Minimum Variance Problem**

It will now be shown that the optimal solution can be found by minimising the variance of two cost terms. By using an innovations signal representation (Grimble, 1984a,b) the future reference and disturbance signals can be modelled as : $(r(t+j) - H_{fj}d(t)) = Y_{fj}\epsilon(t)$, where $\{\epsilon(t)\}$ is a white noise sequence of unity variance. Thus, the control signal can be expressed, using (3.40), in the equivalent form:

$$\hat{u}_j(t+j|t) = M_j Y_{fj}\epsilon(t) \tag{3.45}$$

Observe that the $I_{jo}$ cost term does not depend upon the controller. Hence by reference to (3.44) and (3.45), the optimal control is clearly equivalent to that for the minimum-variance problem, where the signal to be minimised:

$$\theta_j(t) = Y_{cj}\hat{u}_j(t+j|t) - \left(Y_{cj}^{*-1}\Phi_h Y_{fj}^{*-1}\right)\epsilon(t) \tag{3.46}$$

The optimisation will now proceed through the minimisation of the variance of this signal, rather than considering (3.44) directly. This is different to the approach of Kučera (1980), and is valuable for demonstrating the link to GPC control. Note that the minimum value of the cost-function will equal the variance of this signal plus the term due to $I_{jo}$ in (3.44).

**Polynomial Equations**

The generalised spectral-factors may now be given their polynomial forms:

$$Y_{cj}^* Y_{cj} = \frac{D_{cj}^* D_{cj}}{(AA_{wj})^*(AA_{wj})}$$

$$= \frac{B^* B_{qj}^* B_{qj} B + A^* B_{rj}^* B_{rj} A - B^* B_{pj}^* B_{fj} A - A^* B_{fj}^* B_{pj} B}{(AA_{wj})^*(AA_{wj})} \tag{3.47}$$

$$Y_{fj} Y_{fj}^* = \frac{D_{fj} D_{fj}^*}{AA^*} = (H_j H_j^* + E_r E_r^*)/(AA^*) \tag{3.48}$$

and hence identify $Y_{cj} = D_{cj}/(AA_{wj})$ and $Y_{fj} = D_{fj}/A$. Here the polynomial spectral factors $D_{cj}$ and $D_{fj}$ can, from the system and cost-weighting assumptions, be taken to be strictly Schur.

**Partial Fraction Computations**

The various terms in the above expression (3.46) may now be simpified by substituting from the polynomial system models in Section 3.4.2 and the spectral factor results given above. It will first be necessary to introduce the diophantine equations, which replace the need for partial fraction expansions. To motivate the form of the diophantine equations, consider the polynomial form of the following transfers:

$$Y_{cj}^{*-1}\Phi_h Y_{fj}^{*-1} = Y_{cj}^{*-1}(W^* Q_{cj} - G_{cj}^*)\Phi_{f_j f_j} Y_{fj}^{*-1} = \frac{(B^* B_{qj}^* B_{qj} - A^* B_{fj}^* B_{pj})D_{fj}}{D_{cj}^* AA_{wj}}$$

The above transfer must now be split into causal and non-causal components with denominators ($AA_{wj}$ and $D_{cj}^*$). To accomplish this separation, the following diophantine equations must be introduced. The solution ($G_{oj}, H_{oj}, F_{oj}$) is required, with $F_{oj}$ of smallest degree, of the coupled equations:

$$D_{cj}^* G_{oj} z^{-g_1} + F_{oj} AA_{wj} = (B^* B_{qj}^* B_{qj} - A^* B_{fj}^* B_{pj}) D_{fj} z^{-g_j} \qquad (3.49)$$

$$D_{cj}^* H_{oj} z^{-g_1} - F_{oj} BA_{wj} = (A^* B_{rj}^* B_{rj} - B^* B_{pj}^* B_{fj}) D_{fj} z^{-g_j} \qquad (3.50)$$

The integer $g_j$ is chosen for each of the $j$ coupled equations to ensure these only contain polynomials in $z^{-1}$. The smallest value of $g_j$ which satisfies this requirement is selected. The first equation (3.49) gives the desired partial fraction expansion:

$$G_{oj}/(AA_{wj}) + z^{gj} F_{oj}/D_{cj}^* = (B^* B_{qj}^* B_{qj} - A^* B_{fj}^* B_{pj}) D_{fj}/(D_{cj}^* AA_{wj})$$

**Implied Equation Determining Stability Properties**

Adding equations (3.49) and (3.50) multiplied by $B$ and $A$, respectively, and invoking the spectral-factor relationship (3.47), obtain the *implied equation* :

$$G_{oj} B + H_{oj} A = D_{cj} D_{fj} \qquad (3.51)$$

The solution of the optimal control problem which follows uses the time-domain expression (3.46), which is rather different to the usual frequency domain argument.

**Expression for the $\theta_j(t)$ Time-function**

The polynomial expressions may now be used to find an equation for the time-function $\theta_j(t)$ which depends upon the disturbance input. First note from equation (3.41) that the component of the error $\hat{e}_j(t+j|t)$, due to the reference and disturbance signal, may be represented (using the innovations model discussed earlier) as: $\hat{e}_j(t+j|t) = S_j(r(t+j) - H_{fj}d(t)) = S_j Y_{fj}\epsilon(t)$. From equations (3.46) and (3.49) obtain:

$$\theta_j(t) = \frac{D_{cj}}{AA_{wj}} \hat{u}_j(t+j|t) - \left(\frac{G_{oj}}{AA_{wj}} + z^{gj}\frac{F_{oj}}{D_{cj}^*}\right)\epsilon(t)$$

$$= \frac{D_{cj}}{AA_{wj}} \hat{u}_j(t+j|t) - \frac{G_{oj}}{AA_{wj}}(1+WC_j)\frac{A}{D_{fj}}e(t+j|t) - z^{gj}\frac{F_{oj}}{D_{cj}^*}\epsilon(t)$$

Substituting from the *implied equation* (3.51), now obtain:

$$\theta_j(t) = \frac{1}{AA_{wj}D_{fj}}(G_{oj}B + H_{oj}A)\hat{u}_j(t+j|t)$$

$$- \frac{G_{oj}}{A_{wj}D_{fj}}(1+WC_j)e_j(t+j|t) - z^{gj}\frac{F_{oj}}{D_{cj}^*}\epsilon(t)$$

but from (3.45), $\hat{u}_j(t+j|t) = M_j Y_{fj}\, \epsilon(t) = C_j \hat{e}_j(t+j|t)$, and hence substituting, using this result:

$$\theta_j(t) = \frac{1}{A_{wj}D_{fj}}(H_{oj}\hat{u}_j(t+j|t) - G_{0j}\hat{e}_j(t+j|t)) - z^{gj}\frac{F_{oj}}{D^*_{cj}}\epsilon(t)$$

**Minimisation Procedure**

Consider the minimization of the variance of the sequence $\{\theta_j(t)\}$ where

$$\theta_j(t) = \theta_{ja}(t) + \theta_{jb}(t) = \left[\frac{1}{A_{wj}D_{fj}}\left((H_{oj}\hat{u}_j(t+j|t) - G_{0j}\hat{e}_j(t+j|t))\right)\right]$$

$$+\left(-z^{gj}\frac{F_{oj}}{D^*_{cj}}\epsilon(t)\right) \tag{3.52}$$

The term within the square brackets $\theta_{ja}(t) = [.]$ includes only stable operators. The final term, in the round brackets $\theta_{jb}(t) = (.)$, involves only a strictly unstable system. From the diophantine equation, note that $deg(F_{oj}) < g_j$. Thus, the operators in this equation can be represented as convergent sequences in $z^{-1}$ for the term $[.]$, and a convergent sequence in terms of $z$ for the final term $(.)$. It follows that the final term involves only future values of $\{\epsilon(t)\}$, and the first two terms involve only past values of $\{\epsilon(t)\}$. Since these values are assumed to be independent (white noise) sequences, the variance of $\{\theta_j(t)\}$ can clearly be written as:

$$J_{\theta j} = E\{\theta_j(t)^2\} = E\{\theta_{ja}(t)^2\} + E\{\theta_{jb}(t)^2\} \tag{3.53}$$

where from (3.52) : $\theta_{ja}(t) = [.]$ and $\theta_{jb}(t) = -z^{gj}(F_{oj}/D^*_{cj})\epsilon(t)$.

The second term in (3.52), representing the variance of the signal $\{\theta_{jb}(t)\}$, does not depend upon the control signal (real or predicted). Thence, the optimal control and predicted control signals, to minimise the variance of $\{\theta_j(t)\}$, must minimise the variance of the first term $\{\theta_{ja}(t)\}$. The optimal predicted control is therefore clearly that which sets the square bracketed term to zero, giving:

$$\hat{u}(t+j|t) = H_{0j}^{-1}G_{oj}\hat{e}_j(t+j|t) \tag{3.54}$$

**Closed-loop Stability**

The closed-loop controllers, for the $j$th prediction, follow as:

$$C_j = C_{jd}^{-1}C_{jn} = (H_{oj})^{-1}G_{oj} \tag{3.55}$$

where the $j$th controller $C_j$ is written as: $C_j = C_{jn}/C_{jd}$. It is easily confirmed that the characteristic-equations for the plant shown in Fig. 3.1, or the predictor loops shown in Fig. 3.2 are, for each $j$:

$$AC_{jd} + BC_{jn} = AH_{oj} + BG_{oj} = D_{cj}D_{fj} \tag{3.56}$$

(after substituting from the implied equation (3.51)). The polynomials on the right of this equation are strictly Schur, given the weighting and system assumptions.

**Minimum Value of the Cost-index**

From (3.44) and (3.53), the minimum value of the cost index integrand follows as:

$$I_{\min} = \sum_{j=0}^{n_1} \left\{ \left( \frac{F_{oj} F_{oj}^*}{D_{cj} D_{cj}^*} \right) + I_{j0} \right\} \tag{3.57}$$

**Frequency-domain Solution**

The optimal control solution is normally derived from a frequency-domain argument. However, to draw the analogy with GPC control, a time-domain derivation was provided above. For completeness, a brief sketch of the z-domain solution will now be given to confirm the link to the Kucera polynomial approach. From the diophantine equation (3.49):

$$\left( Y_{cj} M_j Y_{fj} - Y_{cj}^{*-1} \Phi_h Y_{fj}^{*-1} \right) = Y_{cj} M_j Y_{fj} - \left( \frac{G_{oj}}{AA_{wj}} + z^{gj} \frac{F_{oj}}{D_{cj}^*} \right)$$

$$= \left[ \frac{(H_{oj} C_{jn} - G_{oj} C_{jd})}{A_{wj}(AC_{jd} + BC_{jn})} \right] - \left( z^{gj} \frac{F_{oj}}{D_{cj}^*} \right)$$

It is simple to show (Grimble and Johnson, 1988), using the Residue theorem, that the term within the square brackets must be set to zero to define the optimal controller. The optimal controller then follows as:

$$C_j = C_{jd}^{-1} C_{jn} = H_{oj}^{-1} G_{oj}$$

**Theorem 3.1** : *Linear Quadratic Gaussian Predictive Control*

The optimal predictive control law to minimise the cost-function (3.39), for the prediction model based upon Assumption Ass(1) and shown in Fig. 3.2, can be computed from the following spectral factorization and diophantine equations:

**Spectral factorization** : The solutions $D_{cj}, D_{fj}$, for $j = 0, 1, ..., n_1$ of the following equations are required:

$$D_{cj}^* D_{cj} = B^* B_{qj}^* B_{qj} B + A^* B_{rj}^* B_{rj} A - B^* B_{pj}^* B_{fj} A - A^* B_{fj}^* B_{pj} B \tag{3.58}$$

$$D_{fj} D_{fj}^* = E_r E_r^* + H_j H_j^* \tag{3.59}$$

**Diophantine equations** : The solutions $(G_{oj}, H_{oj}, F_{oj})$ for $j = 0, 1, ..., n_1$, with $F_{oj}$ of smallest degree, of the following equations are required:

$$D_{cj}^* G_{oj} z^{-gj} + F_{oj} A A_{wj} = (B^* B_{qj}^* B_{qj} - A^* B_{fj}^* B_{pj}) D_{fj} z^{-gj} \tag{3.60}$$

$$D_{cj}^* H_{oj} z^{-gj} - F_{oj} B A_{wj} = (A^* B_{rj}^* B_{rj} - B^* B_{pj}^* B_{fj}) D_{fj} z^{-gj} \qquad (3.61)$$

where $g_j$ are the smallest positive integers which ensure the equations involve only powers of $z^{-1}$.

**Feedback and prediction model controllers :** For $j = 0, 1, 2, ..., n_1$.

$$C_j = C_{0d}^{-1} C_{0n} = H_{0j}^{-1} G_{oj} \qquad (3.62)$$

The closed-loop predictors and systems are stable, and the characteristic polynomial for the actual system is given by:

$$\rho_0 = A C_{0d} + B C_{0n} = D_{c0} D_{f0} \qquad (3.63)$$

and the minimum cost $J_{min}$ is determined by (3.57).

**Proof :** By collecting the results preceding the theorem.

### 3.4.6   Vector-matrix Computation of Future Controls

The future predicted control sequence can be found by using a similar approach to that in GPC algorithms. The analogy with GPC algorithms suggests the manner in which constrained optimisation problems should also be solved.

**Separation of the Term** $(H_{oj}/(A_{wj} D_{fj}))$

The term $(H_{oj}/(A_{wj} D_{fj})) \, \hat{u}_j(t + j|t)$ in equation (3.52) can be separated into terms which include future values of the control signal and those in the past (at time $t - 1, t - 2, ...$). This requires the introduction of a further diophantine equation in terms of the unknown polynomials $(S_j, W_j)$, where $W_j$ is of smallest degree:

$$W_j A_{wj} D_{fj} + z^{-j-1} S_j = H_{oj} \qquad (3.64)$$

giving

$$H_{oj}/(A_{wj} D_{fj}) = W_j + z^{-j-1} S_j/(A_{wj} D_{fj}) \qquad (3.65)$$

The polynomial $W_j$ has order $j$ and hence this equation provides the desired decomposition.

**Expression for** $\theta_{ja}(t)$

The two components of the signal $\theta_j(t) = \theta_{ja}(t) + \theta_{jb}(t)$ were defined in (3.52). From this equation obtain the expression for $\{\theta_{ja}(t)\}$ as:

$$\theta_{ja}(t) = \frac{1}{A_{wj} D_{fj}} \left( H_{0j} \hat{u}_j(t + j|t) \right) - G_{oj} \hat{e}_j(t + j|t)$$

$$= W_j \hat{u}_j(t + j|t)) - \frac{1}{A_{wj} D_{fj}} (G_{oj} \hat{e}_j(t + j|t) - S_j \hat{u}_j(t - 1|t - j - 1)) \qquad (3.66)$$

and note that $\hat{u}_j(t-1|t-j-1)$ can be replaced by the actual values of the control signal $u(t-1)$. Introduce the signal $f_{ja}(t)$, dependent upon known signals at time $t$, as:

$$f_{ja}(t) = (-1/(A_{wj}D_{fj}))\,(G_{oj}\hat{e}_j(t+j|t) - S_j u(t-1))$$

Then equation (3.66) may be written as:

$$\theta_{ja}(t) = W_j\hat{u}_j(t+j|t)) + f_{ja}(t) \tag{3.67}$$

**Computation of $\hat{e}_j(t+j|t)$**

From equation (3.34) for the predicted output signal:

$$\hat{y}_j(t+j|t)) = H_{fj}d(t) + W\hat{u}_j(t+j|t)$$

and hence $\hat{e}_j(t+j|t) = r(t+j) - \hat{y}_j(t+j|t)$ depends on quantities which are all known at time $t$ and can be computed from these results.

**Vector of Signals $\theta_{ja}(t)$**

The signal $\theta_{ja}(t)$ for $j = 0, 1, .., n_1$ can be stacked to form the following vector:

$$\theta_a = [\theta_{0a}(t), \theta_{1a}(t), ..., \theta_{n1a}(t)]^T$$

where the elements $\theta_{ja}(t)$ are defined by (3.67).

**Vector-matrix Form of the Cost-function**

Note that the cost-function is summed over $n_1$ steps. Collecting the above results, it is clear that the terms in the cost-function, representing the variance of $\theta_j(t)$, can be written in the vector-matrix form:

$$J_\theta = E\{\theta_a^T(t)\theta_a(t) + \theta_b^T(t)\theta_b(t)\} = E\{(XU+F)^T(XU+F)\} + J_b \tag{3.68}$$

where

$$
U = \begin{bmatrix} \hat{u}_{n1}(t+n_1|t) \\ \hat{u}_{n_1-1}(t+n_1-1|t) \\ \cdot \\ \cdot \\ \cdot \\ \hat{u}_1(t+1|t) \\ u(t) \end{bmatrix}, \quad
F = \begin{bmatrix} f_{n_1 a}(t) \\ f_{(n_1-1)a}(t) \\ \cdot \\ \cdot \\ \cdot \\ f_{0a}(t) \end{bmatrix}
$$

$$
X = \begin{bmatrix}
w_0^{n_1} & w_1^{n_1} & w_2^{n_1} & \cdot & \cdot & \cdot & w_{n_1}^{n_1} \\
0 & w_0^{n_1-1} & \cdot & & & & \cdot \\
 & 0 & & & & & \cdot \\
\cdot & & & & & & \cdot \\
\cdot & & & & w_0^2 & w_1^2 & w_2^2 \\
\cdot & & & & & w_0^1 & w_1^1 \\
0 & 0 & \cdot & \cdot & \cdot & 0 & w_0^0
\end{bmatrix} \tag{3.69}
$$

and

$$J_b = E\{\theta_b^T(t)\theta_b(t)\} = \sum_{j=0}^{n_1}\left\{\frac{1}{2\pi j}\oint_{|z|=1}\left(\frac{F_{oj}F_{oj}^*}{D_{cj}D_{cj}^*}\right)\frac{dz}{z}\right\}$$

**Optimal Control Law**

The optimal control vector (Grimble, 1988c) may be found to set the first term in (3.68) to zero, since X is square and non-singular, so that,

$$U = -X^{-1}F$$

Note that $u(t)$ given by the final element in $U$ is the same as that obtained for the optimal control given by (3.54) when $j = 0$. Although the assumption Ass(2) affects the future predicted controls, the final equation is unaffected (since X is upper triangular), and the feedback control is therefore the same as that found previously. The stability properties of the feedback system will therefore also be the same.

**Interpretation of the Polynomial $W_j$**

The physical significance of the term $W_j$ is interesting, and the following explanation provides an alternative method of computing $W_j$. From the diophantine equation (3.64):

$$W_j = H_{oj}/(A_{wj}D_{fj}) - z^{-j-1}S_j/(A_{wj}D_{fj})$$

but from the implied equation (3.51) :

$$H_{oj}/(A_{wj}D_{fj}) = ((D_{cj}D_{fj})/A - G_{oj}(B/A))/(A_{wj}D_{fj})$$

The expression for $W_j$ therefore becomes:

$$W_j = \frac{D_{cj}}{AA_{wj}} - z^{-j}\left(G_{oj}\frac{B}{A}z^j + z^{-1}S_j\right)/(A_{wj}D_{fj})$$

$$= D_{cj}/(AA_{wj}) - z^{-j}\left(G_{oj}B_0z^{-k+j} + z^{-1}S_jA\right)/(AA_{wj}D_{fj})$$

It follows from the above result that if $k > j$ (Ass(1a)), the polynomial $W_j$ includes the first $(j + 1)$ Markov parameters $W_i$ of the transfer function $D_{cj}/(AA_{wj})$. If the weightings are chosen to be the same for each $j$, which is a reasonable assumption, the $W_j$ polynomials will have the same build up of coefficients. Thence,

$$W_0 = w_0$$
$$W_1 = w_0 + w_1z^{-1}$$
$$\vdots \qquad \vdots$$
$$W_j = w_0 + w_1z^{-1} + w_2z^{-2} + ... + w_jz^{-j}$$

The matrix X, defined in (3.69), then has a particularly simple form, since it consists of the same elements as in the first row, but displaced one step for each row.

### 3.4.7  Vector-matrix Form of the LQGPC Algorithm

The alternative method of computing the control law in vector matrix form which emphasises the link to GPC controllers, may now be summarised in the following lemma.

**Lemma 3.2 :** *Vector-matrix Form of LQGPC Calculation*

The LQGPC control law for the system shown in Fig. 3.1 can be found to minimise the cost-function (3.39), which may be written as:

$$J = E\{(XU + F)^T (XU + F)\} + J_{min} \tag{3.70}$$

where

$$J_{min} = \sum_{j=0}^{n_1} \left\{ \frac{1}{2\pi j} \oint_{|z|=1} \left\{ \left( \frac{F_{oj} F_{oj}^*}{D_{cj} D_{cj}^*} \right) + I_{jo} \right\} \frac{dz}{z} \right\}$$

and $I_{j0}$ is defined in (3.44). The matrix $X$ and vectors, $U, F$ are defined as:

$$X = \begin{bmatrix} w_0^{n_1} & w_1^{n_1} & w_2^{n_1} & \cdots & & w_{n_1}^{n_1} \\ 0 & w_0^{n_1-1} & w_1^{n_1-1} & & & \\ & 0 & w_0^{n_1-2} & & & \vdots \\ \vdots & \vdots & & \ddots & & \\ & & \vdots & & w_0^1 & w_1^1 \\ 0 & 0 & 0 & & 0 & w_0^0 \end{bmatrix},$$

$$F = \begin{bmatrix} f_{n_1 a}(t) \\ \cdot \\ \cdot \\ \cdot \\ f_{1a}(t) \\ f_{0a}(t) \end{bmatrix}, \quad U = \begin{bmatrix} u(t+n_1) \\ \cdot \\ \cdot \\ \cdot \\ u(t+1) \\ u(t) \end{bmatrix} \tag{3.71}$$

The elements of the vector $F$ can be defined using:

$$f_{ja}(t) = (-1/(Aw_{2j} D_{fj}))(G_{oj} \hat{e}_j(t+j|t) - S_j u(t-1)) \tag{3.72}$$

The polynomial $S_j$ is obtained from the solution $(S_j, W_j)$, where $W_j$ is a polynomial of smallest degree, of the following diophantine equation $(j = 0, 1, 2, ..., n_1)$:

$$W_j A_{wj} D_{fj} + z^{-j-1} S_j = H_{oj} \tag{3.73}$$

and $H_{oj}$ follows from the coupled diophantine equations (3.49) and (3.50).

**Optimal and Predicted Controls:**

$$U = -X^{-1}F \tag{3.74}$$

The optimal control is stable and satisfies the characteristic equation (3.63).

**Proof :** The proof involves collecting the above results.

**Remarks**

(i) The optimal control and predicted controls can alternatively be calculated using the loop controllers $C_j = H_{oj}^{-1}G_{oj}$, where $G_{oj}$ and $H_{oj}$ are given by (3.49) and (3.50), and the prediction models are as shown in Fig. 3.2.

(ii) Note that the procedure under (i) does not require Assumption Ass (2) to be invoked, and this is therefore referred to as the optimal case. The assumption was necessary, to derive the vector-matrix solution (3.74).

(iii) If $j = 0$, the controller $C_j$ for the actual closed-loop system is the same as a usual (single cost term problem) LQG controller (Grimble, 1984a,b).

(iv) Each of the predictive loops with $C_j$ computed, as in (i), satisfy the following prediction loop characteristic equations:

$$G_{oj}B + H_{oj}A = D_{cj}D_{fj}$$

**Example 3.1 :** *Aero Engine Gas Turbine Control Problem*
The design of the feedback loop controller ($C_0$), and the computation of the future predicted controls for a gas turbine is considered below. The continuous time gas turbine model was sampled using a sample time of $T_s = 0.005$ seconds. The model represents the dynamics of the gas turbine, and is not strictly proper, since the constant through term represents very fast modes which are approximated. When the sampled model is calculated, a discrete transport delay is added for physical computational realizability.

**Continuous-time Gas Turbine Model**
**Plant:**

$$W = \frac{B_0}{A_0} = \frac{(13.448s^3 - 7774.4s^2 + 1.34078 \times 10^6 s + 9.5257 \times 10^6)}{(s^3 + 206.1676s^2 + 1634.8s + 2277.4)}$$

**Discrete-time Models**
**Plant:**

$$W = \frac{B_0}{A_0} = \frac{13.448(1 - 0.965844z^{-1})[(1 - 2.462349z^{-1})^2 + 0.673668^2]}{(1 - 0.01061z^{-1})(1 - 0.9679839z^{-1})(1 - 0.991017z^{-1})}$$

**Disturbance:**

$$W_d = \frac{C_d}{A_d} = 0.05/(1 - 0.6z^{-1})$$

The common denominator $A = A_0 A_d$ and for simplicity the reference model $W_r = W_d$.

**Computed Feedback Loop Controller:**

$$C_0 = \frac{\begin{array}{l}9.05536 \times 10^{-4}(1 - 0.010161z^{-1})(1 - 0.59805z^{-1}) \\ \times(1 + 0.60985z^{-1})(1 - 0.967984z^{-1})(1 - 0.991017z^{-1})\end{array}}{\begin{array}{l}(1 + 0.23016z^{-1})[(1 - 0.23033z^{-1})^2 + 0.338974^2 z^{-2}] \\ \times(1 - 0.74954z^{-1})(1 - 0.965836z^{-1})(1 - 0.999z^{-1})\end{array}}$$

**Discussion of Results**

The unit step response of the output of the system is shown in Fig. 3.3 for both the actual loop dynamics and the predicted step response, (up to six steps ahead). The prediction calculation in this case will be referred to as the optimal predictor, since the calculations are based directly on (3.34).

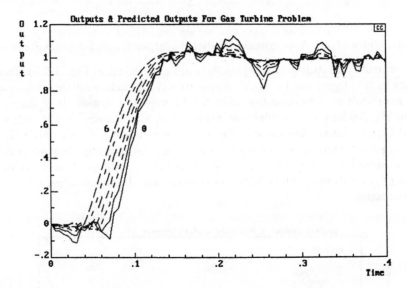

**Fig. 3.3** : System Outputs and Predicted Outputs
for a Step Input at t = 0.05 seconds

The computation of the future predicted controls, based upon the matrix inverse method in Eqn. (3.74), will be referred to as the approximate method, since an assumption must be invoked (Ass. (2)) for this method to be valid. However, the predicted output, six steps ahead ($j = 6$), shown in Fig. 3.4, reveals that there is little to choose between the two computational methods. Note that if the bandwidth of the system is pushed up substantially, the error in computing the future predicted outputs by this method increases, although not for the components of the output due to the reference changes. It is only in the components of the predicted output and control signals, which stem from the disturbance signal, that errors are introduced by the aforementioned approximation. It follows that when the disturbance changes are small relative to the reference changes, either method of computation can be used.

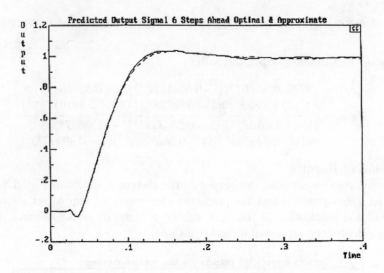

**Fig. 3.4** : Optimal and Approximate Predicted Output Signal for j = 6 Steps Ahead

The control and future predicted controls are also shown in Fig. 3.5. In fact, the responses in this figure are for both computational methods, and the results are so close it is difficult to determine the difference between the signals. It is shown more clearly in Fig. 3.6 for the particular case where $j = 6$. Confirmation has therefore been obtained that the prediction equations give excellent future predictions (particularly for the transients representing the reference change), and secondly the control law can be evaluated by the vector-matrix method form of the equations. This latter point is of course important, since the GPC constrained optimisation algorithms use this type of representation.

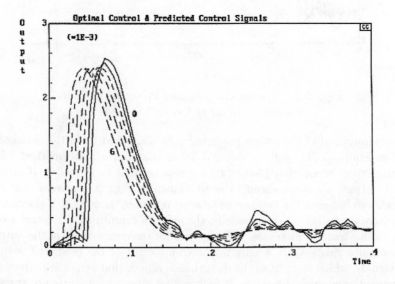

**Fig. 3.5** : Optimal and Approximate Control and Predicted Control Signals

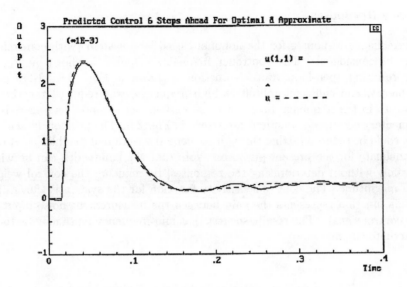

**Fig. 3.6** : Optimal and Approximate Predicted Control Signals
for the Case j=6

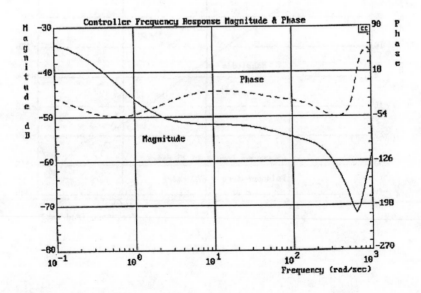

**Fig. 3.7** : Feedback Controller ($C_0$) Frequency Response Magnitude
and Phase Diagram

**Frequency Responses**

The frequency responses for the nominal closed loop system (with controller $C_0$) can now be considered. The controller frequency response is shown in Fig. 3.7, and the resulting open-loop transfer-function is shown in Fig. 3.8. High gain at low frequency, and sufficient roll-off at high frequency, was required together with a bandwidth in the frequency range 14 to 20 radians per second. The sensitivity and complementary sensitivity functions are shown in Fig. 3.9. The peak of the sensitivity function could be reduced further through the careful selection of weightings. However, this is adequate for the present purposes. Note that the bandwidth can be widened substantially without deteriorating the responses, by reducing the control weighting function magnitude. The control sensitivity function for the system is shown in Fig. 3.10. This function represents the gain between the measurement noise input point and the control signal. The results suggest that mid-frequency noise may be too high at the actuator inputs.

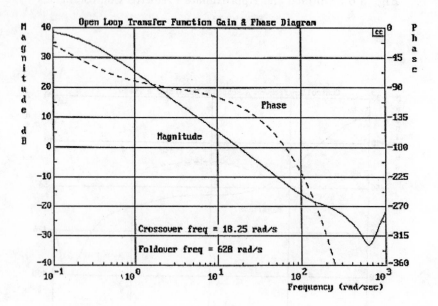

**Fig. 3.8** : Open Loop Transfer Function Gain and Phase Diagram

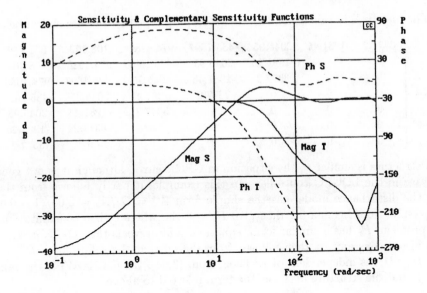

**Fig. 3.9** : Sensitivity and Complementary Sensitivity Function

Gain and Phase

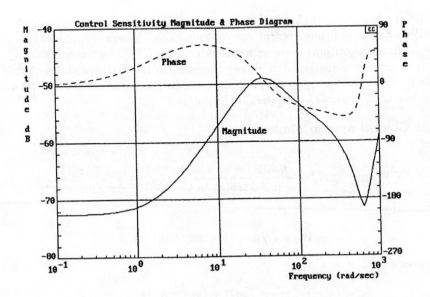

**Fig. 3.10** : Control Sensitivity Function Gain and Phase

The matrix X in this example has the Toeplitz type of structure shown below:

$$X = \begin{bmatrix} 3377.2 & -1981.45 & 2045.95 & -435.1697 & 974.4986 & 108.502 & 631.659 \\ 0 & 3377.2 & -1981.45 & 2045.95 & -435.1697 & 974.4986 & 108.502 \\ 0 & 0 & 3377.2 & -1981.45 & 2045.95 & -435.1697 & 974.4986 \\ 0 & 0 & 0 & 3377.2 & -1981.45 & 2045.95 & -435.1697 \\ 0 & 0 & 0 & 0 & 3377.2 & -1981.45 & 2045.95 \\ 0 & 0 & 0 & 0 & 0 & 3377.2 & -1981.45 \\ 0 & 0 & 0 & 0 & 0 & 0 & 3377.2 \end{bmatrix}$$

This structure is similar to the situation in GPC control, although it is not generally the case in the LQGPC controller. In this example, it partly follows from the fact that the disturbance model has the simple form $W_d = C_d/A_d = 0.05/(1 - 0.6z^{-1})$. The results from Lemma 3.1 then give $H_j = c_j A_0$, where $c_j$ is a constant. Since, in this problem, $E_r$ has a similar form, equation (3.59) reveals that $D_{fj} = c_j A_0$, where $c_j$ is also a constant. It then follows, from (3.61), that the solution for $H_{oj}$ is such that $H_{oj}/D_{fj}$ is independent of $j$. Then, from (3.64), the matrix $W_j$ is the same for each $j$, and the structure of X has the form referred to above.

**Example 6.2** : *Predictive Optimal Control for Flight Autopilot Design*

The following example is based upon Kassapakis and Warwick (1994) for the roll control autopilot of a fighter aircraft. The aim is to demonstrate that the optimal control and future predicted optimal controls can be calculated from similar results to the GPC control law, but with the guarantee of stability and optimality. It will also be shown that the degree of approximation introduced by using the vector matrix form does not cause a marked deterioration in the form of the future predicted control signals.

The previous results indicate that the optimal control laws can be evaluated using the GPC system of future control law computations. Several simplifications can be achieved in the algorithms in the same spirit as in the GPC literature. The rows of the matrix X are, for example, often almost the same as the first row shifted by the row number. Note that this approximation does not by itself give computational savings, but it is a further parallel with the GPC solution.

**Flight Control System Model**
**Plant :**

$$W = \frac{B}{A} = z^{-1} \frac{0.401(1 + 0.043669z^{-1})(1 + 1.42765z^{-1})}{(1 - 0.0060244z^{-1})(1 - 0.4979756z^{-1})(1 - z^{-1})}$$

**Disturbance :**

$$W_d = C/A = (1 - 0.000001z^{-1})/A$$

**Reference :**

$$W_r = E_r/A = 2/(1 - z^{-1})$$

**Cost-function Weightings**

$$Horizon \; n_1 = 7$$

**Error Weighting :**

$$H_q = A_w^{-1} B_q = (1 - z^{-1})^{-1}$$

**Control Weighting:**

$$H_r = A_w^{-1} B_r = 60(1 - 0.5z^{-1})$$

**Computed Controller (for j = 0):**

$$C_0(z^{-1}) = \frac{0.12151(1 - 0.00602z^{-1})(1 - 0.497977z^{-1})(1 - 0.88494z^{-1})}{(1 - 0.005917z^{-1})((1 - 0.34784z^{-1})^2 + 0.2386)^2)(1 - z^{-1})}$$

**Computed X Matrix:**

$$X = \begin{bmatrix}
71.5868 & -14.1732 & 5.68456 & 2.83208 & 1.41030 & 0.70229 & 0.34972 & 0.17415 \\
0 & 71.5868 & -14.1732 & 5.68456 & 2.83208 & 1.41030 & 0.70229 & 0.34972 \\
0 & 0 & 71.5868 & -14.1732 & 5.68456 & 2.83208 & 1.41030 & 0.70229 \\
0 & 0 & 0 & 71.5868 & -14.1732 & 5.68456 & 2.83208 & 1.41030 \\
0 & 0 & 0 & 0 & 71.5868 & -14.1732 & 5.68456 & 2.83208 \\
0 & 0 & 0 & 0 & 0 & 71.5868 & -14.1732 & 5.68456 \\
0 & 0 & 0 & 0 & 0 & 0 & 71.5868 & -14.1732 \\
0 & 0 & 0 & 0 & 0 & 0 & 0 & 71.5868
\end{bmatrix}$$

**Results**

The system and predicted system outputs, for a step input at time $t = 20$ are shown in Fig. 3.11. The predicted system output for $j = 7$ does, of course, reflect the early knowledge of the reference change. The prediction based on the disturbance model is not so effective, although there is clearly some predictive capability. If the reference change variations are much larger than the disturbance effects, then it follows that the predictive capability will be very good. The corresponding control signals for this case are shown in Fig. 3.12. The impact of the step reference change is clearly evident through the peak in control signal variations.

**GPC Matrix Based Control Calculations**

The effectiveness of the predicted controls using the matrix equation (3.74) will now be explored. The control signals for the case $j = 0$ and $j = 7$, as computed by equation (3.74), are shown in Fig. 3.13. Note that the control action for the nominal ($j = 0$) system is the same as calculated from the results of Theorem 3.1. However, there is some small error in the future predicted control action because of the assumption which is invoked. The results shown in Fig. 3.14 illustrate that this approximation is very good.

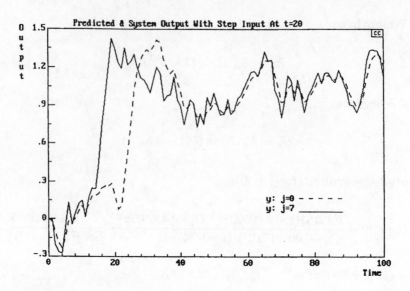

Fig. 3.11 : System Output and Predictive System Output for Step
Reference Change at t = 20

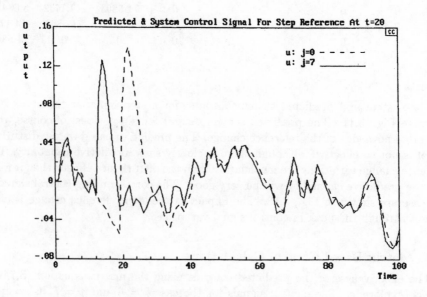

Fig. 3.12 : Control Signal and Predicted Controls for Step
Reference Change at t = 20

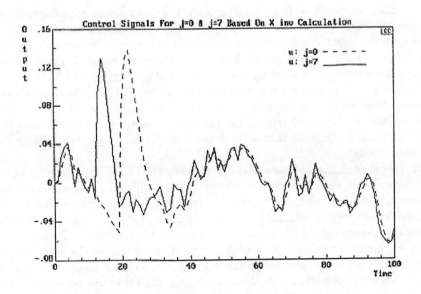

**Fig. 3.13** : System Control Signal and Predictive Control Based upon $X^{-1}$ Calculation

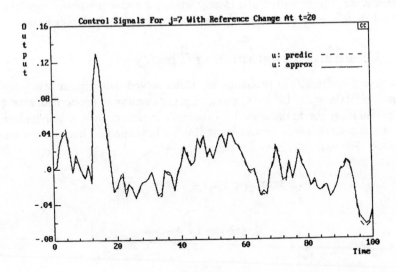

**Fig. 3.14** : Comparison of Control Signals for Exact and Approximate (Matrix Inverse) Calculations

### 3.4.8 The Value of Disturbance Prediction

It is normally assumed in GPC and other algorithms that prediction of the disturbance model is needed. However, it might be questioned whether removing the predictor (by setting $H_{fj} = 1$) would make a significant difference. The first point to note is that, as shown in Fig. 3.2, the predicted disturbance is multiplied by the loop sensitivity functions before obtaining the predicted output. In usual regulating loop designs, the sensitivity function is very small in the very frequency range where the disturbance estimation is good. That is, if the disturbance is dominantly low frequency, the effect of this term on the output would be small, and in fact it is the high frequency disturbance estimates which will be seen in the outputs. Moreover, inspection of results from a disturbance predictor will reveal that, if the disturbance is changing, the predictor will only follow the changes as they occur. It is only during a steady increase or decrease that a predictor will be able to provide reliable future output estimates.

If the reference changes in the system (assumed to be known for future times) dominate the responses, then it seems an obvious conclusion that disturbance prediction is unnecessary. Even in the case where the disturbance and reference changes are of similar magnitude, it is questionable whether to use a predictor, or make the assumption that the disturbance in the future is the same as that at time $t$. As noted above, at times when the disturbance is changing, a predictor will give the same output as the disturbance at time $t$ and hence this assumption will not lead to a large error. Similarly, when the disturbance is slowly changing, the error will be small. There is therefore some evidence to suggest that disturbance component of the prediction is unnecessary in some applications. The consequential numerical savings, in both GPC design and the approach taken here, will be valuable. Moreover, the high frequency prediction error effects which are clearly visible on the predicted outputs will not occur, and these are particularly troublesome for long prediction horizons.

### 3.4.9  Quadratic Programming Theory

Constrained optimisation problems are often solved using quadratic programming techniques (Cottle et al. 1992). Quadratic programming is concerned with the problem of minimising (or maximising) a quadratic function over a polyhedron. These problems can take different forms, depending on how the polyhedral feasible region is represented. For example, consider the problem:

$$\text{minimise } f(x) = c^T x + \frac{1}{2} x^T Q x \tag{3.75}$$

$$\text{subject to } Ax \geq b$$

$$x \geq 0$$

where $Q \in R^{m \times n}$ is symmetric, $c \in R^n$, $A \in R^{m \times n}$ and $b \in R^n$. In some cases, it is more convenient to consider another form,

$$\text{minimise } f(x) = c^T x + \frac{1}{2} x^T Q x$$

$$\text{subject to } Ax \geq b$$

The first problem is readily converted to the second by writing the constraints as:

$$\begin{bmatrix} A \\ I \end{bmatrix} x \geq \begin{bmatrix} b \\ 0 \end{bmatrix}.$$

The aim of a quadratic program with objective function $f$ and feasible region $X$, is to determine a global minimum, that is, a vector $\tau \in X$ , such that

$$f(\bar{x}) \leq f(x), \text{for all } x \in X.$$

On occasions, it is necessary to settle for a local minimum, that is, a vector $\bar{x} \in X$ , such that:

$$f(\bar{x}) \leq f(x), \text{for all } x \in X \cap N(\bar{x})$$

where $N(\bar{x})$ denotes a neighbourhood of $\bar{x}$ .

### Existence of a Global Minimum

A quadratic programming problem has a continuous function as its objective, and a closed set as its feasible region. The Bolzano-Weierstrass theorem guarantees the existence of a global minimum (or maximum) if the feasible region is bounded, since then it is compact. Known as the Frank-Wolfe theorem, the following result is of interest for cases where the feasible region is unbounded. It gives (necessary and) sufficient conditions for the existence of a global minimum in a quadratic programming problem.

**Theorem 3.2 :** If the quadratic function $f$ is bounded below on the nonempty polyhedron $X$, then $f$ attains its infimum on $X$. (That is, if there exists a real number such that $f(x) \geq \gamma$ for all $x \in X$, then there exists a vector $\bar{x} \in X$ such that $f(\bar{x}) \geq f(x)$ for all $x \in X$ ).

## 3.5 Concluding Remarks

The GPC optimal control problem was discussed at the start of the chapter. The main advantage was the simplicity of the algorithm, which has a form suitable for adaptive and non-adaptive applications. Unfortunately the algorithm has poor stability characteristcs for certain classes of system.

The LQGPC approach was investigated in the second half of the chapter. Attempts have been made previously to enhance LQG cost functions by adding in future tracking error and control terms. In one case, the controller employed was restricted to being time-invariant (Grimble, 1997). The LQGPC solution presented here involves a nested set of diophantine equations which allow a time ($j$) dependent solution to be developed for the prediction. In fact, the controller to be implemented at time $t$, which is the controller that determines stability, is time-invariant. However, at each future time step a different control action is required. This has parallels in the solution to the finite time deterministic optimal control problem, where the constant gain matrix is well known to be time-varying.

It is important to note that the feedback controller which is implemented at time $t$ is the same as would be obtained by solving a standard LQG problem with future set point knowledge. In fact, if this was not the case, then the variance of the error and control signals at time $t$ would not be minimised. Other control strategies, which impose a constraint on the controller, to ensure it is used at future time instants, do not enable the true tracking error and control cost to be minimised at the time $t$.

The LQGPC algorithm enables the minimum cost to be achieved at time $t$, and at future predicted times. Other LRPC/GPC algorithms, which may of course have the benefit of simplicity, will not achieve the optimum cost achieved by the LQGPC solution proposed. More importantly, they do not usually have the guaranteed stability properties, for all cost weightings, achieved for the LQGPC controller.

The GPC control law and related LRPC algorithms have established an important place in the toolbox of the process control engineer. However, despite the work of Bitmead et al. (1990), that threw considerable light on the relationship between LQ and GPC control, there have remained several unanswered questions, particularly when the GPC problem is approached from a polynomial systems viewpoint.

The LQGPC stochastic optimal control solution enabled the relationships between LQG and GPC control to be explored. Of particular importance are the assumptions which must be made to ensure a GPC type of solution results. It is important that under quite reasonable conditions the good LQG related stability properties can be assured. At the same time, since the solution is so close to a GPC result, it is clear that constraints can be handled by adopting quadratic programming results which already exist for LRPC and GPC type algorithms. Note that the solution procedure followed, that involves transferring the frequency-domain results back into the time-domain, was the key step in relating the LQG type of results to the familiar GPC type of solution.

The predictive control algorithms may easily be extended to include feedforward control features (Grimble, 1988b ; Sebek et. al., 1988). There is a bright future for applications in a number of industries (Lewis et al. 1991; Prett and Garcia, 1988; Richalet et al. 1978; Ricker 1990, Grimble, 1990).

## 3.6   References

Bitmead, R., Gevers, M. and Wertz, V., 1990, *Adaptive optimal control : the thinking man's GPC*, Prentice Hall Australia, Victoria.

Bitmead, R., Gevers, M. and Wertz, V., 1989a, Adaption and robustness in predictive control, *Proc. 28th Conf on Decision and Control*, Tampa, Florida.

Bitmead, R., Gevers, M. and Wertz, V., 1989b, Optimal control redesign of generalized predictive control, *IFAC Symposium on Adaptive Systems in Control and Signal Proc.*, Glasgow. UK.

Clarke, D.W. and Mohtadi, C., 1989, Properties of generalized predictive control, *Automatica*, Vol. **25**, No.6, pp. 859-875.

Clarke, D.W., Mohtadi, C. and Tuffs, P.S., 1987a, Generalized predictive control - part I : The basic algorithm, *Automatica*, Vol. **23**, No.2, pp. 137-148.

Clarke, D.W., Mohtadi, C. and Tuffs, P.S., 1987b, Generalized predictive control - part II : Extensions and interpretations, *Automatica*, Vol. **23**, No.2, pp. 149-160.

Cottel, R.W., Pang, J.S. and Stone, R.E.,1992, *The linear complementary problem,* Academic Press, San Diego.

Cutler, C.R. and Ramaker, B.L., 1980, Dynamic matrix control - A computer control algorithm, *JACC*, San Francisco.

De Keyser, R.M.,C. Ph, van de Velde, G.A. and Dumortier, F.A.G., 1988 A comparative study of self-adaptive long range predictive control methods, *Automatica,* Vol. **24**, No. 2, pp. 149-163.

Duan, J., 1993, *On weighted predictive control,* PhD thesis, University of Strathclyde.

Duan, J., Grimble, M.J. and Johnson, M.A.,1997, Multivariable weighted predictive control, *J. Process control*, Vol. **7**, No. 3, pp. 219-235.

Garcia, C.E. Prett, D.M. and Morari, M., 1989, Model predictive control : Theory and practice - a survey, *Automatica,* No. **3**, pp. 335-348.

Grimble, M.J., 1984a, Implicit and explicit LQG self-tuning controllers, *Automatica,* Vol. **20**, No. 5, pp. 661-669.

Grimble, M.J., 1984b, LQG multivariable controllers : minimum variance interpretation for use in self-tuning systems, *Int. J. Control,* Vol. **40**, No. 4, pp. 831-842.

Grimble, M.J., 1986a, Observations weighted controllers for linear stochastic systems, *Automatica,* Vol. **22**, No. 4, pp. 425-431.

Grimble, M.J., 1986b, Multivariable controllers for LQG self-tuning applications with coloured measurement noise and dynamic cost weighting, *Int. J. Systems Sci.,* Vol. **17**, No. 4, pp. 543-557.

Grimble, M.J., 1990, LQG predictive optimal control for adaptive applications, *Automatica,* Vol. **26**, No. 6, pp. 949-961.

Grimble, M.J., 1992, Generalized predictive control : An introduction to the advantages and limitations, Int. *J. of Systems Science* Vol. **23**, pp. 85-98.

Grimble, M.J., 1994, Multivariable linear quadratic generalised predictive control, *Transactions of ASME, Dynamic Systems, Measurement and Control.*

Grimble, M.J., 1995, Two degree of freedom linear quadratic Gaussian predictive control, *IEE Proc, Control Theory and Applications,* Vol. **142**, No. 4, pp. 295-306.

Grimble, M.J., 1997, Time-invariant 2 DOF multivariable LQG predictive controllers, *International Journal of Control,* Vol. **68**, No. 1, pp. 1-30.

Grimble, M.J., 1988a, Generalized minimum variance control revisited, *Optimal Control Applications and Methods,* Vol. **9**, pp. 63-77.

Grimble, M.J., 1988b, Two-degrees of freedom feedback and feedforward optimal control of multivariable stochastic systems, *Automatica,* Vol. **34**, No. 6, pp. 809-817.

Grimble, M.J., 1988c, Multi-step $H_\infty$ generalized predictive control, *Dynamics and Control*, Vol. **8**, pp. 303-339.

Grimble, M.J., 1994, *Robust industrial control : An optimal approach for polynomial systems*, Prentice Hall, Hemel Hempstead.

Grimble, M.J. and Johnson, M.A., 1988, *Optimal Control and Stochastic Estimation, Vols I and II*, John Wiley, Chichester.

Kassapakis, E.G., and Warwick, K., 1994, *Predictive control for autopilot design*, in Clarke D.W., (Ed) : Advances in Model Based Predictive Control (Oxford University Press), pp. 458-470.

Kouvaritakis, B., Rossiter J.A. and Chang, A.O.T., 1992, Stable generalized predictive control : an algorithm with guaranteed stability, *IEE Proceedings*, Part-D, Vol. **139**, No. 4.

Kučera, V., 1980, Stochastic multivariable control : A polynomial equation appraoch, *IEEE Trans. on Auto. Control*, Vol. AC-**25**, No. 5, pp. 913-919.

Kwon, W.H. and Pearson, A.E., 1977, A modified quadratic cost problem and feedback stabilization of a linear systems, *IEEE Trans. on Auto. Control*. Vol. AC-**22**, No. 5, pp. 838-842.

Lewis, D.G., Evans, C.C. and Sandoz, P.J., 1991, Application of predictive control techniques to a distillation column, *J. Proc. Contr.* Vol. **1**.

Mohtadi, C. and Clarke, D.W., 1986, Generalized predictive control, *Proc. of 25th Conference on Decision and Control*, Athens, Greece, pp. 1536-1541.

Mosca, E., 1982, *Multivariable adaptive regulators based on multistep quadratic cost functionals*, Publ. in the NATO Workshop Notes, Algarve, Portugal, Edited by R.S. Bucy and J.M.F. Moura (eds). Nonlinear Stochastic Problems, 187-204, D. Reidel Publishing Company.

Peterka, V., 1984, Predictor based self-tuning control, *Automatica*, Vol. **20**, No. 1, pp. 39-50.

Prett, D.M. and Garcia, C.E., 1988, *Fundamental process control*, Butterworth, Stoneham, M.A.

Richalet, J., 1993, Industrial applications of model based predictive control, *Automatica*, Vol. **29**, No. 8, pp. 1251-1274.

Richalet, J., Rault, A., Testud, J.L. and Papon, J., 1978 Model predictive heuristic control : applications to industrial processes, *Automatica*, **14**, 413-428.

Ricker, N.K., 1990, Model predictive control with state estimation, *Ind. Eng. Chem, Res,* **29**, 374-382.

Sebek, M., Hunt, K.J. and Grimble, M.J., 1988, LQG regulation with disturbance measurement feedforward, *Int. J. Control*, Vol. **7**, No. 5, pp. 1497-1505.

Shakoor, S., 1995, *Industrial uses of predictive control*, MPhil Thesis, University of Strathclyde.

Warren, J., 1992, Model based control of catalytic cracking, *C & I Magazine*, pp. 557-581.

# 4

# H$_2$ Multivariable Control

## 4.1  Introduction

The aim in this Chapter is to show that the solutions to the scalar $H_2$ optimal control problem, considered earlier, are of similar form in the multivariable case. The solution to the stochastic linear quadric multivariable optimal control problem, using polynomial system methods, was first obtained by Kucera (1979). The form of the solution was found to be rather more convenient for numerical algorithm construction, than the transfer-function matrix based solutions of Youla et al. (1976).

To make the problem more interesting, it is assumed that potential fault conditions are also to be estimated. This is of practical importance, since there are often greater financial gains from reducing down time, than from production quality improvements. In the following, the Linear Quadratic Gaussian (LQG) optimal control problem is first extended to include fault detection requirements, and is then solved using polynomial matrices. The controller for this more general problem requires the solution of additional diophantine equations, and the spectral factorisation stage is also modified. The general techniques employed in Chapter 2 for solving scalar problems, may be applied in this fault estimation and control case.

A very different solution strategy is considered in the second part of the chapter. The solutions up this point have all been in polynomial or transfer-function matrix closed form for the desired optimal controllers. However, there are often advantages in using an observer structure, similar to that employed in Kalman filtering problems. Until recently neither the Wiener nor the polynomial theories have been used to derive observer forms of estimators and controllers based on the frequency domain forms of models. The second part of the chapter explores this possibility, and a *separation principle* type of solution, for polynomial systems, is derived.

The rather surprising result is established, in the later parts of the chapter, that there are two possible separation type theorems, since in this frequency domain input/ouput model setting the result depends upon the order in which the separate control and estimation problems are solved. The first separation principle result requires the solution of the optimal observer problem, which then defines the disturbance model to be used in the optimal *ideal output* feedback control problem. The second separation principle result involves solving the ideal output feedback control problem, which then defines the weighting for the optimal observer estimation problem. These might be termed dual solutions. Note that there is some slight abuse of terminology in referring to these results as a separation principle, since the control and filter computations are not totally independent. Nevertheless there are strong similarities in the structure of the control law with the usual LQG separation result.

The *separation principle* established has the same advantages and insights as for the traditional state-space solution, namely:

- The role of the optimal linear filter in the solution is clarified and the signal estimates can often be of value.

- The control law calculation $K_c$ does not depend upon the measurement noise properties and the estimator calculation $K_f$ does not depend upon the cost-function control weighting.

- The estimator can be commissioned separately, before attempting to close the feedback control loop.

## 4.2   Combined Fault Estimation and Control

The general approach followed in the rest of the section is the polynomial equivalent of the solution of the combined fault detection and control problem that was considered by Stoustrup, Grimble and Niemann (1997). In this state equation based work, the standard system description was utilised, and an $H_\infty$ solution to the combined problem was obtained. The results below utilise an $H_2$ cost function and a rather more classical system model structure. The main advantage of the approach, presented below, is in the insights that the frequency domain technique provides. It is particularly valuable where stochastic properties dominate and where the frequency domain characteristics are important to the underlying signal processing problem. The technique also provides an integrated control and diagnostics capability that could be extended to nonlinear systems (Robertson and Lee, 1993). The general approach proposed is rather different to the traditional methods of fault monitoring and detection (Patton et al. 1997, 1989, 1998; Frank 1994; Chen and Patton, 1996).

The strategy followed is to represent fault conditions by stochastic fault signals, which is fundamentally different to the so called *reliable control* philosophy due to Veillette and co-workers (1992, 1995). The design approach in *reliable control* involves the representation of failures by uncertain gain elements and is $H_\infty$ design based. The approach below does not provide the same guarantees on stability, but the design philosophy is more suitable when stochastic requirements are important. There are many applications where fault tolerant techniques are necessary, not least in aerospace systems (Joshi, 1986).

The combined fault monitoring and control problem is illustrated in Fig. 4.1. This figure shows a traditional feedback loop, but with faults represented by additional signals. The effective control action is denoted by $u_c$, since this is the actual input to be limited in the range of mechanical operation. Similarly the fault $f_1$ affects the output $y_c$, and this may represent the real plant output to be controlled. The actual plant output does of course differ from the measured output because of dynamics in the measurement system ($W_3$) and the addition of measurement noise $n$. There may also be measurement system faults, represented by the signal $f_3$. The cost-function to be minimised in such a problem can include:

(a) The controller output $u$.

(b) The effective main actuator control signal $u_c$.

(c) The tracking error for the effective plant output $y_c$.

(d) The tracking error for the ideal plant output $y$.

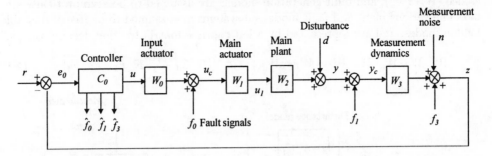

**Fig. 4.1** : Combined Fault Monitoring and Multivariable Control Problem

The control problem will have a traditional solution when the fault signals $f_0, f_1$, and $f_3$ tend to zero. The combined problem will have a more unusual character when one or more of these signals dominate. The characteristics that result in certain asymptotic situations are very different :

$f_0 \rightarrow \infty$  The system will tend to minimum variance control characteristics, with beneficial properties similar to *loop transfer recovery* design (high controller gain resulting).

$f_1 \rightarrow \infty$  This fault signal does not go through the main plant dynamics and the effect on overall robustness is related to that for the signal $f_3$ (next case).

$f_3 \rightarrow \infty$  This signal will have a similar effect to a large coloured noise signal, resulting in a low loop gain in the dominant frequency range for the fault.

Thus, by allowing for certain fault conditions the controller gain will be increased or decreased in particular frequency ranges. If the controller is implemented in state equation form, and the standard system description is employed, the estimates of the fault signals can easily be obtained, with little increase in complexity. This observation suggests the same should be true for the Wiener or polynomial solution, and this is confirmed through the following analysis.

The combined control/fault estimation compensator (C/FE compensator) will generate the optimal control action and also provide estimates of signals representing actuator or measurement system faults. The resulting controller will be aware of the possible fault conditions, and should therefore be optimal for this situation. There will of course be a point beyond which maintenance or preventative action must occur and the plant will need to be shut-down. The estimated fault signals will provide process operators and engineers with the information required to make this judgement. It is of course only those faults which have been defined before-hand, and therefore included in the models, that will be identified correctly.

## 4.2.1   System Description

The discrete-time, multivariable, linear, time-invariant system, illustrated in Fig. 4.1, is shown in polynomial form in Fig. 4.2. The measurement noise $(W_n(z^{-1}))$, reference $(W_r(z^{-1}))$, and fault generation models are assumed to be asymptotically stable. The various plant and fault model subsystems are assumed to be free of unstable hidden modes, and are represented by a left coprime matrix fraction decomposition:

$$[W \ W_d \ W_n \ W_r \ W_{f_1} \ W_{f_2} \ W_{f_3}] = A^{-1}[B \ C_d \ C_n \ E \ F_1 \ F_2 \ F_3] \qquad (4.1)$$

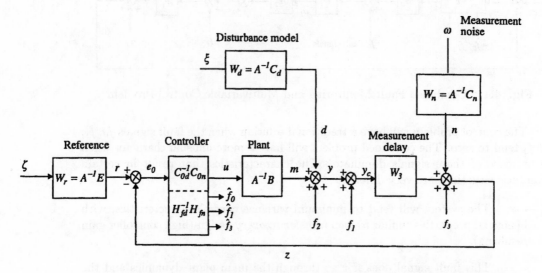

**Fig. 4.2 :** Canonical Feedback Control Problem with Fault Signals
($f_2$ = fault signal, $f_0$ reflected to plant output)

Without loss of generality a common denominator matrix $A(z^{-1})$ is utilised for the system models and to simplify the notation the argument $(z^{-1})$ of these matrices is often omitted.   The system can therefore be represented in polynomial matrix fraction form as:

**Plant :**

$$W = W_2 W_1 W_0 = A^{-1}B$$

**Input disturbance :**

$$W_d = A^{-1}C_d$$

**Measurement noise:**

$$W_n = A^{-1}C_n$$

**Reference:**

$$W_r = A^{-1}E$$

**Controller/fault estimator:**

$$C_0 = C_{0d}^{-1}C_{0n}$$

**Measurement system delay:**

$$W_3 = z^{-k_m}I_r$$

The possible fault signals have the following polynomial system descriptions:
**Measurement fault signal:**

$$f_3(t) = W_{f3}\eta_3(t) = A_{f3}^{-1}F_{f3}\eta_3(t) = A^{-1}F_3\eta_3(t)$$

**Plant output fault signal:**

$$f_1(t) = W_{f1}\eta_1(t) = A_{f1}^{-1}F_{f1}\eta_1(t) = A^{-1}F_1\eta_1(t)$$

**Actuator fault signal:**

$$f_0(t) = W_{f0}\eta_0(t) = A_{f0}^{-1}F_{f0}\eta_0(t)$$

**Reflected actuator fault:**

$$f_2(t) = W_2W_1f_0(t) = A^{-1}F_2\eta_0(t)$$

where $A^{-1}F_{20} = W_2W_1A_{f0}^{-1}$ and hence $F_2 = F_{20}F_{f0}$, $W_2W_1 = A^{-1}B_{21} = A^{-1}F_{20}A_{f0}$ , and $B_{21} = F_{20}A_{f0}$ .

The white driving noise sources for the system $\{\xi(t)\}$, $\{\zeta(t)\}$, $\{\omega(t)\}$, and the faults $\{\eta_0(t)\}$, $\{\eta_1(t)\}$ and $\{\eta_3(t)\}$, are assumed to be zero-mean and statistically independent. The covariance matrices for these signals are, without loss of generality, taken to be the identity matrix. The $r$-output and $m$-input system equations can easily be obtained, by reference to Fig. 4.2, as:
**Plant output equation:**

$$y(t) = m(t) + d(t) + f_2(t) \tag{4.2}$$

**Observations:**

$$z(t) = W_3(z^{-1})\left(y(t) + f_1(t)\right) + W_n(z^{-1})\omega(t) + f_3(t) \tag{4.3}$$

**Error:**

$$e(t) = r(t) - y(t) \tag{4.4}$$

**Error for output fault:**

$$e_c(t) = r(t) - y_c(t) = r(t) - f_1(t) - y(t) = e(t) - f_1(t) \tag{4.5}$$

**Controller input signal:**

$$e_0(t) = r(t) - z(t)$$

**Reference generation:**

$$r(t) = W_r(z^{-1})\zeta(t)$$

The output and input *sensitivity* and *complementary sensitivity* matrices are defined, respectively, as:

$$S_r = (I_r + W_3 W C_0)^{-1}, \quad S_m = (I_m + C_0 W_3 W)^{-1}$$

and

$$T_r = I_r - S_r = W_3 W C_0 S_r, \quad T_m = I_m - S_m = C_0 W_3 W S_m$$

and the control sensitivity operator $M = C_0 S_r$ . These matrices determine the power spectra of the following signals:

$$e_0 = r - z = S_r (r - f_3 - n - W_3(d + f_1 + f_2))$$

$$\Phi_{e_0 e_0} = S_r \Phi_{cc} S_r^* \tag{4.6}$$

where $\Phi_{cc}$ denotes the spectrum of the so-called *innovations signal*:

$$c = r - f_3 - n - W_3(d + f_1 + f_2)$$

and hence the innovations spectrum :

$$\Phi_{cc} = \Phi_{rr} + \Phi_{f_3 f_3} + \Phi_{nn} + W_3(\Phi_{dd} + \Phi_{f_1 f_1} + \Phi_{f_2 f_2})W_3^* \tag{4.7}$$

The total innovations spectrum $\Phi_{cc}(z^{-1})$ may be assumed to be positive definite on the unit circle $|z| = 1$ in the complex plane.

**Control signal and spectrum:**

$$u = C_0 S_r(r - z) = C_0 S_r (r - f_3 - n - W_3(d + f_1 + f_2))$$

$$\Phi_{uu} = C_0 S_r \Phi_{cc} S_r^* C_0^* \tag{4.8}$$

**Undisturbed plant output and spectrum:**

$$m = Wu = W C_0 S_r(r - z) = W C_0 S_r (r - f_3 - n - W_3(d + f_1 + f_2))$$

$$\Phi_{mm} = T_r \Phi_{cc} T_r^* \tag{4.9}$$

**Plant output:**

$$y = m + d + f_2 = (I_r - WC_0S_rW_3)(d + f_2) + WC_0S_r(r - f_3 - n - W_3f_1)$$

**Tracking error and spectrum:**

$$e = r - y = (I_r - WC_0S_r)r - (I_r - WC_0S_rW_3)(d + f_2) + WC_0S_r(f_3 + n + W_3f_1)$$

$$\Phi_{ee} = (I_r - WC_0S_r)\Phi_{rr}(I_r - WC_0S_r)^* + (I_r - WC_0S_rW_3)\Phi_{d_fd_f}(I_r - WC_0S_rW_3)^*$$

$$+(WC_0S_r)(\Phi_{nn} + \Phi_{f_3f_3} + W_3\Phi_{f_1f_1}W_3^*)(WC_0S_r)^* \qquad (4.10)$$

where $d_f = d + f_2$ and $\Phi_{d_fd_f}$ denotes the spectrum of the signal $d_f$.

**Fault Estimation Errors:**

Let the estimator for the fault signal $f_0$, have the form $\hat{f}_0 = -H_{f0}(e_0 + W_3Wu)$ $= H_{f0}(-r + z - W_3Wu)$, where the known signal $W_3Wu$ is added to $e_0$ to simplify the expression. Thence, the fault estimation error expressions become:

$$\tilde{f}_0 = f_0 - \hat{f}_0 = f_0 - H_{f0}(-r + n + f_3 + W_3(f_1 + W_2W_1f_0 + d))$$

$$= (I - H_{f0}W_3W_2W_1)f_0 - H_{f0}(-r + n + f_3 + W_3(f_1 + d))$$

Similarly, the remaining fault estimates are obtained as:

$$\tilde{f}_3 = f_3 - \hat{f}_3 = f_3 - H_{f_3}(-r + n + f_3 + W_3(f_1 + f_2 + d))$$

$$= (I - H_{f_3})f_3 - H_{f_3}(-r + n + W_3(f_1 + f_2 + d))$$

and

$$\tilde{f}_1 = f_1 - \hat{f}_1 = f_1 - H_{f1}(-r + n + f_3 + W_3(f_1 + f_2 + d))$$

$$= (I - H_{f1}W_3)f_1 - H_{f1}(-r + n + f_3 + W_3(f_2 + d))$$

**Fault Spectra:**

The power spectra $\Phi_0$, $\Phi_1$ and $\Phi_3$ of the following signals, may be defined:

$$\phi_0 = -r + n + f_3 + W_3(f_1 + d)$$

$$\phi_1 = -r + n + f_3 + W_3(f_2 + d)$$

$$\phi_3 = -r + n + W_3(f_1 + f_2 + d)$$

The spectra of these fault signals follow as:

$$\Phi_{\tilde{f}_0\tilde{f}_0} = (I - H_{f0}W_3W_2W_1)\Phi_{f_0f_0}(I - H_{f0}W_3W_2W_1)^* + H_{f0}\Phi_0H_{f0}^* \qquad (4.11)$$

$$\Phi_{\tilde{f}_1\tilde{f}_1} = (I - H_{f1}W_3)\Phi_{f_1f_1}(I - H_{f1}W_3)^* + H_{f1}\Phi_1H_{f1}^* \qquad (4.12)$$

$$\Phi_{\tilde{f}_3\tilde{f}_3} = (I - H_{f3})\Phi_{f_3f_3}(I - H_{f3})^* + H_{f3}\Phi_3H_{f3}^* \qquad (4.13)$$

Note that the estimates could have been defined to be of the form: $H_f(z - W_3Wu)$ in which case the reference $r$ would have been absent from these last three equations. The so-called filter generalised spectral factor : $Y_f$, obtained in the computation of the LQG controller, would then have been different to the fault estimation spectral-factors, leading to a slightly larger computational burden.

It will be seen from the viewpoint of the control design that the fault signal $f_0$ (or $f_2$ reflected to the output) acts in the same manner as the disturbance $d$, whereas the fault signals $f_1$ and $f_3$ enter the equations for control design, in the same manner as the measurement noise $n$.

**Innovations Signal Spectrum:**

The physical interpretation of the innovations signal spectrum is explained by reference to Fig. 4.3. The spectrum $\Phi_{cc}$ can be spectrally factored into the form $\Phi_{cc} = Y_fY_f^*$ , where $Y_f$ has the polynomial matrix form $Y_f = A^{-1}D_f$. If $\{\varepsilon(t)\}$ denotes white noise with identity covariance matrix and zero mean, then it may easily be shown that the system model can be represented in the very simple *innovations signal form*, shown in Fig. 4.3.

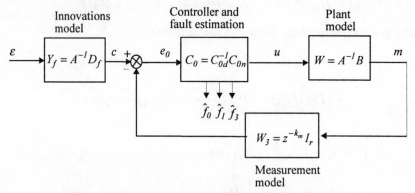

**Fig. 4.3** : Innovations Signal Model Description

## 4.2.2   LQG Sensitivity Reduction and Fault Estimation

The canonical feedback system, shown in Fig. 4.2, is very general since it allows for the presence of fault signals and coloured measurement noise. The cost-function defined below is also more general than that usually considered, but the solution may be obtained following the strategy developed by Kučera (1979).

The criterion (Grimble, 1986, 1998a) defined below allows normal LQG control and error signals to be minimised, but also enables fault estimation error signal terms to be costed. Let the extended cost-function be defined as:

$$J = \frac{1}{2\pi j} \oint_{|z|=1} trace\{Q_c\Phi_{ee}\} + trace\{R_c\Phi_{uu}\} + trace\{Q_{c1}\Phi_{e_ce_c}\}$$

$$+trace\{R_{c1}\Phi_{u_c u_c}\} + trace\{U_{f0}\Phi_{\tilde{f}_0 \tilde{f}_0}\} + trace\{U_{f1}\Phi_{\tilde{f}_1 \tilde{f}_1} + U_{f3}\Phi_{\tilde{f}_3 \tilde{f}_3}\}\frac{dz}{z} \quad (4.14)$$

The first term denotes the tracking error spectrum, and the second term includes the control signal weighting. The third and fourth terms represent the error and actuator output signals, but with the addition of the output or actuator faults, respectively. The final three terms denote the weighted fault estimation error spectra. The error weightings $Q_c$ and $Q_{c1}$ are assumed to be positive semi-definite and at least one of the control weightings $R_c$ and $R_{c1}$ are assumed to be positive definite on the unit-circle of the z-plane, $|z| = 1$. Note that in general it may not be necessary to use both weightings $(Q_c, R_c)$ and $(Q_{c1}, R_{c1})$, but it is helpful here to illustrate the difference that arises due to the presence of the fault signals.

### Dynamic Weighting Functions

The dynamic cost-function weighting terms (Grimble, 1998a) have the following polynomial matrix representations:

$$Q_c = A_q^{*-1}Q_n A_q^{-1} \quad \text{and} \quad R_c = A_r^{*-1}R_n A_r^{-1} \quad (4.15)$$

$$Q_{c1} = A_q^{*-1}Q_{n1} A_q^{-1} \quad \text{and} \quad W_0^* R_{c1} W_0 = A_r^{*-1}R_{n1} A_r^{-1} \quad (4.16)$$

where $A_q(0) = I_r$, $A_r(0) = I_m$ and $A_q$, $A_r$ are strictly Schur. The combined polynomial weightings $\tilde{Q}_n = Q_n + Q_{n1}$ and $\tilde{R}_n = R_n + R_{n1}$. The weightings on the fault signal $U_{f0}, U_{f1}$ and $U_{f3}$ will be assumed constant full rank matrices.

The right-coprime decomposition of the system $A_q^{-1}A^{-1}B$ is defined using:

$$A_q^{-1}A^{-1}B = B_1 A_1^{-1} \in R^{r \times m}(z^{-1}) \quad (4.17)$$

and similarly write:

$$A_r^{-1}A_1 = A_{10} A_{r0}^{-1} \in R^{m \times m}(z^{-1}) \quad (4.18)$$

and define:

$$A_c = A_r A_{10} = A_1 A_{r0} \in P^{m \times m}(z^{-1}). \quad (4.19)$$

### System Assumptions

(i) The system model is free of unstable hidden modes.

(ii) The measurement noise $W_n$, reference $W_r$ and fault signals models $W_{f0}, W_{f1}$, $W_{f3}$ are assumed to be asymptotically stable.

(iii) The control and filter spectral factors $D_c$ and $D_f$ are Schur by definition and depend upon the cost-function weightings and noise/disturbance models, respectively. Assume that these weightings and models are chosen so that these spectral factors are free of unit-circle zeros; that is $D_c$ and $D_f$ are strictly Schur.

(iv) If estimates of measurement and output faults ($f_1$, $f_3$) are required, then in this approach the system must be assumed open-loop stable. This assumption is not necessary if there are only actuator faults $f_2$.

**Theorem 4.1** : *Multivariable Optimal Controller and Fault Estimation*

Consider the system shown in Fig. 4.2 and assume that the cost function (4.14) is to be minimised, with the cost weights (4.15) and (4.16). Define the strictly Schur spectral factors $D_c$ and $D_f$ using:

$$D_c^* D_c = A_{r0}^* B_1^* \tilde{Q}_n B_1 A_{r0} + A_{10}^* \tilde{R}_n A_{10} \tag{4.20}$$

$$D_f D_f^* = EE^* + C_d C_d^* + C_n C_n^* + F_1 F_1^* + F_2 F_2^* + F_3 F_3^* \tag{4.21}$$

**Control Diophantine Equations**

The following diophantine equations must be solved for the smallest degree solution $(H_0, G_0, Z_0)$, with respect to $Z_0 \in P^{m \times r}(z^{-1})$ :

$$D_c^* z^{-g} G_0 + Z_0 A_2 = A_{r0}^* B_1^* \tilde{Q}_n D_2 z^{-g+k_m} \tag{4.22}$$

$$D_c^* z^{-g} H_0 - Z_0 B_2 z^{-k_m} = A_{10}^* \tilde{R}_n D_3 z^{-g} \tag{4.23}$$

The right-coprime decompositions are defined as:

$$D_f^{-1} A A_q = A_2 D_2^{-1} \in R^{r \times r}(z^{-1}) \quad \text{and} \quad D_f^{-1} B A_r = B_2 D_3^{-1} \in R^{r \times m}(z^{-1})$$

The following measurement noise and fault signal diophantine equation must be solved for the smallest-degree solution $(L_0, P_0)$ with respect to $P_0$:

$$D_{fq}^* D_c^* z^{-g_0} L_0 + P_0 A_{qf} = L z^{-g_0} \tag{4.24}$$

where $D_{fq}$, $A_{qf}$ (strictly Schur), and $L$ are coprime and satisfy:

$$D_{fq}^{*-1} L A_{qf}^{-1} = A_{r0}^* B_1^* \tilde{Q}_n A_q^{-1} A^{-1} \left[ EE^* (z^{k_m} - 1) + (C_n C_n + F_3 F_3^*) z^{k_m} \right] D_f^{*-1}$$

$$+ [A_{r0}^* B_1^* Q_n A_q^{-1} A^{-1} F_1 F_1^* - A_c^* W_0^* R_{c1} A_{f0}^{-1} F_{f0} F_{f0}^* F_{20}^*] W_3^* D_f^{*-1}$$

**Fault Detection Diophantine Equations**

The fault estimation diophantine equations that must be solved for the smallest degree solutions $(X_0, Y_0), (X, Y)$ and $(X_1, Y_1)$, with respect to $Y_0, Y, Y_1$ become:

**Fault estimate $\hat{f}_0$**

$$X_0 D_f^* z^{-h_0 + k_m} + A_{f0} Y_0 = F_{f0} F_2^* z^{-h_0 + k_m} \tag{4.25}$$

$$V_0 D_f^* z^{-h_0} - B_{21} Y_0 = (D_f D_f^* - F_2 F_2^*) z^{-h_0 + k_m} \qquad (4.26)$$

**Fault estimate $\widehat{f_1}$**

$$X_1 D_f^* z^{-h_1 + k_m} + A_{f_1} Y_1 = F_{f_1} F_1^* z^{-h_1 + k_m} \qquad (4.27)$$

**Fault estimate $\widehat{f_3}$**

$$X_3 D_f^* z^{-h_3} + A_{f_3} Y_3 = F_{f_3} F_3^* z^{-h_3} \qquad (4.28)$$

where scalar weightings have been assumed, and $h_0, h_1, h_3$ are the smallest positive integers that make these equations polynomials in $z^{-1}$.

The optimal controller and fault estimators can be computed from the solution of the above equations using the following results.

**Feedback controller:**

$$C_0 = C_{0d}^{-1} C_{0n} = C_{1n} C_{1d}^{-1}$$

$$C_0 = (H_0 D_3^{-1} A_r^{-1} + L_0 A_{qf}^{-1} D_f^{-1} B z^{-k_m})^{-1} (G_0 D_2^{-1} A_q^{-1} - L_0 A_{qf}^{-1} D_f^{-1} A) \qquad (4.29)$$

**Fault estimators :**

$$H_{f0} = A_{f0}^{-1} X_0 D_f^{-1} A \quad \text{(actuator faults)} \qquad (4.30)$$

$$H_{f1} = A_{f1}^{-1} X_1 D_f^{-1} A \quad \text{(output faults)} \qquad (4.31)$$

$$H_{f3} = A_{f3}^{-1} X_3 D_f^{-1} A \quad \text{(measurement faults)} \qquad (4.32)$$

Both the controller $C_0(z^{-1})$ and the fault estimators : $\bar{H}_f = [H_{f0}^T \ H_{f1}^T \ H_{f3}^T]^T$ must be realised in a minimal form.

**Proof :** The proof is similar to the solutions provided in Kučera (1979) or Grimble (1994), and the main steps are summarised in the following section.

**Remarks**

- The optimal problems for control and fault estimation can be decoupled in this case, where uncertainty in the system models was neglected.

- Since some of the calculations are common to both the control and fault estimation terms, the problem complexity is not increased as much as would be expected.

- This latter point is important if an adaptive version is to be produced, requiring on-line evaluation of the controller/estimator.

- Because the controller is computed knowing about the possible fault condition, it can be accused of hiding the fault, since the regulating action will tend to compensate for the problems arising. However, the plant operator should be fully aware of the presence of the faults if the fault estimator triggers alarms.

- Fault signals $f_1$ and $f_3$ enter the control equations like the measurement noise (see the signal $n_1 = n + f_3 + W_3 f_1$ ).

- Fault signal $f_2$ enters the control equations like the disturbance signal $d$.

- The consequence of the last two remarks is that the larger the variance of the fault signals $f_1$ and $f_3$ the smaller will be the controller gain, and the larger the variance of $f_2$ the greater will be the gain of the controller.

- The optimal controller defined in (4.29) has the form of a Youla parameterisation (1976). The stable gain term $L_0(D_f A_{qf})^{-1}$ depends upon the measurement noise and the fault signals.

**Lemma 4.1** : *Optimal System Properties*
The stability of the closed loop system is determined by the zeros of the return difference matrix :

$$(I + C_0 W z^{-k_m}) = A_r D_3 H_0^{-1} D_c (A_r A_{10})^{-1} \qquad (4.33)$$

**Proof** : Presented in the following section.

The degree of stability of the closed-loop system is determined by the control spectral factor $D_c$ and the filter spectral factor $D_f$ (from (4.20) and (4.21)).

## 4.2.3   Minimisation of the Combined Criterion

There now follows an outline of the solution of the optimal control and fault estimation problem, using a traditional completing the squares method. Define the closed-loop transfer-function matrix $M$ as:

$$M = C_0(I_r + W_3 W C_0)^{-1} = (I_m + C_0 W_3 W)^{-1} C_0$$

The control, plant output, tracking error and controller input signals can now be listed, using the results in §4.2.1, as:

$$u = M(r - f_3 - n - W_3(d + f_1 + f_2))$$

$$y = WM(r - f_3 - n - W_3 f_1) + (I_r - WMW_3)(d + f_2)$$

$$e = (I_r - WM)r - (I_r - WMW_3)(d + f_2) + WM(f_3 + n + W_3 f_1)$$

$$e_0 = (I_r - W_3 WM)(r + f_3 - n - W_3(d + f_1 + f_2))$$

The actuator output and error signals, including faults, were defined as:

$$u_c = W_0 u + f_0 \quad \text{and} \quad e_c = r - y - f_1 = e - f_1$$

and the fault *estimation error* signals become:

$$\tilde{f}_0 = (I - H_{f0}W_3W_2W_1)f_0 - H_{f0}[-r - n + f_3 + W_3(f_1 + d)] \qquad (4.34)$$

$$\tilde{f}_1 = (I - H_{f1}W_3)f_1 - H_{f1}[-r - n + f_3 + W_3(f_2 + d)] \qquad (4.35)$$

$$\tilde{f}_3 = (I - H_f)f_3 - H_{f1}[-r - n + f_3 + W_3(f_1 + f_2 + d)] \qquad (4.36)$$

where $\Phi_0, \Phi_1$ and $\Phi_3$ denote the power spectra of the signals within the square brackets, on the inputs of the estimator, in the equations (4.34) to (4.36), respectively.

**Cost-function Expression**

The combined control/fault estimation criterion was defined in (4.14) as:

$$J = \frac{1}{2\pi j} \oint_{|z|=1} (trace\{Q_c\Phi_{ee}\} + trace\{R_c\Phi_u\} + trace\{Q_{c1}\Phi_{e_c e_c}\}$$

$$+ trace\{R_{c1}\Phi_{u_c u_c}\} + trace\{U_{f0}\Phi_{\tilde{f}_0 \tilde{f}_0}\} + trace\{U_{f1}\Phi_{\tilde{f}_1 \tilde{f}_1} + U_{f3}\Phi_{\tilde{f}_3 \tilde{f}_3}\})\frac{dz}{z}$$

and after substitution for $\tilde{Q}_c = Q_{c1} + Q_{c1}$ and $\tilde{R}_c = R_c + W_0^* R_{c1} W_0$ obtain:

$$J = \frac{1}{2\pi j} \oint_{|z|=1} (trace\{\tilde{Q}_c[(I_r - WM)\Phi_{rr}(I_r - WM)^*\}$$

$$+ (I_r - WMW_3)\Phi_{d_f d_f}(I_r - WMW_3)^* + WM\Phi_{n_1 n_1}M^*W^*]\}$$

$$+ trace\{\tilde{R}_c[M\Phi_{cc}M^*]\} + trace\{Q_{c1}[\Phi_{f_f f_f} - WM_3\Phi_{f_f f_f} - \Phi_{f_f f_f}W_3^*M^*W^*]\}$$

$$+ trace\{R_{c1}[\Phi_{f_0 f_0} - W_0 M W_3 W_2 W_1 \Phi_{f_0 f_0} - \Phi_{f_0 f_0}W_1^*W_2^*W_3^*M^*W_0^*]\}$$

$$+ trace\{U_{f0}[(I - H_{f0}W_3W_2W_1)\Phi_{f_0 f_0}(I - H_{f0}W_3W_2W_1)^* + H_{f0}\Phi_0 H_{f0}^*]\}$$

$$+ trace\{U_{f1}[(I - H_{f1}W_3)\Phi_{f_f f_f}(I - H_{f1}W_3)^* + H_{f1}\Phi_1 H_{f1}^*]$$

$$+ U_{f3}[(I - H_{f3})\Phi_{f_3 f_3}^*(I - H_{f3})^* + H_{f3}\Phi_3 H_{f3}^*]\})\frac{dz}{z} \qquad (4.37)$$

where $d_f = d + f_2$, and the following signals are defined:

$$n_1 = f_3 + n + W_3 f_1 \quad \text{and} \quad c = r - f_3 - n - W_3(d + f_1 + f_2) \qquad (4.38)$$

The integrand of the combined cost-function may therefore be written, after simplification, as:

$$I_c = trace\{(W^*\widetilde{Q}_cW + \widetilde{R}_c)M\Phi_{cc}M^* - W^*\widetilde{Q}_c(\Phi_{rr} + \Phi_{d_fd_f}W_3^*)M^*$$

$$- (W^*Q_{c1}\Phi_{f_1f_1}W_3^* + W_0^*R_{c1}\Phi_{f_0f_0}W_1^*W_2^*W_3^*)M^*$$

$$- M(\Phi_{rr} + W_3\Phi_{d_fd_f})\widetilde{Q}_cW - M(W_3\Phi_{f_1f_1}Q_{c1}W + W_3W_2W_1\Phi_{f_0f_0}R_{c1}W_0)$$

$$+ \widetilde{Q}_c(\Phi_{rr} + \Phi_{dd} + \Phi_{f_2f_2}) + Q_{c1}\Phi_{f_1f_1}\} + trace\{R_{c1}\Phi_{f_0f_0}\}$$

$$+ trace\{U_{f0}[\Phi_{f_0f_0} - H_{f0}MW_3W_2W_1\Phi_{f_0f_0} - \Phi_{cc}W_1^*W_2^*W_3^*H_{f0}^* + H_{f0}\Phi_{cc}H_{f0}^*]\}$$

$$+ trace\{U_{f1}[\Phi_{f_1f_1} - H_{f1}W_3\Phi_{f_1f_1} - \Phi_{f_1f_1}W_3^*H_{f1}^* + H_{f1}\Phi_{cc}H_{f1}^*]\}$$

$$+ trace\{U_{f3}[\Phi_{f_3f_3} - H_{f3}\Phi_{f_3f_3} - \Phi_{f_3f_3}H_{f3}^* + H_{f3}\Phi_{cc}H_{f3}^*]\} \qquad (4.39)$$

**Transfer Function Spectral Factors**

Introduce the control and filter generalised spectral factors $Y_c$ and $Y_f$ (Shaked, 1976), which are strictly Schur, and satisfy:

$$Y_c^*Y_c = W^*\widetilde{Q}_cW + \widetilde{R}_c = W^*(Q_c + Q_{c1})W + R_c + W_0^*R_{c1}W_0 \qquad (4.40)$$

$$Y_fY_f^* = \Phi_{cc} = \Phi_{rr} + \Phi_{nn} + \Phi_{f_3f_3} + W_3(\Phi_{dd} + \Phi_{f_1f_1} + \Phi_{f_2f_2})W_3^* \qquad (4.41)$$

Substituting for the spectral factors in the cost integrand (4.39) obtain:

$$I_c = trace\{Y_c^*Y_cM\Phi_{cc}M^* - \Phi_{rd}M^* - M\Phi_{dr} + \widetilde{Q}_c(\Phi_{rr} + \Phi_{dd} + \Phi_{f_2f_2})$$

$$+ Q_{c1}\Phi_{f_1f_1}\} + trace\{R_{c1}\Phi_{f_0f_0}\}$$

$$+ trace\{U_{fo}[\Phi_{f_0f_0} - H_{f0}W_3W_2W_1\Phi_{f_0f_0} - \Phi_{f_0f_0}W_1^*W_2^*W_3^*H_{f0}^* + H_{f0}Y_fY_f^*H_{f0}^*]\}$$

$$+ trace\{U_{f1}[\Phi_{f_1f_1} - H_{f1}W_3\Phi_{f_1f_1} - \Phi_{f_1f_1}W_3^*H_{f1}^* + H_{f1}Y_fY_f^*H_{f1}^*]\}$$

$$+ trace\{U_{f3}[\Phi_{f_3f_3} - H_{f3}\Phi_{f_3f_3} - \Phi_{f_3f_3}H_{f3}^* + H_{f3}Y_fY_f^*H_{f3}^*]\}$$

where

$$\Phi_{rd} = W^*\widetilde{Q}_c(\Phi_{rr} + \Phi_{d_fd_f}W_3^*) + (W^*Q_{c1}\Phi_{f_1f_1} + W_0^*R_{c1}\Phi_{f_0f_0}W_1^*W_2^*)W_3^*$$

**Completing the Squares**

The weightings for the fault detection criterion are written: $U_{f0} = H_{u0}H_{u0}^*$, $U_{f1} = H_{u1}H_{u1}^*$ and $U_{f3} = H_{u3}H_{u3}^*$. Thence, following the completing the squares procedure first introduced by Kucera (1979), the cost-function may be written as:

$$I_c = trace\{(Y_c M Y_f - Y_c^{*-1}\Phi_{rd}Y_f^{*-1})(Y_f^* M^* Y_c^* - Y_f^{-1}\Phi_{dr}Y_c^{-1})$$

$$-Y_c^{*-1}\Phi_{rd}Y_f^{*-1}Y_f^{-1}\Phi_{dr}Y_c^{-1}$$

$$+\widetilde{Q}_c(\Phi_{rr} + \Phi_{dd} + \Phi_{f_2 f_2}) + Q_{c1}\Phi_{f_1 f_1}\} + trace\{R_{c1}\Phi_{f_0 f_0}\}$$

$$+trace\{(H_{u0}H_{f0}Y_f - H_{u0}\Phi_{f_0 f_0}W_1^*W_2^*W_3^*Y_f^{*-1})(Y_f^* H_{f0}^* H_{u0}^* - Y_f^{-1}W_3 W_2 W_1 \Phi_{f_0 f_0}H_{u0}^*)$$

$$-H_{u0}\Phi_{f_0 f_0}W_1^*W_2^*W_3^*Y_f^{*-1}Y_f^{-1}W_3 W_2 W_1 \Phi_{f_0 f_0}H_{u0}^* + H_{u0}\Phi_{f_0 f_0}H_{u0}^*\}$$

$$+trace\{(H_{u1}H_{f1}Y_f - H_{u1}\Phi_{f_1 f_1}W_3^*Y_f^{*-1})(Y_f^* H_{f1}^* H_{u1}^* - Y_f^{-1}W_3 \Phi_{f_1 f_1}H_{u1}^*)$$

$$-H_{u1}\Phi_{f_1 f_1}W_3^*Y_f^{*-1}Y_f^{-1}W_3 \Phi_{f_1 f_1}H_{u1}^* + H_{u1}\Phi_{f_1 f_1}H_{u1}^*\}$$

$$+trace\{(H_{u3}H_f Y_f - H_{u3}\Phi_{f_3 f_3}Y_f^{*-1})(Y_f^* H_{f3}^* H_{u3}^* - Y_f^{-1}\Phi_{f_3 f_3}H_{u3}^*)$$

$$-H_{u3}\Phi_{f_3 f_3}Y_f^{*-1}Y_f^{-1}\Phi_{f_3 f_3}H_{u3}^* + H_{u3}\Phi_{f_3 f_3}H_{u3}^*\} \tag{4.42}$$

Observe that the *squared terms* depend upon the control sensitivity function $M$ or estimator transfers $H_{f0}, H_{f1}$ and $H_{f3}$, respectively but the remainder of the terms, although influencing the minimum value of the cost-function, do not affect the actual optimal functions. To minimise the criterion it is only therefore necessary to find the smallest cost as measured by the squared terms.

## 4.2.4   Wiener Solution of the Combined Problem

Although the polynomial solution is actually required, a brief look at the Wiener transfer-function solution to the problem will be provided. In fact the solution is clear from the cost integrand expression (4.42) and it is obtained by equating the causal terms (Grimble and Johnson, 1988):

$$M = Y_c^{-1}\{Y_c^{*-1}(W^*\widetilde{Q}_c(\Phi_{rr} + \Phi_{d_f d_f}W_3^*)$$

$$+(W^*Q_{c1}\Phi_{f_1 f_1} + W_0^* R_{c1}\Phi_{f_0 f_0}W_1^*W_2^*)W_3^*)Y_f^{*-1}\}+Y_f^{-1} \tag{4.43}$$

$$H_{f0} = H_{u0}^{-1}\{H_{u0}(\Phi_{f_0 f_0} W_1^* W_2^* W_3^*) Y_f^{*-1}\}_+ Y_f^{-1} \tag{4.44}$$

$$H_{f1} = H_{u1}^{-1}\{H_{u1}(\Phi_{f_1 f_1} W_3^*) Y_f^{*-1}\}_+ Y_f^{-1} \tag{4.45}$$

$$H_{f3} = H_{u3}^{-1}\{H_{u3}(\Phi_{f_3 f_3}) Y_f^{*-1}\}_+ Y_f^{-1} \tag{4.46}$$

where $\{f(z^{-1})\}_+$ denotes the z-transfer function of the positive-time component of the time-response, corresponding to the 2-sided z-transform $f(z^{-1})$.

There are some observations that can be made, given this Wiener solution:

- The control and fault estimation problems are decoupled in the sense that the solution obtained above is the same as would have been derived if the problems had been solved separately (for the same system, disturbance and fault signals). This confirms the observations of Stoustrup et al. (1997), following a state-space based solution.

- The fault estimators, $H_{f0}, H_{f1}$ and $H_{f3}$ are different because of the different paths for the fault signals to the observed outputs. The expressions for the estimators are the solutions to the appropriate Wiener deconvolution estimation problems (Grimble, 1994).

## 4.2.5  Polynomial Solution of the Combined Problem

The Wiener-Hopf transfer function expressions are not very suitable for the implementation of simple and efficient numerical algorithms. The polynomial forms of system operators, defined in Section 2, can be introduced to obtain the desired numerically efficient polynomial algorithm.    The same types of argument utilised in Chapter 2 may be used to obtain the polynomial solution. That is, the spectral factors and diophantine equations listed in Theorem 4.1 must be introduced, and after simplification the control related terms in the cost-function can be expanded into the form:

$$(Y_c M Y_f - Y_c^{*-1}\Phi_{rd}Y_f^{*-1})(Y_f^* M^* Y_c^* - Y_f^{-1}\Phi_{rd}^* Y_c^{-1}) = (T_1 + T_2)(T_1^* + T_2^*)$$

$$= T_1 T_1^* + T_2 T_2^* + T_1 T_2^* + T_2 T_1^* \tag{4.47}$$

The contour integral of the final term $\oint trace\{T_2 T_1^*\}dz/z$ is zero, because poles of $T_2 T_1^*$ are all outside the unit-circle of the z-plane. The contour integral of the term $T_1 T_2^*$ is also therefore null, since it has a value of (-1) times the integral of the $T_2 T_1^*$ term. Now the term involving $T_2 T_2^*$ does not involve the controller $C_0$, and hence it does not enter into the minimisation argument. Clearly the cost is therefore minimised by setting the first term involving $T_1 T_1^*$ to zero. The optimal controller therefore follows by setting $T_1$ to zero, and the controller becomes:

$$C_0 = C_{0d}^{-1} C_{0n} = C_{1n} C_{1d}^{-1}$$

$$= (H_0 D_3^{-1} A_r^{-1} + L_0 A_{qf}^{-1} D_f^{-1} B z^{-k_m})^{-1} (G_0 D_2^{-1} A_q^{-1} - L_0 A_{qf}^{-1} D_f^{-1} A)$$

Following a similar argument, the optimal fault estimators are obtained by setting the relevant cost terms to zero to obtain the expressions for the fault estimators:

$$H_{f0} = A_{f0}^{-1} X_0 D_f^{-1} A, \quad H_{f1} = A_{f1}^{-1} X_1 D_f^{-1} A, \quad H_{f3} = A_{f3}^{-1} X_3 D_f^{-1} A$$

**System Properties**

The stability of the closed loop system depends upon the zeros of the return difference matrix, and after substitution:

$$I + C_0 W_3 W = I + A_r D_3 H_0^{-1} G_0 D_2^{-1} A_q^{-1} W_3 A^{-1} B$$

$$= (A_r D_3 H_0^{-1})(H_0 D_3^{-1} + G_0 W_3 D_2^{-1} B_1 A_1^{-1} A_r) A_r^{-1}$$

$$= (A_r D_3 H_0^{-1}) D_c (A_r A_{10}^{-1})^{-1}$$

where $D_c$ is strictly Schur.

## 4.2.6 Scalar Continuous-time Results

A quick review of the solution reveals that the results for the continuous-time case are almost identical to those for the discrete-time problem. It only requires the $z^{-g}$ type terms be set to unity to recover the continuous time solution. For completeness, the main results for the scalar continuous-time problem, that are much simpler than for the multivariable case, are summarised below. The system models have the form:

$$W = W_2 W_1 W_0 = A^{-1} B,$$

$$W_d = A^{-1} C_d, \quad W_n = A^{-1} C_n, \quad W_r = A^{-1} E$$

$$f_0(t) = A_{f0}^{-1} F_{f0} \eta_0(t), \quad f_1(t) = A^{-1} F_1 \eta_1(t), \quad f_2(t) = A^{-1} F_2 \eta_0(t)$$

$$f_3(t) = A^{-1} F_3 \eta_3(t), \quad A^{-1} F_1 = A_{f1}^{-1} F_{f1}, \quad A^{-1} F_3 = A_{f3}^{-1} F_{f3}$$

$$W_{21} = W_2 W_1 = A^{-1} B_{21}, \quad W_0 = A_0^{-1} B_0$$

The continuous-time criterion to be minimised is defined as:

$$J = \frac{1}{2\pi j} \oint_D (Q_c \Phi_{ee} + Q_{c1} \Phi_{e_c e_c} + R_c \Phi_{uu} + R_{c1} \Phi_{u_c u_c}$$

$$+ U_{f0} \Phi_{\tilde{f}_0 \tilde{f}_0} + U_{f1} \Phi_{\tilde{f}_1 \tilde{f}_1} + U_{f3} \Phi_{\tilde{f}_3 \tilde{f}_3}) \, ds$$

where the weightings:

$$Q_c = A_q^{*-1} Q_n A_q^{-1}, \quad Q_{c1} = A_q^{*-1} Q_{n1} A_q^{-1}, \quad R_c = A_r^{*-1} R_n A_r^{-1}$$

and

$$W_0^* R_{c1} W_0 = A_r^{*-1} R_{n1} A_r^{-1}, \quad \tilde{Q}_n = Q_n + Q_{n1}, \quad \tilde{R}_n = R_n + R_{n1}$$

and $A_1 = A_q A$.

**Algorithm 4.1** : *Scalar Continuous Time Fault Estimation and Control*

1. *Spectral factorisation:*

$$D_c^* D_c = A_r^* B^* \tilde{Q}_n B A_r + A_q^* A^* \tilde{R}_n A A_q$$

$$D_f D_f^* = E E^* + C_d C_d^* + C_n C_n^* + F_1 F_1^* + F_2 F_2^* + F_3 F_3^*$$

2. *Control diophantine equations:*

$$D_c^* G_0 + Z_0 A A_q = A_r^* B^* \tilde{Q}_n D_f$$

$$D_c^* H_0 - Z_0 B A_r = A_q^* A^* \tilde{R}_n D_f$$

$$D_{fq}^* D_c^* L_0 + P_0 A_{qf} = L$$

where $D_{fq}, A_{qf}$ and $L$ satisfy:

$$D_{fq}^{*-1} L A_{qf}^{-1} = \frac{1}{A_0^* D_f^*} A_0^* A_r^* B^* \tilde{Q}_n (C_n C_n^* + F_3 F_3^*) A_{f0}$$

$$+ [A_0^* A_r^* B^* Q_n F_1 F_1^* A_{f0} - A_q A A^* A_q^* B_0^* R_{c1} F_{f0} F_{f0}^* F_{20}^*] \frac{1}{A_{f0} A_q A}$$

3. *Estimator diophantine equations:*

$$X_0 D_f^* + A_{f0} Y_0 = F_{f0} F_2^*$$

$$V_0 D_f^* - B_{21} Y_0 = (D_f D_f^* - F_2 F_2^*)$$

$$X_1 D_f^* + A_{f_1} Y_1 = F_{f_1} F_1^*$$

$$X_3 D_f^* + A_{f_3} Y_3 = F_{f_3} F_3^*$$

4. *Feedback controller :*

$$C_0 = (H_0 A_r^{-1} + L_0 A_{qf}^{-1} B)^{-1} (G_0 A_q^{-1} - L_0 A_{qf}^{-1} A)$$

5. Fault estimators :

$$H_{f0} = A_{f0}^{-1} X_0 D_f^{-1} A, \quad H_{f1} = A_{f1}^{-1} X_1 D_f^{-1} A, \quad H_{f3} = A_{f3}^{-1} X_3 D_f^{-1} A$$

**Design Hints**

1. If a single fault source is considered, the best fault estimate of that fault will be obtained. The problem that arises when two or more fault sources are considered possibilities, is that the fault signal which has not been estimated becomes part of the effective measurement noise model.

2. Including fault models also has a deleterious affect on the control action. Clearly the solution when no faults are present will be sub-optimal, since the controller is designed allowing for the presence of the fault model terms.

3. A good control design that allows for fault conditions should not affect the regulating action excessively, and yet it should be able to accommodate the faults when they arise. Clearly, this is a difficult trade-off design situation.

4. Parametric uncertainty may be allowed for in the system description, as in Grimble (1982).

### 4.2.7 Marine Roll Stabilisation Example

The model for the following example was based upon the roll dynamics of a ship (Chapter 12). In this type of problem the requirement is to minimise, in some sense, the transfer function between the wave force inputs and the roll of the vessel. In control engineering terms this is of course the minimisation of the sensitivity function, which is referred to in marine systems as the *roll reduction ratio* frequency response. However, the example which follows is more intended to show the effects of combined fault monitoring and control design, rather than a realistic study in roll suppression. The system model is defined below.

**System Description**
**Input actuator :**

$$W_0 = B_0/A_0 = 1/(1 + s/40)$$

**Main actuator :**

$$W_1 = B_1/A_1 = 1/(1 + s/10)$$

**Plant :**

$$W_2 = B_2/A_2 = s/(s^2 + 0.0698s + 0.12188), \quad W_3 = 1$$

**Actuator fault :**

$$W_{f0} = F_{f0}/A_{f0} = 0.01/s^2$$

**Output fault :**

$$W_{f1} = F_{f1}/A_{f1} = 0.051/s$$

**Measurement fault :**

$$W_{f3} = F_{f3}/A_{f3} = 0.051/s$$

**Cost weightings :**

$$Q_c = Q_n/(A_q^* A_q) = (s + 0.05)(-s + 0.05)/(-s^2)$$

$$R_c = R_n/(A_r^* A_r) = 0.1, \quad Q_{c1} = Q_{n1}/(A_q^* A_q) = Q_c$$

$$W_0^* R_{c1} W_0 = R_{n1}/(A_r^* A_r) = 0.1/((1 + s/40)(1 - s/40))$$

**Example Strategy**

The example will begin by considering a system without fault models. A design will be completed for the situation where no faults are modelled, and a sensitivity function will be computed that can be used as a base against which to judge the modified sensitivity functions when faults are included. Recall that the sensitivity function is a measure of the roll reduction performance. When fault models are present, the resulting design will of course be sub-optimal for the case when there is no fault.

The next case to be considered is when one output fault is modelled. The third case involves both of the output fault conditions $f_1$ and $f_3$. Finally, the example when all three faults are included is considered. The example will be evaluated using the scalar results detailed in Section 4.2.6.

**Plant Frequency Responses**

The plant frequency responses are shown in Fig. 4.4. This includes the plant transfer function, measurement noise and disturbance models, and the fault transfer function models. Observe that the plant transfer function represents the lightly damped second order roll response of a ship. The disturbance model is taken to be the same as the ship dynamics approximately, since in practice it is the product of the roll model and the ship model.

**Case (i) : No fault conditions modelled**

The cost weightings are chosen in a traditional manner (Grimble, 1994). That is, the error weighting includes an integrator with a lead term so that errors are penalised heavily at low frequency, but not so heavily in mid frequencies. The control weighting is chosen to be a simple constant. A rule of thumb is that the intersection of the frequency responses of $W^* Q_c W$ and $R_c$ will be in the region of the unity gain crossover frequency for the system. This enables the relative size of the two weightings to be chosen. The result is indicated in Fig. 4.5. Observe that the intersection between the weighted error term and the control term lies at a frequency of 3.0 radian/sec and the point at which the resulting open loop transfer function crosses the zero dB's line is at 1.0 radian/sec.

The frequency response of the controller where no faults are modelled is shown in Fig. 4.6. In fact, this is similar to the classical frequency response of the roll stabilisation controller, which normally involves a PID term that is cut off at high frequencies.

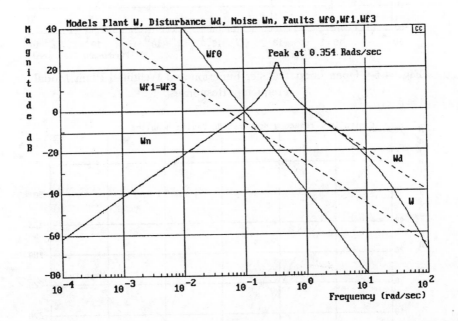

Fig. 4.4 : Plant Model, Disturbance, Noise and Fault Model Frequency Responses

If the roll angle sensor includes a bias signal the controller cannot include pure integral action at low frequencies. This complicates the design problem, since a high gain is needed to reduce the sensitivity function. Thus, the controller characterteristic in this case must have zero gain at zero frequency and reasonable gain in the frequency ranges where the sensitivity function peaks.

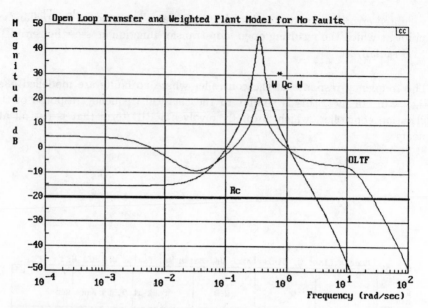

Fig. 4.5 : Open Loop Transfer Function and Weighted Plant Model
Frequency Responses

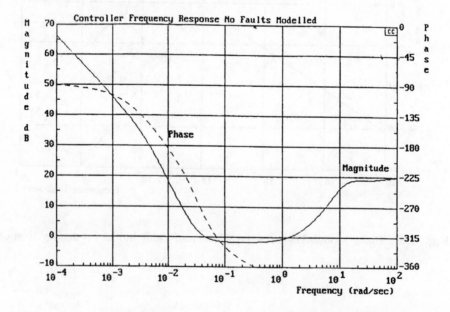

Fig. 4.6 : Controller Frequency Responses when no Faults are Modelled

In the roll stabilisation problem, the sensitivity function determines the performance of the system. This represents the transfer function between the wave input and the roll angle of the vessel, and the lower the sensitivity the greater is the roll reduction. One of the problems is to reduce the two peaks which occur at low and high frequencies, but this is not the main objective in this example. The sensitivity function frequency response is shown in Fig. 4.7, and excellent attenuation is obtained in the mid frequency range and at low frequencies.

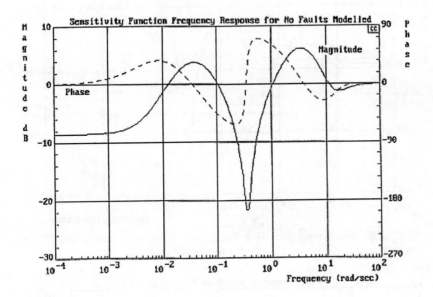

**Fig. 4.7** : Sensitivity Function Frequency Responses when No Fault Condition is Modelled

### Case (ii) : Inclusion of one fault condition $f_1$

In this case, the presence of either fault $f_1$ or $f_3$ is considered. Since in this problem they enter the system at the same point, and the fault models are the same, it does not matter which of the two faults is considered, but only one of them is assumed present.

The sensitivity function frequency response is shown in Fig. 4.8. There is some deterioration in the frequency response at the higher frequencies. The frequency response of the fault estimator is shown in Fig. 4.9. Note that at low frequencies the gain of the estimator is unity, which is compatible with the signal to be estimated, that is represented by white noise feeding into an integrator. The equivalent deterministic situation is for the estimation of a constant signal and hence a low frequency filter gain of unity is required. At higher frequencies, the other signals in the system, which represent both measurement noise and disturbances, act like an effective measurement noise for the fault estimator, and hence the gain decreases rapidly in the mid frequency range.

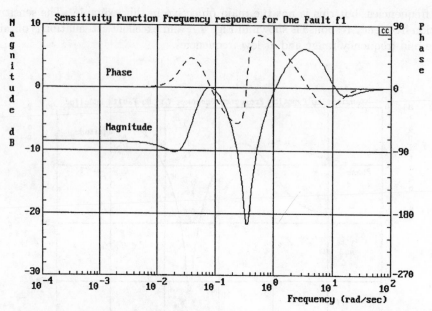

**Fig. 4.8** : Sensitivity Function Frequency Responses when One Fault
Condition is Modelled

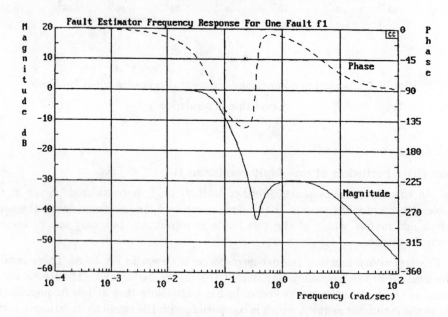

**Fig. 4.9** : Fault Estimator Frequency Response when One Fault
Condition is Modelled

**Case (iii) : Inclusion of two faults $f_1$ and $f_3$**

The controller frequency response has changed slightly in this case because of
the presence of the two fault conditions. The sensitivity function frequency response

also changes, but the most significant difference lies in the fault estimator frequency response, shown in Fig. 4.10. Note that, in this case, both the faults $f_1$ and $f_3$ involve the same models, and yet the results change relative to the previous case. Note that, at low frequencies, the gain of the estimator is no longer unity (compare Figs. 4.9 and 4.10). The reason this occurs is that, when designing the estimator for fault condition $f_1$, all other noise sources (including the model for fault $f_3$) appear like measurement noise in the equations. Since fault $f_3$ has a significant low frequency component, the fault estimator zero frequency response is no longer unity. This is, of course, only the ideal value when fault $f_3$ is not present.

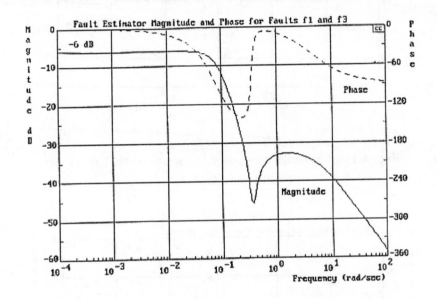

Fig. 4.10 : Fault Estimator Frequency Response when Two Fault
Conditions are Included

## Case (iv) : All three fault conditions $f_0$, $f_1$, $f_3$ included

There is a significant change from the first case in the controller frequency response, shown in Fig. 4.11, and in the sensitivity function frequency response, Fig. 4.12.

The frequency responses of the estimators for three faults are shown in Fig. 4.13. The responses for the estimators for faults $f_1$ and $f_3$ are, of course, the same in this case, but the response of the actuator fault model $f_0$ has a high gain at low frequencies. In fact, the fault model for the actuators involves a double integrator, which in deterministic terms would indicate a ramp fault signal. It may be helpful to consider the transfer functions between the white noise inputs to the fault models and the estimated signals. These transfer function frequency responses are shown in Fig. 4.14. Observe that in the case of the estimator for the actuator faults (labelled 0), the approximate transfer appears as $1/s^2$, and in the case of the output faults ($f_1$ and $f_3$), it has approximate frequency responses of $1/s$. Thus, these responses approximate the actual fault signals well in the low frequency range, where most energy is concentrated.

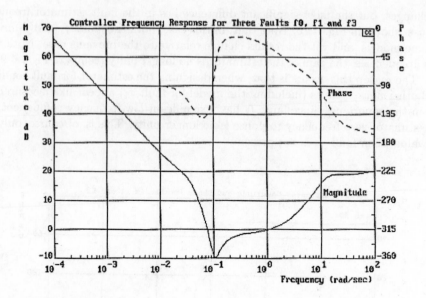

Fig. 4.11 : Controller Frequency Responses when Three Fault
Conditions are Modelled

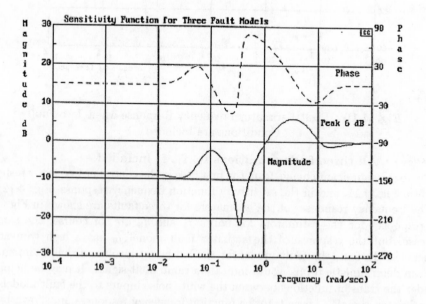

Fig. 4.12 : Sensitivity Function Frequency Response when Three Fault
Conditions are Modelled

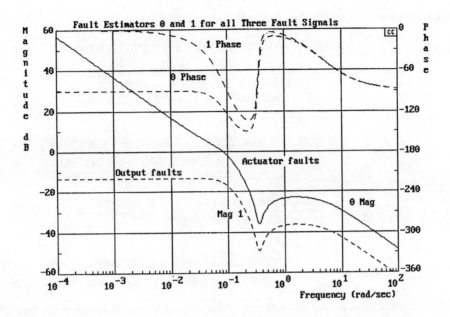

Fig. 4.13 : Fault Estimator Frequency Responses when All Three Fault
Conditions are Modelled
(Estimators for Faults 0 and 1, 3)

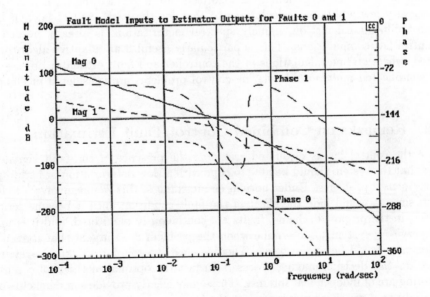

Fig. 4.14 : Frequency Responses between Fault Model Inputs and
Fault Estimator Outputs for Faults 0 and 1

**Computed Controller and Fault Estimators Case (iv)**

$$C_0(s) = \frac{\begin{array}{c}-10.50493(s + 0.0355113)[(s + 0.017675)^2 + 0.1010306^2] \\ \times (s - 1.996054)(s + 10)(s + 40)\end{array}}{\begin{array}{c}s[(s + 0.0238678)^2 + 0.03206884^2] \\ [(s + 7.868753)^2 + 10.12384^2](s + 58.33688)\end{array}}$$

$$H_{f0}(s) = \frac{0.3723922(s + 0.06722878)[(s + 0.0349)^2 + 0.3473643^2]}{s[(s + 0.0688281)^2 + 0.06867264^2](s + 4.694794)}$$

$$H_{f1}(s) = \frac{0.0793638[(s + 0.0349)^2 + 0.3473643^2]}{[(s + 0.0688288)^2 + 0.06867264^2](s + 4.694794)}$$

$$H_{f3}(s) = \frac{0.0793638[(s + 0.0349)^2 + 0.3473643^2]}{[(s + 0.0688288)^2 + 0.06867264^2](s + 4.694794)}$$

The example has drawn attention to some difficulties the designer must face. The fault estimators depend upon the number of faults expected. This also applies to the control law, which tries to suppress the worst effects of the faults. For this reason, a design to allow for all possible fault conditions may be unrealistic, unless there is a high probability that these faults can occur at the same time.

A much more realistic situation is where combined control and only one fault condition is involved. The numerical algorithm for implementing such a solution involves very few additional calculations to that for the basic control law. In fact the most difficult calculation, namely spectral factorisation, is already needed for the control law computations. This is particularly useful if an adaptive algorithm is required, where on-line calculations of the controller and fault estimator are needed. More complicated multivariable marine control problems are considered in Chapter 12.

## 4.2.8  Remarks on Combined Control/Fault Estimation

Since the control law is evaluated assuming the presence of the fault signals, it follows that the system should be able to cope with a slow deterioration or build up of actuator/sensor problems. Under normal circumstances, this would involve a hidden danger, since it would hide the effect of the fault condition until it became serious. However, in the proposed scheme, faults are continuously monitored, and it is up to the supervisory system to determine when the problem is so urgent that immediate maintenance should be undertaken. There is clearly a role for an expert system to be used for this decision making process. In fact the optimal methods of condition monitoring are of independent interest. These may clearly provide some benefits over more traditional condition monitoring methods.

The example revealed that there remain many questions to be considered. Should, for example, the controller try to compensate for the adverse effects of the fault models? This may indeed provide for improved performance when the fault is present, but in normal operation there will be some degradation. There is also a difficulty when including a number of fault models, as the example indicated, when both faults $f_1$ and $f_3$ were present, and both entered the system at the same point and had the same models. These fault estimators changed, relative to the case when only one fault was present. The explanation was that a particular estimator sees all other fault signals as effective measurement noise. Thus, a fault estimator which is ideal for estimating a single fault, will be different to the equivalent fault estimator when more than one fault is present. If it is unlikely that two faults can occur at the same time, it may not be realistic to include them both, since suboptimality of the estimator will occur.

## 4.3 Separation Principle for Polynomial Systems

The separation principle of stochastic optimal control theory has often been utilised for systems represented in state equation form until recently. However, no such results have been established for systems represented in transfer-function or polynomial matrix form (Grimble, 1998b). The benefits of this type of solution are similar to those for systems represented in state equations. That is, the decomposition of the LQG controller into a Kalman filter and control gain is physically justifiable, and the signal estimates are often useful indicators of important physical variables.

The frequency domain approach to optimal control and estimation was initiated by Norbert Wiener (1949), but two seminal contributions later established the main tools for synthesis. These contributions were undertaken in the same period by Youla and Bongiorno and co-workers (1976) in New York, and Kucera in Prague (1979). The Youla et al approach introduced the idea of a stabilising controller parameterisation and the Kucera method also included a technique to ensure closed-loop stability was achieved.

The separation principle that is well known in state-space LQG synthesis (Kwakernaak and Sivan, 1972) was not used in the frequency-domain solutions, although Kucera (1978, 1979) later provided independent solutions of the LQ state feedback control and the Kalman filtering problems (1961). In this work, by Kucera, if the polynomial models are related back to a system described in state equation form, it is possible to use the polynomial solutions to calculate the constant control and filter gains. The state-space separation principle results can then be invoked to obtain the LQG output feedback controller. The separation principle was not, however, established using a polynomial setting. Moreover, there was no attempt to generalise these results to the case where the control law feedback included a reduced set of variables, such as plant output estimates. The major objective of the analysis that follows is to use frequency domain models and analysis to establish a separation principle result for systems represented in frequency domain polynomial matrix-fraction form.

To establish the separation result, it is necessary to introduce the filter in an observer form. That is, using a model of the process with an observer feedback loop to ensure what is termed pseudo-state estimate tracking. If the filter is not represented in an observer form the decomposition of the LQG controller into a filter and control gain is not straightforward.

**Solution Strategy**

To generate the polynomial equivalent of the state-space LQG separation principle results the following strategy will be followed:

(i) The system must first be placed in polynomial form, with internal variables referred to as pseudo states, and the observer structure for the estimator must be introduced, based on this model.

(ii) The gain for the estimator $K_f$ should be found so that the estimates of the signal $h$ are orthogonal to the estimation error.

(iii) The output feedback control problem can then be expressed (using the orthogonality of signals in the cost-function) in terms of a pseudo state estimate feedback control problem, where the estimates are available for feedback control.

(iv) The pseudo state estimate feedback control problem can then be solved to obtain the control gain $K_c$, and this completes the solution.

## 4.3.1   Polynomial System Description

The linear, time-invariant, discrete-time, multivariable, finite-dimensional system of interest is illustrated in Fig. 4.15. The noise free system output sequence is denoted by $\{y(t)\}$ , where $y(t) \epsilon R^r$, and the observations signal is denoted by $\{z(t)\}$. In this simple case the pseudo states are equal to the noise free output vector. The white driving noise signals, $\{\xi(t)\}$ and $\{v(t)\}$, represent the disturbance, and coloured measurement noise signals, respectively. These signals are statistically independent, and the covariance matrices are given as:

$$cov\{\xi(t), \xi(\tau)] = I_q \delta_{t\tau} \quad \text{and} \quad cov\{v(t), v(\tau)] = R_f \delta_{t\tau}$$

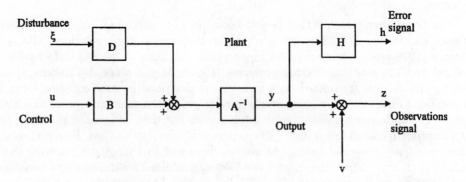

**Fig. 4.15** : Discrete Polynomial Matrix Representation of the System Model

The time-invariant linear system of interest has the following polynomial matrix form (Kailath, 1980):

**Plant:**

$$G(z^{-1}) = \begin{bmatrix} G_{11}(z^{-1}) & G_{12}(z^{-1}) \\ G_{21}(z^{-1}) & G_{22}(z^{-1}) \end{bmatrix} = \begin{bmatrix} HA^{-1}D & HA^{-1}B \\ A^{-1}D & A^{-1}B \end{bmatrix}$$

$$= \begin{bmatrix} H \\ I_r \end{bmatrix} A^{-1} \begin{bmatrix} D & B \end{bmatrix}$$

**Signals:**

$$\begin{bmatrix} h(t) \\ y(t) \end{bmatrix} = G \begin{bmatrix} \xi(t) \\ u(t) \end{bmatrix} = \begin{bmatrix} HA^{-1}D\xi(t) + HA^{-1}Bu(t) \\ A^{-1}D\xi(t) + A^{-1}Bu(t) \end{bmatrix} \qquad (4.48)$$

**Observations:**

$$z(t) = (y(t) + v(t)) \epsilon R^r \qquad (4.49)$$

The last signal in the observations represents white measurement noise.

Introduce the right coprime factorisation, $(A_1, B_1)$, of the plant-transfer-function model, to satisfy:

$$B_1 A_1^{-1} = A^{-1} B \epsilon R^{r \times m}(z^{-1}) \qquad (4.50)$$

and note that these matrices may easily be defined to include transport-delay elements.

**Assumptions**

The assumptions may be listed as follows:

- The system is assumed to be free of unstable hidden modes.

- The disturbance polynomial matrix $D \epsilon R^{r \times q}(z^{-1})$ is assumed to be full rank, $q \geq r$ and $DD^*$ is also therefore full rank.

- The control $D_c \epsilon R^{m \times m}(z^{-1})$ and filter $D_f \epsilon P^{r \times r}(z^{-1})$ spectral-factors, introduced in the following, will depend upon the cost weightings and noise covariances, respectively. It will be assumed that these are chosen so that $D_c$ and $D_f$ are strictly-Schur polynomial matrices.

## 4.3.2  Noise Free Output Feedback Control Problem

The noise free output feedback control problem is illustrated in Fig. 4.16. For simplicity the regulating problem is considered, where the output measurement is ideal. That is, the measurement noise is assumed to be null. The initial results obtained are similar to those of the usual state feedback LQ control problem and are unlikely to be of direct benefit for real applications. However, these results are needed to establish the *separation* result that follows. The properties of the resulting feedback system are also of interest.

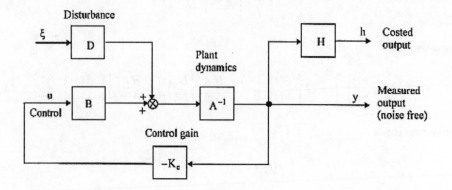

**Fig. 4.16**: Plant Model and Controller for the Ideal Output Zero
Measurement-Noise Output Feedback Control Problem

The cost-function to be minimised may be represented, using Parseval's theorem, in the complex frequency domain as:

$$J_c = \frac{1}{2\pi j} \oint_{|z|=1} (trace\{Q_c\Phi_{yy}\} + trace\{R_c\Phi_{uu}\})\frac{dz}{z} \qquad (4.51)$$

where the contour is evaluated around the unit-circle of the complex frequency plane, and $Q_c = H^*H$. The weighting $H$ is assumed to be of normal full rank, and $H\epsilon P^{r\times r}(z^{-1})$. The optimal control gain transfer-function matrix $K_c$, that gives the noise-free output feedback, can be computed from the theorem which follows below. This is required as the first part of the separation principle result.

**Theorem 4.2** : *Discrete Ideal Output Feedback Control Problem Solution*
    Consider the system shown in Fig. 4.16, where perfect measurements of the output are available, and assume that the variance (4.51) of the weighted output and control signal is to be minimised. Then, the output feedback controller, when measurements are ideal, can be computed from the solution of the following spectral factors and diophantine equations.

**Spectral factors:**
    The strictly Schur control spectral factor : $D_c\epsilon P^{m\times m}(z^{-1})$, satisfies:

$$D_c^*D_c = B_1^*Q_cB_1 + A_1^*R_cA_1 \qquad (4.52)$$

The spectral factor $D_s\epsilon R^{q\times q}(z^{-1})$ is Schur and satisfies:

$$D_sD_s^* = DD^* \qquad (4.53)$$

**Diophantine equations:**
    The smallest degree solution $(G_{c0}, H_{c0}, F_{c0})$, in terms of the polynomial matrix $F_{c0}\epsilon P^{m\times r}(z^{-1})$, satisfies:

$$D_c^*G_{c0}z^{-g} + F_{c0}A_d = B_1^*Q_cD_dz^{-g} \qquad (4.54)$$

$$D_c^* H_{c0} z^{-g} - F_{c0} B_{d0} = A_1^* R_c D_{d0} z^{-g} \tag{4.55}$$

where $D_d, A_d$ and $B_{d0}, D_{d0}$ are right coprime and satisfy:

$$D_d A_d^{-1} = A^{-1} D_s \quad \text{and} \quad B_{d0} D_{d0}^{-1} = D_s^{-1} B_1 \tag{4.56}$$

and $g$ is the smallest positive integer chosen to ensure (4.54) and (4.55) are polynomial equations in $z^{-1}$.

**Optimal control and controller gain:**

$$u(t) = -K_c y(t) \quad \text{and} \quad K_c = D_{d0} H_{c0}^{-1} G_{c0} D_d^{-1} \tag{4.57}$$

and the controller should be implemented in its minimal order form.

**Proof:** The proof is given in the following section.

**Lemma 4.2:** *Ideal Output Discrete Optimal Feedback System Properties*
     The properties of the optimal ideal output feedback control problem, when measurements are perfect, are summarised below. The so called *implied equation* and the *return difference* matrix, that determine the degree of stability of the system, are respectively:

$$G_{c0} D_d^{-1} B_1 + H_{c0} D_{d0}^{-1} A_1 = D_c \tag{4.58}$$

and

$$I_m + K_c A^{-1} B = D_{d0} H_{c0}^{-1} D_c A_1^{-1}$$

The minimum value of the *cost function integrand* $I_{cmin}$ can be computed from the functions $T_2$ and $T_3$ as:

$$T_{2c} = D_c^{*-1} F_{c0} z^g$$

$$T_{3c} = D_s^* A^{*-1} Q_c (I_r - B_1)(D_c^* D_c)^{-1} B_1^* Q_c) A^{-1} D_s$$

$$I_{c\,min} = trace\{T_{2c}^* T_{2c} + T_{3c}\} \tag{4.59}$$

The optimal *sensitivity functions* are given as:

$$S_c = (I + K_c A^{-1} B)^{-1} = A_1 D_c^{-1} H_{c0} D_{d0}^{-1}$$

and

$$M_c = S_c K_c = A_1 D_c^{-1} G_{c0} D_d^{-1} \tag{4.60}$$

**Proof:** The proof is given in the following section.

**Solution of the Ideal Output Feedback Control Problem**
     There now follows a solution to the zero measurement noise, output feedback control problem. This uses the completing the squares optimisation argument, established by Kučera (1979). The solution is straightforward, and the results are therefore summarised briefly.
     The proof is divided into the following stages:

- The cost integrand is expressed in terms of signal power spectra.

- Introducing spectral factors enables the cost spectra to be expressed in a squared form depending upon the control sensitivity matrix $M_c$.

- Diophantine equations are introduced to enable the squared term to be expanded in terms of stable and unstable components.

- The cost function is then expanded, and it is shown that the integral of the cross terms between stable and unstable components is null.

- The cost is then minimised when the only term in the cost expression which is control law dependent is set to zero.

**Summary of Control and Costed Output Signals**

$$u = -K_c y = -K_c A^{-1}(Bu + D\xi) \tag{4.61}$$

$$h = HA^{-1}(Bu + D\xi) \tag{4.62}$$

Thence, after some obvious manipulation:

$$u = -(I + K_c A^{-1}B)^{-1}K_c A^{-1}D\xi$$

$$h = HA^{-1} - (-B(I + K_c A^{-1}B)^{-1}K_c A^{-1} + I)D\xi$$

$$= H(I + A^{-1}BK_c)^{-1}A^{-1}D\xi$$

**Sensitivity Function Definitions: The Control Problem**

Introduce the definition of the sensitivity and control sensitivity functions, for the output feedback control case, respectively, as:

$$S_c = (I + K_c A^{-1}B)^{-1} \quad \text{and} \quad M_c = S_c K_c$$

and by reference to Fig. 4.16, substituting into (4.61) and (4.62) gives:

$$h = Hy = HA^{-1}(B(-M_c A^{-1}D) + D\xi$$

$$u = -M_c A^{-1}D\xi$$

**Power Spectra**

The *power spectra* of these signals may therefore be written as:

$$\Phi_{hh} = HA^{-1}(-BM_c A^{-1} + I_r)DD^*(-A^{*-1}M_c^* B^* + I_r)A^{*-1}H^*$$

and

$$\Phi_{uu} = M_c A^{-1}DD^* M_c A^{*-1}M_c^* \tag{4.63}$$

Thence, the cost integrand (4.51) may be expressed (noting $trace\{XY\} = trace\{YX\}$) as:

$$I_c = trace\{Q_c\Phi_{yy}\} + trace\{R_c\Phi_{uu}\}$$

$$= trace\{D^*((-A^{*-1}M_c^*B^* + I)A^{*-1}Q_cA^{-1}(-BM_cA^{-1} + I)$$

$$+A^{*-1}M_c^*R_cM_cA^{-1})D\} \tag{4.64}$$

**Generalised Control Spectral Factors**

This expression may be simplified by introducing the generalised control spectral factor $Y_c$, that is defined to satisfy (Shaked , 1976):

$$Y_c^*Y_c = B^*A^{*-1}Q_cA^{-1}B + R_c \tag{4.65}$$

and noting $A^{-1}B = B_1A_1^{-1}$ the rational spectral factor may be written as:

$$Y_c = D_cA_1^{-1}\epsilon R^{m\times m}(z^{-1})$$

where the Schur polynomial spectral factor $D_c$ satisfies:

$$D_c^*D_c = B_1^*Q_cB_1 + A_1^*R_cA_1$$

The weightings are assumed to be chosen so that $D_c$ is also free of unit-circle zeros. Also, introduce the spectral factor $D_s$ that satisfies:

$$D_sD_s^* = DD^*$$

and assume that the disturbance model is such that this matrix is full rank. Thence, a Schur spectral factor $D_s$ exists.

**Completing the Squares**

There now follow the usual steps (Kucera, 1979) to complete the squares in the cost integrand expression. Substituting (4.65) into (4.64) obtain:

$$I_c = trace\{D^*A^{*-1}(M_c^*Y_c^*Y_cM_c - M_c^*B^*A^{*-1}Q_c - Q_cA^{-1}BM_c + Q_c)A^{-1}D\}$$

$$= trace\{(D_s^*A^{*-1}M_c^*Y_c^* - D_s^*A^{*-1}Q_cA^{-1}BY_c^{-1})$$

$$\times(Y_cM_cA^{-1}D_s - Y_c^{*-1}B^*A^{*-1}Q_cA^{-1}D_s)$$

$$+D_s^*A^{*-1}Q_cA^{-1}(I - BY_c^{-1}Y_c^{*-1}B^*A^{*-1}Q_cA^{-1})D_s\} \tag{4.66}$$

**Diophantine Equations**

Two diophantine equations (4.54) and (4.55) are essential to the proof of optimality, and stability and a third equation that depends upon these first two is referred to as the implied equation, which determines the stability margins of the solution.     To obtain the so called implied equation for this problem, first note from (4.50) and (4.56), the following result:

$$A_d D_d^{-1} B_1 = D_s^{-1} A B_1 = D_s^{-1} B A_1 = B_{d0} D_{d0}^{-1} A_1$$

Right multiplying the diophantine equation (4.54) by $D_d^{-1} B_1$ and the equation (4.55) by $D_{d0}^{-1} A_1$ and adding these equations obtain:

$$B_1^* Q_c B_1 + A_1^* R_c A_1 = D_c^* D_c$$

and from (4.52) the *implied equation* follows as:

$$G_{c0} D_d^{-1} B_1 + H_{c0} D_{d0}^{-1} A_1 = D_c$$

**Cost-function Simplification**

The so called *squared term* in the cost-function integrand (4.66) may be simplified using the implied equation:

$$Y_c M_c A^{-1} D_s - Y_c^{*-1} B^* A^{*-1} Q_c A^{-1} D_s = Y_c M_c A^{-1} D_s - (G_{c0} A_d^{-1} + D_c^{*-1} F_{c0} z^g)$$

$$Y_c K_c (A + BK_c)^{-1} D_s - G_{c0} A_d^{-1} - D_c^{*-1} F_{c0} z^g$$

$$= [H_{c0} D_{d0}^{-1} K_c - G_{c0} A_d^{-1} D_s^{-1} A](A + BK_c)^{-1} D_s - D_c^{*-1} F_{c0} z^g \qquad (4.67)$$

To simplify the expression for the cost integrand let,

$$T_{1c} = (H_{c0} D_{d0}^{-1} K_c - G_{c0} D_d^{-1})(A + BK_c)^{-1} D_s$$

$$T_{2c} = D_c^{*-1} F_{c0} z^g$$

The above cost term (4.67) may therefore be written as:

$$Y_c M_c A^{-1} D_s - Y_c^{*-1} B^* A^{*-1} Q_c A^{-1} D_s = T_1 - T_2$$

The final term in (4.66) is independent of the choice of controller and is denoted as:

$$T_{3c} = D_s^* A^{*-1} Q_c A^{-1} (I_r - B Y_c^{-1} Y_c^{*-1} B^* A^{*-1} Q_c A^{-1}) D_s$$

Collecting results, the cost integrand (4.66) may therefore be written as:

$$I_c = trace\{(T_{1c}^* - T_{2c}^*)(T_{1c} - T_{2c})\} + trace\{T_{3c}\}$$

It is important to observe, for the optimisation argument, that $T_{1c}$ includes the following terms:

- $D_{d0}$ related to $D_s$ through (4.56), and this is by definition Schur.

- $D_d$ related to $D_s$ through equation (4.56).

- The zeros of $D_s$ on the unit-circle will cancel with the zeros of $D_d$ and $D_{d0}$.

- $(A + BK_c)$ that must be strictly Schur, since the cost $J_c$ is assumed finite and the closed-loop must necessarily be stable.

The term $T_{1c}$ is therefore asymptotically stable, and the term $T_{2c}$ is strictly unstable (since the control spectral factor is strictly Schur, and $D_c^*$ is strictly non-Schur).

**Solution of the Optimisation Problem**

The cost function (4.51) may now be written in the expanded form:

$$J_c = \frac{1}{2\pi j} \oint_{|z|=1} (trace\{T_{1c}^* T_{1c} + T_{2c}^* T_{2c} - T_{1c}^* T_{2c} - T_{2c}^* T_{1c}\} + trace\{T_{3c}\}) \frac{dz}{z} \quad (4.68)$$

and by assumption the integral converges $J_c < \infty$. The integral of the term $T_{2c}^* T_{1c}$ which is analytic in and on the *unit-circle*, is zero by the residue theorem of complex analysis. Similarly, for the term $T_{1c}^* T_{2c}$. Since the unknown final term $T_{3c}$ does not depend upon the controller, the cost is clearly minimised when $T_{1c} = 0$.

**Feedback controller:**

$$K_c = D_{d0} H_{c0}^{-1} G_{c0} D_d^{-1}$$

**Control sensitivity**:

$$M_c = (I_m + K_c A^{-1} B)^{-1} K_c = A_1 D_c^{-1} G_{c0} D_d^{-1}$$

This latter expression may be obtained more directly from the first equation in (4.67).

## 4.3.3  Discrete-time Output Estimation Problem

The second step in establishing the separation principle requires the solution of the optimal observer based estimation problem, where the filter structure is chosen to be as shown in Fig. 4.17. The cost-function to be minimised in the unweighted output optimal estimation problem is defined as:

$$J = E\{(y(t) - \widehat{y}(t))^T (y(t) - \widehat{y}(t))\}$$

In later analysis, the estimator orthogonality property that is required refers to weighted estimation error signals (the weighting $H$ being due to $Q_c$ in the cost index (4.51)). Thus, a more general weighted estimation error criterion will be minimised, where the weighting $D_{ho}$ is assumed Schur and $D_{h0} \epsilon P^{r \times r}(z^{-1})$. The weighted output estimate, generated by the observer in Fig. 4.17, is defined to have the form : $\widehat{y} = D_{h0} \widehat{y}$. The weighted estimation error is defined as $\widetilde{y}_h(t) = (D_{h0} y)(t) - \widehat{y}_h(t)$, and the estimation error criterion:

$$J_{fh} = E\{\widetilde{y}_h(t)^T \widetilde{y}_h(t)\} = E\left\{ \left( (D_{h0}(y(t) - \widehat{y}(t)))^T \right) (D_{h0}(y(t) - \widehat{y}(t))) \right\}$$

The *weighted cost-function* may therefore be written, invoking Parseval's theorem, as:

$$J_{fh} = \frac{1}{2\pi j} \oint_{|z|=1} (trace\{\Phi_{\widetilde{y}_h \widetilde{y}_n}\}) \frac{dz}{z} = \frac{1}{2\pi j} \oint_{|z|=1} trace\{D_{h0} \Phi_{\widetilde{y}_h \widetilde{y}_n} D_{h0}^*\} \frac{dz}{z} \quad (4.69)$$

where $\Phi_{\widetilde{y}\widetilde{y}}$ denotes the power spectrum of the estimation error signal $\widetilde{y} = y - \widehat{y}$ .

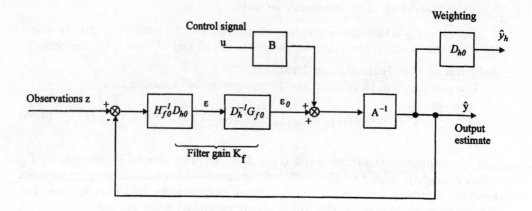

Fig. 4.17 : Discrete Output Estimator in an Observer Polynomial Matrix Form

**Theorem 4.3 :** *Discrete Output Optimal Observer Based Estimator*
Consider the optimal discrete-time observer, shown in Fig. 4.17, where the system is represented in polynomial matrix form, and the estimation error criterion (4.69) is to be minimised. The optimal estimates of the outputs may then be computed from the solution of the following spectral factor and diophantine equations.

**Spectral factors:**
The strictly Schur control spectral factor $D_f \epsilon P^{r \times r}(z^{-1})$, satisfies:

$$D_f D_f^* = DD^* + AR_f A^* \qquad (4.70)$$

**Diophantine equations:**
The smallest degree solution $(G_{f0}, H_{f0}, F_{f0})$, in terms of the polynomial matrix $F_{f0} \epsilon P^{r \times r}(z^{-1})$, satisfies:

$$G_{f0} D_f^* z^{-g_0} + A_h F_{f0} = D_h DD^* z^{-g_0} \qquad (4.71)$$

$$H_{f0} D_f^* z^{-g_0} - F_{f0} = D_{h0} R_f A^* z^{-g_0} \qquad (4.72)$$

where $g_0$ is the smallest positive integer chosen to ensure (4.71) and (4.72) are polynomial equations in $z^{-1}$, and the left-coprime factors $A_h, D_h$ satisfy:

$$A_h^{-1} D_h = D_{h0} A^{-1} \qquad (4.73)$$

**Optimal estimate and filter gain:**

$$\hat{y}(t) = A^{-1}(K_f(z(t) - \hat{y}(t)) + Bu(t)) = (A + K_f)^{-1} K_f z(t) + Bu(t))$$

and

$$K_f = D_h^{-1} G_{f0} H_{f0}^{-1} D_{h0} \qquad (4.74)$$

**Proof :** By collecting the results established in the following section.

**Lemma 4.3 :** *Properties of the Discrete Observer Based Output Estimator*

The properties of the pseudo state-estimator, in observer form, are summarised below. The so called *implied equation* and the *return difference* matrix, that determine the degree of stability of the system, are respectively:

$$D_h^{-1}G_{f0} + AD_{h0}^{-1}H_{f0} = D_f \tag{4.75}$$

and

$$I + A^{-1}K_f = A^{-1}D_fH_{f0}^{-1}D_{h0}$$

The *minimum value* of the weighted cost function integrand (4.69) can be computed from the functions $T_{2f}$ and $T_{3f}$ as:

$$I_{f\min} = trace\{T_{2f}^*T_{2f} + T_{3f}\} \tag{4.76}$$

where

$$T_{2f} = F_{f0}D_f^{*-1}z^{g_0}$$

$$T_{3f} = D_{h0}A^{-1}DD^*(I - D_fD_f^*)^{-1}DD^*)A^{*-1}D_{h0}^*$$

The optimal *sensitivity* and *filter sensitivity* functions are given, respectively, as:

$$S_f = (I_r + A^{-1}K_f)^{-1} = D_{h0}^{-1}H_{f0}D_f^{-1}A$$

and

$$M_f = K_fS_f = D_h^{-1}G_{f0}D_f^{-1}A \tag{4.77}$$

The following *orthogonality property* applies to the weighted output estimation error signal:

$$E\{\tilde{y}_h^T(t)\hat{y}_h(t)\} = 0 \tag{4.78}$$

and the spectrum of the *weighted pseudo state* $y = \tilde{y}_h - \hat{y}_h$ can be written as:

$$\Phi_{y_yy_y} = \Phi_{\tilde{y}_h\tilde{y}_h} + \Phi_{\hat{y}_h\hat{y}_h}$$

**Proof :** By collecting results in the following solution.

**Remarks**

The optimal observer properties have some similarities with Kalman filter properties but there are also distinct differences. Notice that, in Fig. 4.17, the signal $\{\varepsilon(t)\}$ is white noise but this is not true of the signal $(z - \hat{y})$ which would represent the *innovations signal* in the Kalman filter.

**Solution of the Discrete-time Optimal Output Observer Problem**

The solution of the optimal output estimation problem (Grimble, 1994) will now be obtained for the filter in observer form. The expression for the estimation error may be obtained from the following expressions for the plant and estimator, shown in Fig. 4.17:

**System output:**

$$y = A^{-1}(D\xi + Bu)$$

**Estimated output:**

$$\widehat{y} = A^{-1}(K_f(z - \widehat{y}) + Bu) \tag{4.79}$$

**Observations signal:**

$$z = A^{-1}Bu + A^{-1}D\xi + v$$

Substituting from the above results, the output estimation error:

$$\widetilde{y} = y - \widehat{y} = A^{-1}(D\xi + Bu) - A^{-1}(K_f(z - \widehat{y}) + Bu)$$

$$= A^{-1}D\xi - A^{-1}K_f(z - \widehat{y}) = (I_r + A^{-1}K_f)^{-1}(A^{-1}D\xi - A^{-1}K_f v) \tag{4.80}$$

**Sensitivity Functions for the Estimation Problem**

Let the *sensitivity functions* for the estimator be defined as:

$$S_f = (I_r + A^{-1}K_f)^{-1} \quad \text{and} \quad S_{f1} = (I_m + K_f A^{-1})^{-1}$$

and

$$M_f = (I_r + K_f A^{-1})^{-1}K_f \; = K_f(I_r + A^{-1}K_f)^{-1}$$

Then the filter sensitivities $M_f$ and $S_f$ satisfy: $A^{-1}M_f + S_f = I_r$ and the estimator gain also follows from (4.77) as:

$$K_f = M_f(I_r + A^{-1}M_f)^{-1}$$

These sensitivity functions play a similar role to the sensitivity $S_c$ and the control sensitivity $M_c$ functions in the previous control problem.

Using (4.79) and (4.80), and the above definitions for the estimator sensitivity functions, obtain the *estimate* and *estimation error* as:

$$\widehat{y} = S_f A^{-1}K_f(A^{-1}D\xi + v) + A^{-1}Bu = A^{-1}M_f \,(A^{-1}D\xi + v) + A^{-1}Bu \tag{4.81}$$

$$\widetilde{y} = S_f A^{-1}D\xi - A^{-1}M_f \, v = (I_r - A^{-1}M_f)A^{-1}D\xi - A^{-1}M_f v$$

$$= A^{-1}D\xi - A^{-1}M_f(A^{-1}D\xi + v) \tag{4.82}$$

and the optimisation can proceed in terms of the function $M_f$.

Let the signal: $\varepsilon_f = A^{-1}D\xi + v$ and denote the spectrum of this combined signal and noise as: $\Phi_{\varepsilon_f \varepsilon_f}$ . Recall that the process noise source sequence $\{\xi(t)\}$ is assumed to be of zero mean and identity covariance and the measurement noise source $v\{t\}$ is assumed to be zero mean and of covariance matrix $R_f$.

**Estimator Spectral Factors**

The estimator spectral factor $Y_f = A^{-1}D_f$ may then be introduced to satisfy:

$$Y_f Y_f^* = \Phi_{\varepsilon_f \varepsilon_f} = A^{-1}D_f D_f^* A^{*-1} \quad \text{where} \quad D_f D_f^* = DD^* + AR_f A^*$$

**Completing the Squares**

The spectrum of the weighted output estimation error signal $\tilde{y}_h = D_{ho}\tilde{y} = D_{ho}(y - \hat{y})$ can now be expressed, from (4.82), using the definition of $Y_f$ , as:

$$\Phi_{\tilde{y}_h \tilde{y}_h} = D_{ho}A^{-1}(DD^* + M_f Y_f Y_f^* M_f^* - DD^* A^{*-1}M_f^* - M_f A^{-1}DD^*)A^{*-1}D_{h0}^*$$

Following the *completing the squares* argument used previously, obtain :

$$\Phi_{\tilde{y}_h \tilde{y}_h} = (D_{ho}A^{-1}M_f Y_f - D_{h0}A^{-1}DD^* A^{*-1}Y_f^{*-1})$$

$$\times (Y_f^* M_f^* A^{*-1}D_{h0}^* - Y_f^{-1}A^{-1}DD^* A^{*-1}D_{h0}^*)$$

$$+ D_{ho}A^{-1}(DD^* - DD^* A^{*-1}Y_f^{*-1}Y_f^{-1}A^{-1}DD^*)A^{*-1}D_{h0}^*$$

This equation may be written in a more convenient form by introducing the left coprime factors $A_h, D_h$ where,

$$A_h^{-1}D_h = D_{ho}A^{-1}$$

Thence, the *estimation error spectrum* which enters the cost integrand, becomes:

$$\Phi_{\tilde{y}_h \tilde{y}_h} = (D_{ho}A^{-1}M_f Y_f - A_h^{-1}D_h DD^* D_f^{*-1})$$

$$\times (Y_f^* M_f^* A^{*-1}D_{h0}^* - D_f^{-1}DD^* D_h^* A_h^{*-1})$$

$$+ D_{ho}A^{-1}(DD^* - DD^* D_f^{*-1}D_f^{-1}DD^*)A^{*-1}D_{h0}^* \tag{4.83}$$

**Diophantine Equations**

Introduce the following filter diophantine equations in terms of the solution polynomial matrices $(G_{f0}, H_{f0}, F_{f0})$, where $F_{f0}$ is of smallest degree,

$$G_{f0}D_f^* z^{-g_0} + A_h F_{f0} = D_h DD^* z^{-g_0}$$

$$H_{f0}D_f^*z^{-g_0} - F_{f0} = D_{h0}R_fA^*z^{-g_0}$$

where $g_0$ is the smallest positive integer that ensures these equations are polynomials in $z^{-1}$.

**Implied Diophantine Equation**

The filter implied equation now follows, by appropriately multiplying and adding (4.71) and (4.72), to obtain:

$$D_h^{-1}(G_{f0}D_f^* + A_hF_{f0}) + AD_{h0}^{-1}(H_{f0}D_f^* - F_{f0}) = D_fD_f^*$$

Thence, obtain the implied equation for the estimator as:

$$D_h^{-1}G_{f0} + AD_{h0}^{-1}H_{f0} = D_f$$

This equation determines the stability of the output estimator in its observer form.

**Cost Function Simplification**

Using equation (4.83), the squared term in the weighted cost integrand expression (4.69), becomes:

$$(D_{h0}A^{-1}M_fY_f - D_{h0}A^{-1}DD^*A^{*-1}Y_f^{*-1}) = (D_{h0}A^{-1}M_fY_f - A_h^{-1}D_hDD^*D_f^{*-1})$$

$$= D_{h0}A^{-1}M_fY_f - A_h^{-1}G_{f0} - F_{f0}D_f^{*-1}z^{g_0} = T_{1f} - T_{2f} \qquad (4.84)$$

where these cost terms are defined as:

$$T_{1f} = D_{h0}A^{-1}M_fY_f - A_h^{-1}G_{f0} \quad \text{and} \quad T_{2f} = F_{f0}D_f^{*-1}z^{g_0}$$

The term $T_{3f}$ may also be defined, which is independent of the filter gain:

$$T = D_{h0}A^{-1}DD^*A^{*-1}(I_n - Y_f^{*-1}Y_f^{-1}A^{-1}DD^*A^{*-1})D_{h0}^*$$

$$= D_{h0}A^{-1}(DD^* - DD^*(D_fD_f^*)^{-1}DD^*)A^{*-1}D_{h0}^*$$

Observe that the term $T_{1f}$ can be simplified, using the implied equation (4.75), as:

$$T_{1f} = D_{h0}(I_r + A^{-1}K_f)^{-1}[A^{-1}K_fY_f - (I_r + A^{-1}K_f)D_{h0}^{-1}A_h^{-1}G_{f0}]$$

$$= D_{h0}(A + K_f)^{-1}[K_fD_{h0}^{-1}H_{f0} - D_h^{-1}G_{f0}]$$

The location of the poles of the $T_{1f}$ and $T_{2f}$ cost terms may now be considered. Note that the spectral factor $D_f$ is assumed to be strictly Schur, and the estimator must necessarily be asymptotically stable, since the minimum cost is assumed to be finite. Thence, the term $T_{1f}$ is asymptotically stable, and the term $T_{2f}$ is strictly unstable.

**Solution of the Optimisation Problem**

The cost integrand (4.83) may now be used to obtain:

$$J_c = \frac{1}{2\pi j} \oint_{|z|=1} (trace\{T_{1f}^* T_{1f} + T_{2f}^* T_{2f} - T_{1f}^* T_{2f} - T_{2f}^* T_{1f}\} + trace\{T_{3c}\}) \frac{dz}{z}$$

and by assumption $J_f < \infty$. As in the previous case, the integral of the term $T_{1f} T_{2f}^*$ which is analytic outside and on the unit-circle is zero by the residue theorem. It follows that the integral of the term $T_{2f} T_{1f}^*$ is also null, and the criterion is therefore minimised by setting the term $T_{1f}$ to zero. Thence, obtain:

**Filter gain:**

$$K_f = D_h^{-1} G_{f0} H_{f0}^{-1} D_{h0} \tag{4.85}$$

**Filter sensitivity:**

$$M_f = K_f (I_r + A^{-1} K_f)^{-1} = D_h^{-1} G_{f0} D_f^{-1} A$$

This last expression can also be obtained from the $T_{1f}$ term in (4.84).

**Return Difference Matrix and Stability**

The implied equation (4.75) may be written as:

$$A^{-1}(I_r + D_h^{-1} G_{f0} H_{f0}^{-1} D_{h0} A^{-1}) A = A^{-1} D_f H_{f0}^{-1} D_{h0}$$

or, using this result, the *return difference* matrix becomes:

$$F_f = (I_r + A^{-1} K_f) = A^{-1} D_f H_{f0}^{-1} D_{h0} \tag{4.86}$$

The left hand side of this equation represents the estimator return-difference matrix $F_f$, and since $D_f$ is strictly Schur, the asymptotic stability of the estimator is established. The sensitivity function:

$$S_f = (I_r + A^{-1} K_f)^{-1} = D_{h0}^{-1} H_{f0} D_f^{-1} A$$

From the return difference equation (4.86), and the spectral factor relationship for $Y_f$, obtain:

$$F_f D_{h0}^{-1} H_{f0} H_{f0}^* D_{h0}^{*-1} F_f^* = R_f + A^{-1} D D^* A^{*-1}$$

The second group of terms on the right hand side of this equation are positive semi-definite on the unit-circle of the z-plane, and since these terms are also strictly proper:

$$|\det(F_f)|^2 \geq |\det(R_f)| / |\det(H_{f0})|^2$$

The eigenvalue loci of the estimator return-difference matrix must lie outside the region defined by this inequality.

**Estimator White Noise Property**

The stochastic signal at the input to block $A^{-1}$ in the estimator becomes:

$$\varepsilon_0 = M_f(v + A^{-1} D\xi) = (I_r + K_f A^{-1})^{-1} K_f(v + A^{-1} D\xi)$$

and

$$\varepsilon_0 = D_h^{-1}G_{f0}D_f^{-1}(Av + D\xi) \tag{4.87}$$

and the spectrum becomes:

$$\Phi_{\varepsilon_0\varepsilon_0} = D_h^{-1}G_{f0}D_f^{-1}(AR_fA^* + DD^*)D_f^{*-1}G_{f0}^*D_h^{*-1} = D_h^{-1}G_{f0}G_{f0}^*D_h^{*-1} \tag{4.88}$$

If the gain matrix is expressed as $K_f = D_h^{-1}G_{f0}H_{f0}^{-1}D_{h0}$, then the signal input to $G_{f0}$ is white noise with unity power spectrum (see Fig. 4.17). This is important in the proof of the separation principle, since the estimator may be portrayed as being similar to the plant, but with inputs $Bu$ and $\varepsilon_0 = D_h^{-1}G_{f0}\varepsilon$, where the signal $\{\varepsilon(t)\}$ denotes white noise of identity covariance matrix.

## 4.3.4   Output Feedback Problem and Separation Principle

The solutions of the optimal noise-free output feedback control problem, and the output observer estimation problem, are utilised below to generate the desired *separation principle* result (Grimble and Johnson, 1988). This provides the desired solution to the output feedback control problem, where these outputs are corrupted by measurement noise.

The same cost-function employed earlier, (4.51) will be minimised, and for simplicity, the weightings $Q_c = H^*H$ and $R_c$ will be assumed to be constant matrices. Also assume that the estimator weighting $D_{ho}$ is defined to be equal to $H \epsilon R^{r\times r}$ ($D_{ho} = H$). The estimator is optimal, and hence the following orthogonality property is satisfied $\Phi_{y_h y_h} = \Phi_{\hat{y}_h \hat{y}_h} + \Phi_{\tilde{y}_h \tilde{y}_h}$; or

$$H\Phi_{yy}H^* = H(\Phi_{\hat{y}\hat{y}} + \Phi_{\tilde{y}\tilde{y}})H^*$$

The cost-function can therefore be written, using this result, as the sum of two terms:

$$J_c = J_{c0} + J_{c1}$$

where the first term (independent of the choice of control action):

$$J_{c0} = \frac{1}{2\pi j}\oint_{|z|=1} (trace\{H\Phi_{\tilde{y}\tilde{y}}H^*\})\frac{dz}{z} = \frac{1}{2\pi j}\oint_{|z|=1} (trace\{T_{2f}^*T_{2f} + T_{3f}\})\frac{dz}{z}$$

and the second term (dependent upon the optimal control) satisfies:

$$J_{c1} = \frac{1}{2\pi j}\oint_{|z|=1} (trace\{Q_c\Phi_{\hat{y}\hat{y}}\} + (trace\{R_c\Phi_{uu}\})\frac{dz}{z} \tag{4.89}$$

It is clear that the cost-index $J_c$ is optimised by minimising $J_{c1}$.

The first term in the criterion (4.89) represents the predicted output signal spectrum, and from (4.81):

$$\hat{y} = A^{-1}(K_f(y + v - \hat{y}) + Bu) = S_fA^{-1}(K_f(A^{-1}Bu + A^{-1}D\xi + v) + Bu)$$

$$= A^{-1}M_f(A^{-1}D\xi + v) + A^{-1}Bu = A^{-1}D_h^{-1}G_{f0}\varepsilon + A^{-1}Bu \qquad (4.90)$$

where the last step involves substitution from the white noise estimator property (4.88). This represents the output of a system with noise free measurable outputs $\widehat{y}$ and stochastic inputs $D_h^{-1}G_{f0}\varepsilon$, and control input $Bu$.

The problem to be solved is therefore equivalent to that in §4.3.3, but with the disturbance subsystem $A^{-1}D_h^{-1}G_{f0}\varepsilon$ replacing $A^{-1}D\xi$, where $\varepsilon$ represents white noise of identity covariance matrix (as was the case for $\xi$). Recall that, in this problem, $D_h$ is a real full rank constant matrix ($D_{h0} = H$). Thus $D_h^{-1}G_{f0}$ is an $r \times r$ polynomial matrix that can be taken to be full rank, since for usual full rank signal models, $G_{f0}$ must be full rank for the estimates to be unbiased (see the expression for $M_f$ obtained from (4.77)).

Thence, the problem of minimising the cost-function (4.51), with the behaviour of $\widehat{y}(t)$ described by (4.90), is a stochastic linear regulating problem, where the complete noise free output $\widehat{y}(t)$ can be measured. It follows from Theorem 4.2 that the optimal linear solution of this output feedback stochastic regulator problem is the linear control law:

$$u(t) = -K_c(z^{-1})\widehat{y}(t) = -K_c(A + K_f + BK_c)^{-1}K_f z(t)$$

where the control gain $K_c(z^{-1})$ is given by (4.57), and the filter gain $K_f(z^{-1})$ follows by substituting from (4.74). The desired separation principle solution is illustrated in Fig. 4.18, and is summarised in Theorem 4.4.

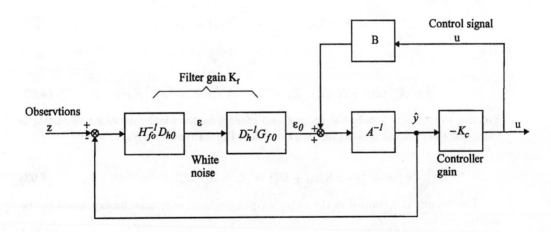

**Fig. 4.18** : Discrete Polynomial Matrix Implementation of $H_2$ Optimal Controller

The two cases for the system in Fig. 4.18 may be summarized as:

- *Case I : Solve the optimal observer problem with weighting $H$ and the ideal output control problem with disturbance polynomial matrix $D_h^{-1}G_{f0}$*

- *Case II : Solve the ideal output control problem with disturbance polynomial $D$ and the optimal observer problem with weighting $G_{c0}D_d^{-1}$*

**Theorem 4.4 :** *Output Feedback Control Problem and Separation Result I*

The solution of the optimal stochastic linear output feedback regulator problem, with constant pseudo-state and control weightings $Q_c = H^*H$ and $R_c$, where the measurements are corrupted by measurement noise, can be found from the solution of the stochastic optimal ideal output feedback regulator problem (Theorem 4.2), and the optimal observer problem (Theorem 4.2). The observer problem must be solved with the observer weighting : $D_{h0} = H$ and the noise free regulator problem must then be solved (with the same plant model) using the disturbance polynomial matrix $D = D_h^{-1}G_{f0}$. The optimal control signal can then be generated using:

$$u(t) = -K_c(z^{-1})\widehat{y}(t) \tag{4.91}$$

where $K_c$ is given by (4.57) and $\widehat{y}(t)$ is the output of the optimal observer shown in Fig. 4.17.

**Proof :** By collecting results preceding the theorem.

## 4.3.5   Alternative Separation Principle Result

Unlike the solution of the state-space LQG problem, there are two possible solutions of separation principle type, since the polynomial matrix case depends on which problem is solved first. In the present case, the noise free regulator problem will be assumed to be solved first, so that $K_c$ is available. If it is therefore assumed that the control signal is generated as $u = -K_c\widehat{y}$, then the system and estimator equations will have the form:

$$y = A^{-1}(Bu + D\xi) = A^{-1}(-BK_c\widehat{y} + D\xi)$$

and

$$\widehat{y} = A^{-1}(Bu + K_f(z - \widehat{y})) = (I_r + A^{-1}BK_c)^{-1}A^{-1}K_f(z - \widehat{y}) \tag{4.92}$$

Let the subscript 0 denote the output and control in the ideal case when the noise is zero and the optimal system (as in §4.3.2) has the form $u_0 = -K_c y_0$, where:

$$y_0 = A^{-1}(-BK_c y_0 + D\xi) = (I_r + A^{-1}BK_c)^{-1}A^{-1}D\xi \tag{4.93}$$

The output and control in the noisy output problem may now be expressed in the form:

$$y = A^{-1}(-BK_c(\widehat{y} - y_0)) + A^{-1}(-BK_c y_0 + D\xi)$$

$$u = -K_c(\widehat{y} - y_0) - K_c y_0$$

In terms of the error $(H)$ and control $(H_r)$ weightings $(Q_c = H^*H$ and $R_c = H_r^*H_r)$, the signals in the cost function (4.51) may be written in the vector form:

$$p = p_0 + p_1 = \begin{bmatrix} HA^{-1}(-BK_c y_0 + D\xi) \\ -H_r K_c y_0 \end{bmatrix} + \begin{bmatrix} HA^{-1}B \\ H_r \end{bmatrix} K_c \widetilde{y}_0$$

where $\widetilde{y}_0 = y_0 - \widehat{y}$ denotes the estimation error, calculated as the difference between the system output with ideal control $u_0$, and the estimator output utilizing the proposed control $u = -K_c\widehat{y}$. From the above results, (4.92) and (4.93), note that,

$$\widetilde{y}_0 = (I_r + A^{-1}BK_c)^{-1}A^{-1}(D\xi - K_f(z - \widehat{y}))$$

and

$$K_c\widetilde{y}_0 = M_c(A^{-1}(D\xi - K_f(z - \widehat{y})) = M_c\widetilde{y} \tag{4.94}$$

where $\widetilde{y} = y - \widehat{y}$ is the estimation error for the actual noisy system.

The optimal cost index for this case may be written using the above vector equation, as :

$$J = E\{p^T(t)p(t)\} = E\{(p_0(t) + p_1(t))^T(p_0(t) + p_1(t))\} \tag{4.95}$$

Observe that the first term:

$$J_0 = E\{p_0^T(t)p_0(t)\}$$

represents the cost-function for an *ideal output* feedback control problem, and it is a minimum if $K_c$ is defined as in Theorem 4.2.

The second term in the cost expression (4.95):

$$J_1 = E\{p_1^T(t)p_1(t)\}$$

is dependent upon the estimation error $\widetilde{y}_0$, and represents the difference between the absolute minimum cost (using optimal control $-K_c y_0(t)$) and the actual cost (using the constrained control $-K_c\widehat{y}(t)$). It follows from the optimality of $J_0$ that the vectors $p_0$ and $p_1$ are orthogonal. The criterion may therefore be written as:

$$J = E\{p_0^T(t)p_0(t)\} + E\{(p_1^T(t)p_1(t)\}$$

The term $J_0$ is independent of the choice of the estimator, and to minimise this criterion the second term must therefore be minimised:

$$J_1 = E\{p_1^T(t)p_1(t)\}$$

$$= \frac{1}{2\pi j} \oint_{|z|=1} (trace\{(H_r^* H_r + B^* A^{*-1} H^* H A^{-1} B)K_c\Phi_{\widetilde{y}_0\widetilde{y}_0}K_c^*\})\frac{dz}{z}$$

$$= \frac{1}{2\pi j} \oint_{|z|=1} (trace\{Y_c^* Y_c K_c\Phi_{\widetilde{y}\widetilde{y}}K_c^*\})\frac{dz}{z}$$

but, from (4.94) $K_c\widetilde{y}_0 = M_c\widetilde{y}$, and substituting for $M_c$ from (4.60), obtain:

$$J_1 = \frac{1}{2\pi j} \oint_{|z|=1} (trace\{G_{c0}D_d^{-1}\Phi_{\widetilde{y}\widetilde{y}}G_{c0}^*\})\frac{dz}{z}$$

Assume for simplicity that $D_d$ is a constant matrix, then the cost-function $J$ is clearly minimised when the criterion $J_1$, is minimised. This is of course a weighted estimation error problem, where the weighting $D_{h0}$ is a polynomial matrix $D_{h0} = G_{c0}D_d^{-1}$. The alternative form of the separation principle result may now be summarised in Theorem 4.5 below.

**Theorem 4.5 :** *Output Feedback Control Problem and Separation Result II*
The solution of the optimal stochastic linear *output feedback* regulator problem, with cost weightings $Q_c = H^*H$ and $R_c = H_r^*H_r$, where the measurements are corrupted by measurement noise, can be found from the solution of the *ideal output feedback regulator problem* (Theorem 4.2) and the optimal observer problem (Theorem 4.3). The control problem must be solved for a constant disturbance model matrix $D$, and the optimal observer problem must then be solved with the estimator cost weighting $D_{h0} = G_{c0}D_d^{-1}$. The optimal control signal is obtained as:

$$u(t) = -K_c(z^{-1})\widehat{y}(t)$$

where $K_c$ is given by (4.57), and $\widehat{y}(t)$ denotes the output of the optimal observer shown in Fig. 4.17.

**Proof :** By collecting results preceding the theorem.

The Fig. 4.18 shows the optimal controller in its observer form. The main results for Cases I and II are summarised below this figure, and the symmetry of the results is clearly evident.

## 4.3.6    Closed-loop System Stability

Assume that the measurement noise is null for the present stability proof. Then the plant and estimator equations become:

$$y = A^{-1}D\xi + A^{-1}Bu = A^{-1}D\xi - A^{-1}BK_c\widehat{y}$$

$$\widehat{y} = A^{-1}(K_f(y - \widehat{y}) - BK_c\widehat{y})$$

If the estimation error is denoted as $\widetilde{y} = y - \widehat{y}$, then the equations may be written as:

$$y = A^{-1}D\xi - A^{-1}BK_c(y - \widetilde{y})$$

$$\widetilde{y} = -A^{-1}K_f\widetilde{y} + A^{-1}D\xi$$

thence,

$$\begin{bmatrix} y \\ \widetilde{y} \end{bmatrix} = \begin{bmatrix} -A^{-1}BK_c & A^{-1}BK_c \\ 0 & -A^{-1}K_f \end{bmatrix}\begin{bmatrix} y \\ \widetilde{y} \end{bmatrix} + \begin{bmatrix} I \\ I \end{bmatrix}A^{-1}D\xi$$

or

$$\begin{bmatrix} (I + A^{-1}BK_c) & -A^{-1}BK_c \\ 0 & (I + A^{-1}K_f) \end{bmatrix}\begin{bmatrix} y \\ \widetilde{y} \end{bmatrix} + \begin{bmatrix} I \\ I \end{bmatrix}A^{-1}D\xi$$

The closed-loop system matrix is therefore upper block triangular, and hence the stability of the closed-loop system depends upon the separate characteristic equations for the ideal output feedback control and optimal observer problems, $(I + A^{-1}BK_c = 0$ and $(I + A^{-1}K_f) = 0$, respectively). The closed-loop stability then follows from the *strictly Schur* properties of the spectral-factors in equations (4.58) and (4.75), respectively.

**Alternative Structures**

A more general, but more complicated, result occurs if the estimator is assumed to provide estimates of so called pseudo-states, rather than outputs. A so called pseudo-state is a set of variables, which can be states or inferred outputs, that are to be estimated. A pseudo-state estimator structure is similar to a Kalman filter. The continuous-time version of this case is explored in Grimble (1999), and it has the advantage that the polynomial equivalent of the usual state estimate feedback control problem can be considered. However, it loses the structural simplicity of the case considered here, where in effect the pseudo states are equivalent to outputs. This is the simplest problem to consider.

Note that the optimal structure, which involves a type of separation result, need not be utilised in full. It would be possible to replace the optimal controller $K_c$ with one designed using QFT (see Chapter 9). This would provide a robust solution (via the QFT solution) and good noise rejection properties through the use of the optimal observer.

Note that the structure assumed for the observer does not result in sub-optimality relative to the usual Kalman filtering state-space solution. Say, for example, that the optimal controller $C_0(z^{-1})$ is obtained by Wiener or Kalman means, and then consider the equivalence:

$$K_c(A + K_f + BK_c)^{-1}K_f = C_0$$

If the polynomial estimator gain is chosen, then after simple manipulation :

$$K_c = C_0(K_f - BC_0)^{-1}(A + K_f)$$

and under the appropriate conditions on the inverse, there exists a dynamic gain $K_c$ to give the same optimal controller transfer-function. Thus, the assumption that is inherent in the assumed structure for the observer should not result in performance deterioration relative to the usual LQG solutions.

## 4.3.7  Flight Control Example

A simple flight control problem will now be considered, involving the pitch control for a military aircraft. This example will illustrate the type of results obtained from the algorithm. The system description now follows:

**Sample time :**

$$T_s = 0.2 \text{ seconds}$$

**Plant model:**

$$W = \frac{B}{A} = \frac{0.01z^{-1}(0.401 + 0.59z^{-1} + 0.025z^{-2})}{(1 - 0.0060244z^{-1})(1 - 0.49797z^{-1})(1 - 0.998z^{-1})}$$

**Disturbance model :**

$$W_d = \frac{D}{A}$$

$$= \frac{0.1175(1 - 1.679656z^{-1} + 0.725549z^{-2})(1 - 0.496219z^{-1})(1 - 0.0060244z^{-1})}{(1 - 0.0060244z^{-1})(1 - 0.49797z^{-1})(1 - 0.998z^{-1})}$$

**Weightings:**

$$H = 10, \quad Q_c = H^*H \quad \text{and} \quad H_r = 1, \quad R_c = H_r^*H_r = 1$$

**Measurement noise :**

$$R_f = 0.1$$

**Computed Control and Observer Polynomials**

$$D_c(z^{-1}) = 1.092674(1 - 0.8066577z^{-1})(1 - 0.516139z^{-1})1 - 6.02296 \times 10^{-3}z^{-1})$$

$$D_f(z^{-1}) = 10.3478048(1 - 0.9828991z^{-1})(1 - 0.4980659z^{-1})$$

$$\times (1 - 8.392025 \times 10^{-2}z^{-1})(1 - 0.0060244z^{-1})$$

$$G_{c0}(z^{-1}) = 4.749469 \times 10^{-2}(1 - 0.5035845z^{-1} + 2.9975 \times 10^{-3}z^{-2})$$

$$F_{c0}(z^{-1}) = 2.492673 \times 10^{-4}(1 - 76.06039z^{-1} + 208.4406z^{-2})$$

$$H_{c0}(z^{-1}) = 1.092674(1 - 1368988 \times 10^{-2}z^{-1})(1 - 0.3480339z^{-1})$$

$$G_{f0}(z^{-1}) = 4.765199 \times 10^{-2}(1 - 0.4983207z^{-1} + 2.962688 \times 10^{-3}z^{-2})$$

$$F_{f0}(z^{-1}) = 2.952951 \times 10^{-4}(1 - 170.8778z^{-1} + 819.8825z^{-2}$$

$$-1471.738z^{-3} + 1185.931z^{-4} - 338.6443z^{-5})$$

$$H_{f0}(z^{-1}) = 3.430396(1 - 8.381467 \times 10^{-2}z^{-1})$$

**Plant Frequency Response**

   The magnitude and phase frequency responses for the plant and disturbance models are shown in Fig. 4.19. Note that these frequency responses are basically low-pass characteristics up to a point where the foldover frequency occurs (15.71 radians per second). The computed controller for the ideal outputs, zero measurement noise case has the gain transfer function:

$$K_c = \frac{0.41032(1 - 0.4964138z^{-1})}{(1 - 0.4944437z^{-1})(1 - 1.509903z^{-1} + 0.6100211z^{-2})}$$

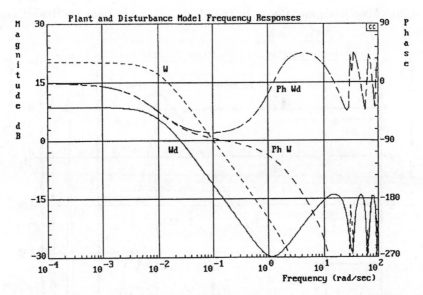

**Fig. 4.19** : Plant and Disturbance Model Frequency Responses, Magnitude and Phase Diagrams

### Ideal Output Feedback System Response

The frequency response for this controller is shown in Fig. 4.20. Note that the response, which approximates a constant gain and a high frequency lead term, is unrealistic, but then measurement noise has been neglected at this point. The resulting sensitivity function frequency response and the closed-loop transfer-function responses are shown in Fig. 4.21, and these indicate that good results will be obtained for both disturbance rejection and tracking.

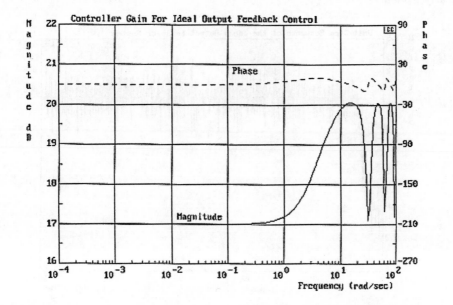

**Fig. 4.20** : Magnitude and Phase for the Ideal Output Optimal Feedback Controller

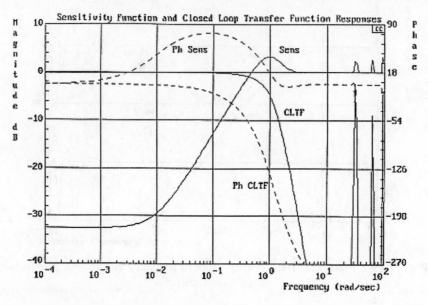

**Fig. 4.21** : Sensitivity and Closed-loop Transfer Function
Frequency Responses

If the output can be measured without noise, the unit step-response for the system is as shown in Fig. 4.22. Note that the unit step is not injected into the system until after 4 seconds. Measurement noise is not included, but the effect of the disturbance is clearly evident.

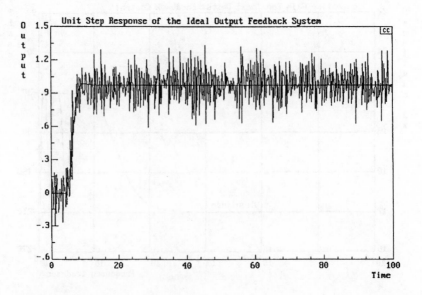

**Fig. 4.22** : Unit Step Responses for Ideal Output System With and Without
Disturbance Signal Component

### Estimator Frequency and Time Responses

The discrete-time transfer-function for the optimal observer gain, and the total
transfer-function for the estimator, are obtained as follows:

**Optimal observer gain:**

$$K_f = \frac{0.01389(1 - 0.49832z^{-1} + 0.0029627z^{-2})}{(1 - 0.08381467z^{-1})}$$

**Optimal observer transfer function:**

$$H_f = \frac{0.0137(1 - 0.4923z^{-1})}{(1 - 0.08392z^{-1})(1 - 0.49807z^{-1})(1 - 0.982899z^{-1})}$$

The signal and noise time responses, when only the estimation problem is considered, are shown in Fig. 4.23. Note that the noise has a much larger variance than the
signal variance. The signal and its estimates in the presence of measurement noise
are shown in Fig. 4.24. The estimator acts, as expected, as a low pass filter in this
high noise case. The frequency responses of the optimal observer are shown in Fig.
4.25. The observer has a low pass characteristic that only changes near the foldover
frequency.

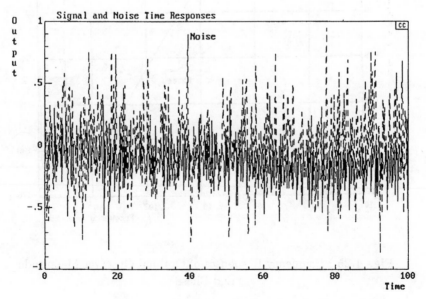

**Fig. 4.23** : Time Responses for the Signal to be Estimated
and Noise Signal

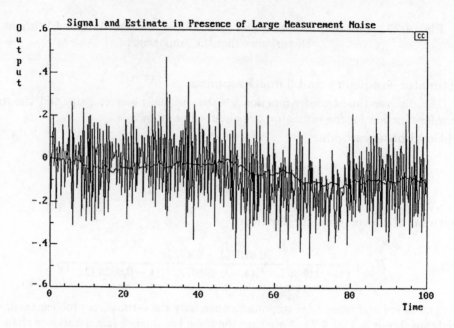

**Fig. 4.24** : Signal and Estimated Signal in the Presence of
Measurement Noise

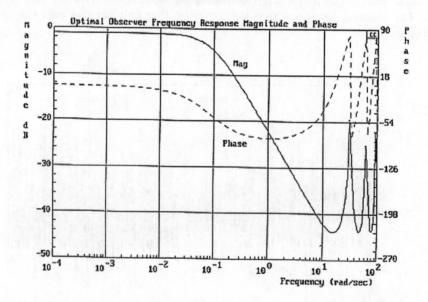

**Fig. 4.25** : Frequency Responses of Optimal Observer Magnitude
and Phase

**Response in the Presence of Measurement Noise**

The case where measurement noise is included and only noisy outputs are available for feedback will now be considered. By application of the separation principle results, the computed controller gain, observer gain, and observer transfer functions are respectively:

**Control Gain:**

$$K_c = \frac{9.1216(1 - 0.0060244z^{-1})(1 - 0.49756z^{-1})}{(1 - 0.01368988z^{-1})(1 - 0.3480339z^{-1})}$$

**Observer Gain:**

$$K_f = \frac{0.01389(1 - 0.49832z^{-1} + 0.0029627z^{-2})}{(1 - 0.08381467z^{-1})}$$

**Filter Transfer:**

$$H_f = \frac{0.0137(1 - 0.006018z^{-1})(1 - 0.4923z^{-1})}{(1 - 0.006025z^{-1})(1 - 0.08392z^{-1})(1 - 0.498066z^{-1})(1 - 0.982899z^{-1})}$$

**Controller Transfer:**

$$C_0 = \frac{0.1249735(1 - 0.49756z^{-1})}{(1 - 0.08388919z^{-1})(1 - 0.5213z^{-1})(1 - 0.7870235z^{-1})}$$

**Output Feedback System Frequency and Time Responses**

The optimal controller frequency response is shown in Fig. 4.26. This clearly has a dominantly low pass characteristic. The individual responses of the closed loop system to the step change in the reference signal ($t = 2$ seconds), and to the measurement noise and disturbances, are shown in Fig. 4.27.

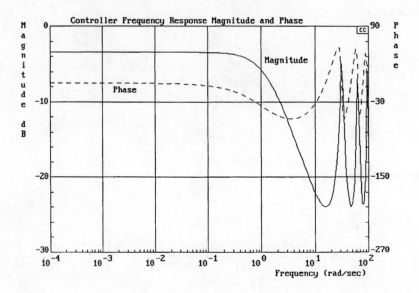

**Fig. 4.26** : Frequency Responses of the Output Feeedback Optimal Controller
Magnitude and Phase

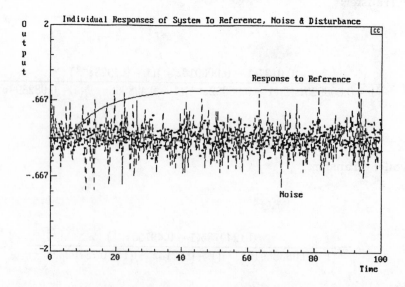

**Fig.4.27** : Individual Responses of the Closed-loop System due to Reference, Noise
and Disturbance Input Signals

The output of the optimal system is shown in Fig. 4.28. This figure also shows the response of the system, without stochastic inputs, due to the reference change only. The control signal corresponding to the unit step reference change is shown in Fig. 4.29, both for the cases where the stochastic signals are present and where they are absent.

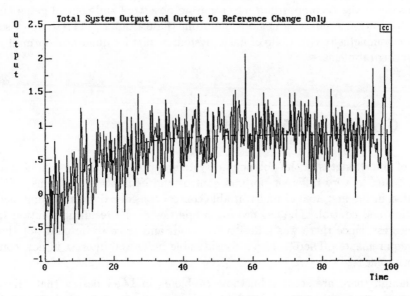

**Fig. 4.28** : Output of the Optimal Systems and Output When Only the Reference
Change is Present

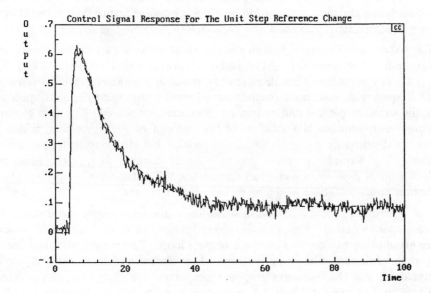

**Fig. 4.29** : Control Signal Time Responses When All Inputs Present and Only the
Reference Change Occurs

**Remarks**

The separation principle of stochastic optimal control theory is well known in terms of state-space system models. However, there has been no such result for systems represented in transfer-function matrix or polynomial systems form. In such cases, the frequency-domain solutions for the $H_2$ or LQG optimal control problems have not considered the decomposition into separate observers and control gains. Indeed, such a separation result was not obvious from these results. This type of solution therefore completes the portfolio of linear system results for quadratic optimal control and filtering problems.

## 4.4   Concluding Remarks

One of the strengths of the polynomial system approach is that a similar approach can be taken to very different optimal control and estimation problems. This was illustrated in the first area of multivariable control considered; that of combined fault estimation and control. The engineering properties of the resulting solution were of most interest, since there was a design trade-off, and price to be paid, for the combined requirements. There is also a considerable industrial interest in new condition monitoring ideas.

Although there are some robustness problems in LQG design that arise when plants are uncertain (Kwakernaak, 1983), it does have some valuable features, like the relationship to internal model control structures (Grimble et al. 1989). Robustness can also be improved (Grimble and Owens, 1986). Taken together with a combined fault detection algorithm, the solution provides a possible future fault tolerant control strategy (Patton, 1997; Jacobson and Nett, 1991).

The second novel concept was to require an observer structure be utilised in the construction of the controller. The rather surprising result was established, in the later parts of the chapter, that there are two possible separation type theorems, since in this frequency domain input/output model setting the result depends upon which order the separate control and estimation problems are solved. The first separation principle result requires the solution of the *optimal observer problem,* which then defines the disturbance model to be used in the optimal ideal output feedback control problem. The second separation principle result involves solving the *ideal output feedback control problem,* which then defines the weighting for the optimal observer estimation problem. These might be termed dual solutions.

The *separation principle* established has the same advantages and insights as for the state-space solution. For example, the estimator can be commissioned separately before attempting to close the feedback control loop. The results obtained here also have some additional interesting features. The control and filter gains $K_c$ and $K_f$ are dynamic, and the frequency response properties are clear from the polynomial equations in Theorems 4.2 and 4.3, respectively. The full implications for the use of the structures obtained still have to be determined. Obvious extensions, like the tracking problem solution (Grimble, 1995) may easily be derived.

## 4.5  References

Chen, J. and Patton, R.J., 1996, Robust fault detection and isolation systems, *Control and Dynamic Systems*, Vol. **74**, pp. 171-223.

Frank, P.M., 1994, Enhancement of robustness in observer-based fault detection, *Int. J. Control*, Vol. **59**, No. 4, pp. 955-981.

Grimble, M.J., 1982, Optimal control of linear uncertain multivariable system, *IEE Proc.*, Pt. D, Vol. **129**, No. 6, pp. 263-270.

Grimble, M.J., 1986, Dual criterion stochastic optimal control problem for robustness improvement, *IEEE Trans* AC-**31** pp. 1891-1895.

Grimble, M.J., 1998a, Combined fault monitoring detection and control, *IEEE CDC Conference*, Tampa, Florida.

Grimble, M.J., 1998b, Stochastic control of discrete systems: a separation principle approach, *IEEE CDC Conference*, Tampa, Florida.

Grimble, M.J., 1994, *Robust Industrial Control: Optimal design approach for polynomial systems*, Prentice Hall, Hemel Hempstead.

Grimble, M.J., 1995, Two-degree-of-freedom linear quadratic Gaussian predictive control, *IEE Proc., Control. Theory of Appl.* Vol. **142**, No. 4, pp. 295-306.

Grimble, M.J., 1999, A separation principle for continuous-time polynomial systems, *European Control Conference*, Karlsruhe.

Grimble, M.J. and Owens, T.J.. 1986, On improving the robustness of LQG regulators, *IEEE Trans Automatic Control* AC-**31** No. 1 pp. 54-55.

Grimble, M.J. and Johnson,M.A., 1988, *Optimal Control and Stochastic Estimation, Vols I and II*, John Wiley, Chichester.

Grimble, M.J. de la Salle, S. and Ho, D., 1989, Relationship between internal model control and LQG controller structure, *AUTOMATICA*, Vol. **25**, No. 1 pp. 41-53.

Jacobson, C.A., and Nett, C.N., 1991, An integrated approach to controls and diagnostics using the four parameter control, *IEEE Control Systems Magazine*, 22-29.

Joshi, S.M., 1986, Failure-accommodating control of large flexible space-craft, *American Control Conference*, Seattle, USA.

Kailath, T., 1980, *Linear systems*, Prentice Hall, Englewood Cliffs, NJ.

Kalman, R.E., 1961, New methods in Wiener filtering theory, *Proc. of the Symposium on Engineering Applications of Random Function Theory and Probability*, pp. 270-388.

Kučera, V., 1979, *Discrete linear control*, John Wiley and Sons, Chichester.

Kučera, V., 1978, Transfer-function solution of the Kalman Bucy filtering problem, *Kybernetika*, Vol. **14**, No. 2, pp. 110-121.

Kwakernaak, H., 1983, Robustness optimization of linear feedback systems, *22nd CDC*, San Antonio, Texas.

Kwakernaak, H., and Sivan, R., 1972, *Linear optimal control systems*, Wiley, New York.

Patton, R.J., 1997, Fault tolerant control : the 1997 situation, I*FAC Symposium, Safeprocess97*, pp. 1033-1055, University of Hull.

Patton, R.J. and Chen, J., 1997, Observer-based fault detection and isolation : robustness and applications, *IFAC Journal : Control Engineering Practice*, Vol. **5**, No. 5, pp. 671-682.

Patton, R.J., Frank, P.M. and Clark, R.N., (Eds), 1989, *Fault diagnosis in dynamic systems : theory and application*, Prentice Hall, UK.

Patton R.J., Frank, P.M. and Clark, R.N., (Eds), 1998, *Advances in fault diagnosis in dynamic systems*, Springer Verlag, London.

Robertson, D., and Lee, J.H., 1993, Integrated state estimation, fault detection and diagnosis for nonlinear systems, *Proc of ACC*, San Francisco, pp. 389-392.

Shaked, U., 1976, A general transfer-function approach to the steady-state linear quadratic Gaussian stochastic control problem, *Int. J. Control,* **24**, 6, pp. 771-800.

Stoustrup, J., Grimble, M.J. and Niemann, H., 1997, Design of integrated systems for the control and detection of actuator/sensor faults, *Sensor Review*, Vol. **17**, No. 2, pp. 138-149.

Veillette, R.J., 1995, Reliable linear-quadratic state feedback control, *Automatica*, Vol. **131**, No. 1, pp. 137-143.

Veillette, R.J., Medanic, J.V. and Perkins, W.R., 1992, Design of reliable control systems, *IEEE Trans. Automatic Control*, Vol. **37**, No. 3, pp. 290-304.

Wiener, N., 1949, *Extrapolation, interpolation and smoothing of stationary time series, with engineering applications*, New York : Technology Press and Wiley (Oringally issued in February 1942 as a classified nat. Defence Res. Council Report).

Youla, D.C., Jabr, H.A. and Bongiorno, J.J., 1976, Modern Wiener-Hopf design of optimal controllers - Part II : the multivariable case, *IEEE Trans. on Aut. Contr.*, Vol. AC-**21**, No. 3, pp. 319-338.

# 5

# $H_\infty$ Optimal Control Laws

## 5.1  Introduction

The $H_\infty$ control synthesis problem was first proposed by George Zames (1981) and was aimed at systems containing uncertainty. His main thesis was that the linear quadratic optimal control design methods were based upon accurate system and disturbance models and that the mathematical framework for this problem did not allow plant uncertainties to be taken into consideration rigorously. The mathematical setting introduced by Zames for the solution of robust control and filtering problems was more appropriate for the design of robust systems (Zames, 1981; Zames and Francis, 1981). That is, he proposed the use of the Hardy space $H_\infty$ which in scalar systems has a norm that can be obtained from the peak value of the frequency response. Such a measure is valuable when assessing robustness and stability properties (Zhou et al.1996). For example, if the distance of the open-loop frequency response, from the critical point for stability is to be maxmised, then this can be achieved by minimisation of the $H_\infty$ norm of the sensitivity function.

Early research in $H_\infty$ design concentrated on the development of suitable state space (Doyle et al. 1989), or polynomial based, computational algorithms (Kwakernaak, 1985, 1986; Grimble, 1985, 1986, 1987c). That is, methods of synthesising and calculating $H_\infty$ controllers were of prime interest. More recently attention has turned to the design of $H_\infty$ optimal systems (Glover and McFarlane, 1989) and the industrial problems which might benefit from its use. Simple introductions to the $H_\infty$ robust design approach were given in Francis (1987) and in Grimble and Johnson (1991).

Some of the benefits of the $H_\infty$ design approach are as follows:

1. The synthesis problem has well defined stability and robustness properties and these can be determined once the system is specified.

2. Design iterations which enables trade offs to be achieved can easily be accomplished.

3. For certain given classes of uncertainty, robustness margins can be guaranteed and the steps in going from the uncertainty description to the optimal problem are straightforward.

4. Software is readily available in most commercial packages for calculating multivariable $H_\infty$ controllers and users do not need a high degree of skill, at least to calculate a stabilising solution

5. The links to LQG control solutions often enable stochastic properties to be optimised in addition to robustness, (Fragopoulos et al. 1991 and Saeki et al. 1987).

There are also some disadvantages with $H_\infty$ design, although these do not outweigh the advantages:

1. Robust solutions may not give adequate transient response, or other properties and hence robustness requirements may have to be relaxed. The reason for using the $H_\infty$ approach is therefore less obvious in this case.

2. Although it is not necessary to understand the theory to be able to use the commercial packages, it is a daunting prospect to try to understand the underlying theory without the help of formal courses.

3. Although $H_\infty$ problems are theoretically tractable it is not clear that the physical robustness requirements match the theoretical problems posed.

4. The maximisation of stability and robustness margins is a rather more complicated problem than the $H_\infty$ sensitivity minimisation design problem suggests.

The $H_\infty$ design approach has the advantage that it is relatively easy to use to obtain a reasonable initial design. There is no need for control designers to fully understand the numerical algorithms, but they should understand the design process which involves the cost weighting definition. There are practical limitations and disadvantages, including the fact that high-order controllers are often obtained, whereas the classical control paradigm utilises low order controllers. However, $H_\infty$ controllers can be model reduced and the results obtained from many researchers have been so attractive that this approach is destined to become one of the options regularly taken by the control designer. The polynomial systems approach (Grimble and Kucera, 1996) to $H_\infty$ design is introduced in this chapter, and the state-space approach is discussed later in Chapter 7.

The chapter begins with the SISO $H_\infty$ optimal control problem, which includes integral action. This parallels the work of Chapter 2. The SIMO optimal control problem is then considered and both $H_2$ and $H_\infty$ results are obtained. The $H_\infty$ equivalent of the LQGPC control law, introduced in Chapter 3, is then discussed. For each of the $H_\infty$ problems the solution is shown to be a special case of a weighted $H_2$ optimisation problem.

# 5.2  Scalar $H_\infty$ Feedback/Feedforward Control

The solution of the scalar $H_\infty$ optimal control problems are derived in this section. This particular solution involves feedback, tracking and feedforward control action. The solution is obtained by constructing a ficticious or auxiliary LQG problem that has the desired solution for the $H_\infty$ problem.    The results of Chapter 2 can then be applied to obtain the optimal control, for the system described in §2.4 and shown in Fig. 2.17.   The solution of this $H_\infty$ problem is given in Grimble (1999a) and the results given here will apply to the 1, 2, $2\frac{1}{2}$, 3 and $3\frac{1}{2}$ DOF $H_\infty$ design problems, depending on the choice of the parameters in Table 2.1.

A stochastic approach will be taken to motivate the $H_\infty$ problem, since this modelling philosophy has been very successful in applications. The independence of the noise sources enables the separate contributions of the feedback, tracking and feedforward controllers to be determined. The optimal control problem posed enables the tracking error and control terms for each of these components to be minimised in either an $H_2$ or an $H_\infty$ sense (Mosca et al. 1990). Thus, for example, the cost-function can include an $H_\infty$ criterion for the feedback controller design and $H_2$ cost terms for the tracking and feedforward components. Any of the alternative $H_2/H_\infty$ permutations will also be possible.

### 5.2.1   LQG Embedding Principle

It will be shown that a particular $H_2$ or LQG optimal control cost-index can be minimised that actually provides the solution to an $H_\infty$ minimisation problem. That is, a fictitious LQG problem that involves a dynamic weighting function can be solved to obtain the solution of an $H_\infty$ optimisation problem. This approach was developed by Grimble (1986) and is referred to as LQG embedding and utilises a key Lemma due to Kwakernaak (1985).

The components $J_0, J_1$ and $J_2$ in the $H_2$ cost problem in Chapter 2, correspond to the functions of the feedback, tracking and feedforward controllers. These depend upon the dynamic weightings $\Sigma_0$, $\Sigma_1$ and $\Sigma_1$, respectively.    If the $i$ th controller is to be $H_\infty$ optimal this can be achieved by appropriate choice of the weighting $\Sigma_i$. In fact the strategy will be to compute the particular $\Sigma_i$ that ensures $H_\infty$ optimality and to then simply invoke Theorem 2.4. This LQG embedding technique (Grimble, 1986) involves solving an $H_\infty$ problem within a rather artificial $H_2$ cost problem. The general problem is considered below where a $H_2$ cost $J$ is formed from $p_1 + 1$ cost terms which are dependent upon $p_1+ 1$ control elements $C_j(z^{-1})$. The following lemma reveals how $p_1 - p_0 + 1$ of these controllers can also minimise a mixed $H_2$ and $H_\infty$ criterion.

**Lemma 5.1** : *LQG Embedding and Mixed $H_2/H_\infty$ Cost Minimisation*

Consider the problem of choosing compensators $C_j(z^{-1})$ for $j = 0,...,p_1$ which minimize the $H_2$ cost-function:

$$J = \sum_{j=0}^{p_1} J_j = \sum_{j=0}^{p_1} \{ \frac{1}{2\pi j} \oint_{|z|=1} \{I_j(z^{-1})\Sigma_j(z^{-1})\} \frac{dz}{z} \} \tag{5.1}$$

Suppose that for some real rational functions: $\Sigma_i(z^{-1}) = \Sigma_i^*(z^{-1}) > 0$ , $i = p_0,...,p_1$ the criterion $J$ is minimised by functions $I_j^o(z^{-1})$ , for which $p_0 - p_1 + 1$ are constants $(I_i^o(z^{-1}) = \lambda_i^2$ , $i = p_0,...,p_1)$. Also assume that the minimum value of the $j$th cost term is dependent only upon the choice of $C_j$. Then the optimal compensators $C_j^0$ ,

$j = 0, ..., p_1$ also minimise:

$$J_{2\infty} = \sum_{j=0}^{p_0-1}\{\frac{1}{2\pi j}\oint_{|z|=1}\{I_j(z^{-1})\Sigma_j(z^{-1})\}\frac{dz}{z}\} + \sum_{j=p_0}^{p_1}\|I_j(z^{-1})\|_\infty$$

and the minimum value of the cost-function is given as:

$$J_{2\infty\,min} = \sum_{j=0}^{p_0-1} J_{j\,min} + \sum_{j=p_0}^{p_1}\lambda_j^2$$

**Proof of Lemma 5.1**

A proof by contradiction will be presented. Let $z = e^{j\omega}$, then the cost integral J may be written as:

$$J = \sum_{j=0}^{p_1}\{\frac{1}{2\pi}\int_{-\pi}^{\pi}\{I_j(e^{-j\omega})\Sigma_j(e^{-j\omega})\}\ d\omega\} = \frac{1}{2\pi}\int_{-\pi}^{\pi} X(e^{-j\omega})\ d\omega \qquad (5.2)$$

where the integrand:

$$X(z^{-1}) = \sum_{j=0}^{p_1} I_j(z^{-1})\Sigma_j(z^{-1})$$

Let $X^o(z^{-1})$ denote the function at the LQG optimum of (5.1). Suppose now there exists a set of compensatorsm $C_i$, such that $X(e^{-j\omega})$ has the property, that for some $i$ ( $p_0 \leq i \leq p_1$), the compensator $C_i$ ensures:

$$\sup_\omega\{I_i(e^{-j\omega})\} < \sup_\omega\{I_i^0(e^{-j\omega})\} = \lambda_i^2$$

Then necessarily, $I_i(e^{-j\omega}) < \lambda_i^2 = I_i^0(e^{-j\omega})$ , for all $\omega \in [-\pi,\ \pi]$ , and since by assumption the choice of $C_i$ does not affect the minimum of the cost terms, for $j \neq i$ , then

$$\sum_{j=0}^{p_1}\int_{-\pi}^{\pi}\{I_j(e^{-j\omega})\Sigma_j(e^{-j\omega})\}\ d\omega < \sum_{j=0}^{p_1}\int_{-\pi}^{\pi}\{I_j^0(e^{-j\omega})\Sigma_j(e^{-j\omega})\}\ d\omega$$

This contradicts the assumption that $X^0(e^{-j\omega})$ minimises (5.2), and the result follows.

## 5.2.2   Mixed $H_2$ and $H_\infty$ Cost Minimisation

To apply the above lemma the proof should first be reviewed. This reveals that although the controller $C_0$ enters the expressions for the tracking $J_1$ and feedforward $J_2$ cost terms, it does not affect the minimum values of these quantities $J_{1min}$ and $J_{2min}$. Similarly neither of the compensator terms $C_{01}$ nor $C_{02}$, introduced in Chapter 2, affect $J_0$. In the present problem $p_1 = 3$ and allowing for the possibility of reordering in the above Lemma it is clear that the following alternative cost-minimisation problems can be considered:

(i)

$$J = J_0 + J_1 + J_2 \qquad\qquad (H_2 \text{ cost-index})$$

(ii)

$$J_{2\infty} = \|I_0\|_\infty + J_1 + J_2, \quad J_{2\infty} = J_0 + \|I_1\|_\infty + J_2,$$

$$J_{2\infty} = J_0 + J_1 + \|I_2\|_\infty \qquad\qquad (\text{Mixed-cost})$$

(iii)

$$J_{2\infty} = \|I_0\|_\infty + \|I_1\|_\infty + J_2, \quad J_{2\infty} = \|I_1\|_\infty + J_1 + \|I_2\|_\infty,$$

$$J_{2\infty} = J_0 + \|I_1\|_\infty + \|I_2\|_\infty \qquad\qquad (\text{Mixed-cost})$$

(iv)

$$J_\infty = \|I_0\|_\infty + \|I_1\|_\infty \quad + \|I_2\|_\infty \qquad\qquad (H_\infty \text{ cost-index})$$

Thus, if suitable robustness weighting terms $\Sigma_0$, $\Sigma_1$, $\Sigma_2$ can be found, any or all of the controller elements can be made $H_\infty$ optimal and the remainder will minimise $H_2$ or LQG criteria. The first case under (ii) is the most practical problem where $C_0$ is to be chosen to minimise an $H_\infty$ criterion but the tracking and feedforward controllers are $H_2$ optimal. However, the following results will enable any of the above combination of cost problems to be solved.

### 5.2.3  Equalising Solutions

The particular values of the robust weighting functions $\Sigma_0$, $\Sigma_1$, $\Sigma_2$ which ensure the respective cost terms $I_0, I_1$ and $I_2$ are minimised in an $H_\infty$ sense, can now be determined.

**Feedback control**

Invoking Lemma 2.2 the expression for the minimum cost (2.121), may be utilised to obtain the equations for the integrand terms at the optimum. That is,

$$F_0 F_0^* / (D_f D_f^* D_{c0} D_{c0}^*) + \Sigma_0 T_0 = \lambda_0^2 \Sigma_0$$

and

$$\Sigma_0 = \frac{B_{\sigma_0} B_{\sigma_0}^*}{A_{\sigma_0} A_{\sigma_0}^*} = \frac{F_0 F_0^*}{(\lambda_0^2 - T_0)(D_f D_{c0})(D_f D_{c0})^*} \qquad (5.3)$$

Similar steps may be followed to obtain the expressions for the tracking and feedforward weightings $\Sigma_1$ and $\Sigma_2$ .

**Tracking**

$$\Sigma_1 = \frac{B_{\sigma_1}B_{\sigma_1}^*}{A_{\sigma_1}A_{\sigma_1}^*} = \frac{F_1F_1^*}{(\lambda_1^2 - T_1)D_{c1}D_{c1}^*} \tag{5.4}$$

**Feedforward**

$$\Sigma_2 = \frac{B_{\sigma_2}B_{\sigma_2}^*}{A_{\sigma_2}A_{\sigma_2}^*} = \frac{F_2F_2^*}{(\lambda_2^2 - T_2)D_{c2}D_{c2}^*} \tag{5.5}$$

These results enable expressions to be derived for $B_{\sigma_0}, B_{\sigma_1}, B_{\sigma_2}$ and $A_{\sigma_0}, A_{\sigma_1}, A_{\sigma_2}$ which ensure minimising an $H_2$ cost-index also results in the minimisation of the respective $H_\infty$ cost function terms.

### 5.2.4   The $GH_\infty$ Cost Function and Computations

The class of control laws derived in this section have a particularly simple form and this stems from the choice of cost-index, which leads to a simplification of the equations. Fortunately the cost index is of value practically and it is referred to as a $GH_\infty$ criterion.

The previous results may be used to compute the polynomial $H_\infty$ controllers via the solution of the robustness diophantine equations (2.109) to (2.111) for the given $\Sigma_i$. The class of so called $GH_\infty$ controllers will be obtained where the weightings are assumed to be chosen so that the $T_i$ terms are null (Grimble 1987a,b). The algorithms are simpler in this particular case which is related to the so called Loop Shaping $H_\infty$ design method. This latter approach employs normalised coprime factor system models introduced by McFarlane and Glover (1990(a), 1990(b), 1992). The assumptions which determine the so called $GH_\infty$ case are as follows:

**Assumptions**

(i) The $GH_\infty$ case involves defining a special cost-function that has the weightings $i = 0, 1, 2$, of the form:

$$Q_{ci} = H_{qi}^* H_{qi}, \qquad R_{ci} = H_{ri}^* H_{ri}, \qquad G_{ci} = H_{ri} H_{qi}^* \tag{5.6}$$

where the sensitivity weightings $M_{c0}$ and $N_{c0}$ are null.

(ii) The output disturbance/measurement noise model is assumed to be absent $(W_n = 0)$.

(iii) In the one DOF design the ideal response model $W_i$ is assumed to be set to unity $(\beta = 0$ case$)$.

None of the above assumptions are very restrictive. The first only requires a cross-product term $G_{ci}$ to be present which is normally relatively small for practical choices of $Q_{ci}$ and $R_{ci}$ (usually $|Q_{ci}|$ is large at LF and $|R_{ci}|$ is large at HF). The measurement noise attenuation must be achieved through proper choice of control weightings in $GH_\infty$ design and hence the absence of the measurement noise model

$W_n$ is not serious. In fact all of these assumptions are aimed at simplifying the equations to be solved and do not seriously restrict the types of designs that can be achieved. Before continuing with the derivation of the $GH_\infty$ optimal controller the cost-function being minimised, under the above assumptions, should be restated:

$$J_\infty = \|I_0\|_\infty + \|I_1\|_\infty + \|I_2\|_\infty$$

where the frequency response functions are defined as:

$$I_i = [Q_{ci}\Phi_{ei} + R_{ci}\Phi_{ui} + G_{ci}\Phi_{u_i e_i} + G_{ci}^*\Phi_{e_i u_i}], \qquad for\ i = 1,2,3$$

The sum of cost terms, with the above cost weighting definitions (5.6), involves minimising the maximum values of the power spectral density $\Phi_{\phi_i}$ of the following signals:

$$\phi_i(t) = (H_{qi}e_i(t) - H_{ri}u_i(t)), \qquad for\ \{i = 1,2,3\}$$

**$T_i$ Cost Terms**

To confirm that the cost terms $T_0, T_1,$ and $T_2$ in (5.3) to (5.5) are null, when the above assumptions are made, consider equations (2.123) to (2.125):

$$T_0 = (Q_{c0}(\beta_1^2 W_i\Phi_{rr}W_i^* + \gamma_1^2\Phi_{d_0 d_0} + \Phi_{d_1 d_1})Y_f Y_f^* Y_{c0}^* Y_{c0} - \Phi_{h0}\Phi_{h0}^*)(Y_f Y_f^* Y_{c0}^* Y_{c0})^{-1}$$

where from (2.123) :

$$\Phi_{h0} = (W^* Q_{c0} - G_{c0}^*)(\beta_1^2 W_i\Phi_{rr} + \gamma_1^2\Phi_{d_0 d_0} + \Phi_{d_1 d_1})$$

but with assumption (i):

$$T_0 = Q_{c0}(H_{q0}W - H_{r0})(H_{q0}W - H_{r0})^*(\beta_1^2 W_i\Phi_{rr}W_i^* + \gamma_1^2\Phi_{d_0 d_0} + \Phi_{d_1 d_1})Y_f Y_f^*$$

$$-(\beta_1^2 W_i\Phi_{rr} + \gamma_1^2\Phi_{d_0 d_0} + \Phi_{d_1 d_1})(\beta_1^2 W_i\Phi_{rr} + \gamma_1^2\Phi_{d_0 d_0} + \Phi_{d_1 d_1})(Y_f Y_f^* Y_{c0}^* Y_{c0})^{-1}$$

$$= 0 \text{ (invoking assumptions (i) to (iii))}$$

Similarly, from equations (2.124) and (2.125):

$$T_1 = (Q_{c1}W_i W_i^* Y_{c1}^* Y_{c1} - W_i W_i^*(Q_{c1}W - G_{c1})(W^* Q_{c1} - G_{c1}^*))(Y_{c1}^* Y_{c1})^{-1}\Phi_{rr}$$

$$= 0 \text{ (invoking (i))}$$

$$T_2 = (Q_{c2}W_{di}W_{di}^* Y_{c2}^* Y_{c2} - W_{di}W_{di}^*(Q_{c2}W - G_{c2})(W^* Q_{c2} - G_{c2}^*))(Y_{c2}^* Y_{c2})^{-1}\Phi_{d_2 d_2}$$

$$= 0 \text{ (invoking (i))}$$

It follows that each $T_i$ term is zero. The robust weightings, that ensure the Lemma 5.1 is satisfied, are given by equations (5.3) to (5.5), and these ensure the $H_\infty$

norm is minimised.  If $F_{is}$ is defined to be Schur and to satisfy: $F_{is}F_{is}^* = F_i F_i^*$ for $i = 0, 1, 2$ then the robustness weightings for the three cases clearly satisfy:

$$\frac{B_{\sigma_0}}{A_{\sigma_0}} = \frac{F_{0s}}{\lambda_0 D_f D_{c0}}, \qquad \frac{B_{\sigma_1}}{A_{\sigma_1}} = \frac{F_{1s}}{\lambda_1 D_{c1}}, \qquad \frac{B_{\sigma_2}}{A_{\sigma_2}} = \frac{F_{2s}}{\lambda_2 D_{c2}} \qquad (5.7)$$

**Feedback equation:** The weighting $\Sigma_0$ and equation (5.3) can be simplified, since with $P = 0$, $F_0 = D_f^* z^{-n_f} \widetilde{F}_0$, where $n_f = \deg(D_f)$.  The equation (2.109) to be solved simplifies as:

$$D_{c0}^* N_0 z^{-g} + \lambda_0 D_{c0} \widetilde{F}_0 = -F \widetilde{F}_{0s} \qquad (5.8)$$

where $\widetilde{F}_{0s}$ is Schur and satisfies $\widetilde{F}_{0s}\widetilde{F}_{0s}^* = \widetilde{F}_0 \widetilde{F}_0^*$ and the weighting simplifies as:

$$A_{\sigma_0}^{-1} B_{\sigma_0} = (\lambda_0 D_{c0})^{-1} \widetilde{F}_{0s} \qquad (5.9)$$

**Tracking equation:** The robust tracking equation (2.110) may be written as:

$$D_{c1}^* N_1 z^{-g_1} + \lambda_1 D_{c1} F_1 = Y_1 F_{1s} \qquad (5.10)$$

**Feedforward equation:** The robust feedforward equation (2.111) may be written as:

$$D_{c2}^* N_2 z^{-g_2} + \lambda_2 D_{c2} F_2 = Y_2 F_{2s} \qquad (5.11)$$

The above equations (5.8) to (5.11) reduce to simple linear eigenvalue problems, in terms of the unknowns, $(N_0, \widetilde{F}_0, \lambda_0)$, $(N_1, F_1, \lambda_1)$ and $(N_2, F_2, \lambda_2)$, respectively.

## 5.2.5   Summary of Scalar $H_2/H_\infty$ Results

The following lemmas stem directly from the previous results and enable any one of the three controllers to be chosen to minimise an $H_\infty$ rather than a $H_2$ norm.

**Lemma 5.2 :** $H_\infty$ **Feedback Controller**

If the measurement noise is null, $M_{c0} = N_{c0} = 0$, then the $H_\infty$ controller to minimise the $H_\infty$ cost term, $\|I_0\|_\infty$ , with generalised weightings $H_{q0}$ and $H_{r0}$, can be computed by invoking Theorem 2.3 but with the solution of (5.8), in terms of $(N_0, \widetilde{F}_0, \lambda_0)$, replacing equation (2.109).

**Lemma 5.3 :** $H_\infty$ **Tracking Controller**

The tracking controller to minimise the $H_\infty$ cost term $\|I_1\|_\infty$ , with generalised weightings $H_{q1}$ and $H_{r1}$, can be computed from Theorem 2.3 but with the solution of (5.10), in terms of $(N_1, F_1, \lambda_1)$, replacing equation (2.110).

**Lemma 5.4 :** $H_\infty$ **Feedforward Controller**

The feedforward controller to minimise the $H_\infty$ cost term $\|I_2\|_\infty$ , with generalised weightings $H_{q2}$ and $H_{r2}$, can be computed from Theorem 2.3 but with the solution of (5.11) in terms of $(N_2, F_2, \lambda_2)$ replacing equation (2.111).  Numerical methods of solving these equations are well established (Grimble 1987a, 1989a).

**$H_\infty$ Marine Control Design Example**

Application of these results in the $H_\infty$ design of ship positioning and roll stabilisation systems is described in the Marine Control Design, Chapter 12.

# 5.3    Single-input Multi-output Control Design

The design of $H_2$ and $H_\infty$ robust controllers for systems with a single-input and many outputs is now considered. Several simplifications can be achieved over the more general multivariable problem which is important in, for example, power generation self-tuning control applications. A separate tracking controller enables the future values of the set point to be taken into account. This can only be utilised in a 2 (or higher) DOF control solution. The tracking problem allows a model following capability to be introduced. The optimisation is based upon an extended cost-function which allows separate costing of the terms due to the tracking and feedback controllers.

In the application of interest it is important to be able to tune the feedback loop robustness properties, independently of the set point following response requirements. The cost-function enables the components of the error and control signals, due to the feedback and tracking controllers, to be costed separately. This enables the feedback loop controller, in the 2 or $2\frac{1}{2}$ DOF designs, to be tuned independently of the tracking controller.

A solution of the $H_2$ or Linear Quadratic Gaussian (LQG) optimal control problem, for systems represented in SIMO linear polynomial matrix form is first considered.

## 5.3.1    SIMO System Description

The system considered is assumed to be discrete-time, linear and time-invariant. The 2 or $2\frac{1}{2}$ DOF system diagram is shown in Fig. 5.1 and the plant includes $r$ outputs and a single input. A two degrees-of-freedom controller structure may be chosen (by defining $\alpha = 0$ and $\beta = 1$) so that future reference signal $\{r(t)\}$ information may be used. The $2\frac{1}{2}$ DOF control structure may be employed (by defining $\alpha = 1$ and $\beta = 1$). Any measurement system delay can be is absorbed into the subsystem model $H_f$.

The system includes an input disturbance signal $\{d(t)\}$ which must be countered by control action and an output disturbance or measurement noise signal $\{n(t)\}$ which must not lead to excessive control activity. The signal $\{d(t)\}$ represents an input disturbance since it directly affects the output $\{y(t)\}$ to be controlled. The signal $\{n(t)\}$ is termed on output-disturbance or noise, since it is added to the system after the point representing the output-signal $\{y(t)\}$. The independent zero-mean, white driving noise signals for the disturbance, measurement noise and reference signal and noise subsystems are denoted by $\{\xi(t)\}$, $\{\omega(t)\}$, $\{\zeta(t)\}$, and $\{\eta(t)\}$, respectively.

The plant output $y(t) = m(t) + d(t)$ is formed from the controlled output $m(t) = W(z^{-1})u(t)$, together with the input disturbance signal: $d(t) = W_d(z^{-1})\xi(t)$. The various controller structures (Grimble, 1994) are obtained by appropriately defining the scalars in Fig. 5.1. This is clarified in Table 5.1, where $\beta_1 = 1 - \beta$.

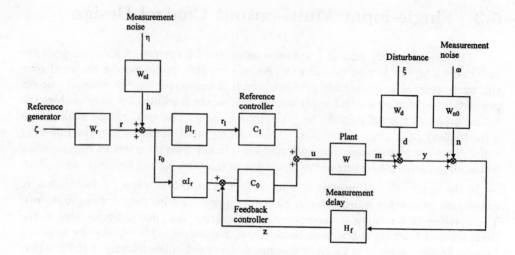

**Fig. 5.1** : Multi-DOF Multi-output Single-input Canonical System

**Table 5.1** : Controller Structure Path Switch Settings

| Controller structure | $\alpha$ | $\beta$ | $\beta_1$ |
|---|---|---|---|
| 1 | 1 | 0 | 1 |
| 2 | 0 | 1 | 0 |
| $2\frac{1}{2}$ | 1 | 1 | 0 |

## System Assumptions

(i) The reference $W_r$, ideal response $W_i$, output disturbance $W_{n0}$ and measurement noise $W_{n1}$, models are assumed to be asymptotically stable. With little loss of generality $W_r$ is assumed to be strictly minimum-phase.

(ii) The plant has no unstable hidden modes.

(iii) The measurement system dynamics are assumed to be absorbed into the system models but the measurement system delay can be included in the dynamic model $H_f$ which is of full rank.

(iv) The covariance matrices for the noise sources may, without loss of generality, be taken to be the identity matrix.

(v) The various noise sources are assumed to be mutually statistically independent.

(vi) For the 2 DOF solution $p$ steps of future reference signal information are assumed to be available ($p \geq 0$). Future reference or setpoint knowledge cannot be incorporated into a single DOF solution.

## System Equations

The plant equations, for the SIMO system shown in Fig. 5.1, may be listed as:
**Observations** :

$$z(t) = H_f(y(t) + n(t))$$

**Output :**

$$y(t) = d(t) + Wu(t)$$

**Input disturbance :**

$$d(t) = W_d \xi(t)$$

**Measurement noise :**

$$n(t) = W_{n0} \omega(t) \quad \text{and} \quad h(t) = W_{n1} \eta(t)$$

**Reference generation :**

$$r(t) = W_r \xi(t)$$

**Ideal response :**

$$y_i(t) = W_i r(t)$$

**Control signal :**

$$u(t) = C_1 \beta r_o(t) - C_0 z(t) + C_0 \alpha r_o(t)$$

**Tracking error :**

$$e(t) = W_i r(t) - y(t)$$

The following output and input sensitivity functions may be defined:

**Sensitivity :**

$$S_r = (I_r + H_f W C_0)^{-1}, \quad S = (1 + C_0 H_f W)^{-1}$$

**Complementary sensitivity :**

$$T_r = I_r - S_r = S_r H_f W C_0 = H_f W C_0 S_r = H_f W M$$

$$T = 1 - S = S C_0 H_f W = C_0 H_f W S = M H_f W$$

**Control sensitivity :**

$$M = S C_0 = C_0 S_r \tag{5.12}$$

**Closed-loop Equations**

The cost function to be introduced will include tracking error terms which are normally taken to be of the form $e'(t) = r(t) - y(t)$ . However, in model following problems the output is required to follow an *Ideal Response Model* $W_i r(t)$. From the system equations and the sensitivity function definitions :

$$z = H_f(d + Wu + n) = H_f (d + n + W C_1 \beta r_0 + W C_0 (\alpha r_0 - z))$$

$$= S_r H_f (d + n + WC_1\beta r_0 + WC_0\alpha r_0)$$

$$u = C_1\beta r_0 + C_0\alpha r_0 - C_0 S_r H_f (d + n + WC_1\beta r_0 + WC_0\alpha r_0)$$

$$= (C_0\alpha + C_1\beta)r_0 - C_0 S_r H_f W(C_0\alpha + C_1\beta)r_0 - C_0 S_r H_f (d + n)$$

$$e = W_i r - y = W_i r - WS(C_0\alpha + C_1\beta)r_0 - (I_r - WMH_f)d + WMH_f n$$

The above expressions can be simplified by substituting for the compensator function term:

$$C_{01} = C_0\alpha + C_1\beta$$

to obtain:

$$u = SC_{01}r_0 - MH_f(d + n)$$

$$e = (W_i - WSC_{01})r - WSC_{01}h - (I_r - WMH_f)d + WMH_f n$$

Observe that the compensator term $C_{01}$ depends upon the feedback and tracking controllers. This is not a physical controller to be implemented but it is easier to solve for $C_{01}$ than to work in terms of the individual controller elements $C_0$ and $C_1$. These equations enter the cost-function and the terms are statistically independent. This property is maintained if the equations are rewritten in an equivalent form but one which allows the 1, 2 and $2\frac{1}{2}$ DOF solutions to be obtained in a single completing the squares argument. Thence, from these equations obtain:

$$u = (SC_{01}\beta + M\beta_1)r + (SC_{01}\beta + M\beta_1)h - MH_f(d + n) \qquad (5.13)$$

$$e = (W_i - WSC_{01})\beta r + (W_i - WM)\beta_1 r - WSC_{01}\beta h - WM\beta_1 h$$

$$-(I_r - WMH_f)d + WMH_f n \qquad (5.14)$$

Let the error and control signal terms in (5.13) and (5.14) be separated into the following statistically independent components:

$$u = u_0 + u_1 \text{ and } e = e_0 + e_1$$

where the error and control signal components may be defined as follows:

$$u_0 = M(\beta_1 r + \beta_1 h - H_f d - H_f n) \qquad (5.15)$$

$$u_1 = SC_{01}\beta(r + h) \qquad (5.16)$$

and

$$e_0 = W_i\beta_1 r - d - WM(\beta_1 r + \beta_1 h - H_f d - H_f n) \qquad (5.17)$$

$$e_1 = (W_i - WSC_{01})\beta r - WSC_{01}\beta h \qquad (5.18)$$

These signals have the following physical significance:

$(e_0, u_0)$ : Component of the error and control signals due to the feedback controller action including that for the one DOF tracking problem.

$(e_1, u_1)$ : Tracking error and control signals resulting from the reference signal input for the two or higher DOF designs.

The above observations suggest that each of the terms should be costed separately, providing the significant advantage of *independent tuning knobs* for the feedback and tracking control actions.

### Polynomial System Description

The system models shown in Fig. 5.1 may be represented in polynomial matrix form, where the system transfer-functions are written as follows:

**Plant (r×1) :**

$$W = A^{-1}B$$

**Output disturbance or measurement noise (rxr) :**

$$W_{n0} = A^{-1}C_{n0} \text{ and } W_{n1} = A^{-1}C_{n1}$$

**Input disturbance (r×r) :**

$$W_d = A^{-1}C_d$$

**Measurement subsystem (r×r) :**

$$H_f \text{ (maximum measurement system delay } k_f)$$

**Reference generator (r×r) :**

$$W_r = A^{-1}E \qquad \text{(E normal full rank)}$$

**Ideal response model (r×r) :**

$$W_i = B_i A_i^{-1}$$

**Feedback controller (1×r) :**

$$C_0 = C_{0d}^{-1}C_{0n}$$

**Reference controller (1×r) :**

$$C_1 = C_{1d}^{-1}C_{1n}$$

The various system polynomial matrices are assumed to be free of unstable hidden modes and to be represented by the left coprime matrix fraction decomposition:

$$[W, W_d, W_r, W_{n0}, W_{n1}] = A^{-1}[B, C_d, E, C_{n0}, C_{n1}] \tag{5.19}$$

There is no loss of generality in assuming the use of a common denominator $A(z^{-1})$ *polynomial matrix* in the plant model. The very reasonable assumption will, also be made that $A$ and $H_f$ commute. The stable reference model may also be expressed in left coprime matrix form, writing: $[W_r, \quad W_{n1}] = A_r^{-1}[E_r, \quad C_{nr}]$.

It can be assumed that the future values of the reference trajectory are known at least p-steps ahead. In this case it only makes physical sense to consider a 2 DOF solution where the tracking measurement noise $\eta \to 0$ . The one and $2\frac{1}{2}$ DOF control structures cannot utilise future set point knowledge, since the controller $C_0$ then includes an input of $\alpha r_0(t) - z(t)$.

## 5.3.2   The $H_2$ SIMO Control Problem

The LQG or $H_2$ criterion (Grimble and Johnson, 1988) involves the minimisation of the tracking error and control signal variances. The criterion $J$ is defined, in terms of the power and cross-power spectra, for the error and control signal components, as:

$$J = \frac{1}{2\pi j} \oint_{|z|=1} \{[trace\{Q_{c0}\Phi_{e0} + G_{c0}\Phi_{u_0 e_0} + \Phi_{e_0 u_0}G_{c0}^*\} + R_{c0}\Phi_{u0}]\Sigma_0$$

$$+ [trace\{Q_{c1}\Phi_{e1} + G_{c1}\Phi_{u_1 e_1} + \Phi_{e_1 u_1}G_{c1}^*\} + R_{c1}\Phi_{u1}]\Sigma_1\}dz/z \qquad (5.20)$$

Let the index $i = 0, 1$ correspond to the regulating and tracking controller signal components, respectively. Then the dynamic weightings (Grimble, 1994) in the above criterion have the following significance:

$Q_{ci}, R_{ci}, G_{ci}$:   Error, control and cross-weighting terms.

$\Sigma_i$ :   Common *robustness improvement* weighting elements.

From an obvious identification of terms, the criterion (5.20) can clearly be written as the sum of two components:

$$J = J_0 + J_1 = \frac{1}{2\pi j} \oint_{|z|=1} \{I_0(z^{-1})\Sigma_0(z^{-1}) + I_1(z^{-1})\Sigma_1(z^{-1})\}\frac{dz}{z} \qquad (5.21)$$

where $I_0$ and $I_1$ correspond to the respective square bracketed terms in (5.20). The polynomial matrix forms of the weightings in the cost index may be defined, for $i = \{0, 1\}$, as:

$$Q_{ci} = A_{qi}^{*-1}Q_{ni}A_{qi}^{-1}, \qquad R_{ci} = A_{ri}^{*-1}R_{ni}A_{ri}^{-1}, \qquad G_{ci} = A_{qi}^{*-1}G_{ni}A_{ri}^{-1}$$

and the scalar weighting:

$$\Sigma_i = W_{\sigma i}^*W_{\sigma i} = A_{\sigma i}^{*-1}B_{\sigma i}^*B_{\sigma i}A_{\sigma i}^{-1}$$

Also introduce the right coprime weighted plant model pairs $(B_0, A_0)$ and $(B_1, A_1)$ as:

$$A_{q0}^{-1}A^{-1}BA_{r0} = B_0A_0^{-1} \qquad and \qquad A_{q1}^{-1}A^{-1}BA_{r1} = B_1A_1^{-1} \qquad (5.22)$$

There is little loss in generality in assuming that the denominators of these weightings are strictly Schur. The assumption will also be made, to simplify the analysis, that $H_f$ and $AA_{q0}$ commute.

**Theorem 5.1** : *Robust Feedback SIMO $H_2$ Controller*

The $H_2$ controller to minimise the $H_2$ dual-criterion (5.20), for the SIMO system described in Section 5.3.1, may be found as follows. Compute the strictly-Schur spectral factors $D_{c0}, D_{c1}, D_f$ and $D_r$ using:

$$D_{c0}^*D_{c0} = B_0^*Q_{n0}B_0 + A_0^*R_{n0}A_0 - B_0^*G_{n0}A_0 - A_0^*G_{n0}^*B_0 \qquad (5.23)$$

$$D_{c1}^* D_{c1} = B_1^* Q_{n1} B_1 + A_1^* R_{n1} A_1 - B_1^* G_{n1} A_1 - A_1^* G_{n1}^* B_1 \qquad (5.24)$$

$$D_f D_f^* = H_f (C_d C_d^* + C_{n0} C_{n0}^*) H_f^* + \beta_1^2 (EE^* + C_{n1} C_{n1}^*) \qquad (5.25)$$

$$D_r D_r^* = E_r E_r^* + C_{nr} C_{nr}^* \qquad (5.26)$$

**Regulating loop equations** : Compute $(G_0, H_0, F_0)$, with $F_0$ of smallest degree:

$$D_{co}^* G_0 z^{-g} + F_0 A_2 = (B_0^* Q_{n0} - A_0^* G_{n0}^*) H_f^{-1} D_2 z^{-g} \qquad (5.27)$$

$$D_{co}^* H_0 z^{-g} - F_0 B_3 = (A_0^* R_{n0} - B_0^* G_{n0}) D_3 z^{-g} \qquad (5.28)$$

where the right $r \times r$ and $r \times 1$ coprime pairs $(A_2, D_2)$ and $(A_3, D_3)$ satisfy:

$$D_2 A_2^{-1} = (AA_{q0})^{-1} D_f \quad \text{and} \quad B_3 D_3^{-1} = D_f^{-1} H_f B A_{r0}$$

**Measurement noise equation** : Compute $(L_0, P_0)$, with $P_0$ of smallest degree:

$$D_{co}^* L_0 D_f^* z^{-g_0} + A_{t0} P_0 = C_{t0} z^{-g_0} \qquad (5.29)$$

where the left coprime pair $(A_{t0}, C_{t0})$ satisfy :

$$A_{t0}^{-1} C_{t0} = (B_0^* Q_{n0} - A_0^* G_{n0}^*) A_{q0}^{-1} A^{-1} H_f^{-1} (H_f C_{n0} C_{n0}^* H_f^*$$

$$+ \beta_1^2 (EE^* + C_{n1} C_{n1}^* - H_f A B_i A_i^{-1} A^{-1} EE^*))$$

**Reference equation** : Compute $(X_1, Y_1)$, with $Y_1$ of smallest degree:

$$D_{c1}^* X_1 D_r^* z^{-g_1} + A_{t1} Y_1 = C_{t1} z^{-g_1 - p} \qquad (5.30)$$

where the left coprime pair $(A_{t1}, C_{t1})$ satisfy :

$$A_{t1}^{-1} C_{t1} = (B_1^* Q_{n1} - A_1^* G_{n1}^*) A_{q1}^{-1} B_i A_i^{-1} A_r^{-1} E_r E_r^*$$

**Robustness feedback equation** : Compute $(N_0, M_0)$, with $M_0$ of smallest degree:

$$D_{c0}^* N_0 D_f^* z^{-g_0} + A_{\sigma0} M_0 = B_{\sigma0} (P_0 - F_0 D_f^* z^{g - g_0}) \qquad (5.31)$$

where $F_0$ and $P_0$ are from the solution of (5.27), (5.28) and (5.29), respectively.

**Robustness tracking equation**: Compute $(N_1, M_1)$, with $M_1$ of smallest degree:

$$D_{c1}^* N_1 D_r^* z^{-g_1} + A_{\sigma1} M_1 = B_{\sigma1} Y_1 \qquad (5.32)$$

The scalars, $g, g_o, g_1$ are chosen to be the smallest positive integers which ensure the above equations are polynomial in the indeterminate $z^{-1}$, that is :

$$g \geq \deg(D_{c0}), \quad g_1 \geq \deg(D_{c1} D_r), \quad g_0 \geq \deg(D_f D_{c0})$$

**Feedback Controller:**

$$C_0 = A_{r0}D_3(A_{t0}B_{\sigma 0}H_0 + [B_{\sigma 0}L_0 + A_{t0}N_0]B_3)^{-1}$$

$$\times (A_{t0}B_{\sigma 0}G_0 - [B_{\sigma 0}L_0 + A_{t0}N_0]A_2)D_2^{-1}A_{q0}^{-1} \tag{5.33}$$

**Tracking Controller:**

$$C_1\beta = C_{01} - C_0\alpha \tag{5.34}$$

where

$$C_{01} = (D_{c1}A_{t1}B_{\sigma 1}C_{0d})^{-1}(C_{0d}\tilde{A}_0 + C_{0n}H_f\tilde{B}_0)\tilde{A}_{q1}A_{r1}[B_{\sigma 1}X_1 + A_{t1}N_1]D_r^{-1}A_r z^p$$

**Proof :**  The proof employs a completing the squares type of solution procedure explained in Chapter 2, and given in Grimble (1999a).

The properties of the optimal controller for this SIMO problem are summarised in the following lemma.

**Lemma 5.5 :** *Robust Weighted SIMO $H_2$ Controller Properties*

The *characteristic polynomial* and the related *implied equation* which determine closed-loop stability, are given respectively as:

$$\rho_c = D_3 B_{\sigma 0} A'_{t0} D_{c0}$$

and

$$G_0 D_2^{-1} A_{q0}^{-1}(H_f\tilde{B}_0) + H_0 D_3^{-1}A_{r0}^{-1}(\tilde{A}_o) = D_{c0}A_{r0}^{-1}\tilde{A}_{q0}^{-1} \tag{5.35}$$

where $A'_{t0}$ is obtained from $A_{t0}$ after cancellation of common terms in $B_3$, and $\tilde{B}_0, \tilde{A}_0$ represent a right coprime model of the plant $\tilde{B}_0\tilde{A}_0^{-1} = A^{-1}B$ , and $\tilde{A}_{q0}$ is a scalar polynomial defined as

$$\tilde{A}_{q0} = \tilde{A}_0^{-1}A_0.$$

**Sensitivity function:**

$$S = A_0(B_{\sigma 0}A_{t0}D_{c0})^{-1}(B_{\sigma 0}A_{t0}H_0 + [B_{\sigma 0}L_0 + A_{t0}N_0]B_3)D_3^{-1}$$

**Control sensitivity:**

$$M = A_0 A_{r0}(B_{\sigma 0}A_{t0}D_{co})^{-1}(B_{\sigma 0}A_{t0}G_0 - [B_{\sigma 0}L_0 + A_{t0}N_0]A_2)D_2^{-1}A_{q0}^{-1}$$

**Complementary sensitivity:**

$$T = H_f\tilde{B}_0\tilde{A}_{qo}A_{ro}(B_{\sigma 0}A_{to}D_{co})^{-1}(B_{\sigma 0}A_{to}G_0 - [B_{\sigma 0}L_0 + A_{t0}N_0]A_2)D_2^{-1}A_{q0}^{-1}$$

The *minimum value* for the cost-function can be computed using:

$$J_{\min} = \frac{1}{2\pi j} \oint_{[z]=1} \{X_{\min}(z^{-1})\} \frac{dz}{z} \tag{5.36}$$

where the cost integrand at the minimum:

$$X_{\min} = [D_{c0}^{*-1}M_0(D_fD_f^*)^{-1}M_0^*D_{c0}^{-1} + T_0\Sigma_0] + [D_{c1}^{*-1}M_1(D_rD_r^*)^{-1}M_1^*D_{c1}^* + T_1\Sigma_1]$$

and the cost integrand functions terms :

$$T_0 = trace\{Q_{c0}(\beta_1^2 W_i\Phi_{rr}W_i^* + \Phi_{dd}) - (Q_{co}W - G_{c0})(Y_{c0}^*Y_{c0})^{-1}(W^*Q_{c0} - G_{c0}^*)$$

$$\times(\Phi_{dd}H_f^* + \beta_1^2 W_i\Phi_{rr})(Y_fY_f^*)^{-1}(H_f\Phi_{dd} + \beta_1^2\Phi_{rr}W_i^*)\}$$

and

$$T_1 = trace\{Q_{c1}W_i\ \Phi_{rr}W_i^*$$

$$-(Q_{c1}W - G_{c1})(Y_{c1}^*Y_{c1})^{-1}(W^*Q_{c1} - G_{c1}^*)(W_i\Phi_{rr}(Y_{r0}Y_{r0}^*)^{-1}\Phi_{rr}W_i^*)\}$$

where the spectral factors : $Y_{c0} = D_{c0}(A_{r0}A_0)^{-1}$, $\quad Y_{c1} = D_{c1}(A_{r1}A_1)^{-1}$,

$$Y_f = A^{-1}D_f \quad \text{and} \quad Y_{r0} = A_r^{-1}D_rz^{-p}$$

**Proof :** These results also follow from the analysis in Chapter 2 and are derived in Grimble (1999a).

**Remarks:**

(i) Only a small set of the above diophantine equations must be solved in any particular case. A null solution to (5.29) is obtained in the one DOF case if measurement noise is null. The solutions to (5.31) and (5.32) are null if the weightings $\Sigma_0$ and $\Sigma_1$ are constants.

(ii) The only terms in the tracking controller defined by (5.34), which may be unstable are included within the feedback controller denominator term $C_{0d}$ and can therefore be implemented in a common element within the feedback loop. This is a necessary condition for the tracking control to be stable.

### 5.3.3 The $H_\infty$ SIMO Control Problem

The $H_\infty$ solution that is derived in the following is rather simpler than in the more general polynomial solution of the multivariable control problem, where in fact the weightings $\Sigma_i$ are introduced via the disturbance models. Approximations may also be made if a simple numerical algorithm is required and for the applications problems of interest this is very appropriate. The relationship between the $H_2$ and $H_\infty$ problems will first be explored. The $H_\infty$ SIMO power generation case study, described in Chapter 10, will illustrate the theoretical results generated below.

The components $J_0$ and $J_1$ in the $H_2$ cost problem (5.21) correspond to functions of the feedback and tracking controllers. These depend upon the dynamic weighting functions $\Sigma_0$ and $\Sigma_1$, respectively. It will be shown below that if the $i$th controller is to be $H_\infty$ optimal this can be achieved by appropriate choice of the weighting function $\Sigma_i$. In fact the strategy will be to compute the particular $\Sigma_i$ which ensures $H_\infty$ optimality and to then invoke Theorem 5.1. This technique was referred to in §5.2.1 as LQG embedding (Grimble, 1986).

The general LQG embedding problem was considered in Lemma 5.1, where a $H_2$ cost $J$ is formed from $p_1 + 1$ cost terms which are dependent upon $p_1 + 1$ control elements $C_j(z^{-1})$.

## 5.3.4   Mixed $H_2$ and $H_\infty$ Cost Minimisation Problems

To apply the above Lemma the proof of Theorem 5.1 should first be reviewed. This reveals that although the controller $C_0$ enters the expression for the tracking cost term $J_1$ , it does not affect the minimum value $J_{1min}$. Similarly the controller $C_{01}$ does not affect $J_{0min}$.   In the present problem $p_1 = 1$ and allowing for the possibility of reordering in the above Lemma, it is clear that the following different cost-minimisation problems can be considered:

(i)  $J = J_0 + J_1$ ($H_2$ cost-index)

(ii)  $J_{2\infty} = \|I_0\|_\infty + J_1, \qquad J_{2\infty} = J_0 + \|I_1\|_\infty$

(iii)  $J_\infty = \|I_0\|_\infty + \|I_1\|_\infty$ ($H_\infty$ cost-index)

Thus, if suitable robustness weighting terms $\Sigma_0$, $\Sigma_1$ can be found, either or both of the controller elements can be made $H_\infty$ optimal and the remaining will minimise a $H_2$ or LQG criterion. The first case under (ii) is the most practical problem where $C_0$ is to be chosen to minimise an $H_\infty$ criterion but the tracking controller is $H_2$ optimal.

### Equalising Solutions

The particular values of the robust weighing functions $\Sigma_0$, $\Sigma_1$ which ensure the respective cost terms $I_0$ and $I_1$ are minimised, in an $H_\infty$ sense, can now be determined.

### Feedback control:

Invoking Lemma 5.1 and utilising (5.36) obtain the conditions, for the feedback control, at the optimum:

$$D_{c0}^{*-1} M_0 (D_f D_f^*)^{-1} M_0^* D_{c0}^{-1} + T_0 \Sigma_0 = \lambda_0^2 \Sigma_0$$

and

$$\Sigma_0 = \frac{B_{\sigma_0} B_{\sigma_0}^*}{A_{\sigma_0} A_{\sigma_0}^*} = D_{c0}^{*-1} M_0 (D_f D_f^*)^{-1} M_0^* D_{c0}^{-1} / (\lambda_0^2 - T_0) \qquad (5.37)$$

The same steps may be followed to obtain the tracking weighting $\Sigma_1$.

### Tracking:

$$\Sigma_1 = \frac{B_{\sigma_1} B_{\sigma_1}^*}{A_{\sigma_1} A_{\sigma_1}^*} = D_{c1}^{*-1} M_1 (D_r D_r^*)^{-1} M_1^* D_{c1}^{-1} / (\lambda_1^2 - T_1) \qquad (5.38)$$

These results enable expressions to be derived for $B_{\sigma_0}, B_{\sigma_1}$ and $A_{\sigma_0}, A_{\sigma_1}$ which ensure minimising an $H_2$ cost-index gives the solutions to the respective $H_\infty$ cost terms.

## 5.3.5   Computation of the $GH_\infty$ Controllers

The previous results may be used to generate the $H_\infty$ controllers via the solution of the robustness diophantine equations (5.31), (5.32). As in §5.2.4 the class of so called generalised $H_\infty$ controllers will be obtained, where the weightings and system are such that the $T_i$ terms, in Lemma 5.5, are null (Grimble, 1987a,b). Recall that the $GH_\infty$ criterion involves defining weightings $i = 0, 1$ that have the following structure:

$$Q_{ci} = H_{qi}^* H_{qi}, \qquad R_{ci} = H_{ri}^* H_{ri}, \qquad G_{ci} = H_{qi}^* H_{ri} \tag{5.39}$$

The scalar $GH_\infty$ problem is considered below but the multivariable results are similar and are described in Grimble (1989a). In computing the robust weighting $\Sigma_0$, for the feedback loop, it is reasonable to consider the limiting case where the measurement noise is null. In this case the solution of (5.29) for one DOF designs is trivial, that is $L_0 = 0$ and $P_0 = 0$. In fact this will also be the case for 2 and $2\frac{1}{2}$ DOF designs, when $C_{n1} = 0, H_f = W_i = I$. The weightings can then be chosen with robustness requirements in mind and measurement noise attenuation can be determined by the choice of the frequency response of $R_{c0}$. If the measurement noise is null and $L_0 = 0$ and $P_0 = 0$ then (5.31) becomes:

$$D_{c0}^* N_0 D_f^* z^{-g_0} + A_{\sigma 0} M_0 = -B_{\sigma 0} F_0 D_f^* z^{g-g_0} \tag{5.40}$$

Clearly $M_0$ can be written in the form $M_0 = \tilde{M}_0 D_f^* z^{-n_f}$, where $n_f = \deg(D_f)$. Let the Schur polynomial $\tilde{M}_{0s}$ now be defined to satisfy $\tilde{M}_{0s} \tilde{M}_{0s}^* = \tilde{M}_0 \tilde{M}_0^*$. Then, from equation (5.37), the weighting $\Sigma_0$ satisfies:

$$\Sigma_0 = A_{\sigma 0}^{-1} B_{\sigma 0} B_{\sigma 0}^* A_{\sigma 0}^{*-1} = D_{c0}^{*-1} \tilde{M}_0 \tilde{M}_0^* D_{c0}^{-1} / \lambda_0^2$$

and hence the required weighting for the feedback loop has the form:

$$A_{\sigma 0}^{-1} B_{\sigma 0} = (\lambda_0 D_{c0})^{-1} \tilde{M}_{0s}$$

Observe from (5.30) that $C_{t1}$ may be written as $C_{t1}' = C_{t1} E_r^*$. Then if the noise term $C_{n1} = 0$, $D_r^* = E_r^*$ and (5.30) reveals $Y_1$ may be written as $Y_1 = Y_1' E_r^* z^{-n_r}$, where $n_r = \deg(D_r)$ and $Y_1'$ satisfies:

$$D_{c1}^* X_1 z^{-g_1'} + A_{t1} Y_1' = C_{t1} z^{-g_1'-p}$$

The robustness tracking equation (5.32) may also be written as:

$$D_{c1}^* N_1 E_r^* z^{-g_1} + A_{\sigma 1} M_1 = B_{\sigma 1} Y_1' E_r^* z^{-n_r} \tag{5.41}$$

Clearly, $M_1$ can be written in the form $M_1 = \tilde{M}_1 E_r^* z^{-n_r}$ where $n_r = \deg(E_r)$. Let the Schur polynomial $\tilde{M}_{1s}$ now be defined to satisfy $\tilde{M}_{1s} \tilde{M}_{1s}^* = \tilde{M}_1 \tilde{M}_1^*$. Then, from equation (5.38), the weighting $\Sigma_1$ satisfies:

$$\Sigma_1 = A_{\sigma 1}^{-1} B_{\sigma 1} B_{\sigma 1}^* A_{\sigma 1}^{*-1} = D_{c1}^{*-1} \tilde{M}_1 \tilde{M}_1^* D_{c1}^{-1} / \lambda_1^2$$

and hence the required weighting for the tracking controller becomes:

$$A_{\sigma 1}^{-1} B_{\sigma 1} = (\lambda_1 D_{c1})^{-1} \tilde{M}_{1s}$$

The two equations to the solved to compute $\tilde{M}_0$ and $\tilde{M}_1$ and hence the weightings, can be summarised, from (5.40) and (5.41), as:

$$D_{c0}^* N_0 z^{-g} + \lambda_0 D_{c0} \tilde{M}_0 = -\tilde{M}_{0s} F_0 \tag{5.42}$$

and

$$D_{c1}^* N_1 z^{-g_1'} + \lambda_1 D_{c1} \tilde{M}_1 = \tilde{M}_{1s} Y' \tag{5.43}$$

These equations are in terms of the scalars $\lambda_0$ and $\lambda_1$ and the polynomial matrices $(N_0, \tilde{M}_0)$ and $(N_1, \tilde{M}_1)$. The polynomials $\tilde{M}_{0s}$ and $\tilde{M}_{1s}$ are Schur and satisfy:

$$\tilde{M}_{0s} \tilde{M}_{0s}^* = \tilde{M}_0 \tilde{M}_0^* \quad \text{and} \quad \tilde{M}_{1s} \tilde{M}_{1s}^* = \tilde{M}_1 \tilde{M}_1^*$$

Note that if $T_0$ and $T_1$ are constant functions in Lemma 5.5, or are approximated by constants, with the same maximum values, the equations to be solved also have the above simplified forms ((5.42) to (5.43)). The numerical algorithms are simpler and converge much faster when $T_0$ and $T_1$ are constants.

### Summary of $H_2/H_\infty$ Results

The following lemmas stem directly from the previous results and enable any one of the feedback, tracking and feedforward controllers to be chosen to minimise a $H_\infty$ rather than a $H_2$ norm, for that component of the cost index.

**Lemma 5.6** : $H_\infty$ *Feedback Controller*

If the measurement noise terms are null, then the $H_\infty$ controller to minimise the $H_\infty$ cost term, $\|I_0\|_\infty$ , with generalised weightings $H_{q0}$ and $H_{r0}$, can be computed by invoking Theorem 5.1 but with the solution of (5.42), in terms of $(N_0, \tilde{M}_0, \lambda_0)$, replacing equation (5.31).

**Lemma 5.7:** $H_\infty$ *Tracking Controller*

The tracking controller to minimize the $H_\infty$ cost term $\|I_1\|_\infty$ , with generalised weightings $H_{q1}$ and $H_{r1}$, can be computed from Theorem 5.1 but with the solution of (5.43) in terms of $(N_1, \tilde{M}_1, \lambda_1)$, replacing equation (5.32).

### $H_\infty$ Power System Design Example

The theoretical framework established was motivated by the type of problem found in multi-loop machine control systems. The particular problem which motivated the work was the control of alternator terminal voltage and the use of stabilising power and speed feedback loops. This power system example is considered in the power generation Chapter 10. The additional feedback from the alternator speed and power loops, in the voltage regulating system, provides damping for the electro-mechanical oscillations. However, this is gained at a price. Additional feedback from the electro-mechanical variables increases the interactions between the voltage and electro-mechanical loops. Thus, the damping of the lightly damped electro-mechanical mode is improved, but the voltage loop regulating action deteriorates. This arises because additional feedbacks change the electro-magnetic poles which are the dominant terms in the voltage loop. The additional feedback gains should therefore be of limited magnitude, to provide some damping but to limit the aforementioned interaction. The SIMO $H_\infty$ design approach enables the interactions between loops to be taken into account in a very natural manner, as shown in the simulation results of Chapter 10.

## 5.4  $H_\infty$ Predictive Optimal Control Laws

An $H_\infty$ Generalised Predictive Control law is derived in the following which has properties in common with both the $H_\infty$ and GPC control laws. The stability and robustness properties are the same as for a $GH_\infty$ optimal controller but a cost on future predictive error and control action is dealt with in the same manner as for the GPC, or model based predictive control laws. This type of long range predictive control algorithm has stability and robustness problems which can be improved through the $H_\infty$ formalism. The first attempt to introduce robust $H_\infty$ properties into the GPC algorithm was made by Bentsman and coworkers (see Tse et al. 1992, 1993 a,b) who developed both adaptive and non-adaptive control solutions.

The optimal control solution enables the relationship between the $GH_\infty$ predictive and the more traditional GPC control law to be explored. The assumptions which must be made to ensure a GPC type of solution results are of particular importance. It is also valuable that under quite reasonable conditions the usual $H_\infty$ type of stability properties can be achieved. At the same time, since the solution is so close to a GPC result, it is clear how constraints can be treated by adopting many of the results which already exist for LRPC and GPC type algorithms.

The strategy followed below is to solve a multi-step Linear Quadratic Gaussian Predictive Control (LQGPC) control problem, which has a cost-function including both cross-product and common dynamic weighting terms. By careful choice of this common weighting function the multi-step $H_\infty$ controllers are obtained. This is the so called LQG embedding solution procedure.

### 5.4.1   Plant and Disturbance Description

The single-input single-output one degree-of-freedom discrete-time system is assumed to be the same as that described in Section 3.4.2 and shown in the Fig. 3.1. The $H_\infty$ problem will be motivated by a stochastic system description, although once the solution is obtained the stochastic disturbance and reference models can be given their usual $H_\infty$ weighting interpretations. The external white noise sources drive colouring filters which represent the reference $W_r(z^{-1})$ and disturbance $W_d(z^{-1})$ subsystems. The system equations and the system assumptions are the same as in Section 3.4.2.

The following expressions may be derived for the output, error and control signals:

$$y(t) = WC_0Sr(t) + Sd(t) \qquad (5.44)$$

$$e(t) = r(t) - y(t) = (1 - WC_0S)r(t) - (1 - WC_0S)d(t) \qquad (5.45)$$

$$u(t) = SC_0(r(t) - d(t)) \qquad (5.46)$$

### 5.4.2   Review of the LQGPC Control Law

The LQG Predictive Optimal Control law introduced in Chapter 3 minimised a cost-function which was related to both LQG and GPC cost-function terms. This is a necessary precursor to solving the $H_\infty$GPC control law in the next section. The weighting functions in the criterion can be dynamic and may be set to zero to represent different error and control prediction intervals. The basic LQGPC cost-function is as described in (3.39). However, in this present section additional common weighting elements $\Sigma_j$ are included (for $j = 0, 1,...,n_1$). These are used when invoking the LQG embedding procedure, later in §5.4.4

For greater generality a cross-product and a common weighting term may be introduced into the cost-function which may then be written in the complex-frequency domain (Grimble, 1995, 1998) as:

$$J = \frac{1}{2\pi j} \oint_{|z|=1} \left\{ \sum_{j=0}^{n_1} \left\{ Q_{cj}\Phi_{\widehat{e}\widehat{e}} + R_{cj}\Phi_{\widehat{u}\widehat{u}} + G_{cj}\Phi_{\widehat{u}\widehat{e}} + \Phi_{\widehat{e}\widehat{u}}G_{cj}^* \right\} \Sigma_j \right\} \frac{dz}{z} \qquad (5.47)$$

where the weightings are again assumed to have the $GH_\infty$ form:

$$Q_{cj} = H_{qj}^*H_{qj}\,, \quad R_{cj} = H_{rj}^*H_{rj}\,, \quad G_{cj} = H_{pj}^*H_{fj} \text{ and } \Sigma_j = W_{\sigma j}W_{\sigma j}^*$$

and these have the following polynomial descriptions:

$$H_{qj} = B_{qj}/A_{wj}, \quad H_{rj} = B_{rj}/A_{wj}$$

$$H_{pj} = B_{pj}/A_{wj}, \quad H_{fj} = B_{fj}/A_{wj} \quad and \quad W_{\sigma j} = B_{\sigma j}/A_{\sigma j}$$

**Remarks**

(i) Some of the $j$ dependent weightings can be zero, as in GPC control, so that the number of steps over which the error is costed can be different to those over which the future controls are penalised.

(ii) In GPC cost problems the weights are constants but in the present case the weightings are allowed to be dynamic. The dynamic weighting functions $W_j$, $j = 0, 1, ..., n_1$, will be computed in later sections, so that the control obtained is the desired multi-step $H_\infty$ predictive control.

Let the form of the predictor equation be based upon assumption Ass(1), (Section 3.4.3) then the optimal predicted controls can be found from a similar solution to that given in the previous chapter.

**Theorem 5.2** : *Linear Quadratic Generalised Predictive Control Law*

The optimal control and predictive control laws to minimise the multi-step cost-function (5.47), for the prediction model based upon Assumption Ass (1), and shown in Fig. 3.2, can be computed from the following spectral factorisation and diophantine equations:

**Spectral factorisation** : The solutions $D_{cj}$, $D_{fj}$, for $j = 0, 1, 2, ..., n_1$ of the following equations are required:

$$D_{cj}^*D_{cj} = B^*B_{qj}^*B_{qj}B + A^*B_{rj}^*B_{rj}A - B^*B_{pj}^*B_{fj}A - A^*B_{fj}^*B_{pj}B \qquad (5.48)$$

$$D_{fj}D_{fj}^* = E_rE_r^* + H_jH_j^* \qquad (5.49)$$

**Diophantine equations** : The solutions $(G_{oj}, H_{oj}, F_{oj})$ for $j = 0, 1, 2, ..., n_1$, with $F_{oj}$ of smallest degree, of the following equations are required:

$$D_{cj}^* G_{oj} z^{-gj} + F_{oj} A A_{wj} A_{\sigma j} = (B^* B_{qj}^* B_{qj} - A^* B_{fj}^* B_{pj}) D_{fj} B_{\sigma j} z^{-gj} \tag{5.50}$$

$$D_{cj}^* H_{oj} z^{-gj} - F_{oj} B A_{wj} A_{\sigma j} = (A^* B_{rj}^* B_{rj} - B^* B_{pj}^* B_{fj}) D_{fj} B_{\sigma j} z^{-gj} \tag{5.51}$$

where $g_j$ are the smallest positive integers which ensure the equations involve only powers of $z^{-1}$.

**Feedback and Prediction Model Controllers for** $j = 0, 1, 2, ..., n_1$.

$$C_j = C_{0d}^{-1} C_{on} = H_{oj}^{-1} G_{oj} \tag{5.52}$$

The closed-loop system and predictors are stable and the characteristic polynomials for the plant and predictor loops are given (for $j = 0, 1, 2, .., n_1$) by:

$$\rho_j = A C_{jd} + B C_{jn} = D_{cj} D_{fj} B_{\sigma j} \tag{5.53}$$

and the minimum-cost $J_{min}$ is given by:

$$J_{\min} = \sum_{j=0}^{n_1} \left\{ \frac{1}{2\pi j} \oint_{|z|=1} \left\{ \left( \frac{F_{0j} F_{oj}^*}{D_{cj} D_{cj}^*} + W_{\sigma j} I_{j1} W_{\sigma j}^* \right) \right\} \frac{dz}{z} \right\} \tag{5.54}$$

where

$$I_{j1} = Y_{fj}^* \left( Q_{cj} - Y_{cj}^{*-1} \left( (W^* Q_{cj} - G_{cj}^*)(Q_{cj} W - G_{cj}) \right) Y_{cj}^{-1} \right) Y_{fj}$$

and

$$Y_{cj} = D_{cj}/(A A_{wj}) \quad \text{and} \quad Y_{fj} = D_{fj}/A$$

**Proof** : Presented in Grimble (1995, 1998) and only a slight extension of the results in Theorem 3.1.

## 5.4.3 Vector-matrix Form of LQGPC Algorithm

An alternative method of computing the control law in vector matrix form emphasises the link to GPC controllers and provides the form of equations where quadratic programming may be used for constrained optimization. This form of the equations is summarised in the following lemma.

**Lemma 5.8:** *LQGPC Controller Calculation*

The LQGPC control law, for the system shown in Fig. 3.1, can be found to minimise the cost-function (5.47) which may be written, invoking the assumptions Ass(1) and (2), as:

$$J = E\{(XU + F)^T (XU + F)\} + J_{min} \tag{5.55}$$

where

$$J_{min} = \sum_{j=0}^{n_1} \left\{ \frac{1}{2\pi j} \oint_{|z|=1} \left\{ \left( \frac{F_{0j} F_{oj}^*}{D_{cj} D_{cj}^*} + W_{\sigma j} I_{j1} W_{\sigma j}^* \right) \right\} \frac{dz}{z} \right\}$$

and $I_{j1}$ is defined by reference to $I_{j1}$. The matrix X and vectors, $U, F$, are defined as:

$$X = \begin{bmatrix} w_0^{n_1} & w_1^{n_1} & w_2^{n_1} & \cdots & & w_{n_1}^{n_1} \\ & w_0^{n-1} & w_1^{n_1-1} & & & \\ & & w_0^{n1-2} & \ddots & & \vdots \\ \vdots & \vdots & \vdots & & w_0^1 & w_1^1 \\ 0 & 0 & 0 & & 0 & w_0^0 \end{bmatrix},$$

$$F = \begin{bmatrix} f_{n_1 a}(t) \\ \vdots \\ f_{1a}(t) \\ f_{0a}(t) \end{bmatrix}, \qquad U = \begin{bmatrix} u(t+n_1) \\ \vdots \\ u(t+1) \\ u(t) \end{bmatrix}$$

The elements of $F$ can be defined using:

$$f_{ja}(t) = (-1/(A_{wj} A_{\sigma j} D_{fj})) (G_{oj} \hat{e}_j(t + j|t) - S_j u(t - 1)) \tag{5.56}$$

where $S_j$ is obtained from the solution $(S_j, W_j)$, where $W_j$ is of smallest degree, of the diophantine equation $(j = 0, 1, 2, ..., n_1)$ :

$$W_j A_{wj} A_{\sigma j} D_{fj} + z^{-j-1} S_j = H_{oj} \tag{5.57}$$

and $H_{oj}$ follows from the coupled diophantine equations (5.50) and (5.51).

**Predicted and optimal controls:**

$$U = -X^{-1} F \tag{5.58}$$

The optimal control and predicted controls are stable and satisfy the characteristic equations (5.53).

**Proof :** The proof is given in Grimble (1998) and is only a small extension of the results in Lemma 3.2.

### 5.4.4   The $H_\infty$ Predictive Control Problem

The $H_\infty$ robust feedback and predictive controllers can be derived directly from the LQGPC polynomial system results by computing the common cost function weightings $W_{\sigma j}$ which link the two problems. This method is the LQG embedding (Grimble, 1986) approach introduced in §5.2.1 and requires an extension of the lemma due to Kwakernaak (1985). The key idea is to find the weighting functions $W_{\sigma j}$ so that the $H_\infty$ cost terms to be minimised have an optimal equalising solution.

**Lemma 5.9:** *Multi-step $H_\infty$ Cost-function minimisation*

Consider the auxiliary problem of minimising the multistep $H_2$ cost-function:

$$J = \sum_{j=0}^{n_1} \left\{ \frac{1}{2\pi j} \oint_{|z|=1} \{I_j(z^{-1})\Sigma_j(z^{-1})\} \frac{dz}{z} \right\} \tag{5.59}$$

where each of the functions $I_j(z^{-1})$ are only dependent upon a controller $C_j(z^{-1})$ for $j = 0, 1, 2, ..., n_1$. Suppose that for some real-rational set of weightings $\Sigma_j(z^{-1}) = \Sigma_j^*(z^{-1}) > 0$ the criterion $J$ is minimised by a set of functions $I_j(z^{-1}$ for which $I_j(z^{-1}) = \lambda_j^2$ (real constants for each $j$) on $|z| = 1$. Then this function also minimises:

$$J_\infty = \sup_{|z|=1} \left\{ \sum_{j=0}^{n_1} I_j(z^{-1}) \right\} = \sum_{j=0}^{n_1} \|I_j(z^{-1})\|_\infty \tag{5.60}$$

**Proof :** A proof by contradiction follows. Let $z = e^{j\omega}$ then the integral expression in equation (5.59) may be written as:

$$J = \sum_{j=0}^{n_1} \left\{ \frac{1}{2\pi j} \int_{-\pi}^{\pi} I_j(e^{-j\omega})\Sigma_j(e^{-j\omega}) e^{-j\omega} d(e^{j\omega}) \right\}$$

$$= \sum_{j=0}^{n_1} \left\{ \frac{1}{2\pi} \int_{-\pi}^{\pi} I_j(e^{-j\omega})\Sigma_j(e^{-j\omega}) \right\} d\omega$$

Suppose that there exists a set of feedback and predictive compensators $\{C_j(z^{-1})\}$ and that one (or more) of the functions $I_j(z^{-1})$ has the property that,

$$\sup_\omega \{I_j(e^{-j\omega})\} < \sup_\omega \{I_j^0(e^{-j\omega})\} = \lambda_j^2$$

where $I_j^0(z^{-1})$, for $j = 0, 1, 2, ..., n_1$, denote the $H_2$ optimal solutions to (5.59). Then necessarily,

$$I_j(e^{-j\omega}) < \lambda_j^2 = I_j^0(e^{-j\omega})$$

for all $\omega \epsilon [-\pi, \pi]$, which implies that,

$$\int_{-\pi}^{\pi} I_j(e^{-j\omega})\Sigma_j(e^{-j\omega}) d\omega < \int_{-\pi}^{\pi} I_j^0(e^{-j\omega})\Sigma_j(e^{-j\omega}) d\omega$$

but this contradicts the assumption that the functions $I_j^0(e^{-j\omega})$, for $j = 0, 1, 2, ..., n_1$ minimise the cost function (5.59).

The function to be minimised in an $H_\infty$ sense involves the same cost terms as in the integrand of (5.47). Although these terms are the basis of the traditional LQG or $H_2$ cost-function they can easily be shown to be physically important in an $H_\infty$ problem. This is due to the fact that the signal spectra are dependent upon the sensitivity functions which must be minimised. Thus let,

$$I_c(z^{-1}) = \sum_{j=0}^{n_1} (Q_{cj}\Phi_{\widehat{e}\widehat{e}} + R_{cj}\Phi_{\widehat{u}\widehat{u}} + G_{cj}\Phi_{\widehat{u}\widehat{e}} + \Phi_{\widehat{e}\widehat{u}}G_{cj}^*) \tag{5.61}$$

Then the cost-function to be minimised may be expressed:

$$J_\infty = \sup_{|z|=1}\{|I_c(z^{-1})|\} = \sup_{|z|=1}\{|I_c(e^{-j\omega})|\} = \sum_{j=0}^{n_1}\lambda_j^2 \tag{5.62}$$

## 5.4.5   The Equalising $H_\infty$GPC Solution

It is now a straightforward step to obtain the expressions which determine the $H_\infty$ equivalent of the preceding LQGPC problem. The system is the same as in Fig. 3.1 and the function to be minimised is defined by (5.61).

To invoke Lemma 5.9 note that the set of weightings $\Sigma_j$ must be determined so that an equalising solution for $j = 0, 1, 2, ..., n_1$ is obtained. Equating the integrand of (5.59) with the expression for the optimum $H_\infty$ function integrand (5.54) obtain:

$$\sum_{j=0}^{n_1}\{\lambda_j^2\Sigma_j(z^{-1})\} = \sum_{j=0}^{n_1}\{F_{0j}F_{0j}^*/(D_{cj}D_{cj}^*) + \Sigma_j(z^{-1})I_{j1}\} \tag{5.63}$$

Thus, if the conditions of Lemma 5.9 are to be satisfied the dynamic weightings must satisfy the above equation (5.63), or each $\Sigma_j$ must satisfy:

$$\Sigma_j = \frac{B_{\sigma j}B_{\sigma j}^*}{A_{\sigma j}A_{\sigma j}^*} = \frac{F_{0j}F_{0j}^*}{(D_{cj}^*D_{cj})(\lambda_j^2 - I_{j1})} \tag{5.64}$$

To obtain the expressions for $B_{\sigma j}$ and $A_{\sigma j}$ from this equation first define the Schur polynomials $F_{sj}$ from the spectral factorisation:

$$F_{sj}F_{sj}^* = F_{0j}F_{0j}^* \tag{5.65}$$

Also define the polynomials $T_{nj}$ and $T_{dj}$ so that after the cancellation of common factors:

$$T_{nj}T_{nj}^*/(T_{dj}T_{dj}^*) = D_{cj}^*D_{cj}I_{jl}$$

where $T_{dj}$ is a Schur polynomial.    Substituting in equation (5.64), for the robust weightings, obtain:

$$B_{\sigma j}B_{\sigma j}^*/(A_{\sigma j}A_{\sigma j}^*) = F_{sj}F_{sj}^*/(\lambda_j^2 D_{cj}^*D_{cj} - T_{nj}T_{nj}^*/(T_{dj}T_{dj}^*)) \tag{5.66}$$

and hence the Schur polynomials $B_{\sigma j}$ and $A_{\sigma j}$ satisfy:

$$B_{\sigma j} = F_{sj} T_{dj}$$

and

$$A_{\sigma j} A_{\sigma j}^* = \lambda_j^2 D_{cj}^* D_{cj} T_{dj} T_{dj}^* - T_{nj} T_{nj}^* \qquad (5.67)$$

This completes the derivation of the expression for the weighting function polynomials $W_{\sigma j} = A_{\sigma j}^{-1} B_{\sigma j}$ which ensure the conditions of Lemma 5.9 are satisfied. The $H_\infty$ predictive controller follows directly by substituting for these weighting terms in the results of Theorem 5.1.

## 5.5   Need for Adaptation and $H_\infty$ Robustness

Adaptive action may be necessary even when a controller is designed by $H_\infty$ techniques. That is, robustness alone may not be sufficient. The reasons for possibly including adaptive or self-tuning action are as follows:

1. Robust controllers have constant coefficients and are designed to deal with a range of system modelling errors. However, there are some systems which have such a large class of uncertainty or where the changes are so significant (even if slow) that no fixed controller could stabilise them. There is therefore a class of problems which could be compensated by an adaptive controller but where system uncertainties/variations make it impossible for one fixed robust controller to stabilise the total system.

2. Even if a robust controller can stabilise an uncertain system the price paid is often very low gains and poor performance. An adaptive controller can reduce the amount of uncertainty and hence widen the bandwidth of the controlled system. Thus, if high performance is required in addition to good robustness it may be necessary to have both adaptation and robust design.

3. Robust design procedures are excellent for aerospace application where good models of the real world system are available and the uncertainty which may result from using a lower order controller. In such cases, the modelling error is reasonably well defined and is readily available from extensive simulation and modelling data. However, as the growth of the adaptive process controls market illustrates, there is a need for controllers which will tune themselves without the need for extensive modelling or design work. Many food processing and small pharmaceutical companies fall into this category. In such cases plant engineers may wish to have an adaptive or self-tuning control but they may also need this to be robust. Since in such cases there is a need for self-tuning action some mechanism must be provided to ensure it is robust and $H_\infty$ design provides one such option.

## 5.6   Concluding Remarks

Three very different types of $H_\infty$ problem have been considered. The scalar problems first discussed enabled mixed $H_2$ and $H_\infty$ criteria to be minimised, for feedback, tracking and feedforward terms. The main idea was that of the *Embedding Principle*, which enables $H_\infty$ problems to be solved using the previous $H_2$ polynomial solutions (in Chapters 2 and 3). The $GH_\infty$ cost function was also introduced which involves a special relationship between the cost terms. This provides a practical cost index which has the merit of simplifying the polynomial equations to be solved.

The single-input and multi-output control problems that have power system applications were then considered. Although this is a special case of the multivariable solution (Grimble 1989a) it is of numerical value since there are considerable simplifications which occur. Finally, $H_\infty$ predictive optimal control laws were introduced. This range of problems demonstrates the great flexibility the $H_\infty$ polynomial approach provides (Casavola and Mosca, 1989).

The high order of $H_\infty$ controllers is sometimes a problem for systems where on-line tuning is desirable. By restricting the plant and cost function descriptions, Grimble (1991) derived a PID controller that minimised an $H_\infty$ criterion. More recently the topic of *restricted structure optimal control* has been introduced (Grimble, 1999b). This involves direct minimisation of $H_2$ or $H_\infty$ criteria for controllers of a pre-specified (low-order) structure. An alternative method of obtaining a simple $H_\infty$ control law is to use the so-called *Observations Weighted concept* (Grimble, 1991). There are several variants of the basic $H_\infty$ problem still to be fully explored, such as different criteria for multivariable systems (Grimble, 1988) and mixed $H_2/H_\infty$ problems (Grimble, 1989b; Yeh et al. 1991a,b; Schaper et al. 1990; Zhou et al. 1990).

The $H_\infty$ synthesis problems have an important role in the toolkit of the design engineer. The unit step responses of such a system are normally much better than might be expected. This also applies to the interaction in multivariable systems, which can often be limited without much difficulty. In fact the $H_\infty$ design approach provides a true multivariable solution, whereas many other methods often equate to a sequence of SISO designs. Robustness issues are addressed in later sections, although real robustness can still be difficult to achieve, since mathematical models can be unrepresentative of the actual situation. Nevertheless, the meritorious behaviour of $H_\infty$ designs suggests a long and valuable future for the approach.

## 5.7   References

Casavola, A. and Mosca, E., 1989, On the polynomial solution of the $H_\infty$ generalised sensitivity minimisation problem, *Proc. 28th CDC*, Tampa, Florida, pp. 1500-1505.

Doyle J.C., Glover, K., Khargonekar, P.P. and Francis, B.A., 1989, State-space solutions to standard $H_2$ and $H_\infty$ control problems, *IEEE Trans. Automatic Control.* Vol. **34**, No. 2, pp. 831-846.

Fragopoulos, D., Grimble M.J. and Shaked, U., 1991, $H_\infty$ controller design for the SISO case using a Wiener approach, *IEEE Transactions on Automatic Control,*

Vol. **36**, No. 10.

Francis, B.A., 1987, A course in control theory, *Lecture notes in control and information science,* Springer-Verlag.

Glover, K. and McFarlane, D., 1989, Robust stabilization of normalised coprime factor plant descriptions with $H_\infty$ bounded uncertainty, *IEEE Trans Automatic Control.* Vol. **34**, No. 8, pp. 821-830.

Grimble, M.J., 1985, $H_\infty$ and LQG robust design methods for uncertain linear system in *Model Error Concepts and Compensation,* Proc. of IFAC Workshop, Boston, USA, pp. 91-96.

Grimble, M.J., 1986, Optimal $H_\infty$ robustness and the relationship to LQG design problems, *Int. J. Control,* Vol. **43**, No. 2, pp. 351-372.

Grimble, M.J., 1986, Observations weighted controllers for linear stochastic sysetms, *Automatica,* Vol. **22**, No. 4, pp. 425-431.

Grimble, M.J., 1987a, $H_\infty$ robust controller for self-tuning applications, part 1 : Controller design, *Int. J. Control,* Vol. **46**, No. 4, pp. 1429-144.

Grimble, M.J., 1987b, $H_\infty$ robust controller for self-tuning applications, part 2 : Self tuning robustness, *Int. J. Control,* Vol. **46**, No. 5, pp. 1819-1840

Grimble, M.J., 1987c, Simplification of the equation in the paper optimal $H_\infty$ robustness and the relationship to LQG design, problems, *Int. J. Control.* Vol. **46**, No. 5, pp. 1841-1843.

Grimble, M.J., 1988, Minimax design of optimal stochastic multivariable systems, *IEE Proc.,* Vol. **135**, Pt.D. No. 6, pp. 436-440.

Grimble, M.J., 1989a, Generalised $H_\infty$ multivariable controllers, *Proc. IEE,* 136, (6), pp. 285-297.

Grimble, M.J., 1989b, minimisation of a combined $H_\infty$ and LQG cost-function for at two degrees of freedom control design, *Automatica,* Vol. **15**, No. 3, pp. 635-638.

Grimble, M.J., 1991, $H_\infty$ controller with a PID structure, *Journal of Dynamic Systems, Measurement and Control,* Vol. **112**

Grimble, M.J., 1994, *Robust Industrial Control,* Prentice Hall, Hemel Hempstead.

Grimble, M.J., 1995, Multivariable linear quadratic generalised predictive control, *IEEE CDC Conference,* New Orleans.

Grimble, M.J., 1998, Multistep $H_\infty$ generalised predictive control, *Dynamics and Control,* **8**, pp. 303-339.

Grimble, M.J., 1999a, Polynomial solution of the $3\frac{1}{2}$DOF $H_2/H_\infty$ feedforward control problem, *IEE Proc. Control Theory Appl.,* Vol. **146**, No. 6, pp. 549-560.

Grimble, M.J., 1999b, Restricted structure feedforward and feedback stochastic optimal control, *IEEE CDC Conference*, Phoenix, Arizona.

Grimble, M.J. and Johnson, M.A., 1988, *Optimal control and stochastic estimation, Vols I and II*, John Wiley, Chichester.

Grimble, M.J. and Johnson, M.A., 1991, $H_\infty$ robust control design - a tutorial review, *IEE Computing and Control Engineering Journal*. Vol. **2**, No. 6, pp. 275-281.

Grimble, M.J. and Kucera, V., (Eds), 1996, *Polynomial methods for control systems design*, Springer-Verlag.

Kwakernaak, H., 1985, Minimax frequency domain performance and robustness optimisation of linear feedback systems, *IEEE Trans. On Auto. Control.*, AC-**30**, 10, pp. 994-1004.

Kwakernaak, H., 1986, *A polynomial approach to $H_\infty$ optimization of control system*, in NATO ASI Series, Series F - Vol. 34, Modeling, Robustness and Sensitivity Reduction in Control Systems, Ed. by R. Curtain, Springer-Verlag, pp. 83-94.

McFarlane, D.C., Glover, K. and Vidyasagar, M., 1990a, Reduced-order controller design using coprime factor model reduction, *IEEE Trans. Automat. Contr.*, Vol. **35**, No. 3, pp. 369-373.

McFarlane, D.C. and Glover, K., 1990b, *Robust controller design using normalised coprime factor plant descriptions*, Vol. 138, Lecture Notes in Control and Information Sciences, Springer-Verlag.

McFarlane, D.C. and Glover, K., 1992, A loop shaping design procedure using $H_\infty$ synthesis, *IEEE Trans. Automat. Contr.*, Vol. **37**, No. 6, pp. 759-769.

Mosca, E., Casavola, A. and Giarre, L., 1990, Minimax LQ stochastic tracking and servo problems, *IEEE Trans. AC,* Vol. **35**, No. 1.

Saeki, M., Grimble, M.J., Kornegoor, E. and Johnson, M.A., 1987, $H_\infty$ optimal control, LQG polynomial systems techniques and numerical procedures, *NATO ASI Series*, Vol. **F34**.

Schaper C.D., Mellichamp, D.A. and Seborg, D.E., 1990, Stability robustness relations for combined $H_\infty$/LQG control, In Proceedings *American Control Conference*, pages 3050-305, San Diego, California.

Tse, J., Bentsman, J. and Miller, N., 1992, Minimax predictive control, *Proc of the 31st Conference on Decision and Control*, Tucson, Arizona, pp. 2165-2170.

Tse, J., Bentsman, J. and Miller, N., 1993a, Mixed $H_2/H_\infty$ predictive control, *IFAC World Congress,* Sydney, Australia, Vol. 2, pp. 467-470.

Tse, J., Bentsman, J. and Miller, N., 1993b, Properties of the self-tuning minimax predictive control (MPC), *Proc of the ACC,* San Francisco, California, pp. 1721-1725.

Yeh, H-H, Banda, S.S. and Heise, S.A., 1991a, Robust control of uncertain systems with combined $H_\infty$ and LQG optimisations,International Journal of System Science, 22(1):85-96.

Yeh, H-H, Banda, S.S., Sparks, A.G. and Ridgely, D.B., 1991b, Loop shaping in mixed $H_2/H_\infty$ optimal control, In Proceedings American Control Conference, pages 11651170, Boston, Massachusetts.

Zames, G., 1981, Feedback and optimal sensitivity : Model reference transformations, multiplicative seminorms and approximate inverses, *IEEE Trans Automatic Control.* Vol. **26**, No. 2, pp 301-320.

Zames, G., and Francis, B.A., 1981, A new approach to classical frequency methods : feedback and minimax sensitivity, *CDC*, San Diego, California.

Zhou, K., Doyle, J.C., Glover, K. and Bodenheimer, B., 1990, Mixed $H_2$ and $H_\infty$ control. In Proceedings American Control Conference, pages 2502-2507, San Diego, California.

Zhou, K., Doyle, J.C. and Glover, K., 1996, *Robust and optimal control*, Prentice Hall, New York.

# 6

# H$_2$ and H$_\infty$ Filtering and Prediction

## 6.1 Introduction

The aim of this chaper is to introduce the polynomial approach to $H_2$ and $H_\infty$ estimation problems. These estimators are suitable for stationary noise and time-invariant systems. They provide an alternative to the Kalman filter, that in its steady-state version has a transfer-function matrix form (where the initial time $t_0 \to -\infty$). In fact polynomial approaches may also be used for non-stationary and time-varying systems (Grimble, 1985b) but these are not explored further here.

The return-difference properties of the stationary Kalman filter were established by MacFarlane (1971). The solution of the Wiener filtering problem, using these relationships for the state-space based Kalman filter (Kalman, 1960), was obtained by Barrett (1977). The useful frequency domain properties of the Kalman filter were identified by Grimble and Åström (1987). However, the polynomial systems approach to optimal control, introduced in previous sections, can also be applied to optimal linear estimation problems, and it is sometimes more convenient numerically (as illustrated in Grimble 1985, 1988). The advantages of using polynomial methods in control problems apply equally to the $H_2$ and $H_\infty$ estimation problems. The advantages of polynomial systems include greater insights and more suitable structure for adaptive applications (Grimble and Kucera 1996, Moir and Grimble 1984).

The frequency-domain or polynomial systems approach to robust linear multichannel estimation and the prediction of time-series is taken in this Chapter, and both $H_2$ and $H_\infty$ problems are considered. The solutions of the first $H_\infty$ filtering and prediction problems were obtained in Grimble (1987) and the $H_\infty$ fixed lag smoothing problem was solved in Grimble (1991). The $H_\infty$ filtering problem was developed further by Elsayed and Grimble (1989), Elsayed (1988) and Grimble and Elsayed (1990). The filter was derived using a polynomial systems model and a frequency-domain based solution. The importance of the $H_\infty$ filter in signal processing problems was recognised early in its development. However, the role of the $H_\infty$ filter in the robust control problem was not realised until more recently, when a state-space solution of the $H_\infty$ control problem was obtained (Doyle et al. 1989). The state-space solution of the $H_\infty$ filtering problem was obtained by Shaked (1990) and by Yaesh and Shaked (1992). This is discussed further in Chapter 7.

The system and noise models in the following can include uncertain system elements, which can be represented by either $H_\infty$ frequency weighted bounds or by linear models with probabilistic parameter deviations. In either case, the optimal robust filter, smoother or predictor can be obtained from the results of a frequency weighted estimation problem. Both of the uncertainty descriptions lead to a similar $H_\infty$ filtering problem and solution:

(a) Problems where a $H_\infty$ filter is required to minimise the peak of the estimation error spectrum but the system includes uncertain probabilistic elements.

(b) Problems where a filter must be computed to be robust to system uncertainties, described both probabilistically and by frequency bounding functions. The latter implies the need for a $H_\infty$ filter.

The $H_2$ deconvolution filtering problem is discussed first, since this can be specialised to the usual smoothing, filtering and prediction problems. The deconvolution filtering problem can involve the estimation of the signal input to a communication channel, where only noisy output measurements are available. The channel can be represented by a dynamical system which distorts the gain and phase characteristics of the signal. The measurement noise can be correlated with the signal sources and can also be coloured. This is more general than the usual white noise filtering problem, whose results can be recovered from those presented (Oppenheim et al. 1983).

A common application for deconvolution smoothers is in reflection seismology where the primary reflectogram is assumed to be white and the distorting system is the seismic wavelet (Mendel, 1977). A polynomial based solution to the deconvolution smoothing problem was proposed by Moir (1986). The solution of smoothing problems, using a polynomial approach, was described by Elsayed et al. (1989). The $H_2$ multichannel deconvolution problem has also been considered by Chisci and Mosca (1992), for the uncorrelated noise source case.

The solution of the $H_2$ deconvolution filtering problem is first derived and the $H_\infty$ filter is then obtained from the equivalent of the LQG embedding procedure described in Chapter 5. The $H_\infty$ norm minimisation problem provides a valuable method of tackling the robustness problem. The use of a probabilistic model for the system parameters treats the parametric uncertainty, while the $H_\infty$ norm is useful for the unmodelled dynamics that have norm bounded uncertainty.

In the second half of the chapter attention turns to the solution of optimal estimation problems, represented by a standard system description. This is an even more general estimation problem construction that can be used for both $H_2$ and $H_\infty$ estimation error cost minimisation. It is a particularly valuable setting for constructing standard computer aided design packages.

## 6.2   $H_2$ Estimation Using Probabilistic Models

Optimal linear estimation problems are considered for systems including probabilistic uncertainty models. The solution of the Wiener and Kalman filtering problems for systems with probabilistic models was obtained by Grimble (1984) and by Sternad and Ahlen (1990, 1993a,b). However, the system model and the uncertainty description is different to that introduced below. The least squares estimation problems for known systems are then considered before attention turns to the related $H_\infty$ filtering, smoothing and prediction problems.

## 6.2.1   Uncertain Signal Model and Noise Descriptions

The system, signal and noise models are assumed to be time-invariant and discrete-time, and to be represented in transfer function or polynomial matrix form.    The signal processing problem is illustrated in Fig. 6.1. The signal to be estimated $\{s(t)\}$ is assumed to enter a transmission channel $W_c(z^{-l})$ and to then be corrupted by coloured measurement noise $\{n(t)\}$. The resulting observations signal $\{z(t)\}$ is input to the linear filter $H_f(z^{-1})$ to obtain the estimate of the signal. The signal source $W_s(z^{-1})$ and coloured noise $W_n(z^{-1})$ models are driven by the white noise sources of $\{\xi(t)\}$ and $\{\omega(t)\}$, respectively. All noise sources are assumed to be zero mean and white noise. The measurement noise $\{n(t)\}$ may be correlated with the signal source, and the covariance matrix for the signal $[\xi(t)^T, \omega^T(t)]^T$ is defined as:

$$E \left\{ \begin{bmatrix} \xi(t) \\ \omega(t) \end{bmatrix} [\xi^T(\tau),\ \omega^T(\tau)] \right\} = \begin{bmatrix} I & G \\ G^T & I \end{bmatrix} \delta_{t\tau} \qquad (6.1)$$

where $\delta_{t\tau}$ is the Kronecker delta ($\delta_{t\tau} = 1$ for $t = \tau$ and $\delta_{t\tau} = 0$ for $t \neq \tau$).    There is no loss of generality in assuming the white noise sources $\{\xi(t)\}$ and $\{\omega(t)\}$, in equation (6.1), are normalised to the identity matrix.

**System equations**
By inspection of Fig. 6.1 the following system equations apply:

**Signal model:**

$$s = W_s\xi$$

**Channel output:**

$$y = W_c s$$

**Measurement noise:**

$$n = W_n\omega$$

**Observations signal:**

$$z = W_n\omega + W_c W_s\xi \qquad (6.2)$$

where $s(t) \in R^{n_s}$, $y(t) \in R^{n_y}$, and $z(t) \in R^{n_z}$ and $n_y = n_z$ denotes the number of measured outputs.

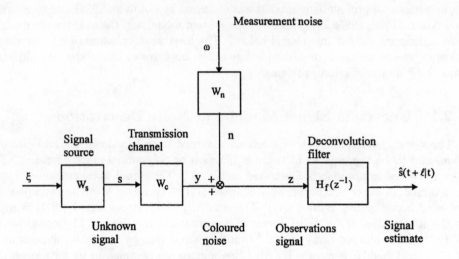

Fig. **6.1** : Optimal Deconvolution Filtering Problem System Model

The optimal linear filter output estimate will be represented in the form $\hat{s}(t|t-\ell) = (H_f(z^{-1})z^{-\ell})z(t)$ , where the filtering, prediction and fixed-lag smoothing problems correspond with $\ell = 0,\ \ell > 0$ and $\ell < 0$ , respectively.

After manipulation the estimation and estimation error equations become as follows:

**Estimate:**

$$\hat{s} = (H_f z^{-\ell})z = (H_f z^{-\ell})(W_n\omega + W_c W_s \xi)$$

**Estimation error:**

$$e = s - \hat{s} = W_s \xi - (H_f z^{-\ell})(W_n\omega + W_c W_s \xi)$$

$$= (I - H_f z^{-\ell} W_c)W_s \xi - (H_f z^{-\ell})W_n\omega$$

**Weighted error:**

$$e_0 = W_p(s - \hat{s}) = W_p W_s \xi - W_p(H_f z^{-\ell})(W_n\omega + W_c W_s \xi)$$

$$= W_p(I - H_f z^{-\ell} W_c)W_s \xi - W_p(H_f z^{-\ell})W_n\omega$$

**Summary of Assumptions**

(1i) Noise sources are assumed to be zero mean.

(1ii) The signal channel $W_c$ and the coloured measurement noise model $W_n$ can be assumed to be asymptotically stable.

(1iii) There can be no unstable hidden modes in the individual plant subsystems or in the cascade system $W_c W_s$.

## Probabilistic Models for Uncertainty

The system models $W_c, W_s$ and $W_n$ are assumed to be uncertain (Grimble, 1992). They have the following uncertain system description:

$$W_c = \overline{W}_c \delta W_c$$

$$W_s = \overline{W}_s \Delta W_s = \overline{W}_s \delta W_s \Delta W_{si}$$

$$W_n = \overline{W}_n \Delta W_n = \overline{W}_n \delta W_n \Delta W_{ni}$$

where $\delta W_c, \delta W_s$ and $\delta W_n$ are linear in the random parameters but are otherwise completely determined and $\Delta W_{si}$ and $\Delta W_{ni}$ are only determined by an upper ($H_\infty$) norm bound.

This type of uncertainty description was proposed by Grimble (1992). It has one advantage over the traditional multiplicative ($\tilde{W} = \overline{W}(I + \Delta W)$) and additive ($\tilde{W} = \overline{W} + \Delta W$) uncertainty models. That is, if say a continuous-time system has a strictly proper perturbation $\Delta W$ then the frequency responses of $\tilde{W}$ and $\overline{W}$ will be identical at high-frequencies. The proposed structure enables extra roll-off to be introduced by the uncertainty at high frequencies. This is a useful model for both continuous and discrete-time systems.

Note that the probabilistic models for the uncertain system elements can be written as:

$$\delta W_c = 1 + \delta W_{c1}, \quad \delta W_s = 1 + \delta W_{s1}, \quad \delta W_n = 1 + \delta W_{n1}$$

where by definition $E_\theta\{\delta W_{c1}\} = E_\theta\{\delta W_{s1}\} = E_\theta\{\delta W_{n1}\} = 0$. Here the expectation, with respect to the unknown parameters (represented by a vector of unknowns $\theta$), is denoted by $E_\theta\{.\}$.

The uncertain $H_\infty$-norm bounded elements $\Delta W_{si}$ and $\Delta W_{ni}$ will be assumed to be set to unity for the generation of the $H_2$ filtering problem results which follow. The uncertainty in these elements will be considered when the $H_\infty$-norm minimisation problem is discussed later.

## Polynomial Matrix Descriptions

The transfer function models for the system are assumed to have the following polynomial matrix representations (Kailath, 1980):

$$[W_c \ W_n] = A^{-1}[C_c \ C_n] \tag{6.3}$$

$$W_s = A_s^{-1} C_s \tag{6.4}$$

where $A, C_c, C_n, A_s$ and $C_s$ are polynomial matrices in $z^{-l}$ of compatible dimensions. Through the assumption concerning the random parameters the probabilistic uncertainty affects the polynomial terms $C_c, C_n$ and $C_s$. Introduce the left-coprime pair $C_1$ and $A_1$ which satisfy:

$$C_c A_s^{-1} = A_1^{-1} C_1 \qquad (6.5)$$

and assume, without loss of generality, $A_1(0) = I$ and $A_s(0) = I$. Note for later use that from Assumption (1iii) the model $C_c A_s^{-1}$ can include no unstable hidden modes.

**Spectral Factorisation**

The solution of the deconvolution filtering problem requires the introduction of an uncertainty averaged spectral-factor $Y_f$ (Shaked, 1976) which satisfies the following equation:

$$Y_f Y_f^* = E_\theta \{ W_c W_s W_s^* W_c^* + W_n W_n^* + W_c W_s G W_n^* + W_n G^T W_s^* W_c^* \} \qquad (6.6)$$

Replacing the system models in equation (6.6) by their polynomial matrix counterparts, in (6.3) to (6.5), the generalised spectral factor $Y_f \epsilon R^{n_y \times n_y}(z^{-1})$ may be written as:

$$Y_f = (A_1 A)^{-1} D_f \qquad (6.7)$$

where $D_f$ is a polynomial matrix left spectral-factor defined using,

$$D_f D_f^* = E_\theta \{ C_1 C_s C_s^* C_1^* + A_1 C_n C_n^* A_1^* + C_1 C_s G C_n^* A_1^* + A_1 C_n G^T C_s^* C_1^* \} \qquad (6.8)$$

The Schur spectral-factor $D_f$ exists, if and only if,

$$rank[\{C_1 C_s, \quad A_1 C_n] \begin{bmatrix} I & G \\ G^T & I \end{bmatrix} \} = n_y$$

The spectral factor $D_f$ is strictly Schur, if and only if, $[C_1 C_s, \quad A_1 C_n] \begin{bmatrix} I & G \\ G^T & I \end{bmatrix}$ has no left common factors with zeros on the unit-circle in the z-plane.

**Assumption (1iv)** : The noise and signal source models are assumed to be such that $D_f$ exists and is strictly Schur.

## 6.2.2  The $H_2$ Cost Minimisation Problem and Theorem

In this section the problem of the minimisation of the variance of the weighted estimation error is considered (Grimble 1994a and Ahlen and Sternad 1991, 1993). These results are needed before attention can turn to the $H_\infty$ optimal estimation problem for an uncertain system (Grimble, 1996). The $H_2$ optimal deconvolution problem involves the minimisation of the estimation error:

$$\hat{e}(t|t - \ell) = s(t) - \hat{s}(t|t - \ell) \qquad (6.9)$$

where $\hat{s}(t|t-\ell)$ denotes the optimal linear estimate of the signal$\{s(t)\}$ at time $t$, given observations $\{z(t)\}$ up to time $(t-\ell)$ . The scalar $\ell$ may be positive or negative (for the smoothing case see Elsayed et al. 1989):

$\ell = 0 :$ *Filtering*
$\ell < 0 :$ *Fixed-lag smoothing*
$\ell > 0 :$ *Prediction*

The estimation error cost-function which is to be minimised has the form:

$$J = trace\{E\{\hat{e}(t|t-\ell)\hat{e}^T(t|t-\ell)\}\} \tag{6.10}$$

where $E\{.\} = E_\theta\{E_s\{.\}\}$ denotes the unconditional expectation operator and $E_\theta, E_s$ denotes the expectations with respect to the random parameters and the stochastic noise signals, respectively.

If, for example, there is one uncertain parameter, having $n$ possible values $\{\alpha_1, \alpha_2 ,.., \alpha_n\}$, with probabilities $\{p_1, p_2, ..., p_n\}$, the variance of the estimation error can be computed as:

$$J = E_\theta\{E_s\{\hat{e}^T(t|t-\ell)\hat{e}(t|t-\ell)\}\} = \sum_{j=1}^{n_1} p_j E_s\{\hat{e}^T(t|t-\ell)\hat{e}(t|t-\ell)\}|_{\alpha_j}$$

This criterion takes into account the likelihood of different modelling errors.

It is useful to introduce a dynamic weighting function if the error is to be penalised in particular frequency ranges. Thus, introduce the asymptotically stable $n_s$ square weighting function matrix: $W_p = A_p^{-1} B_p$ , where $A_p$ and $B_p$ denote square polynomial matrices. Then the uncertainty averaged $H_2$ cost-function to be minimised becomes:

$$J = trace\{E\{W_p\hat{e}(t|t-\ell)(W_p\hat{e}(t|t-\ell))^T\}\}$$

and this can be written in the complex integral form:

$$J = trace\{\frac{1}{2\pi j} \oint_{|z|=1} E_\theta\{W_p\Phi_{ee}W_p^*\}\frac{dz}{z}\} \tag{6.11}$$

where the adjoint $W_p(z^{-1})^* = W_p^T(z)$ .

**Theorem 6.1 :** *$H_2$ Optimal Deconvolution Estimator for an Uncertain System*

Consider the signal processing system and the assumptions described in §6.2.1 and illustrated in Fig. 6.1. The optimal deconvolution filter to minimise the variance of the estimation error (6.11), can be calculated from the following uncertainty averaged spectral factor and diophantine equations.

**Spectral factor:**

$$D_f D_f^* = E_\theta\{C_1 C_s C_s^* C_1^* + A_1 C_n C_n^* A_1^* + C_1 C_s GC_n^* A_1^* + A_1 C_n G^T C_s^* C_1^*\} \tag{6.12}$$

where $C_c A_s^{-1} = A_1^{-1} C_1$ and from the assumptions $D_f$ is strictly Schur.

**Diophantine equations:**

The polynomial matrix solution $(F_0, G_0, S_0)$, with $F_0$ of smallest degree, is required:

$$A_s F_0 + G_0 D_f^* z^{-g} = E_\theta \{C_s (C_s^* C_1^* + G C_n^* A_1^*) z^{\ell-g}\} \tag{6.13}$$

$$-C_c F_0 + S_0 D_f^* z^{-g+\ell_0} = E_\theta \{(C_n C_n^* A_1^* + C_n G^T C_s^* C_1^*) z^{\ell-g}\} \tag{6.14}$$

where for filtering and prediction $\ell_0 = \ell$ and for smoothing $\ell_0 = 0$, and where $g$ denotes the smallest positive integer, which ensures these equations involve polynomials in the indeterminate $z^{-1}$.

**Robustness weighting diophantine equation:**

If the cost-weighting $W_p$ is dynamic the diophantine equation solution $(F_1, N)$, with $F_1$ of smallest degree, is required using:

$$A_p F_1 + N D_f^* z^{-g} = B_p F_0 \tag{6.15}$$

**Optimal estimator:**

$$H_f = (A_s^{-1} G_0 + B_p^{-1} N) D_f^{-1} A_1 A \tag{6.16}$$

**Proof** : Presented in the following section.

### 6.2.3   Solution of Uncertain $H_2$ Deconvolution Problems

A solution is presented in the following for the discrete multivariable deconvolution filtering problem, using the form of solution procedure developed by Kučera (1979), which involves a completing the squares argument.

The estimation error spectrum (to be minimised in a weighted $H_2$ sense) can be expanded using the results in §6.2.1, as follows:

$$\Phi_{ee} = (I - H_f z^{-\ell} W_c) W_s W_s^* (I - W_c^* z^\ell H_f^*) + H_f W_n W_n^* H_f^*$$

$$-(I - H_f z^{-\ell} W_c) W_s G W_n^* z^\ell H_f^* - H_f z^{-\ell} W_n G^T W_s^* (I - W_c^* z^\ell H_f^*) \tag{6.17}$$

Taking the expectation with respect to the uncertain system parameters, and substituting from the spectral-factor expression (6.6):

$$E_\theta \{\Phi_{ee}\} = E_\theta \{W_s W_s^* - H_f z^{-\ell} W_c W_s W_s^* - W_s W_s^* W_c^* z^\ell H_f^*$$

$$-H_f z^{-\ell} W_n G^T W_s^* - W_s G W_n^* z^\ell H_f^*\} + H_f Y_f Y_f^* H_f^*$$

$$= H_f Y_f Y_f^* H_f^* + E_\theta \{-H_f z^{-\ell} (W_c W_s + W_n G^T) W_s^*$$

$$-W_s (W_s^* W_c^* + G W_n^*) z^\ell H_f^* + W_s W_s^*\} \tag{6.18}$$

**Completing the Squares for the Error Spectrum**

Using a completing the squares argument the expression (6.18) may be written as:

$$E_\theta\{\Phi_{ee}\} = (H_f Y_f - E_\theta\{W_s(W_s^* W_c^* - GW_n^*)z^\ell Y_f^{*-1}\})$$

$$\times(Y_f^* H_f^* - E_\theta\{Y_f^{-1}z^{-\ell}(W_c W_s + W_n G^T)W_s^*\})$$

$$+E_\theta\{W_s W_s^*\} - E_\theta\{W_s(W_s^* W_c^* + GW_n^*)\}Y_f^{*-1}Y_f^{-1}E_\theta\{(W_c W_s + W_n G^T)W_s^*\} \quad (6.19)$$

The last two terms of this expression can be denoted by $T_0$.

**Introduction of the Diophantine Equations**

A pair of coupled diophantine equations must be introduced with unknowns $(G_0, S_0, F_0)$, where the optimal solution to the deconvolution problem requires the degree of $F_0$ to be a minimum. Based on these equations a third diophantine equation is required, which is known as the implied equation in the unknowns $G_0$ and $S_0$. This equation is needed in the proof of the stability of terms when the signal model is unstable.

The two basic equations may now be introduced, in terms of the unknowns $(G_0, S_0, F_0)$, with $F_0$ of smallest degree:

$$A_s F_0 + G_0 D_f^* z^{-g} = E_\theta\{C_s(C_s^* C_1^* + GC_n^* A_1^*)z^{\ell-g}\} \quad (6.20)$$

$$-C_s F_0 + S_0 D_f^* z^{-g+\ell_0} = E_\theta\{(C_n C_n^* A_1^* + C_n G^T C_s^* C_1^*)z^{\ell-g}\} \quad (6.21)$$

where for $\ell \geq 0$, $\ell_0 = \ell$ and for $\ell < 0$, $\ell_0 = 0$. The *implied equation* is obtained by adding these equations and using (6.12), to obtain:

$$C_1 G_0 D_f^* z^{-g} + A_1 S_0 D_f^* z^{-g+\ell_0} = D_f D_f^* z^{\ell-g}$$

Thence, the implied equation, which is needed in proving the stability of the estimation error equation (6.19), follows as:

$$C_1 G_0 z^{-\ell} + A_1 S_0 z^{-\ell+\ell_0} = D_f \quad (6.22)$$

Note that the above definitions of $\ell_0$ are chosen to ensure the pair of coupled diophantine equations involve only powers of $z^{-1}$. If $\ell \geq 0$ then $\ell_0 = \ell$, since $g = \deg(D_f) + \ell$. If $\ell < 0$, let $\ell_0 = 0$, since $g = \deg(D_f)$.

**Simplification of the Squared Cost-function Term**

Substituting for the polynomial system models, using (6.5), obtain

$$W_s(W_s^* W_c^* + GW_n^*)z^\ell Y_f^{*-1} = A_s^{-1}C_s(C_s^* A_s^{*-1}C_c^* + GC_n^*)A_1^* D_f^{*-1}z^\ell$$

$$= A_s^{-1}C_s(C_s^* C_1^* + GC_n^* A_1^*)D_f^{*-1}z^\ell$$

This term must now be separated into causal and non-causal components using the equivalent of a partial fraction expansion. This requires the introduction of the diophantine equation (6.13), so that:

$$E_\theta\{W_s(W_s^*W_c^* + GW_n^*)z^\ell Y_f^{*-1}\} = F_0 D_f^{*-1} z^g + A_s^{-1} G_0$$

The *squared term* in (6.19) therefore becomes:

$$(H_f Y_f - E_\theta\{W_s(W_s^*W_c^* + GW_n^*)z^\ell Y_f^{*-1}\})$$

$$= [H_f(A_1 A)^{-1} D_f - A_s^{-1} G_0] - F_0 D_f^{*-1} z^g \qquad (6.23)$$

To prove that the squared term within the square brackets has all its poles strictly within the unit-circle, the implied polynomial equation (6.22) is required which gives:

$$(A_1 A)^{-1} D_f = A^{-1} A_1^{-1} C_1 G_0 z^{-\ell} + A^{-1} S_0 z^{-\ell+\ell_0}$$

Thence, obtain the squared bracketed term in (6.23) as:

$$[H_f(A_1 A)^{-1} D_f - A_s^{-1} G_0] = [(H_f A^{-1} C_c z^{-\ell} - I) A_s^{-1} G_0 + H_f A^{-1} S_0 z^{-\ell+\ell_0}]$$

If the estimation error is to be mean square bounded the terms in this expression must necessarily involve stable operators. This is clear by reference to expression (6.17), which represents the integrand of the unweighted cost-function and includes both $H_f$ and the term:

$$(I - H_f z^{-\ell} W_c) A_s^{-1} = (I - H_f z^{-\ell} A^{-1} C_c) A_s^{-1} \qquad (6.24)$$

Thence, if the $H_2$ cost-function (6.11) is to be finite, the filter $H_f$ and the component of estimation error due to the signal (depending upon (6.24)), must both involve stable transfer functions.

### Robustness Diophantine Equation

It now follows that (6.23) represents an asymptotically stable term [.] and a strictly unstable term $F_0 D_f^{*-1} z^g$ . Thence, the uncertainty averaged, weighted estimation error, can be expressed as:

$$E_\theta\{\Phi_{e_0 e_0}\} = W_p\{([H_f Y_f - A_s^{-1} G_0] - F_0 D_f^{*-1} z^g)$$

$$\times ([H_f Y_f - A_s^{-1} G_0] - F_0 D_f^{*-1} z^g)^* + T_0\} W_p^*$$

$$= (W_p[H_f Y_f - A_s^{-1} G_0] - W_p F_0 D_f^{*-1} z^g)(W_p[H_f Y_f - A_s^{-1} G_0] - W_p F_0 D_f^{*-1} z^g)^*$$

$$+ W_p T_0 W_p^* \qquad (6.25)$$

If $W_p$ is a constant matrix the condition for optimality is that the stable term $[H_f Y_f - A_s^{-1} G_0]$ should be null. However, in the general case considered here $W_\sigma$ is a rational transfer-function matrix. The term $W_\sigma F_0 D_f^{*-1} z^g$ must in this case be separated into causal (stable) and non-causal (unstable) terms using a further diophantine equation (6.15). That is, the term $A_p^{-1} B_p F_0 D_f^{*-1} z^g$ can be written, using (6.15) as:

$$A_p^{-1} B_p F_0 D_f^{*-1} z^g = F_1 D_f^{*-1} z^g + A_p^{-1} N$$

The weighted estimation error criterion, equation (6.25), may now be written as:

$$E_\theta\{\Phi_{e_o e_o}\} = ([W_p(H_f Y_f - A_s^{-1} G_o) - A_p^{-1} N] - F_1 D_f^{*-1} z^g)$$

$$\times ([W_p(H_f Y_f - A_s^{-1} G_o) - A_p^{-1} N] - F_1 D_f^{*-1} z^g)^* + W_p T_0 W_p^* \qquad (6.26)$$

From the previous discussion and recalling that the weighting function $W_p$ can be assumed to be asymptotically stable, the term within the square brackets is strictly Schur and the term $F_1 D_f^{*-1} z^g$ is strictly non-Schur.

**Optimisation Procedure**

The cost-integrand may therefore be written in the concise form:

$$E_\theta\{\Phi_{e_0 e_0}\} = (T^+ + T^-)(T^+ + T^-)^* + W_p T_0 W_p^*$$

where $T^+$ denotes the strictly stable term $T^+ = [.]$ and $T^- = -F_1 D_f^{*-1} z^g$ represents a strictly unstable term in (6.26). In the evaluation of the contour integral the cross-product of the terms $T^-(T^+)^*$ and $(T^+)(T^-)^*$ arise. By the argument in Chapter 2 that employed the residue theorem, it may easily be shown that the integral of such cross terms is null. The cost-function therefore follows from (6.26) as:

$$J = trace\{\frac{1}{2\pi j} \oint_{|z|=1} E_\theta\{\Phi_{e_0 e_0}(z^{-1})\}\frac{dz}{z}\}$$

$$= trace\{\frac{1}{2\pi j} \oint_{|z|=1} \{T^+(T^+)^* + T^-(T^-)^* + W_p T_0 W_p^*\}\frac{dz}{z}\}$$

The terms $T^-$ and $T_0$ do not depend upon the estimator transfer function and hence this equation can be minimised by setting $T_0^+ = 0$, giving:

$$T^+ = [W_p(H_f Y_f - A_s^{-1} G_0) - A_p^{-1} N] = 0$$

After substituting in this equation for the polynomial forms of the transfer functions obtain:

$$H_f = (A_s^{-1} G_0 + B_p^{-1} N) Y_f^{-1} = (A_s^{-1} G_0 + B_p^{-1} N) D_f^{-1} A_1 A \qquad (6.27)$$

The expression for the minimum-cost follows by substituting for $T^-$.

### 6.2.4   Properties of the $H_2$ Estimator

The properties of the $H_2$ deconvolution estimator are summarised in the following lemma and an example is then presented.

**Lemma 6.1**: *Properties of the $H_2$ Deconvolution Filter, Predictor and Smoother*

The properties of the optimal deconvolution filter, defined in Theorem 6.1, can be detailed as follows.

**Minimal cost:**

$$J_{\min} = trace\{\frac{1}{2\pi j} \oint_{|z|=1} \{F_1 D_f^{*-1} D_f^{-1} F_1^* + W_p T_0 W_p^*\}\frac{dz}{z}\} \qquad (6.28)$$

where

$$T_0 = E_\theta\{W_s W_s^*\} - E_\theta\{W_s(W_s^* W_c^* + GW_n^*)\}(Y_f Y_f^*)^{-1} E_\theta\{(W_c W_s + W_n G^T)W_s^*\}$$

**Implied equation determining stability:**

$$C_1 G_0 z^{-\ell} + A_1 S_0 z^{-\ell+\ell_0} = D_f \qquad (6.29)$$

The $H_2$ optimal deconvolution filter, predictor or smoother and corresponding estimation error models, are asymptotically stable for both stable and unstable signal model cases.

**Proof :** By collecting the preceding results.

**Remarks:**

(i) The dynamic characteristics of the deconvolution filter are determined mainly by the right-hand side of the implied equation, that is, the spectral factor $D_f$.

(ii) For $\ell < 0$ the smoothing filter obtained is of fixed-lag smoothing filter form.

(iii) The robustness weighting function $W_p$ adds a term to the estimator and if $W_p$ is a constant matrix $N$ and $F_1$ are null. Thus, only dynamic weighting functions affect the optimal solution and these result in an additional parallel term (Fig. 6.2).

(iv) If $W_p$ is a constant matrix the minimal-degree solution of (6.15) gives $N = 0$ and the optimal estimator (6.16) simplifies accordingly.

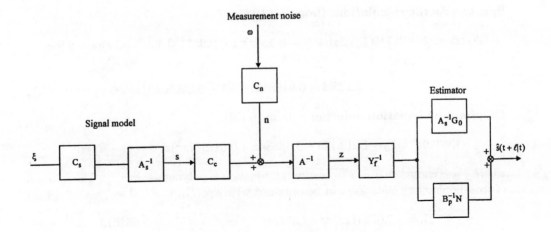

**Fig. 6.2** : Optimal Deconvolution Filter and Uncertain, Noise,
Channel and Signal Models
(for a traditional filtering problem $C_c = A$)

**Example 6.1** : $H_2$ *Discrete Deconvolution Filtering Problem*

Consider the deconvolution filtering problem (where $\ell = 0$), where the transmission channel has both delays and non-minimum phase behaviour. The system and noise models only include parametric uncertainty and are defined as:

**System:**

$$W_s = 1/(1 - 0.9z^{-1}), \quad W_c = 0.25(1 + \alpha)z^{-2}(1 - 2z^{-1})/(1 - 0.1z^{-1}), \quad W_p = 1$$

**Noise:**

$$G = 0, \quad W_n = 0.2(1 + \beta)$$

**Parameters:**

$$E_\theta\{\alpha\} = E_\theta\{\beta\} = 0, \quad E_\theta\{\alpha^2\} = 3, \quad E_\theta\{\beta^2\} = 1$$

The polynomial definitions, of the signal and noise follow from the system transfers as:

$$A_s = 1 - 0.9z^{-1}, \quad C_s = 1, \quad C_n = 0.2(1 + \beta)A$$

$$C_c = 0.25(1 + \alpha)z^{-2}(1 - 2z^{-1}), \quad A = 1 - 0.1z^{-1}$$

$$A_1^{-1}C_1 = C_c A_s^{-1} = 0.2(1 + \alpha)z^{-2}\frac{(1 - 2z^{-1})}{(1 - 0.9z^{-1})}$$

and identify $A_1 = 1 - 0.9z^{-1}$ and $C_1 = 0.2(1 + \alpha)z^{-2}(1 - 2z^{-1})$. Substituting in the results of Theorem 6.1 obtain:

**Spectral factor calculation:** (from (6.12))

$$D_f D_f^* = (1 - 2z^{-1})(1 - 2z) + (1 - 0.1z^{-1})(1 - 0.9z^{-1})0.4(1 - 0.1z)(1 - 0.9z)$$

$$D_f = 2.123(1 - 0.015335z^{-1})(1 - 0.52086z^{-1})$$

**Diophantine equation solution** : (from (6.13))

$$(1 - 0.9z^{-1})F_0 + G_0(2.123 - 1.13834z + 0.016957z^2)z^{-3} = (1 - 2z)z^{-1}$$

where $g$ was chosen as $g = 3$ to ensure the equation is polynomial in $z^{-1}$. The solution of this diophantine equation can be obtained with $deg(F_0) < g = 3$ as:

$$F_0 = -2.079012z^{-1} - 1.3743z^{-2} \quad \text{and} \quad G_0 = -0.582613$$

Note that only a single diophantine equation (6.13) need be solved when the system is stable and the weighting $W_p$ is a constant. In this case, where the weighting is a constant, the solution of (6.15) is $N = 0$ and the filter follows from (6.16), as:

$$H_f = (G_0 A_1 A/(A_s D_f)) = \frac{-0.27443(1 - 0.1z^{-1})}{(1 - 0.52086z^{-1})(1 - 0.015335z^{-1})}$$

# 6.3   $H_\infty$ Optimal Estimation Problems

Attention will now turn to the $H_\infty$ estimation problem. The use of $H_\infty$ estimation is first motivated by considering the treatment of system uncertainty. The embedding principle, outlined in Chapter 5, is then utilised to obtain the $H_\infty$ optimal estimator, given the previous $H_2$ results.

**Frequency Bounded Uncertainty**

The solution of estimation problems, where noise or signal models have additive or multiplicative modelling errors, can be solved in the $H_\infty$ framework. Fu et al. (1990) and Xie (1991) has solved both discrete and continuous-time $H_\infty$ problems, which guarantee certain robustness properties.

## 6.3.1   $H_\infty$ Robust Estimator and Uncertainty

If the uncertainty in the system elements is described by soft (probabilistic) bounds the optimal robust estimator can be found from the solution of the minimum variance estimation problem in the previous section. This was demonstrated by Sternad and Ahlen (1993a,b) who used a different uncertainty description, related to the Wiener solution presented by Grimble (1984).

The dynamically weighted $H_2$ type of estimation problem solved in the previous sections, enables a solution to be obtained to the $H_\infty$ estimation problem discussed below. However, it will first be helpful to consider why an $H_\infty$ problem should be solved to obtain a robust estimator. Recall from §6.2.1 that the uncertain elements $\Delta W_{si}$ and $\Delta W_{ni}$ represent the unknown transfers in the signal and measurement noise paths, respectively. They are determined by upper bounding functions $\delta W_{si}$ and $\delta W_{ni}$, respectively, in the sense:

$$\left\|\delta W_{si}^{-1}\Delta W_{si}\right\|_\infty \le 1 \quad \text{and} \quad \left\|\delta W_{ni}^{-1}\Delta W_{ni}\right\|_\infty \le 1$$

**Signal Processing System**

The objective below is to derive a bound on the estimation error which will motivate the use of an $H_\infty$ cost-function. The signal and noise models may now be introduced that define the output and observations signals system. The system equations are defined as follows:

| | |
|---|---|
| **Signal:** | $s = W_s\xi$ |
| **Noise:** | $n = W_n\omega$ |
| **Output:** | $y = W_c s = W_c W_s\xi$ |
| **Observations:** | $z = n + y = W_n\omega + W_c W_s\xi$ |

The optimal linear filter estimate can be represented in the form:

$$\hat{s} = (H_f z^{-\ell})z = H_f z^{-\ell}(W_n\omega + W_c W_s\xi)$$

and the estimation error:

$$\tilde{s} = s - \hat{s} = (I - H_f z^{-\ell}W_c)W_s\xi - H_f z^{-\ell}W_n\omega$$

The weighted errors are defined as $\tilde{s}_0 = W_{po}(s - \hat{s})$ and hence,

$$\tilde{s}_o = W_{po}(I - H_f z^{-\ell}W_c)W_s\xi - W_{p0}H_f z^{-\ell}W_n\omega$$

**Estimation Error Bound**

Recall that the white noise signals $\{\xi(t)\}$ and $\{\omega(t)\}$ are uncorrelated and that the norm for the uncertain system can be defined as:

$$\|f\|_\theta = \left\{E\{f^T(t)f(t)\}\right\}^{1/2}$$

where the expectation $E\{.\} = E_\theta\{E_s\{.\}\}$ is taken over the ensemble of stochastic signals $(E_\theta)$ and unknown parameters $(E_\theta)$, respectively. Thus obtain:

$$\|\tilde{s}_o\|_\theta^2 = \left\|W_{po}(I - H_f z^{-\ell}W_c)W_s\xi\right\|_\theta^2 + \left\|W_{po}H_f z^{-\ell}W_n\omega\right\|_\theta^2$$

and allowing for the uncertainties:

$$W_c = \bar{W}\delta W_c, \quad W_s = \bar{W}_s\Delta W_s = \bar{W}_s\delta W_s\Delta W_{si}, \quad W_n = \bar{W}_n\Delta W_n = \bar{W}_n\delta W_n\Delta W_{ni}$$

Substituting into the norm of the weighted estimation error:

$$\|\tilde{s}_o\|_\theta^2 = \left\|W_{po}(I - H_f z^{-\ell}\bar{W}_c\delta W_c)\bar{W}_s\Delta W_s\xi\right\|_\theta^2 + \left\|W_{po}H_f z^{-\ell}\bar{W}_n\Delta W_n\omega\right\|_\theta^2$$

$$\le \left\|W_{po}(I - H_f z^{-\ell}\bar{W}_c\delta W_c)\bar{W}_s\delta W_s\Delta W_{si}\right\|_\theta^2 \cdot \|\xi\|^2$$

$$+ \left\| W_{po} H_f z^{-\ell} \bar{W}_n \delta W_n \Delta W_{ni} \right\|_\theta^2 \cdot \|\omega\|^2$$

The $L_2$-induced norm of the operator $G$ is defined as:

$$\|G\| = \sup_{f \neq 0} \|Gf\| / \|f\|$$

and if $G$ is a linear asymptotically stable operator:

$$\|G\| = \left\| G(z^{-1}) \right\|_\infty = \sup_\omega \{ \sigma_{\max} \{ G(e^{-j\omega}) \} \}$$

Since the operators involved in this problem are asymptotically stable:

$$\|\tilde{s}_o\|^2 \le E_\theta \{ \left\| W_{po} (I - H_f z^{-\ell} \bar{W}_c \delta W_c) \bar{W}_s \delta W_s \Delta W_{si} \right\|_\infty^2 \} \cdot \|\xi\|^2$$

$$+ E_\theta \{ \left\| W_{po} H_f z^{-\ell} \bar{W}_n \delta W_n \Delta W_{ni} \right\|_\infty^2 \} \cdot \|\omega\|^2$$

Recall $\left\| \delta W_{si}^{-1} \Delta W_{si} \right\|_\infty \le 1$ and $\left\| \delta W_{ni}^{-1} \Delta W_{ni} \right\|_\infty \le 1$ and $\|AB\|_\infty \le \|A\|_\infty \cdot \|B\|_\infty$. Thence,

$$\|\tilde{s}_o\|^2 \le E_\theta \{ \left\| W_{po} (I - H_f z^{-\ell} \bar{W}_c \delta W_c) \bar{W}_s \delta W_s \delta W_{si} \right\|_\infty^2 \} \cdot \|\xi\|^2$$

$$+ E_\theta \{ \left\| W_{po} H_f z^{-\ell} \bar{W}_n \delta W_n \delta W_{ni} \right\|_\infty^2 \} \cdot \|\omega\|^2$$

Assuming the variances of the zero mean signals $\{\xi(t)\}$ and $\{\omega(t)\}$ are normalised to unity, the above expression may be written as:

$$\|\tilde{s}_o\|^2 \le E_\theta \{ \| W_{po} [(I - H_f z^{-\ell} \bar{W}_c \delta W_c) \bar{W}_s \delta W_s \delta W_{si} \delta W_{si}^* \delta W_s^* \bar{W}_s^* \, (I - H_f z^{-\ell} \bar{W}_c \delta W_c)^*$$

$$+ H_f z^{-\ell} \overline{W}_n \delta W_n \delta W_{ni} \delta W_{ni}^* \delta W_n^* \overline{W}_n^* z^\ell H_f^*] W_{po}^* \|_\infty \}$$

The least upper bound (lub) can clearly be obtained by minimising the expectation of the $H_\infty$-norm on the right-hand side of the above expression. This is clearly the same as the $H_\infty$-norm of the power spectrum of the following weighted estimation error signal:

$$\tilde{s}_0(t) = W_{p0} [(I - H_f z^{-\ell} \overline{W}_c \delta W_c) \overline{W}_s \delta W_s \delta W_{si} \xi(t) - H_f z^{-\ell} \overline{W}_n \delta W_n \delta W_{ni} \omega(t)] \quad (6.30)$$

Note that the uncertain systems $\Delta W_{si}$ and $\Delta W_{ni}$ for which $H_\infty$ norm bounds are known have been replaced by the worst case models $\delta W_{si}$ and $\delta W_{ni}$. The uncertain system elements depending upon the probabilistic descriptions $\delta W_c$, $\delta W_s$ and $\delta W_n$ can be set to unity if there is no uncertainty.

Thus, when such uncertain elements are present the least upper bound on the estimation error can be obtained by minimising the averaged $H_\infty$-norm of the power spectrum of the error signal (6.30). This is simply the weighted estimation error for a system where the uncertain elements $\Delta W_{si}$ and $\Delta W_{ni}$ are replaced by their upper bounds.

The conclusion may therefore be drawn that if the system includes a combination of unmodelled dynamics and signal noise models, including parameters represented by random variables, then an averaged $H_\infty$-norm minimisation problem must be solved to limit the upper bound on the estimation error. The system model in the resulting optimisation problem includes the elements with the random variables ($\delta W_c$, $\delta W_s$, $\delta W_n$) and the bounding functions ($\delta W_{si}$ and $\delta W_{ni}$), replacing the unknown elements $\Delta W_{si}$ and $\Delta W_{ni}$.

## 6.3.2 $H_\infty$ Estimation and Embedding

The $H_\infty$ multichannel estimation problem can be solved using the $H_2$ optimisation results presented in §6.2.4 and the following lemma which links the $H_2$ and $H_\infty$ problems. Note that $W_\sigma$ represents a weighting filter.

**Lemma 6.2** : *Auxiliary $H_2$ Minimisation Problem*
Consider the auxiliary problem of minimising the uncertainty averaged $H_2$ criterion:

$$J_0 = \frac{1}{2\pi j} \oint_{|z|=1} trace\{E_\theta\{W_\sigma(z^{-1})X(z^{-1})W_\sigma^*(z)\}\}\frac{dz}{z} \qquad (6.31)$$

Suppose that for some real-rational matrix: $W_\sigma^*(z^{-1})W_\sigma(z^{-1}) \geq 0$, the cost function $J_0$ is minimised by a function $X(z^{-1}) = X^*(z^{-1})$, for which $E_\theta\{X(z^{-1})\} = \lambda^2 I_r$ (a real constant matrix on $|z| = 1$). Then the function $E_\theta\{X(z^{-1})\}$ also minimises:

$$J_0 = \sup_{|z|=1} E_\theta\{\|X(z^{-1})\|_2\} = \sup_{|z|=1} \{\sigma_{max}\{E_\theta\{X(z^{-1})\}\}\} \qquad (6.32)$$

where $\|X(z^{-1})\|_2$ denotes the spectral norm.

**Proof** : The proof is similar to that given in the $H_2$ embedding Lemmas of Chapter 5 and follows that in Kwakernaak (1984).

A solution was presented in §6.2.2 to a $H_2$ minimisation problem which is similar to that referred to in the above lemma. To relate the two problems let the dynamic weighting function $W_p$ be written as the product of two terms:

$$W_p = W_\sigma W_{p0} \ , \quad W_p \varepsilon R^{n_s \times n_s}(z^{-1}) \qquad (6.33)$$

Then the function $X(z^{-1})$ above can be related to the weighted estimation error as:

$$X = W_{p0}\Phi_{ee}W_{p0}^* \ , \quad X\varepsilon R^{n_s \times n_s}(z^{-1}) \qquad (6.34)$$

The $W_\sigma$ represents the weighting term, still to be determined, which appears in Lemma 6.2. The transfer $W_{po}$ will be assumed to represent the dynamic weighting selected by the designer to frequency shape the estimation error spectrum, so that the cost-function to be minimised is given as:

$$J_0 = \sup_{|z|=1} \{\sigma_{max}\{E_\theta\{X(z^{-1})\}\}\} = \sup_{|z|=1} \{\sigma_{max}\{E_\theta\{W_{p0}\Phi_{ee}W_{p0}^*\}\}\}$$

### 6.3.3    Derivation of the Weighting Filter $W_\sigma$

Assume for the moment that the scalar $\lambda$ in the above lemma is known. To ensure the conditions of the above lemma are satisfied, the $W_\sigma$ must now be found which leads to an equalising solution :

$$E_\theta\{X(z^{-1})\} = \Lambda\Lambda^T = \lambda^2 I_r \text{ on } |z| = 1$$

The matrix $\Lambda$ introduced in the above expression may be defined as $\Lambda = \lambda I$ at the optimum. However, once the maximum singular value has been minimised and a solution for the filter obtained, it is sometimes possible to improve upon the solution by defining a more general $\Lambda$. The improvement can, for example, be in terms of reducing the maximum values of the remaining singular values.

Comparison of equations (6.28) and (6.31), evaluated at the $H_2$ optimum, gives:

$$F_1 D_f^{*-1} D_f^{-1} F_1^* + W_\sigma W_{p0} T_0 W_{p0}^* W_\sigma^* = W_\sigma \Lambda\Lambda^T W_\sigma^T$$

or

$$F_1 D_f^{*-1} D_f^{-1} F_1^* = W_\sigma (\Lambda\Lambda^T - W_{p0} T_0 W_{p0}^*) W_\sigma^* \qquad (6.35)$$

where

$$F_1 \varepsilon P^{n_s \times n_s}(z^{-1})$$

Introduce the stable minimum-phase transfer-function $S \varepsilon R^{n_s \times n_s}(z^{-1})$, which satisfies:

$$SS^* = \Lambda\Lambda^T - W_{p0} T_0 W_{p0}^* \qquad (6.36)$$

and assume that $\Lambda$ is chosen so that the transfer function matrix $S$ is invertible. From (6.35) and (6.36) now obtain:

$$F_1 D_f^{*-1} D_f^{-1} F_1^* = W_\sigma S S^* W_\sigma^* \qquad (6.37)$$

Define the spectral-factor $D_{fo}$ which satisfies:

$$D_{f0}^* D_{f0} = D_f D_f^*$$

and let the left-coprime pair $(F, D)$ satisfy:

$$D^{-1} F = F_1 D_{f0}^{-1} \quad , \quad F \varepsilon P^{n_s \times n_y}(z^{-1})$$

Then, to ensure the equalising solution is obtained, equation (6.37) must be satisfied:

$$D^{-1} F F^* D^{*-1} = W_\sigma S S^* W_\sigma^* \qquad (6.38)$$

Let $F_s$ denote a Schur polynomial matrix which satisfies

$$F_s F_s^* = F F^*$$

Then, from equation (6.38), the desired robustness weighting:

$$W_\sigma = D^{-1} F_s S^{-1} \tag{6.39}$$

To ensure a suitable definition of the matrix $F_s \varepsilon P(z^{-1})^{n_s \times n_s}$ is obtained the following assumption is required:

**Assumption (2i):** The number of signal channels $n_s <$ number of observation channels $n_z = n_y$.

The linear equation which enables $F_s$ to be computed can now be determined. Introduce the left-coprime polynomial matrices $(\overline{S}, \overline{F}_s)$ as:

$$\overline{S}^{-1} \overline{F}_s = F_s S^{-1} W_{p0}$$

Then from this equation and (6.39):

$$W_p = W_\sigma W_{p0} = A_p^{-1} B_p = D^{-1} \overline{S}^{-1} \overline{F}_s \tag{6.40}$$

Hence, identify the desired weighting polynomials which ensure an equalising solution is obtained, as:

$$A_p = \overline{S} D \quad \text{and} \quad B_p = \overline{F}_s \tag{6.41}$$

and if $W_\sigma$ is required this may be found from (6.40) as:

$$W_\sigma = A_p^{-1} B_p W_{p0}^{-1} = (\overline{S} D)^{-1} \overline{F}_s W_{p0}^{-1} \tag{6.42}$$

**Robust Weighting Diophantine Equation**

To calculate $(N, \overline{F}_s)$ a more appropriate form of the robust diophantine equation (6.15) is required. Substituting for the weightings in (6.41) the robustness diophantine equation (6.15) becomes:

$$\overline{S} D F_1 + N D_f^* z^{-g} = \overline{F}_s F_0$$

but $DF_1 = FD_{fo}$ and hence obtain:

$$\overline{S} F D_{f0} + N D_f^* z^{-g} = \overline{F}_s F_0 \tag{6.43}$$

The optimal estimator follows from (6.16) but note that $B_p^{-1}$ enters the expression and hence from (6.41) the matrix $F$ must be of normal full rank. The reason for introducing Assumption (2i) should now be apparent.

## 6.3.4   The $H_\infty$ Estimator and Example

The main theorem that summarises the $H_\infty$ estimation results is introduced below. This is followed by an example to illustrate the calculation procedure and to show typical results.

**Theorem 6.2** : $H_\infty$ *Estimator for Multichannel Uncertain System Estimation*
Consider the system shown in Fig. 6.2, which includes uncertain elements represented both probabilistically and by hard frequency response bounds, and assume that the cost-function:

$$J_0 = \sup_{|z|=1} \{\sigma_{\max}\{E_\theta\{W_{p0}\Phi_{ee}W_{p0}^*\}\}\}$$

is to be minimised where $\Phi_{ee}$ denotes the power spectrum of the estimation error. The $H_\infty$ optimal estimator may be computed from the averaged spectral factor and diophantine equations, calculating $D_f$ from (6.12) and obtaining $(G_0, S_0, F_0, g)$ from (6.13) and (6.14). Define the transfer $T_0$ from (6.28), then the stable minimum phase transfer $S$ is defined as:

$$SS^* = \Lambda\Lambda^T - W_{p0}T_0W_{p0}^* = \lambda^2 I - W_{p0}T_0W_{p0}^* \tag{6.44}$$

The left-coprime pair $(F, D)$ can be found using:

$$D^{-1}F = F_1 D_{f0}^{-1} \text{ where } D_{fo} \text{ satisfies} : D_{f0}^* D_{f0} = D_f D_f^* \tag{6.45}$$

Let $F_s$ denote the Schur spectral-factor satisfying: $F_s F_s^* = FF^*$ and introduce the left-coprime polynomial matrices $(\overline{S}, \overline{F}_s)$ as:

$$\overline{S}^{-1}\overline{F}_s = F_s S^{-1} W_{p0} \tag{6.46}$$

The $H_\infty$ optimal linear estimator can then be found by first calculating the solution $(N, F)$, with $F$ of smallest degree, of the diophantine equation:

$$\overline{S}FD_{f0} + ND_f^* z^{-g} = \overline{F}_s F_0 \tag{6.47}$$

**Optimal estimator:**

$$H_f = (A_s^{-1} G_0 + \overline{F}_s^{-1} N)D_f^{-1} A_1 A \tag{6.48}$$

**Optimum function and minimum cost:**

$$X_{\min} = \lambda_0^2 I \quad \text{and} \quad J_\infty = \lambda_0^2$$

where $\lambda_0^2$ is the smallest scalar $\lambda^2$ such that equations (6.44) to (6.47) are satisfied.

**Proof** : The theorem follows by collecting the previous results and by invoking the results of Theorem 6.1.

**Remarks:**

(i) The polynomial form of the estimator is related to the equivalent state-space versions but there are several insights into the frequency domain properties of the estimator which are more obvious in the polynomial form.

(ii) The robust weighting diophantine equation (6.47) has a non-zero solution for the matrix $N$, since $W_\sigma$, and hence $W_p$ is normally dynamic.

(iii) Xie (1991) has shown that a range of robust filtering problems, where the system is uncertain, has an $H_\infty$ filter as a solution.

**Example 6.2** : *Uncertain System Filtering Problem*

Consider a discrete system with sample interval $T_s = 0.005$ seconds. Define the following signal and noise models:

**Signal:**

$$\overline{W}_s = 0.001246/(1 - 1.987563z^{-1} + 0.9900498z^{-2})$$

**Uncertainty bound:**

$$\delta W_{si} = 802.568(1 - 0.9987544z^{-1})$$

**Noise:**

$$\overline{W}_n = 1 \text{ (white noise of unity variance)}, \quad G = 0$$

**Channel dynamics:**

$$\overline{W}_c = 1$$

The signal model frequency response is shown in Fig. 6.3. The upper bound on the uncertainty in the signal model $\delta W_{si}$ is also shown in this figure. Note that the uncertainty is assumed to be large at high frequencies. For simplicity assume that there is no probabilistic uncertainty, so that $\delta W_c = \delta W_s = \delta W_n = 1$ and let $\Delta W_{ni} = 1$. Thus, as explained at the start of §6.3.1 the robust filter can be found by minimising the $H_\infty$-norm of the weighted estimation error, for a system where the signal model becomes:

$$\overline{W}_s \delta W_{si} = (1 - 0.9987544z^{-1})/(1 - 1.987563z^{-1} + 0.9900498z^{-2})$$

Define the system polynomials from the above transfers as follows:

$$A_s^{-1} C_s = \overline{W}_s \delta W_{si}$$

$$A = C_c = C_n = 1 \Rightarrow A_1 = A_2, \quad C_1 = 1$$

The filter may then be computed from the results of the above Theorem 6.2, assuming the weighting $W_{po} = 1$. Then from (6.48):

$$H_f = (G_0 + \bar{F}_s^{-1} N A_s)/D_f$$

$$= \frac{0.619728(1 - 0.9987544z^{-1})(0.9921 + 0.9924z^{-1})}{(1 - 0.9972141z^{-1})(1 - 0.3813036z^{-1})(1 + 0.9968z^{-1})}$$

The algorithm which solves the equations in Theorem 6.2 is described in Grimble (1991). The minimum cost $\lambda_0^2 = 1.3729$. The time responses for the signal, noise and estimated signal, are shown in Fig. 6.4.

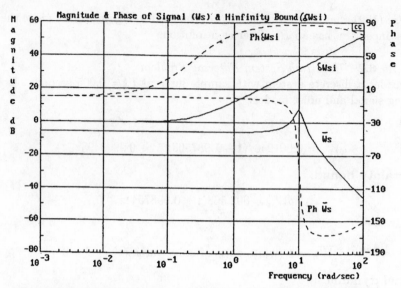

Fig. 6.3 : Signal Model and Amplitude Bound on Uncertainty

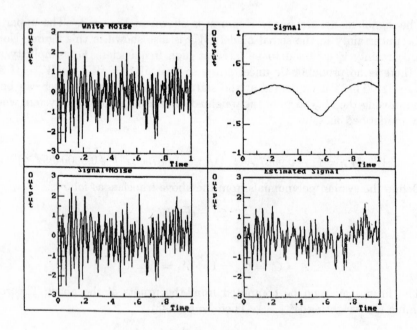

Fig. 6.4 : Signal, Noise and Estimated Signal

## 6.4   Standard System for $H_2$ Estimation

The solution of the optimal $H_\infty$ smoothing problem was considered by Grimble (1991) using a polynomial systems analysis. The polynomial solution of estimation problems, using the standard model approach, has been considered in Grimble (1994a), and this is valuable for the so called inferential estimation problems where the signal to be estimated is not directly measurable (otherwise known as soft sensing problems).

The use of the *standard system model* in control applications is now well established for providing a general solution to many $H_\infty$ problems. Such models can be tailored to different system configurations and hence the optimal control results obtained are generally applicable. Most of the state-space based $H_2$ and $H_\infty$ commercial control packages utilise a standard system model representation. A solution is presented in the following to the $H_\infty$ filtering, smoothing and prediction problems for systems represented in standard system form. The results can be applied to the optimal deconvolution estimation problems, considered earlier and to a wide variety of other problems.

In the solution of most filtering problems by frequency domain methods the output to be estimated is assumed to be measured but corrupted by measurement noise, and possibly distorted by signal path dynamics. However, in some cases the signal to be estimated cannot be measured directly. For example, in inferential control problems, the quantity to be controlled cannot be measured directly. This occurs in many servo mechanism control problems such as coordinate measuring machines and steel rolling mills (Grimble, 1995). The estimation of variables which cannot be measured directly is also called *soft sensing*.

An optimal linear estimator is therefore required which can provide an estimate of the signal of interest using observations of other variables. These observations will of course be corrupted by measurement noise and the signal paths may introduce distortion. It is this very general estimation problem, involving the minimisation of the $H_\infty$ norm of the estimation error which is of interest in the following. The polynomial approach to the solution of these problems is particularly appropriate for signal processing problems, which are often posed in frequency domain terms. It is easier to determine the likely frequency-domain performance characteristics of estimators in polynomial system form, than their state-space equivalents. Both $H_2$ and $H_\infty$ estimation problems are considered but the emphasis is first on the $H_2$ problem.

## 6.4.1 Signal Processing Standard System Description

The standard system, signal and noise models are assumed to be time-invariant and discrete-time and to be represented in transfer-function or polynomial matrix form. The multichannel signal processing problem is illustrated in Fig. 6.5. The signal to be estimated is denoted by $\{s(t)\}$. A secondary measured signal $\{m(t)\}$ is assumed to be corrupted by white $\{v(t)\}$ and coloured $\{n(t)\}$ measurement noise, respectively. In the deconvolution filtering problems considered previously, $m(t)$ is given by the output of a transmission channel which has an input $s(t)$. In inferential control problems the signal processing problem is more difficult, since the signal measured $m(t)$ is not simply a distorted version of the desired signal $s(t)$.

The physical system of interest has a measured output that involves a very different combination of stochastic signals to the output to be estimated. The signal source $W_s(z^{-1})$ and coloured measured noise $W_n(z^{-1})$ models are driven by the white noise sources $\{\xi(t)\}$ and $\{\omega(t)\}$, respectively. All noise sources are assumed to be zero mean. The secondary signal measurements are formed from the output of the subsystem

$W_m(z^{-1})$which is also driven by the signal $\{\xi(t)\}$.

In the multichannel estimation problem the number of signal outputs $(n_s)$ to be estimated can be larger, or smaller, than the number of observations $(n_y)$. However, it is physically reasonable to assume that the power spectrum: $\Phi_{mm} = W_m W_m^*$ of the signal $\{m(t)\}$ which forms the measured output is full rank. Thus, the measured output model $W_m$ can be assumed to be square and full rank.

The covariance matrix for the white measurement noise $\{v(t)\}$ is denoted by $R = R^T = E\{v(t)v^T(t)\} \geq 0$ and $\{v(t)\}$ is statistically independent of the other noise terms. The measurement noise $\{n(t)\}$ may, however, be correlated with the signal source, and the covariance matrix for the signal $[\xi(t)^T, \omega^T(t)]^T$ is assumed to be of the form:

$$E\left\{ \begin{bmatrix} \xi(t) \\ \omega(t) \end{bmatrix} [\xi^T(t), \omega^T(t)] \right\} = \begin{bmatrix} I & G \\ G^T & I \end{bmatrix} \qquad (6.49)$$

There is no loss of generality in assuming that the white noise sources $\{\xi(t)\}$ and $\{\omega(t)\}$ , in equation (6.49), are normalised to the identity matrix.

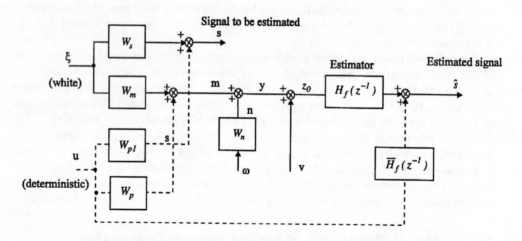

**Fig. 6.5** : $H_\infty$ Signal Processing Problem in Standard Model Form

## Summary of Assumptions

(3i) Noise sources are assumed to be zero mean. The measurement noise source $\{v(t)\}$ is statistically independent of the other noise sources and has a covariance matrix $R$. The covariance matrix indicates there is cross-correlation between the measurement noise and the signal source, which can occur in some control engineering applications.

(3ii) The measurement noise model $W_n$ is assumed to be asymptotically stable.

(3iii) There can be no unstable hidden modes in the individual plant subsystems and any unstable modes in the signal model $W_s$ must also be present in the observations signal model $W_m$ (in state-space terms the system is detectable).

### Optimal Estimator Transfer-function

The optimal linear filter estimate will be represented in the general form $\hat{s}(t|t-\ell) = (H_f(z^{-1})z^{-\ell})z(t)$ , where the filtering, prediction and fixed-lag smoothing problems correspond with $\ell = 0$, $\ell > 0$ and $\ell < 0$, respectively. The estimator will be obtained in z-transfer function form which is simple to implement using difference equations.

The signal $\{u(t)\}$ in Fig. 6.5 is assumed to be a known signal representing, for example, control action or a deterministic disturbance. This signal is transmitted through the transfers $W_p$ and $W_{p1}$. If the optimal filter transfer between the observations signal $\{z_0(t)\}$ and estimate $\{\hat{s}(t)\}$ is denoted by $H_f(z^{-1})$ it is easy to show that for unbiased estimates the optimal additional transfer function between $\{u(t)\}$ and $\hat{s}(t)$ is given by:

$$\overline{H}_f(z^{-1}) = (W_{p1} - H_f W_p)$$

It follows that the deterministic signal can be neglected in the following, since the analysis is simplified. After calculating the optimal filter $H_f$ the transfer can always be recovered from this result.

## 6.4.2  Polynomial Models and System Equations

Assume now that the deterministic signal $\{u(t)\}$ is null. From inspection of Fig. 6.5 the following system equations apply:

Signal model:                     $s = W_s\xi$
Measured (noise free) signal:     $m = W_m\xi$
Measurement noise:                $n = W_n\omega$
Observations:                     $z_0 = v + W_n\omega + W_m\xi$

The system equations may therefore be written as:

**Estimate:**

$$\hat{s} = (H_f z^{-\ell})z_0 = (H_f z^{-\ell})(v + W_n\omega + W_m\xi) \tag{6.50}$$

**Estimation error:**

$$\hat{e} = s - \hat{s} = W_s\xi - (H_f z^{-\ell})(v + W_n\omega + W_m\xi) \tag{6.51}$$

$$= (W_s - H_f z^{-\ell}W_m)\xi - (H_f z^{-\ell})(v + W_n\omega)$$

**Weighted error:**

$$\hat{e}_0 = W_\sigma(s - \hat{s}) = W_\sigma W_s\xi - W_\sigma(H_f z^{-\ell})(v + W_n\omega + W_m\xi)$$

$$= W_\sigma(W_s - H_f z^{-\ell}W_m)\xi - W_\sigma(H_f z^{-\ell})(v + W_n\omega) \tag{6.52}$$

**Polynomial Matrix Descriptions**

The system and noise models have the following polynomial matrix descriptions (Kailath 1980):

**Signal model:**                      $W_s = A_s^{-1} C_s$
**Secondary measurements:**            $W_m = A^{-1} C_m$
**Noise model:**                       $W_n = A^{-1} C_n$

Note that there is no loss of generality in assuming the existence of the left coprime polynomial matrices $A$ and $[C_m \ C_n]$ where,

$$[W_m \ W_n] = A^{-1}[C_m \ C_n]$$

The matrices $A, C_m, C_n$, are polynomial matrices in the indeterminate $z^{-l}$ of compatible dimensions. Also introduce the right coprime matrices $[C_{s1}^T \ C_{m1}^T]^T$ and $A_1$ which satisfy:

$$\begin{bmatrix} W_s \\ W_m \end{bmatrix} = \begin{bmatrix} A_s^{-1} C_s \\ A^{-1} C_m \end{bmatrix} = \begin{bmatrix} C_{s1} \\ C_{m1} \end{bmatrix} A_1^{-1} \tag{6.53}$$

It is convenient to introduce the greatest common right divisor $C_0$, of the polynomial matrices $C_{s1}$, and $C_{m1}$, so that,

$$\begin{bmatrix} C_{s1} \\ C_{m1} \end{bmatrix} = \begin{bmatrix} C_{s0} \\ C_{m0} \end{bmatrix} C_0 \tag{6.54}$$

It follows that the terms $C_0 A_1^{-1}$ are common to the signal and observation channels. Thence, the signal and observations models can be written as:

$$\begin{bmatrix} W_s \\ W_m \end{bmatrix} = \begin{bmatrix} A_s^{-1} C_s \\ A^{-1} C_m \end{bmatrix} = \begin{bmatrix} C_{s0} \\ C_{m0} \end{bmatrix} C_0 A_1^{-1} \tag{6.55}$$

A further assumption is now required, which is not restrictive, since it applies to any system which can be represented in state equation model form. In fact noise and signal models are often an approximation to a more complex nonlinear physical situation and as such they can be treated as design variables, so that the designer can ensure the assumptions are valid.

**Assumption 3(iv)** The system has a signal model polynomial matrix which can be written as $C_{s0} = K C_k$ and $C_k$ is a square strictly Schur polynomial matrix of normal full rank, and $K$ is a constant (possibly nonsquare) matrix.

Note that this assumption is not very restrictive, since any system which can be written in state space form will satisfy this condition, where $C_k = 1$. The signal model can be written as $W_s = A_s^{-1} C_s = C_{s0} C_0 A_1^{-1} = K C_k C_0 A_1^{-1}$ and hence the left coprime matrices $A_{s1}$ and $C_1$ can be introduced as:

$$A_{s1}^{-1} C_1 = C_k C_0 A_1^{-1} \tag{6.56}$$

so that

$$W_s = A_{s1}^{-1} C_s = K A_{s1}^{-1} C_1 \tag{6.57}$$

Note from (6.55) and (6.56) that $W_m$ can also now be written as:

$$W_m = A^{-1}C_m = C_{m0}C_k^{-1}A_{s1}^{-1}C_1 \tag{6.58}$$

**Spectral Factorisation**

The solution of the inferential estimation problem requires the introduction of a filter spectral-factor $Y_f$. This enables the power spectrum of the observations signal to be factored into the form $\Phi_{z_0 z_0} = Y_f Y_f^*$ . The spectral factor satisfies the following equation:

$$Y_f Y_f^* = W_m W_m^* + W_n W_n^* + R + W_m G W_n^* + W_n G^T W_m^* \tag{6.59}$$

Replacing the system models by their polynomial matrix counterparts, note that $Y_f$ may be written as:

$$Y_f = A^{-1}D_f \tag{6.60}$$

where $D_f$ is a Schur polynomial matrix, left spectral factor, defined using

$$D_f D_f^* = C_m C_m^* + C_n C_n^* + ARA^* + C_m G C_n^* + C_n G^T C_m^* \tag{6.61}$$

The Schur spectral factor $D_f$ exists, if and only if, $rank\{[C_m \ \ C_n] \begin{bmatrix} I & G \\ G^T & I \end{bmatrix}^{1/2} \}$

and $AR^{1/2}$ have no left common factors with zeros on the unit-circle of the z-plane.

The following assumption normally holds if realistic system models and noise descriptions are employed.

**Assumption 3(v)** : The signal and noise source models are assumed to be such that $D_f$ exists and is strictly Schur.

### 6.4.3 The Standard $H_2$ Optimal Estimation Problem

The minimisation of the variance of the weighted estimation error may now be considered. The solution of the $H_2$ minimisation problem is required before the solution to the $H_\infty$ problem can be found. The $H_2$ optimal standard model problem involves the minimisation of the estimation error:

$$\hat{e}(t|t - \ell) = s(t) - \hat{s}(t|t - \ell) \tag{6.62}$$

where $\hat{s}(t|t-\ell)$ denotes the optimal linear estimate of the signal $\{s(t)\}$ at time $t$, given observations $\{z(t)\}$ up to time $(t - \ell)$ . The integer $\ell$ may be positive or negative and as noted:

$\ell = 0$ : Filtering

$\ell < 0$ : Fixed lag smoothing

$\ell > 0$ : Prediction

The usual estimation error cost-function which is minimised has the form:

$$J = trace\{E\{\hat{e}(t|t - \ell)\hat{e}^T(t|t - \ell)\}\} \qquad (6.63)$$

It is useful to introduce a dynamic cost function weighting if the error is to be penalised in a particular frequency range. For example, if the errors are particularly important in the low frequency region the transfer $W_p$ should be defined as a low pass filter. Thus, introduce the asymptotically stable weighting function $W_p = A_p^{-1}B_p$, where $A_p$ and $B_p$ denote polynomial matrices, and where $A_p$ is strictly Schur. Then the $H_2$ cost-function to be minimised becomes:

$$J_0 = trace\{E\{W_p\hat{e}(t|t - \ell)(W_p\hat{e}(t|t - \ell))^T\}\}$$

and this can be written in the complex integral form:

$$J_0 = trace\{\frac{1}{2\pi j} \oint_{|z|=1} \{W_p\Phi_{ee}W_p^*\}\frac{dz}{z}\} \qquad (6.64)$$

where the adjoint $W_\sigma(z^{-1})^* = W_\sigma^T(z)$ .

The following theorem summarises the solution of the $H_2$ optimal estimation problem, for a standard system description.

**Theorem 6.3 :** $H_2$ *Optimal Linear Estimator for a Standard System*
The optimal linear filter ($\ell = 0$), predictor ($\ell > 0$) or smoother ($\ell < 0$), to minimise the weighted estimation error (6.64) can be calculated as follows:

**Spectral factorisation :** Compute the strictly Schur spectral factor $D_f$ using:

$$D_fD_f^* = C_mC_m^* + C_nC_n^* + ARA^* + C_mGC_n^* + C_nG^TC_m^* \qquad (6.65)$$

**Diophantine equations :** Compute the solution $(F_0, G_0, S_0)$, with $F_0$ of smallest degree, of the coupled diophantine equations:

$$A_{s1}F_0 + G_0D_f^*z^{-g} = C_1(C_m^* + GC_n^*)z^{\ell-g} \qquad (6.66)$$

$$-C_{m3}F_0 + S_0D_f^*z^{-g+\ell_0} = C_{s3}(C_nC_n^* + ARA^* + C_nG^TC_m^*)z^{\ell-g} \qquad (6.67)$$

where $\begin{bmatrix} W_s \\ W_m \end{bmatrix} = \begin{bmatrix} KC_k \\ C_{m0} \end{bmatrix} C_0A_1^{-1}, A_{s1}^{-1}C_1 = C_kC_0A_1^{-1}$ , $C_{s3}^{-1}C_{m3} = AC_{m0}C_k^{-1}$.
For filtering and prediction problems $\ell_0 = \ell$ and for smoothing problems $\ell_0 = 0$. The integer $g$ is the smallest positive number which ensures these equations are polynomial expressions in the indeterminate $z^{-1}$.

**Robust weighting equation :** The solution $(F_1, N)$, with $F_1$ of smallest degree, is required of the following diophantine equation:

$$A_pF_1 + ND_f^*z^{-g} = B_pKF_0 \qquad (6.68)$$

**Optimal estimator:**

$$H_f = (KA_{s1}^{-1}G_0 + B_p^{-1}N)D_f^{-1}A \qquad (6.69)$$

**Minimal cost:**

$$J_0 = trace\{\frac{1}{2\pi j} \oint_{|z|=1} \{F_1(D_f D_f^*)^{-1}F_1^* + W_p T_0 W_p^*\}\frac{dz}{z}\} \qquad (6.70)$$

where

$$T_0 = W_s \left(I - (W_m^* + GW_n^*)(Y_f Y_f^*)^{-1}(W_m + W_n G^T)\right) W_s^*$$

The *implied equation* determining stability becomes:

$$C_{m3}A_{s1}^{-1}G_0 z^{-\ell} + S_0 z^{\ell_0 - \ell} = C_{s3}D_f \qquad (6.71)$$

**Proof :** From Grimble (1994b).

The above equations are similar to those obtained in the solution of Wiener filtering problems. The spectral factorisation is the same and it relates to the Riccati equation in Kalman filtering problems. The diophantine equations are a simple way of performing the partial-fraction expansion step in Wiener filtering. These are needed for the separation of the causal and non-causal terms in the cost-function. The standard system model encompasses the $H_2$ deconvolution and the usual optimal estimation problems.

## 6.5 Standard System for $H_\infty$ Estimation

The $H_\infty$ filtering problems can again be solved by using the $H_2$ *embedding procedure*. That is, an auxiliary minimum variance filtering problem can be solved for a cost function with dynamic weighting. The Lemma linking the solutions of the $H_2$ and $H_\infty$ problems can then be employed to provide the desired solution. The corresponding $H_\infty$ optimal linear estimators can be derived using these results and the key linking Lemma, introduced below.

### 6.5.1 Relationship between $H_2$ and $H_\infty$ Problems

The $H_\infty$ multivariable estimation problem can be solved using the $H_2$ embedding procedure and the following lemma (Kwakernaak, 1984). The function $W_\lambda$ represents a cost weighting filter.

**Lemma 6.3 :** *Auxiliary $H_2$ Minimisation Problem*
Consider the auxiliary problem of minimising the $H_2$ criterion:

$$J = \frac{1}{2\pi j} \oint_{|z|=1} trace\{W_\lambda(z^{-1})X(z^{-1})W_\lambda^*(z^{-1})\}\frac{dz}{z} \qquad (6.72)$$

Suppose that for some real-rational matrix $W_\lambda^*(z^{-1})W_\lambda(z^{-1}) \geq 0$, the cost function $J$ is minimised by a function $X(z^{-1}) = X^*(z^{-1})$, for which $X(z^{-1}) = \lambda^2 I_r$ (a real constant matrix on $|z| = 1$). Then the function $X(z^{-1})$ also minimises:

$$J_\infty = \sup_{|z|=1} \left\| X(z^{-1}) \right\|_2 = \sup_{|z|=1} \left\{ \sigma_{\max} \{ X(z^{-1}) \} \right\} \tag{6.73}$$

where $\left\| X(z^{-1}) \right\|_2$ denotes the spectral norm.

**Proof :** The proof is similar to that for Kwakernaak (1984) and in Lemma 5.1 of the previous chapter.

Consider the generalised system model discussed in §6.4.1, under the Assumptions 3(i) to 3(iv). To solve the optimal $H_\infty$ estimation problem the $H_2$ embedding procedure will be followed and $W_\lambda$ will be found so that the solution $X(z^{-1})$ is equalising. The objective is to find the weighting function $W_\lambda$ which when substituted into the results of the $H_2$ estimation Theorem 6.3 will ensure that Lemma 6.3 is satisfied. An *equalising* solution will then result which corresponds to a solution $X(z^{-1}) = \lambda^2 I_r$. The $H_\infty$ optimal linear estimator will then follow.

To relate the two problems let the dynamic weighting function $W_p$ in Theorem 6.3, be written as the product of two terms:

$$W_p = W_\lambda W_0 \quad , \quad W_p \varepsilon R^{n_s \times n_s}(z^{-1}) \tag{6.74}$$

One of the weighting terms is necessary to relate the $H_2$ and $H_\infty$ problems ($W_\lambda$). The second is to enable a dynamic weighting function ($W_o$) on the estimation error to be introduced.

The function $X(z^{-1})$ to be minimised can be identified with the weighted estimation error as :

$$X = W_0 \Phi_{ee} W_0 \quad , \quad X \varepsilon R^{n_s \times n_s}(z^{-1}) \tag{6.75}$$

The weighting function $W_0$ represents the dynamic weighting function, selected by the designer, to shape the frequency response of the estimation error spectrum, so that the cost-function to be minimised is given as:

$$J_\infty = \sup_{|z|=1} \left\{ \sigma_{\max} \{ X(z^{-1}) \} \right\} = \sup_{|z|=1} \left\{ \sigma_{\max} \{ W_0 \Phi_{ee} W_0^* \} \right\}$$

The problem is now to determine the function $W_\lambda$ which will ensure the conditions of the Lemma 6.3 are satisfied.

## 6.5.2  Derivation of the Weighting Filter $W_\lambda$

Assume that the scalar $\lambda$ in the above lemma is for the moment known. To ensure the conditions of the above lemma are satisfied, the $W_\lambda$ must now be found which leads to an equalising solution:

$$X(z^{-1}) = \Lambda \Lambda^T = \lambda^2 I_r \quad \text{on} \quad |z| = 1 \tag{6.76}$$

The matrix $\Lambda$ introduced in the above expression may be defined as $\Lambda = \lambda I$ at the optimum. However, once the maximum singular value has been minimised and a solution for the filter has been obtained, it is sometimes possible to improve upon the solution by defining a more general $\Lambda$. The improvement can, for example, be in terms of reducing the maximum values of some of the remaining singular values.

An expression is given in Theorem 6.3 for the minimal value of the cost-function for the $H_2$ optimal estimation problem. The following expression, which must be satisfied if Lemma 6.3 is to be invoked, follows from the integrand of (6.70) and the above equalising result:

$$F_1 D_f^{*-1} D_f^{-1} F_1^* + W_\lambda W_0 T_0 W_0^* W_\lambda^* = W_\lambda \Lambda \Lambda^T W_\lambda^*$$

or

$$F_1 D_f^{*-1} D_f^{-1} F_1^* = W_\lambda (\Lambda \Lambda^T - W_0 T_0 W_0^*) W_\lambda^T \tag{6.77}$$

where $F_1 \varepsilon P^{n_s \times n_y}(z^{-1})$.

Introduce the stable minimum-phase transfer-function $S \varepsilon R^{n_s \times n_y}(z^{-1})$ which satisfies:

$$SS^T = \Lambda \Lambda^T - W_0 T_0 W_0^* \tag{6.78}$$

and assume that $\Lambda$ is chosen so that the transfer $S$ is invertible. From (6.77) and (6.78) now obtain:

$$F_1 D_f^{*-1} D_f^{-1} F_1^* = W_\lambda SS^* W_\lambda^* \tag{6.79}$$

Define the spectral-factor $D_{fo}$ which satisfies:

$$D_{f0}^* D_{f0} = D_f D_f^*$$

and let the left-coprime pair $(F, D)$ satisfy:

$$D^{-1} F = F_1 D_{f0}^{-1} \quad , \quad F \varepsilon P^{n_s \times n_y}(z^{-1}) \tag{6.80}$$

Then, to ensure the equalising solution is obtained, from (6.79) and (6.80):

$$D^{-1} F F^* D^{*-1} = W_\lambda SS^* W_\lambda^* \tag{6.81}$$

Let $F_s$ denote a Schur polynomial matrix which satisfies :

$$F_s F_s^* = FF^* \tag{6.82}$$

Then, from (6.81) and (6.82):

$$W_\lambda = D^{-1} F_s S^{-1} \tag{6.83}$$

**Robust Weighting Diophantine Equation**

The equation which enables the robust weighting function and the minimum value of $\lambda$ to be found will now be derived. To ensure a suitable definition for $F_s \epsilon P^{n_s \times n_s}(z^{-1})$ is obtained, the following assumption is required:

**Assumption 3(vi):** The number of signal channels $n$ equals the number of observation channels $n_z = n_y$.

Introduce the left-coprime polynomial matrices $(\overline{S}, \overline{F}_s)$ as:

$$\overline{S}^{-1}\overline{F}_s = F_s S^{-1} W_0 \qquad (6.84)$$

Then from (6.83) and (6.84) obtain:

$$W_p = W_\lambda W_0 = A_p^{-1} B_p = D^{-1}\overline{S}^{-1}\overline{F}_s$$

Hence, identify the desired $H_2$ weighting polynomials, which ensure an equalising solution is obtained, as:

$$A_p = \overline{S}D \quad\text{and}\quad B_p = \overline{F}_s \qquad (6.85)$$

and if $W_\lambda$ is required this may be found as:

$$W_\lambda = A_p^{-1} B_p W_0^{-1} = (\overline{S}D)^{-1}\overline{F}_s W_0^{-1}$$

To calculate $N$ and $\overline{F}_s$ a more appropriate form of equation (6.68) is required. Substituting for the weightings in (6.85) the robustness diophantine equation (6.68) becomes:

$$\overline{S}DF_1 + ND_f^* z^{-g} = \overline{F}_s K F_0$$

but from (6.80) : $DF_1 = FD_{f0}$ and hence obtain:

$$\overline{S}FD_{f0} + ND_f^* z^{-g} = \overline{F}_s K F_0$$

The optimal estimator follows from (6.69) but note that $B_p^{-1}$ enters the expression and hence from (6.79), (6.80) and (6.85) the matrix $F$ must be of normal full rank.

### 6.5.3   $H_\infty$ Optimal Estimator for the Standard System

The $H_\infty$ optimal linear estimator which has an equalising solution may now be defined. This provides the solution to the filtering, prediction and fixed lag smoothing problems, based on the results in Theorem 6.3.

**Theorem 6.4 :** $H_\infty$ *Estimator for Multichannel Estimation Problems*

Consider the system as shown in Fig. 6.1 and discussed in §6.4.3. Assume that the cost-function:

$$J_\infty = \sup_{|z|=1} \{\sigma_{max}\{W_0 \Phi_{ee} W_0^*\}\} \qquad (6.86)$$

is to be minimised where $\Phi_{ee}$ denotes the spectrum of the estimation error signal $\hat{e}(t|t - \ell) = s(t) - \hat{s}(t|t - \ell)$, where $\ell = 0$, $\ell > 0$ and $\ell < 0$ for filtering, smoothing and prediction problems, respectively. The $H_\infty$ optimal estimator may be computed by first calculating $D_f$ from (6.65) and then obtaining $(G_0, S_0, F_0, g)$ from (6.66) and (6.67). Also define the transfer $T_0$ from (6.70) and the stable minimum-phase transfer $S(z^{-1})$ using:

$$SS^* = \Lambda\Lambda^* - W_0 T_0 W_0^* = \lambda^2 I - W_0 T_0 W_0^* \tag{6.87}$$

The left-coprime pair $(F, D)$ can be found using:

$$D^{-1}F = F_1 D_{f0}^{-1} \text{ where } D_{f0} \text{ satisfies: } D_{f0}^* D_{f0} = D_f D_f^* \tag{6.88}$$

Let $F_s$ denote the Schur spectral factor satisfying $F_s F_s^* = FF^*$ and introduce the left-coprime polynomial matrices $(\overline{S}, \overline{F}_s)$ using:

$$\overline{S}^{-1}\overline{F}_s = F_s S^{-1} W_0 \tag{6.89}$$

The $H_\infty$ optimal linear estimator can then be found by first calculating the solution $(N, F)$, with $F$ of smallest degree, of the linear equation:

$$\overline{S}FD_{f0} + ND_f^* z^{-g} = \overline{F}_s K F_0 \tag{6.90}$$

**Optimal estimator:**

$$H = (KA_{s1}^{-1}G_0 + \overline{F}_s^{-1}N)D_f^{-1}A \tag{6.91}$$

**Optimum function and minimum cost:**

$$X_{\min} = \lambda^2 I \quad \text{and} \quad J_\infty = \lambda^2$$

where $\lambda^2$ is the smallest scalar such that equations (6.87) to (6.90) are satisfied.

**Proof :** The theorem follows by collecting the previous results and by invoking Lemma 6.3.

### 6.5.4 Properties of the Optimal Estimator

The following properties for the $H_\infty$ optimal estimator may be established:

(i) **Limiting form :** If the solution for $N = 0$ the estimator (6.91) reduces to the least-squares estimator of Theorem 6.3, for a constant weighted cost-function

(ii) **Robustness :** The second term in the estimator (6.91) determines the robustness properties. This term distinguishes the least-squares from the $H_\infty$ estimator and can be chosen according to the form of the uncertainty.

(iii) **Stability :** Even when the signal source is unstable the minimal form of the optimal estimator can be shown to be stable. The proof follows from the definitions of $\overline{F}_s$ and $D_f$ and by noting $det(A_{s1}) = det(A_1)$ and $det(A)$ includes any unstable zeros in $det(A_1)$.

(iv) **Optimality** : At the optimum $W_0 \Phi_{ee} W_0^* = \lambda^2 I$, and thus by specifying a desired weighting $W_0(z^{-1})$ frequency response, the estimation error can be frequency shaped. The actual level of the weighted estimation error cannot be specified, since it depends upon the calculated value of $\lambda$. However, the form of the frequency response can be predetermined and this provides a valuable and unusual characteristic for the $H_\infty$ class of estimators.

### 6.5.5   Deconvolution Filtering Problem Example

As discussed in the introduction §6.1 the deconvolution filtering problem involves the estimation of the input signal to a linear system, given an output signal that is corrupted by a measurement noise term. When the $H_\infty$ norm of the weighted estimation error spectrum is minimised the spectral error can be reduced to lower values (in selected frequency ranges) than is possible with a least squares or $H_2$ optimal filter (Grimble 1994b). The optimal deconvolution problem arises in digital communications, underwater acoustics, seismology and control systems. In many applications the system models include uncertainties which are difficult to take into account in a $H_2$ cost minimisation framework. The $H_\infty$ philosophy has, however, been shown to provide robust solutions particularly in equivalent control engineering problems.

**Example 6.3 :** *$H_\infty$ Optimal Linear Filtering*

To illustrate the form of the optimal filter and results consider the following $H_\infty$ filtering problem for the system shown in Fig. 6.1. Assume that the sample interval $T_s = 0.005$ seconds and that the dynamic weighting function and system models are defined as follows.

**Weighting:**

$$W_p = 1.6136 \frac{(1 - 0.997214z^{-1})(1 - 0.38104z^{-1})}{(1 - 0.998755z^{-1})}$$

**Signal model:**

$$W_s = \frac{(1 - 0.998755z^{-1})}{(1 - 0.98756z^{-1} + 0.99005z^{-2})}$$

**Noise and signal channel:**

$$W_n = 1, \quad G = 0, \quad \text{and} \quad W_c = 1$$

The $H_\infty$ filter can be computed as:

$$H_f = \frac{(1 - 0.998755z^{-1})(0.9921 + 0.9924z^{-1})}{1.61336(1 - 0.997214z^{-1})(1 - 0.3813z^{-1})(1 + 0.9968z^{-1})}$$

The estimate of the output follows using : $\hat{s}(t|t-1) = H_f(z^{-1})z_0(t)$.

**Results**

The filter given above may be compared with the smoothing filter defined in Grimble (1996). The output of the signal model for the filtering problem is shown in Fig. 6.6. In this example the measurement noise shown in Fig. 6.7 is much greater than the signal. The observations signal and the estimate of the signal are shown in Fig. 6.8. That the solution is equalizing is demonstrated in Fig. 6.9, since this reveals the weighted error spectrum is a constant.

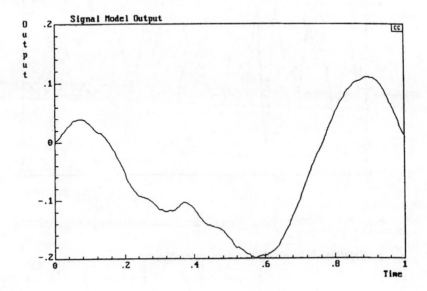

**Fig. 6.6** : Signal Model Time Response

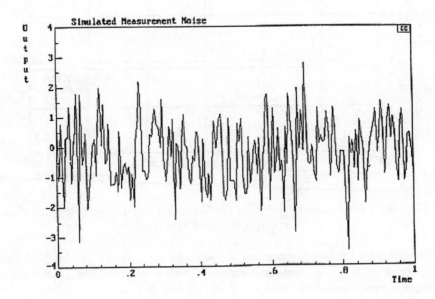

**Fig. 6.7** : Simulated Measurement Noise Signal

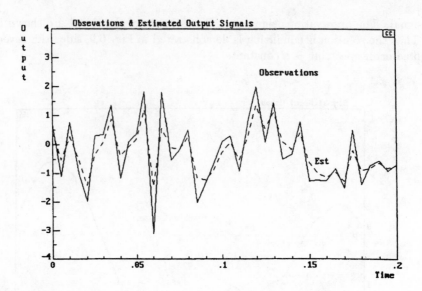

Fig. **6.8** : Observations Signal and Estimated Output Signal

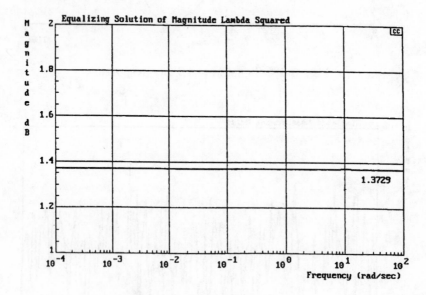

Fig. **6.9** : Frequency Response of the Weighted Error Signal Revealing an
Equalising Solution Obtained

The estimator frequency response is shown in Fig. 6.10. In this problem the
measurement noise is relatively small and the dynamic weighting therefore determines
the frequency response of the estimator fairly closely. In fact it is an unusual property
of $H_\infty$ filters that the estimators error frequency response is directly determined by the
weighting function. The actual magnitude of the error depends upon the calculated
value of $\lambda^2$

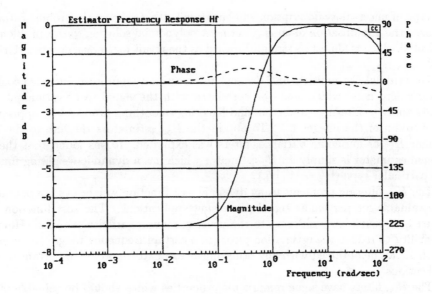

**Fig. 6.10** : Magnitude and Phase of the Filter Frequency Response

## 6.5.6  Multivariable Estimation Problems

The preceding results are specialised to particular problems, are straightforward to implement in the scalar case and provide many insights into the frequency-domain behaviour. However, for the multivariable case it is an advantage to derive a solution which is in standard systems form and can be used to solve a variety of estimation and control problems. The resulting numerical algorithms and software can be used to solve multivariable problems as described and illustrated in Grimble and Kucera (1996). The example presented in this reference is for the multivariable case and it includes a solution of the spectral factorisation and diophantine equations. In this particular work a sub-optimal solution is in fact obtained. This type of algorithm is interesting since it is the direct parallel of the state-space $H_\infty$ filtering results which also obtain sub-optimal solutions (Yaesh and Shaked, 1992). The procedure described above provides an optimum solution. However, each of these approaches has different merits for different application problems.

## 6.6  Concluding Remarks

A solution was presented to the robust linear smoothing, filtering and prediction problems. The $H_2$ optimal linear estimation problems were first considered. When stochastic problems dominate the least squares estimation methods remain the most valuable. They provide signal estimates using the most physically realistic criterion. However, when robustness is the main problem then the $H_\infty$ approach has advantages. The uncertainty in signal and noise models can be represented by both a $H_\infty$ frequency bounding function and a probabilistic system model, although in most problems only one type of uncertainty description will be needed. It was shown that robust estimates required the minimisation of an $H_\infty$ norm. A polynomial solution to this problem was presented which is almost as straightforward as the result for known systems (Grimble 1992).

The optimal $H_\infty$ multichannel deconvolution estimation problem included coloured measurement noise which could be correlated with the signal to be estimated. The results apply to filtering, prediction and fixed-lag smoothing problems by appropriate specification of the integer $\ell$.     To obtain the $H_\infty$ estimators the link to the more familiar $H_2$, or minimum variance, filters was exploited. It was found that the $H_\infty$ optimal estimator is simply an $H_2$ estimator which has a dynamic weighting function of a particular type $(W_p = W_\lambda W_0)$.

The $H_\infty$ filtering problems using dynamic cost function weightings are potentially very valuable for particular signal processing applications. The cost function minimised is of course very different to the usual minimum variance criterion. However, the ability to reduce the estimation error (in a certain frequency range) to lower values than can be obtained with any other linear estimator, may be important in some applications.

The $H_\infty$ filters have some remarkable properties which should be valuable in specialised signal processing applications. For example, the scalar optimal estimation error spectrum is completely characterised by the inverse of the weighting function $W_0$, irrespective of the noise and signal models employed. These models only affect the minimum value of the cost-function $(\lambda^2)$, not the actual shape of the frequency response of the error spectrum.

A generalisation of the $H_2$ and $H_\infty$ estimation problem was also presented and the solution was obtained in standard system polynomial matrix form. This represents a modern approach to the classical Wiener filtering problem. The advantage of the solution is that it may be used to solve a class of problems of practical importance, where currently no transfer-function or polynomial based solution is available. This type of problem involves estimating the output of a system, where the output cannot be measured directly (only a secondary output is available from which the desired output is to be estimated). It is relatively easy to make the polynomial forms of all of the above estimators adaptive (Fung et al. 1984 and Grimble, 1981).

## 6.7   References

Ahlen, A. and Sternad, M., 1991, Wiener filter design using polynomial equations, *IEEE Transactions on Signal Processing*, Vol. **39**, pp. 2387-2399.

Ahlen, A. and Sternad, M., 1993, *Optimal filtering problems*, Chapter 5 Editor K. Hunt, Polynomial Methods in Optimal Control and Filtering, Control Engineering Series, Peter Peregrinus.

Barrett, J.F., 1977, Construction of Wiener filters using the return difference matrix, *Int J Control*, Vol **26**, No 5, 797-803.

Chisci., L. and Mosca, E., 1992, MMSE multichannel deconvolution via polynomial equations, *Automatica*.

Doyle, J.C., Glover, K., Khargonekar, P.P. and Francis, B.A., 1989, State-space solutions to standard $H_2$ and $H_\infty$ control problems, *IEEE Trans. on Auto. Contr.*, **34**, 9, pp. 831-846.

ElSayed, A., 1988, Design of $H_\infty$ filters, PhD thesis, Industrial Control Centre, Department of Electronic and Electrical Engineering, University of Strathclyde.

ElSayed, A. and Grimble, M.J., 1989, A new approach to the $H_\infty$ design of optimal digital linear filters, *IMA Journal of Mathematical Contr., and Information*, **6**, pp. 233-251.

ElSayed, A., Shaked, U. and Grimble, M.J., 1989, Application of spectral factorisation to the linear discrete time fixed-point smoothing problem, *IEEE Trans. on Auto. Contr.*, Vol. **34**, No. 3, pp. 333-335.

Fu, M., de Souza, C.E. and Xie, L., 1990, $H_\infty$ *estimation for uncertain systems*, Technical Report No. EE 9039 University of Newcastle, New South Wales.

Fung, P.T.K., Chen, X.P. and Grimble, M.J. 1984, The adaptive tracking of slowly varying process with coloured noise disturbances, *Trans. Inst. Measurement and Control*, Vol. **6**, No. 6, pp. 299-304.

Grimble, M.J., 1981, Adaptive Kalman filter for control of systems with unknown disturbance, *IEE Proc.* Vol. *128*, Pt. D. No. 6, pp. 263-267.

Grimble, M.J., 1984, Wiener and Kalman filters for systems with random parameters, *IEEE Trans.* Vol. AC-**29**, No. 6, pp. 552-554.

Grimble, M.J., 1985a, Polynomial systems approach to optimal linear filtering and prediction, *Int. J. Control.* Vol. **41**, No. 6, pp. 1545-1564.

Grimble, M.J., 1985b, Time-varying polynomial system approach is multichannel optimal linear filtering, *Proc. of the ACC*, WC6-10:45, pp. 168-174.

Grimble, M.J., 1987, $H_\infty$ Design of Optimal Linear Filter, *MTNS Conference*, Phoenix, Arizona, Published 1988 in C.I. Byrnes, C.F. Martin and R.E. Saeks, Elsevier Science Publishers BV, North Holland, Amsterdam, pp. 553-540.

Grimble, M.J., 1988, Single versus double diophantine equation debate; comments on an approach to Wiener filtering, *Int. J. Control*, Vol. **48**, No. 5, pp. 2161-2165.

Grimble, M.J. and ElSayed, A., 1990, Solution of the $H_\infty$ optimal linear filtering problem for discrete-time systems, *IEEE Trans. on Acoustics Speech and Signal Processing*, Vol. **38**, No. 7, pp. 1092-1104.

Grimble, M.J., 1991, $H_\infty$ fixed-lag smoothing filter for scalar systems, *IEEE Trans. on Signal Processing*, Vol. **39**, No. 9, pp. 1955-1963.

Grimble, M.J., 1992, LQG optimal control design for uncertain system, *IEE Proc.*, Pt. D., Vol. **139**, No. 1, pp. 21-30.

Grimble, M.J., 1994a, *Robust industrial control*, Prentice Hall, Hemel Hempstead.

Grimble, M.J., 1994b, $H_2$ inferential filtering, prediction and smoothing with application to rolling mill gauge estimation, *IEEE Trans. Signal Processing,* Vol. **42**, No. 8, pp. 2078-2093.

Grimble, M.J., 1995, Multichannel optimal linear deconvolution filters and strip thickness estimation from gauge measurements, *ASME Journal of Dynamic Systems, Measurement and Control,* Vol. **117**, pp. 165-174.

Grimble, M.J., 1996, $H_\infty$ optimal multichannel linear deconvolution filters, predictors and smoothers, *Int. J. of Control.* Vol **63**, No. 3, pp. 519-533.

Grimble, M.J. and Kučera, V., 1996, *Polynomial methods for control systems design,* Springer-Verlag, London.

Grimble, M.J. and Åström, K.J., 1987, Frequency domain properties of Kalman filters, *Int. J. Control,* Vol. **45**, No. 3, pp. 907-925.

Kalman, R.E., 1960, A new approach to linear filtering and prediction problems, *Journal of Basic Engineering,* pp. 35-45.

Kailath, T., 1980, *Linear systems,* Prentice Hall, Englewood Cliffs, NJ.

Kucera, V., 1979, *Discrete linear control,* John Wiley and Sons, Chichester.

Kwakernaak, H., 1984, Minimax frequency domain optimization of multivariable linear feedback systems, *Proceedings IFAC World Congress,* Budapest, Hungary.

MacFarlane, A.G.J., 1971, Return-difference matrix properties for optimal stationary Kalman-Bucy filter, *Proc IEE,* **118**, No 2, 373-376.

Mendel, J.M., 1977, White noise estimators for seismic data processing in oil exploration, *IEEE Trans on Auto. Contr.,* AC-**22**, 5, pp. 694-707.

Moir, T.J., 1986, Optimal deconvolution smoother, *IEE Proc.,* **133**, Pt. D, 1, pp. 13-18.

Moir, T.J. and Grimble, M.J., 1984, Optimal self-tuning filtering prediction and smoothing for discrete multivariable processes, *IEEE Trans. on Auto Contr.,* Vol. AC-**29**, No. 2.

Oppenheim, A.V., Wilsky, A.S. and Young, I.T., 1983, *Signals and Systems,* Prentice Hall.

Shaked, U., 1990, $H_\infty$ minimum error state estimation of linear stationary processes, *IEEE Trans. on Auto. Contr.,* **35**, 5, pp. 554-558.

Shaked, U., 1976, A general transfer function approach to linear stationary filtering and steady-state optimal control problems, *Int. J. Control,* **24**, 6, pp. 741-770.

Sternad, M. and Ahlen, A., 1990, The structure and design of realizable decision feedback equalizers for IIR channels with coloured noise, *IEEE Trans. on Information Theory,* Vol. **36**, No. 4, pp. 848-858.

Sternad, M. and Ahlen, A., 1993a, Robust filtering and feedforward control based on probabilistic descriptions of model errors, *Automatica*, Vol. **29**, No. 3, pp. 661-679.

Sternad, M. and Ahlen, A., 1993b, Robust Weiner filtering based on probabilistic descriptions of model errors, *Kybernetika*, Vol. **29**, No. 5, pp. 439-454.

Xie, L., 1991, $H_\infty$ *control and filtering of systems with parameter uncertainty*, PhD Thesis, University of Newcastle, NSW, Australia.

Yaesh, I. and Shaked, U., 1992, Transfer function approach to the problems of discrete-time systems : $H_\infty$ optimal linear control and filtering, *IEEE Trans.* AC. **26**, No. 1, pp. 1264-1271.

# Part II

# State Space and Frequency
# Response Descriptions

# 7

# H$_2$ and H$_\infty$ State-space Control and Filtering

## 7.1   Introduction

The state-space approach to modelling is particularly popular in North America and this is probably due to the strong interest in aerospace applications in the 1960's. In such applications excellent models are often available from detailed design information, whereas in the process industries this is often not the case. In fact, in the process industries simple models are often found from plant test results and experimentation. There is one example for the process industries where state space methods have proved very valuable. This is in the design of supervisory (predictive) control systems which is considered in the next chapter. The main advantage of state-space methods, that they are numerically reliable for large scale systems design, is then important.

The state-space based LQG design approach was probably, up to very recently, the most popular multivariable control design philosophy (leaving aside predictive methods which might be considered a different category). The *separation principle* of stochastic optimal control (Kwakernaak and Sivan, 1972; Grimble and Johnson, 1988; Davis, 1977) provides a very convenient form to implement the LQG controller. This principle reveals that the solution of the output feedback linear quadratic optimal control problem, when the measurements are corrupted by white noise, is obtained from the solution of two problems:

(i)   The state feedback LQ regulator problem, where noise free measurements of states are assumed available.

(ii)   The Kalman filtering state estimation problem using the noisy output measurements and the control signal.

It is therefore possible to solve these two rather simple problems and obtain the solution to the more difficult output feedback control problem. These problems require the solution of the discrete algebraic control and filtering Riccati equations, respectively. These Riccati equations are introduced in §7.3. However, since the separation principle results are reasonably well known a slightly more general problem will be considered.   This problem, discussed in the first part of the chapter, involves a system with explicit transport delay elements on input and output channels. If there are no such explicit delay elements the results correspond with the traditional LQG output feedback control problem solution.

In the second half of the chapter attention turns to the $H_\infty$ optimal control problems. These are probably the basis of the most popular multivariable control design philosophy, at least in terms of the number of research studies undertaken.   It is shown that the state-space $H_\infty$ output feedback optimal controller also satisfies a type of separation principle result that can provide a useful option for implementation.   Moreover, it throws light on the relationship between the $H_2$ and $H_\infty$ solutions. However, before continuing with the optimal control solutions, a relatively old and established classical transport delay compensation technique will be reviewed.

## 7.1.1   Smith Predictor

The optimal control solution will be related to the Smith Predictor structure (Smith, 1959) for time-delay systems, which is well known in the process industries.  The basic idea of the Smith predictor is illustrated in Fig. 7.1, in its continuous-time version. Observe that the transfer function from the control input to the observations signal $\{z(t)\}$ is zero if the model matches the actual plant. The only loop that then remains active is that formed by the compensator and model which do not contain the transport delay. It is therefore straightforward to design the compensator using PID or other design methods.  However, the disturbance feeds into the system at another point and it is regulated by the closed-loop involving the actual plant model in the usual way.  From the viewpoint of the disturbance signal it enters a closed system and if there is sufficient gain in the loop it will be attenuated.   Thus, the Smith predictor generates an ideal control signal for a delay free system and this may be considered to be input to the plant, which is open-loop from the perspective of the control input signal (Marshall, 1974). This is of course the reason that the closed-loop system is unstable, whenever the plant is open-loop unstable.

Some of the major problems with the Smith predictor are overcome in the technique to be presented. The first difficulty is that the Smith predictor provides a structure for the controller and not any guidance on how to design the controller (Normey-Rico, 1997). It is often the case that poor disturbance rejection is achieved because the compensator design usually concentrates on stability, rather than stochastic properties.  The Smith predictor also has a robustness problem, since the basic idea of adding two signals together, which should ideally cancel, is suspect. Finally, there is the problem, referred to already, that the system will be unstable if the plant is open loop unstable. In fact difficulties are likely to arise for lightly damped systems, even if they are marginally stable.

The Wiener frequency-domain form of the optimal transport delay compensator to be derived can be implemented in a different form, which avoids the need to assume open-loop stability and this is the major contribution in the following. The equivalent state-space realisation involves a subsystem with a finite impulse-response and this is the recommended way to implement the controller.  It provides a more robust solution than the Smith predictor structure and is suitable for both open-loop stable and unstable systems.

Transport delays may be present in a plant, or be introduced as part of an approximate model of a high order plant (Fuller, 1968; Malek-Zavris and Jamshidi, 1987). In this latter case the high order plant may be modelled by a low order system with a transport delay. It is well known that small lags in a plant have a combined effect similar to that of a pure delay. In the problem of interest the control inputs to the multivariable plant and observed outputs are assumed to be delayed by a diagonal set of time delays, that may be different in different channels. Marinescu and Bourles (1997) considered the optimal control of this type of state-space system and presented an example involving the control of the voltages in an electrical power transmission network that included significant delays. In this case the control law was computed in a dispatching centre, using measurements of voltages and reactive powers from different points in the electrical network, and the controls were then sent via transmission buses to each regional generator.

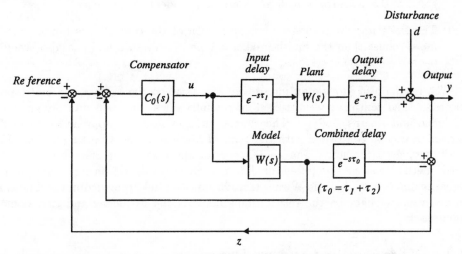

**Fig. 7.1** : Smith Predictor Structure for a Scalar Continuous-time System

## 7.2   Wiener-Hopf Optimal Controller

A discrete-time *Linear Quadratic Gaussian* (LQG) multivariable optimal control problem is first considered where the plant has different time-delays in different data channels. The solution is obtained for the closed-loop optimal controller using a Wiener approach in the z-domain. It is shown later that this controller may be implemented directly in transfer-function matrix, or state-space Kalman predictor form (Kalman and Bucy, 1961; Kleinman, 1969).

The state equation form of the optimal controller will be shown to consist of a Kalman predictor, a diagonal transport delay block and a feedback control gain matrix. This is straightforward to implement, since it depends upon simple finite-dimensional terms and pure delay elements. It is also intuitively reasonable since it can be related to the Smith predictor (Smith, 1959) structure and satisfies a type of separation principle. In this form it has the same limitation as the Smith predictor, namely that the plant must be open-loop stable, if the closed-loop is to be stable. However, the optimal solution can be implemented in a form which stabilises both open loop stable and unstable systems.

The continuous-time case, where the plant included the same delays in each input or each output channel, was considered by Grimble (1979a). Many applications, including the hot strip mill control problem (Chapter 11), involve different delays on different output variables. The solution presented below includes different channel delays and is valid for the more general open loop unstable situation.

The advantages of the approach, that are described in the next section, may be summarised as follows:

(i) The implementation of the controller is very much more economical than the usual LQG solution, having an order equal to the nominal plant (without delays).

(ii) If the delay changes with time the controller can easily be adapted since the gains in the state-space realisation are independent of the delay.

(iii) The finite impulse-response implementation of the controller is fundamentally more robust than the Smith-predictor type of structure and can also stabilise open-loop unstable systems.

The results also provide an interesting link between the Wiener and state-space theories. The controller was utilised successfully on a hot strip mill design study (Grimble and Hearns, 1998). The Wiener transfer function approach used in the derivation can be contrasted with the work of Marinescu and Bourles (1997). This work involved a predictive and an LQ control approach using state-space models. Their method was also proposed for different transport delays on input and output signal channel paths, but the Wiener transfer function link, provided in the following, throws valuable light on the relationships between the frequency and time-domain solutions.

## 7.2.1  Discrete-time System Description

The plant will first be defined using state-space equations, since the z-domain results are to be related to the time-domain. It is assumed that the nominal plant is stabilisable and detectable and contains non-equal transport delays on both input and output channels (Marinescu and Bourles, 1997):

$$(D_1 u)(t) = diag\{u_1(t - \tau_{11}), u_2(t - \tau_{12}), ..., u_m(t - \tau_{1m})\} \text{ for } u(t) \in R^m$$

and

$$(D_2 y)(t) = diag\{y_1(t - \tau_{21}), y_2(t - \tau_{22}), ..., y_r(t - \tau_{2r})\} \text{ for } y(t) \in R^r$$

where $\tau_{ij} \geq 0$ for $j = \{1, ..., m\}$ and $\tau_{2j} \geq 0$ for $j = \{1, ..., r\}$   These delay elements may be separated into synchronous and asynchronous components. Thus, the input delay can be written as $(D_1 u)(t) = (D_{1s} D_{1a})u(t)$ where $(D_{1s}u)(t) = diag\{u_1(t - \tau_1), ..., \{u_m(t - \tau_1)\}$ and $\tau_1 = min\{\tau_{1j}\}$ for $j = \{1, ..., m\}$. That is, the synchronous part is simply the common part of the delay in each signal channel. It may easily be confirmed that if the nominal plant is stabilisable and detectable, then the plant with delays will also be stabilisable and detectable. Let it be assumed that the asynchronous part of the delay be included in the nominal plant state equation model, represented by the triple $(A, B, C)$. Then $(D_1 u)(t) = diag(u_1(t - \tau_1), ..., u_m(t - \tau_1))$.

**State equations**

The generic regulating system is shown in Fig. 7.2 and the constant coefficient plant equations become:

$$x(t + 1) = Ax(t) + B(D_1 u)(t) + D\omega(t) \tag{7.1}$$

$$y(t) = Cx(t) \tag{7.2}$$

$$z(t) = (D_2 y)(t) + (D_2 v)(t) \tag{7.3}$$

where $x(t) \in R^n$, $u(t) \in R^m$, $\omega(t) \in R^q$ and $y(t) \in R^r$. The process noise signal $\{\omega(t)\}$ and the observation noise sequence $\{v(t)\}$ are assumed to be zero-mean, white and stationary with the following process and measurement noise covariance matrices:

$$\left.\begin{array}{l} \text{cov}[\omega(t), \ \omega(\sigma)] = Q\delta_{t\sigma} \\ \text{cov}[v(t), \ v(\sigma)] = R\delta_{t\sigma} \\ \text{cov}[\omega(t), \ v(\sigma)] = 0 \end{array}\right\} \tag{7.4}$$

where $Q \geq 0$ and $R > 0$, and $\delta_{t\sigma}$ denotes the Kronecker delta-function.

Let the following notation be used to denote the z-transforms of the diagonal transport-delay operators:

$$D_1(z^{-1}) = z^{-\Lambda_1} = diag\{z^{-\tau_1}, ..., z^{-\tau_1}\} \in R^{m \times m}(z^{-1})$$

$$D_2(z^{-1}) = z^{-\Lambda_2} = diag\{z^{-\tau_{21}}, ..., z^{-\tau_{2r}}\} \in R^{r \times r}(z^{-1})$$

Also let the resolvent, delay and transfer-function matrices for the plant be defined as:

$$\Phi(z^{-1}) = (zI_n - A)^{-1}, \quad D_i(z^{-1}) = z^{-\Lambda_i}, \quad i = \{1, 2\} \tag{7.5}$$

$$W(z^{-1}) = C\Phi(z^{-1})Bz^{-\Lambda_1}, \quad \overline{W}(z^{-1}) = \Phi(z^{-1})Bz^{-\Lambda_1} \tag{7.6}$$

$$W_0(z^{-1}) = C\Phi(z^{-1})D, \quad \overline{W}_0(z^{-1}) = \Phi(z^{-1})D$$

The optimal *return-difference* matrix $F_0(z^{-1})$ and the *control sensitivity* matrix $M_0(z^{-1})$ can be defined as:

$$F_0(z^{-1}) = I_m + C_0(z^{-1})D_2(z^{-1})W(z^{-1}) \tag{7.7}$$

$$M_0(z^{-1}) = (I_m + C_0(z^{-1})D_2(z^{-1})W(z^{-1}))^{-1}C_0(z^{-1}) \tag{7.8}$$

By reference to Fig. 7.2 the regulating problem equations can be written as:

$$z(z^{-1}) = D_2(z^{-1})(y(z^{-1}) + v(z^{-1}))  \tag{7.9}$$

$$u(z^{-1}) = -C_0(z^{-1})z(z^{-1})  \tag{7.10}$$

$$x(z^{-1}) = \overline{W}(z^{-1})u(z^{-1}) + \overline{W}_0(z^{-1})\omega(z^{-1})  \tag{7.11}$$

Thence, the expressions for the control and state vectors become:

$$u(z^{-1}) = -M_0(z^{-1})D_2(z^{-1})\left(v(z^{-1}) + W_0(z^{-1})\omega(z^{-1})\right)  \tag{7.12}$$

and

$$x(z^{-1}) = \left(\overline{W}_0(z^{-1}) - \overline{W}(z^{-1})M_0(z^{-1})D_2(z^{-1})W_0(z^{-1})\right)\omega(z^{-1})$$

$$-\overline{W}(z^{-1})M_0(z^{-1})D_2(z^{-1})v(z^{-1})  \tag{7.13}$$

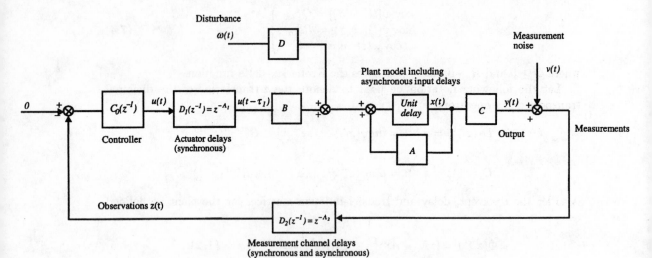

**Fig. 7.2** : LQG Optimal Regulating System with Different Transport
Delays on Input and Output Channels

## 7.2.2   The LQG Stochastic Regulating Problem

The LQG discrete optimal control problem to be considered (Grimble, 1979a,b)
is to determine the output feedback controller $C_0(z^{-1})$, which minimises the usual
time-averaged performance criterion:

$$J(u) = E\{\lim_{T \to \infty} \frac{1}{2T} \sum_{t=-T}^{T} (< x(t), Q_1 x(t) >_{E_n} + < u(t), R_1 x(t) >_{E_m})\}  \tag{7.14}$$

where the constant weighting matrices satisfy $Q_1 \geq 0$ and $R_1 > 0$. The unconditional expectation $E\{.\}$ is taken over the ensemble $\{\omega(t)\}$ and $\{v(t)\}$.

If there are no explicit time-delays the LQG optimal linear controller is normally calculated in the time-domain by utilising the separation principle. The solution for the Wiener-Hopf form of the LQG controller obtained below follows from the conditions for *optimality* (in the form of the Wiener-Hopf equation). The necessary and sufficient condition for optimality is derived in the time-domain by computing the gradient of the cost-function. The resulting gradient equation is then transformed into the z-domain. The frequency domain form of the optimal feedback controller may then be obtained.

**Time-domain Conditions for Optimality**

A brief formal derivation is given below of the necessary and sufficient conditions for optimality (Grimble and Johnson, 1988). First note that the time-average cost-function (7.14) may be written in the form:

$$J(u) = \lim_{T \to \infty} \frac{1}{2T}\{J_T(u)\}$$

where $J_T(u)$ is defined in terms of the $\ell_2(-T, T)$ inner product. Thence,

$$J_T(u) = E\{< x, Q_1 x >_{H_n} + < u, R_1 u >_{H_m}\} \qquad (7.15)$$

The following derivation of the conditions for optimality will initially be to minimise the above finite-time cost-function for a fixed cost summation time $T$. The necessary conditions for minimising the infinite-time cost function are then obtained by taking the limit as $T \to \infty$. Substituting from equations (7.12) and (7.13):

$$u(t) = -(MD_2 v)(t) - (MD_2 W_0 \omega)(t)$$

$$x(t) = \left((\overline{W}_0 - \overline{W} MD_2 W_0)\omega\right)(t) - (\overline{W} MD_2 v)(t)$$

and noting :

$$E\{< \overline{W}_0 \omega, Q_1 \overline{W}_0 \omega >_{H_n}\} = \sum_{t=-T}^{T} [Trace(\overline{W}_0^T Q_1 \overline{W}_0 \phi_{\omega\omega})(t)]_T$$

where $\overline{W}_0^*$ is the adjoint (Grimble, 1978b) of the operator $\overline{W}_0$. Thus the first term in the cost index (7.15) becomes:

$$E\{< x, Q_1 x >_{H_n}\}$$

$$= \sum_{t=-T}^{T} [Trace\left\{(\overline{W}_0 - \overline{W} MD_2 W_0)^* Q_1 (\overline{W}_0 - \overline{W} MD_2 W_0)\phi_{\omega\omega}\right\}$$

$$+ Trace\left\{(\overline{W} MD_2)^* Q_1 \overline{W} MD_2 \phi_{vv}\right\}]_T$$

$$= \sum_{t=-T}^{T} [Trace\left\{(\overline{W}^* Q_1 \overline{W} M \phi_{cc} - 2\overline{W}^* Q_1 \phi_{c0}) M^*\right\} + Trace\left\{Q_1 \overline{W}_0 \phi_{\omega\omega} \overline{W}_0^*\right\}]_{\scriptscriptstyle T}$$

$$(7.16)$$

where the functions:

$$\phi_0(\tau) = (D_2 \phi_{vv} D_2^* + D_2 W_0 \phi_{\omega\omega} W_0^* D_2^*)(\tau)$$

$$\phi_{c0}(\tau) = (\overline{W}_0 \phi_{\omega\omega} W_0^* D_2^*)(\tau)$$

The remaining terms in the cost-function (7.15) are given by:

$$E\{< u, R_1 u >_{H_r}\} = \sum_{t=-T}^{T} Trace[R_1 M \phi_0 M^*]_{\scriptscriptstyle T}$$

The finite-time cost-function (7.15) may therefore be written as:

$$J_T(u) = \sum_{t=-T}^{T} [Trace\left\{(\psi M \phi_0 - 2\overline{W}^* Q_1 \phi_{c0}) M^*\right\} + Trace\{Q_1 \overline{W}_0 \phi_{\omega\omega} \overline{W}_0^*\}]_{\scriptscriptstyle T} \quad (7.17)$$

where the time-domain operator $\psi$ is defined as:

$$\psi = \overline{W}^* Q_1 \overline{W} + R_1$$

### Necessary and Sufficient Conditions

Let the operator $M$ be replaced by $M_0 + \epsilon \tilde{M}$ , where $M_0$ is the optimal control sensitivity operator and $\epsilon \tilde{M}$ is some physically realisable variation about the optimum. A *necessary condition* for optimality in the finite time case is then,

$$\frac{\partial J_T}{\partial \epsilon}\big|_{\epsilon=0} = \sum_{t=-T}^{T} 2[Trace(\chi_T(\lambda - t)\tilde{M}^*)]_{\scriptscriptstyle T} = 0$$

where $\lambda$ is the variable of integration for the adjoint operator $\tilde{M}^*$ and the matrix function $\chi_T(\lambda - t)$ is defined as:

$$\chi_T(\lambda - t) = (\psi M_0 \phi_0 - \overline{W}^* Q_1 \phi_{c0})|_{\scriptscriptstyle T} \quad (7.18)$$

Since $\tilde{M}^*$ is an arbitrary integral operator the matrix $\chi_T(\lambda - t)$ must be null for all $\lambda \in [t, T)$ and all $t \in (-T, T)$ . By defining $\tau = \lambda - t$ the condition for optimality may be written as: $\chi_{\scriptscriptstyle T}(\tau) = 0$ for all $\tau \in [0, 2T)$ .

The necessary condition for optimality for minimising the time-average criterion is obtained by taking the limit as $T \to \infty$ . This statement may be justified as follows. Let the operator $M$ be replaced by $M_0 + \epsilon \tilde{M}$ , then the cost-function $J(u)$ may be expressed as:

$$J(u) = J(u^0) + 2\epsilon \lim_{T \to \infty} \frac{1}{2T} \sum_{t=-T}^{T} [Trace\left\{(\psi M_0 \phi_0 - \overline{W}^* Q_1 \phi_{c0})\tilde{M}^*\right\}]_{\scriptscriptstyle T}$$

$$+\epsilon^2 \lim_{T\to\infty} \frac{1}{2T} \sum_{t=-T}^{T} [Trace(\psi\tilde{M}\phi_0^*\tilde{M})]_{_{\mathrm{T}}}$$

where $J(u^0)$ is the minimum value of the cost-function. A necessary condition for optimality (Grimble, 1978a, 1979b) is then given by:

$$\left(\frac{\partial J}{\partial \epsilon}\right)_{\epsilon=0} = \lim_{T\to\infty} \frac{1}{2T} \sum_{t=-T}^{T} 2[Trace(\chi_T(\lambda-t)\tilde{M}^*)]_{_{\mathrm{T}}} = 0$$

Recall that the output of an asymptotically stable dynamical system, driven by stationary white noise, has a finite (non-zero) average power. Thus, since the operator $\tilde{M}^*$ is arbitrary, the condition for optimality may be written as:

$$\chi_\infty(\tau) = (\psi M_0\phi_0 - \overline{W}^*Q_1\phi_{c0})(\tau) = 0 \qquad (7.19)$$

for all $\tau > 0$. Equation (7.19) also yields a sufficient condition for optimality, since $\frac{\partial^2 J}{\partial \epsilon^2} > 0$ .

The condition (7.19) which determines the optimal controller to minimise the time-average performance criterion is therefore the same as the limiting case of the condition (7.18). That is, the controller which minimises the time average criterion is the same as the controller which minimises the finite time performance criterion when $T$ is arbitrarily large. Expression (7.19) is a form of the well known *Wiener-Hopf equation* (Grimble and Johnson, 1988).

### 7.2.3   Wiener-Hopf Solution for Optimal Controller

The Wiener-Hopf (WH) integral equation, derived above, is first transformed into the z-domain (Grimble, 1979b). Spectral-factorisation is then introduced and the equations are simplified to obtain an expression for the controller $C_0(z^{-1})$. Each term in the WH equation may be transformed into the z-domain, using the two-sided z-transform $\mathcal{Z}_2\left((W^*y)(t)\right) = W^T(z)y(z^{-1})$ (noting Grimble (1978b). The conditions which determine the existence of the two-sided z-transform are detailed by Lepage (1961) and it will be assumed that these conditions are satisfied. That is, attention is restricted here to those cases in which the cross-correlation function in the WH equation is of exponential order. Thus, after transformation, the gradient equation (equation (7.19)) becomes:

$$\chi(z^{-1}) = \Psi(z^{-1})M_0(z^{-1})\Phi_0(z^{-1}) - \overline{W}^T(z)Q_1\Phi_{c0}(z^{-1}) \qquad (7.20)$$

where for optimality $\chi(z^{-1})$ is the transform of a time function which is zero for $\tau \geq 0$ . That is, the gradient must be zero for all time $t \geq 0$ . The matrix transforms $\Psi(z^{-1})$, $\Phi_0(z^{-1})$ and $\Phi_{c0}(z^{-1})$ are defined directly from the time-domain operator versions introduced in the previous section.

### Assumptions

The optimal controller required from this solution, should be as simple to implement as a Smith predictor (Marshall, 1974), from the perspective that it can be realised with finite-dimensional subsystems and pure delay elements (Åström, 1977). However, this requires assumptions which slightly limit the generality of the system and noise descriptions. The main assumption is that a rational transfer-function, minimum-phase, spectral-factor exists, corresponding to each of the power spectrum terms $\Psi(z^{-1})$ and $\Phi_0(z^{-1})$ contained in (7.20).

**Assumption 1 :** Rational transfer-function matrix generalised spectral-factors $Y_1(z^{-1})$ and $Y(z^{-1})$ exist (Shaked, 1976a,b) and are defined to satisfy:

$$Y_1^T(z)Y_1(z^{-1}) = \Psi(z^{-1}) = \overline{W}^T(z)Q_1\overline{W}(z^{-1}) + R_1 \qquad (7.21)$$

$$Y(z^{-1})Y^T(z) = \Phi_0(z^{-1}) = D_2(z^{-1})\left(W_0(z^{-1})QW_0^T(z) + R\right)D_2^T(z) \qquad (7.22)$$

Recall that the asynchronous input delays are absorbed in the plant model. The input delays included in $D_1(z^{-1})$ are therefore synchronous, or the same in each channel ( $\tau_{1i} = \tau_1$ for all $j = \{1, ..., m\}$). The corresponding $z^{-\tau_1}$ and $z^{\tau_1}$ terms in (7.21) cancel (these are involved in $\overline{W}$ and $\overline{W}^*$ ).

**Assumption 2 :** Assume the noise and disturbances are uncorrelated so that the power spectrum $\Phi_0(z^{-1})$ is a diagonal matrix, then the operators $D_2(z^{-1})$ and $D_2^*(z^{-1}) = D_2^T(z) = z^{\Lambda_2}$ cancel in (7.22), and $Y(z^{-1})$ is a diagonal matrix.

A less restrictive assumption that can replace Assumption 2 does not require the noise and disturbance spectrum to be diagonal. This assumption now follows.

**Assumption 2a :** Assume the noise and disturbance models are block diagonal so that the spectrum $\Phi_0(z^{-1})$ is a block diagonal matrix with $p$ blocks that correspond to an output delay operator:

$$D_2(z^{-1}) = z^{-\Lambda_2} = diag\{z^{-\tau_{21}}I_{21},\ z^{-\tau_{22}}I_{22}, ..., z^{-\tau_{2p}}I_{2p}\}$$

Then the operators $D_2(z^{-1})$ and $D_2^*(z^{-1}) = D_2^T(z)$ cancel in (7.22) and the spectral factor $Y(z^{-1})$ is a block diagonal transfer-function matrix.

This latter assumption covers the extreme case when there is only one block, $Y(z^{-1})$ is a full diagonal matrix, and the output delays are the same on each output. An example of a system with a block diagonal disturbance model follows.

**Example 7.1 :** Consider the system shown in Fig. 7.3. For $p = 3$ the state-space block structure has the form referred to above:

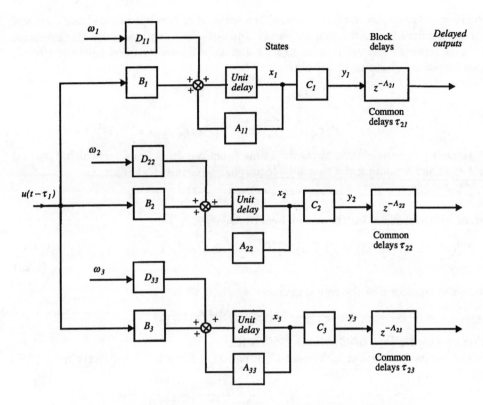

Fig. 7.3 : Discrete Multivariable System with Independent
Disturbances for the Different Output Channels

$$x(t+1) = \begin{bmatrix} A_{11} & 0 & 0 \\ 0 & A_{22} & 0 \\ 0 & 0 & A_{33} \end{bmatrix} x(t) + \begin{bmatrix} B_1 \\ B_2 \\ B_3 \end{bmatrix} u(t - \tau_1)$$

$$+ \begin{bmatrix} D_{11} & 0 & 0 \\ 0 & D_{22} & 0 \\ 0 & 0 & D_{33} \end{bmatrix} \omega(t)$$

$$y(t) = \begin{bmatrix} C_1 & 0 & 0 \\ 0 & C_2 & 0 \\ 0 & 0 & C_3 \end{bmatrix} x(t)$$

and $(D_2 y)(t) = diag\{y_1(t - \tau_{21}), y_2(t - \tau_{22}), y_3(t - \tau_{23})\}$

## Control Sensitivity Expression

If the measurement noise and disturbance models have compatible block diagonal
structures then the filter Riccati equation and Kalman filter gain will also have such a

structure.    Given Assumption 2 (or 2a) the spectral factors are rational matrices and it may therefore be shown that necessary and sufficient conditions for the existence of the *generalised spectral factors* $Y(z^{-1})$ and $Y_1(z^{-1})$ are satisfied (Shaked 1976a). From equations (7.20) to (7.22) obtain:

$$(Y_1^T(z))^{-1}\chi(z^{-1})(Y^T(z))^{-1}$$

$$= Y_1(z^{-1})M_0(z^{-1})Y(z^{-1}) - (Y_1^T(z))^{-1}\overline{W}^T(z)Q_1\Phi_{c0}(z^{-1})(Y^T(z))^{-1}$$

By equating the transforms of positive-time functions (functions for which $f(t) = 0$ for $t < 0$) and by noting the causality requirement (Grimble, 1978a):

$$\{(Y_1^T(z))^{-1}\chi(z^{-1})(Y^T(z))^{-1}\}_+ = 0$$

obtain, the expression for the *optimal control sensitivity* as:

$$M_0(z^{-1}) = (Y_1(z^{-1}))^{-1}\{(Y_1^T(z))^{-1}\overline{W}^T(z)Q_1\Phi_{c0}(z^{-1})(Y^T(z))^{-1}\}_+(Y(z^{-1}))^{-1}$$

$$\tag{7.23}$$

where the transform of the signal response $\phi_{c0}(t)$ in (7.16), gives:

$$\Phi_{c0}(z^{-1}) = \overline{W}_0(z^{-1})QW_0^T(z)D_2^T(z)$$

### Wiener-Hopf Form of Optimal Controller

The optimal controller in Wiener-Hopf equation form follows directly from (7.8) and (7.23) as:

$$C_0(z^{-1}) = M_0(z^{-1})(I_r - D_2(z^{-1})W(z^{-1})M_0(z^{-1}))^{-1}$$

$$= (I_m - M_0(z^{-1})D_2(z^{-1})W(z^{-1}))^{-1}M_0(z^{-1}) \tag{7.24}$$

The resulting optimal closed-loop regulator can be shown to be asymptotically stable (Grimble, 1998). This expression for the *Control Sensitivity Function* $M_0(z^{-1})$ can be simplified by defining:

$$N_c(z^{-1}) = B^T\Phi^T(z)Q_1 \quad \text{and} \quad N_f(z^{-1}) = DQD^T\Phi^T(z)C^T \tag{7.25}$$

Thus, obtain the desired Wiener solution for the optimal control sensitivity as:

$$M_0(z^{-1}) = (Y_1(z^{-1}))^{-1}\{(Y_1^T(z))^{-1}D_1^T(z)N_c(z^{-1})$$

$$\times\Phi(z^{-1})N_f(z^{-1})D_2^T(z)(Y^T(z))^{-1}\}_+(Y(z^{-1}))^{-1} \tag{7.26}$$

The frequency domain version of the optimal controller follows using equations (7.24) and (7.26).    The causal transform in (7.26), denoted by the curly brackets $\{.\}_+$, is the transform of a finite impulse-response term.  The Wiener-Hopf form of the optimal controller also has a finite impulse-response that causes implementation difficulties in the continuous-time solution but is simple to treat in the discrete-time case.  This is considered further in §7.3.2.    This transfer-function solution is not so valuable as the equivalent polynomial results (Kucera, 1980), that provide simpler numerical algorithms.

## 7.3 State-space Form of Optimal Controller

The state-space form of the output feedback LQG optimal controller is derived in this section but first the well known results (Grimble and Johnson, 1988) for the LQ state feedback optimal control and the Kalman filtering problems (Anderson and Moore, 1979) are reviewed. Note that the results for the slightly more general case, when the cost index includes cross-product terms, are summarised in a Theorem in Chapter 8.

### Summary of Results for State-feedback LQ Optimal Control

For the present consider a state-space system, without the presence of explicit input delays and where states are available for feedback (Grimble, 1987b). In this *ideal state feedback* control problem the measurements are not corrupted by noise and the state feedback optimal control has the form:

$$u(t) = -K_1 x(t)$$

where $K_1$ is a constant gain matrix and $x(t) \in R^n$ denotes the state vector, (Grimble and Johnson, 1988). The optimal feedback control gain matrix can be obtained as:

$$K_1 = (R_1 + B^T P_1 B)^{-1} B^T P_1 A \tag{7.27}$$

where $P_1$ satisfies the algebraic Riccati equation:

$$Q_1 = P_1 - A^T P_1 A + K_1^T \tilde{R}_1 K_1 \tag{7.28}$$

and $\tilde{R}_1 = R_1 + B^T P_1 B$ .

### Optimal return-difference matrix

The *control return-difference* matrix (MacFarlane, 1970) is defined as:

$$F_1(z^{-1}) = I_m + K_1 \Phi(z^{-1}) B \tag{7.29}$$

It may easily be shown that the return-difference matrix is related to the spectral factor $Y_l(z^{-1})$, needed in the following, since:

$$F_1^T(z) \tilde{R}_1 F_1(z^{-1}) = \bar{W}^T(z) Q_1 \bar{W}(z^{-1}) + R_1 \tag{7.30}$$

and thus from (7.30), the desired relationship between the return difference matrix and the *control spectral factor* is obtained as:

$$Y_1(z^{-1}) = \tilde{R}_1^{1/2} F_1(z^{-1}) = \tilde{R}_1^{1/2} (I_m + K_1 \Phi(z^{-1}) B) \tag{7.31}$$

### Transfer-function matrix result

Through simple manipulations a result may be obtained which is helpful in simplifying the Wiener-Hopf equation obtained later. Manipulating the Riccati equation (7.28), obtain:

$$Q_1 = P_1 - A^T P_1 A + K_1^T \tilde{R}_1 K_1 + z(A^T P_1 - A^T P_1) + z^{-1}(P_1 A - P_1 A)$$

$$= (z^{-1}I - A^T)P_1(zI - A) + A^T P_1(zI - A) + (z^{-1}I - A)P_1 A + K_1^T \tilde{R}_1 K_1$$

and

$$B^T \Phi^T(z)Q_1\Phi(z^{-1}) = B^T(P_1 + \Phi^* A^T P_1 + PA\Phi + \Phi^* K_1^T \tilde{R}_1 K_1 \Phi)$$

$$= B^T P_1 + B^T \Phi^* A^T P_1 + \tilde{R}_1 K_1 \Phi + B^T \Phi^* K_1^T \tilde{R}_1 K_1 \Phi$$

After utilising the above expression for the gain matrix (7.27) obtain:

$$B^T \Phi^T(z)Q_1\Phi(z^{-1}) = B^T P_1 + B^T \Phi^T(z)A^T P_1 + Y_1^T(z)\tilde{R}_1^{1/2} K_1 \Phi(z^{-1})$$

$$= B^T \Phi^T(z)z^{-1}P_1 + Y_1^T(z)\tilde{R}_1^{1/2} K_1 \Phi(z^{-1}) \qquad (7.32)$$

## Summary of Results for Discrete Estimation Problem

The dual to the LQ state feedback optimal control problem is the Kalman (1961) filtering problem. The main results are summarised below. In fact the so-called single-stage prediction problem, where a one step delay is involved in computing the estimates, will actually be considered, since this is probably the most common. The estimator equations are given as :

$$\hat{x}(t+1|t) = A\hat{x}(t|t-1) + Bu(t) + K\left(z(t) - \hat{y}(t|t-1)\right) \qquad (7.33)$$

$$\hat{y}(t|t-1) = C\hat{x}(t+1|t) \qquad (7.34)$$

The Kalman filter gain matrix for a system, without explicit measurement delays, is given by :

$$K = APC^T(R + CPC^T)^{-1} \qquad (7.35)$$

where $P$ satisfies the algebraic Riccati equation:

$$DQD^T = P - APA^T + K\tilde{R}K^T \qquad (7.36)$$

and $\tilde{R} = R + CPC^T$.

## Optimal estimator return-difference matrix

The *filter return-difference* matrix is defined as:

$$F(z^{-1}) = I_r + C\Phi(z^{-1})K \qquad (7.37)$$

The return-difference matrix is related to the spectral factor $Y(z^{-1})$, needed in the following, since:

$$F(z^{-1})\tilde{R}F^T(z) = W_0(z^{-1})QW_0^T(z) + R \qquad (7.38)$$

and thus the relationship between the *filter spectral factor* and the return difference matrix may be obtained as :

$$Y(z^{-1}) = F(z^{-1})\tilde{R}^{1/2} = (I_r + C\Phi(z^{-1})K)\tilde{R}^{1/2} \qquad (7.39)$$

**Transfer-function matrix result**

Through similar manipulations to those in the previous section, by manipulating the filter Riccati equation, obtain:

$$DQD^T = P - APA^T + K\tilde{R}K^T + z(PA^T - PA^T) + z^{-1}(AP - AP)$$

$$= (zI - A)P(z^{-1}I - A^T) + AP(z^{-1}I - A^T) + (zI - A)PA^T + K\tilde{R}K^T$$

and

$$\Phi(z^{-1})DQD^T\Phi^T(z)C^T = (P + \Phi AP + PA^T\Phi^* + \Phi K\tilde{R}K^T\Phi^*)C^T$$

$$= PC^T + PA^T\Phi^*C^T + \Phi K\tilde{R} + \Phi K\tilde{R}K^T\Phi^*C^T$$

After utilising the filter gain matrix expression (7.35) obtain :

$$\Phi(z^{-1})DQD^T\Phi^T(z)C^T = PC^T + PA^T\Phi^T(z)C^T + \Phi(z^{-1})K\tilde{R}^{1/2}Y^T(z)$$

$$= Pz^{-1}\Phi^T(z)C^T + \Phi(z^{-1})K\tilde{R}^{1/2}Y^T(z) \qquad (7.40)$$

## 7.3.1  LQG Output Feedback Controller

The Wiener Hopf expressions for the optimal controller (equations (7.24) and (7.26)) may now be used to generate the state-space discrete-time optimal control law. The causal term in (7.26), denoted by the curly brackets $\{.\}_+$ involves either a partial-fraction expansion, or diophantine equations, and the use of z and inverse z-transforms. An alternative approach involves relating the results to state-space models, as described below.

The positive-time transform $\{.\}_+$ in (7.26) can be simplified using the state equation results in the previous section. Substituting for $N_c(z^{-1})\Phi(z^{-1})$, noting from (7.32) and (7.25):

$$\{.\}_+ = \{Y_1^T(z)^{-1}D_1^T(z)\left(N_c(z^{-1})\Phi(z^{-1})\right)N_f(z^{-1})D_2^T(z)Y^T(z)^{-1}\}_+$$

$$= \{Y_1^T(z)^{-1}B^T\Phi^T(z)z^{-1}P_1N_f(z^{-1})D_0^T(z)Y^T(z)^{-1}\}_+$$

$$+\{\tilde{R}_1^{1/2}K_1\Phi(z^{-1})N_f(z^{-1})D_0^T(z)Y^T(z)^{-1}\}_+ \qquad (7.41)$$

where $D_0(z^{-1}) = D_2(z^{-1})D_1(z^{-1})$ denotes a combination of the diagonal output delay operator and the synchronous input delay operator $D_1(z^{-1}) = I_m z^{-\tau_1}$. The first transform within the braces in this equation represents an impulse input to a non-causal system (recall that $\Phi^T(z)$ can be expressed as a sequence in $z, z^2, z^3, ...$). The corresponding positive-time transform is therefore zero.

### Causal Transform Calculation

Equation (7.41) may be further simplified by substituting for $\Phi(z^{-1})N_f(z^{-1})$ from (7.40) and (7.25), and recalling that the spectral factor $Y(z^{-1})$ and covariance $R$ are diagonal matrices:

$$\{.\}_+ = \{\tilde{R}_1^{1/2} K_1(Pz^{-1}\Phi^T(z)C^T + \Phi(z^{-1})K\tilde{R}^{1/2}Y^T(z))D_0^T(z)Y^T(z)^{-1}\}_+$$

$$= \{\tilde{R}_1^{1/2} K_1 P z^{-1}\Phi^T(z)C^T D_0^T(z)Y^T(z)^{-1}\}_+ + \{\tilde{R}_1^{1/2} K_1 \Phi(z^{-1})K\tilde{R}^{1/2}D_0^T(z)\}_+$$

The system in (7.1) always includes a unit step delay and hence $D_0$ will include all positive powers of z. Thence, the first term, representing the positive-time transform of a non-causal response, is again zero and thus,

$$\{.\}_+ = \{\tilde{R}_1^{1/2} K_1 \Phi(z^{-1})K\tilde{R}^{1/2}D_0^T(z)\}_+$$

By substituting this expression into equation (7.26) the closed-loop operator becomes:

$$M_0(z^{-1}) = (Y_1(z^{-1}))^{-1}\tilde{R}_1^{1/2} K_1\{\Phi(z^{-1})K\tilde{R}^{1/2}D_0^T(z)\}_+(Y(z^{-1}))^{-1}$$

$$= F_1(z^{-1})^{-1}K_1 N_k(z^{-1})F(z^{-1})^{-1} \qquad (7.42)$$

where the causal transform:

$$N_k(z^{-1}) = \{\Phi(z^{-1})KD_0^T(z)\}_+$$

The significance of the $N_k(z^{-1})$ term is explored below. It is shown that this represents a block diagonal predictor multiplied by the forward path Kalman filter gain $\Phi(z^{-1})K$. That is, the causal transform:

$$N_k(z^{-1}) = A^{\Lambda_0}\Phi(z^{-1})K \qquad (7.43)$$

where the following block diagonal matrix is defined as:

$$A^{\Lambda_0} = diag\{(A_{11}^{\tau_{01}}), ..., (A_{pp}^{\tau_{0p}})\}$$

**Computation of the Causal Transform $N_k(z^{-1})$**

The causal transform, $N_k(z^{-1})$, defined in equation (7.43), is evaluated below. Invoking Assumption 2a write the gain $K$ in column matrix form and the total transform of the delay $D_0(z^{-1})$ as a block diagonal matrix. Then, obtain:

$$K = [K_1, \ K_2 \ ,..., \ K_p] \quad \text{and} \quad D_0(z^{-1}) = diag\{z^{-\tau_{01}}I_1, \ z^{-\tau}I_2 \ ,..., \ z^{-\tau_{0p}}I_p\}$$

where $I_1$ to $I_p$ denote identity matrices corresponding to the block sizes. Under Assumption 2a it may easily be confirmed that the filter gain $K$ will also be block diagonal. Then, the filter gain $K$ may be expressed as:

$$K = diag\{K_{11}, \ K_{22} \ ,..., \ K_{pp}\}$$

The *causal transform* now follows from equation (7.43) as:

$$N_k(z^{-1}) = \{\Phi(z^{-1})KD_0^T(z)\}_+ = diag\{\Phi_{11}(z^{-1})K_{11}z^{\tau_{01}},...,\Phi_{pp}(z^{-1})K_{pp}z^{\tau_{0p}}\}_+$$

where $\Phi_{jj}(z^{-1}) = (zI - A_{jj})^{-1}$ denotes the $j$th block of the resolvent matrix. Recall that the time response corresponding to the resolvent matrix $\Phi(z^{-1}) = (zI - A)^{-1}$ is determined by the transition matrix :

$$\phi(t) = A^t = diag\{A_{11}^t, \ A_{22}^t \ ,..., \ A_{pp}^t\} = diag\{\phi_{11}(t), \ \phi_{22}(t) \ ,..., \ \phi_{pp}(t)\}$$

where $\phi_{jj}(t) = A_{jj}^t$ . The time advanced function $\phi_{jj}(t + \tau_{0j})$ then has the causal transform $\mathcal{Z}_1\left(\phi_{jj}(t + \tau_{oj})\right) = \phi_{jj}(\tau_{0j})\mathcal{Z}_1(\phi_{jj}(t)) = A_{jj}^{\tau_{0j}}\Phi_{jj}(z^{-1})$ . The desired causal transform therefore becomes:

$$N_k(z^{-1}) = diag\{(A_{11}^{\tau_{01}})\Phi_{11}(z^{-1})K_{11} \ ,..., (A_{pp}^{\tau_{0p}})\Phi_{pp}(z^{-1})K_{pp}\}$$

$$= diag\{(A_{11}^{\tau_{01}}) \ ,..., \ (A_{pp}^{\tau_{0p}})\}\Phi(z^{-1})K = A^{\Lambda_0}\Phi(z^{-1})K \quad (7.44)$$

The optimal control sensitivity function may therefore be expressed, using (7.42) and (7.44), as:

$$M_0(z^{-1}) = F_1(z^{-1})^{-1}K_1A^{\Lambda_0}\Phi(z^{-1})KF(z^{-1})^{-1} \quad (7.45)$$

This is a very convenient and simple form for the frequency-domain form of the control sensitivity function. The next step is to obtain a simple expression for the state equation form of the optimal controller.

**The Optimal Control Signal**

From equations (7.8) and (7.10) the transform of the optimal control is given by:

$$u^0(z^{-1}) = -M_0(z^{-1})z(z^{-1}) + M_0(z^{-1})D_2(z^{-1})W(z^{-1})u^0(z^{-1})$$

Substituting from equation (7.45) and simplifying:

$$u^0(z^{-1}) = -K_1N_k(z^{-1})F(z^{-1})^{-1}z(z^{-1})$$

$$+K_1 \left( N_k(z^{-1})F(z^{-1})^{-1}D_2(z^{-1})W(z^{-1}) - \Phi(z^{-1})B \right) u^0(z^{-1})$$

$$= -K_1 \left( \Phi(z^{-1})Bu^0(z^{-1}) + N_k(z^{-1})F(z^{-1})^{-1} \left( z(z^{-1}) - D_2(z^{-1})W(z^{-1})u^0(z^{-1}) \right) \right)$$

The optimal control signal may therefore be expressed, using (7.44), in the form:

$$u^0(z^{-1}) = -K_1(\Phi(z^{-1})Bu^0(z^{-1})$$

$$+A^{\Lambda_0}\Phi(z^{-1})KF(z^{-1})^{-1} \left( z - D_2(z^{-1})W(z^{-1})u^0(z^{-1}) \right)) \qquad (7.46)$$

Recall that $F(z^{-1})^{-1}$ denotes the *return-difference* matrix for the Kalman filter and $D_2(z^{-1})W(z^{-1}) = D_2(z^{-1})C(z^{-1})BD_1(z^{-1})$ . This transfer can be written (under Assumptions 1 and 2) as: $D_2(z^{-1})W(z^{-1}) = D_0(z^{-1})C(z^{-1})B$. The prediction stage of the control law is introduced by the $A^{\Lambda_0}$ term. The control gain matrix $K_1$ is the solution to the equivalent deterministic optimal control problem. These results indicate clearly that a form of the separation theorem holds in this time-delay problem (Davis, 1977; Schmotzer and Blankenship, 1978).

### Limiting Expression for the Optimal Controller

The solution for the optimal controller $C_0(s)$ is simplified below. From equations (7.8) and (7.45):

$$C_0(z^{-1}) = (I_m - F_1^{-1}K_1A^{\Lambda_0}\Phi KF^{-1}z^{-\Lambda_0}C\Phi B)^{-1}F_1^{-1}K_1A^{\Lambda_0}\Phi KF^{-1}$$

$$= K_1(\Phi^{-1}A^{-\Lambda_0} + (\Phi^{-1}A^{-\Lambda_0}\Phi B - KF^{-1}z^{-\Lambda_0}C\Phi B)K_1)^{-1}KF^{-1} \qquad (7.47)$$

but $KF^{-1} = K(I_r + C\Phi(z^{-1})K)^{-1} = (I_n + KC\Phi(z^{-1}))^{-1}K$ and hence:

$$C_0(z^{-1}) = K_1[(I_n + KC\Phi)\Phi^{-1}A^{-\Lambda_0}(I_n + \Phi BK_1)$$

$$-(I_n + KC\Phi)KF^{-1}z^{-\Lambda_0}C\Phi BK_1]^{-1}K$$

$$= K_1[(I_n + KC\Phi)(zI_n - A)A^{-\Lambda_0}(I_n + \Phi BK_1) - Kz^{-\Lambda_0}C\Phi BK_1]^{-1}K$$

The closed-loop controller may therefore be expressed in the transfer-function matrix form:

$$C_0(z^{-1}) = K_1[(zI_n - A + KC)A^{-\Lambda_0}(I_n + \Phi BK_1) - Kz^{-\Lambda_0}C\Phi BK_1]^{-1}K \qquad (7.48)$$

If the delay terms are null $\Lambda_0 = 0$. The controller (7.48) then reduces to the usual transfer-function for the separation principle LQG solution:

$$C_0(s) = K_1[zI_n - A + KC + BK_1]^{-1}K \qquad (7.49)$$

## 7.3.2   Finite-impulse Response Form of Control Law

As noted earlier the Smith predictor type of controller structure, is only suitable for open-loop stable systems. However, this is not a fundamental limitation of the Wiener solution. The analysis below indicates how the state-space solution should be implemented to avoid difficulties on open-loop unstable systems. Consider the final term in (7.46) and note the following:

$$\Phi(z^{-1})KF(z^{-1})^{-1}D_2(z^{-1})W(z^{-1})$$

$$= (I_n + \Phi(z^{-1})KC)^{-1}\Phi(z^{-1})KC.D_0'(z^{-1})\Phi(z^{-1})B$$

where the operator $D_0(z^{-1})$ and $C$ have a compatible block-diagonal matrix structure and hence $D_0'(z^{-1})$ can be defined to satisfy:

$$CD_0'(z^{-1}) = D_0(z^{-1})C \quad \text{and} \quad D_0'(z^{-1}) = z^{-\Lambda_0}$$

where $\Lambda_0$ has the block structure defined above. Thence, obtain:

$$\Phi KF^{-1}D_2W = \left(I_n - (I_n + \Phi KC)^{-1}\right) z^{-\Lambda_0}\Phi B$$

Now consider the finite-impulse response form of the matrices in the first and final terms in equation (7.46):

$$\left(\Phi(z^{-1}) - A^{\Lambda_0}\Phi(z^{-1})KF(z^{-1})^{-1}D_2(z^{-1})C\Phi(z^{-1})D_1(z^{-1})\right) B$$

$$= \left(I_n - A^{\Lambda_0}z^{-\Lambda_0}\right)\Phi(z^{-1})B + A^{\Lambda_0}\left(I_n + \Phi(z^{-1})KC\right)^{-1}z^{-\Lambda_0}\Phi(z^{-1})B \qquad (7.50)$$

### Finite-impulse Response Block Definition

If the plant is open-loop unstable the final term can still be implemented in a form which is free of unstable hidden modes. The first two terms in (7.50) may now be denoted:

$$T_0(z^{-1}, \Lambda_0) = (I_n - A^{\Lambda_0}z^{-\Lambda_0})\Phi(z^{-1}) \qquad (7.51)$$

Note that the impulse response corresponding to the resolvent matrix $\Phi(z^{-1})$ is given by:

$$\mathcal{Z}^{-1}(\Phi(z^{-1})) = A^{t-1}U(t-1)$$

where $U(t)$ denotes the unit step function. The impulse response of the system which implements $T_0(z^{-1}, \Lambda_0)$ is therefore given by,

$$T_0(t, \Lambda_0) = A^{t-1}U(t-1) - A^{\Lambda_0}A^{t-1}A^{-\Lambda_0}diag\{U(t-1-\tau_{01})I, ..., U(t-1-\tau_{0p})I\}$$

$$= A^{t-1}diag\{(U(t-1) - U(t-1-\tau_{01}))I, ..., (U(t-1) - U(t-1-\tau_{0p}))I\} \qquad (7.52)$$

Clearly the jth term is null for $t > \tau_{0j}$ for $j = \{1, ..., p\}$.

The output of this system depends only upon the inputs $Bu^0(t)$ over the previous $\tau_{0j}$ seconds. Such a system may be realized easily. Thus, for unstable plants the signal $T_0(z^{-1}, \Lambda_0)Bu^0(t)$ must be generated by a single subsystem having the above impulse responses and not by the two blocks, $\Phi(z^{-1})$ and $A^{\Lambda_0}z^{-\Lambda_0}\Phi(z^{-1})$, in which the cancellation of the unstable modes occurs. It follows that stability can be ensured if the control related terms in (7.46) are implemented using the above finite impulse response term.

**Control Law**

The first and third terms in (7.46) may therefore be summarised as:

$$\Phi(z^{-1})B - A^{\Lambda_0}\Phi(z^{-1})KF(z^{-1})^{-1}D_2(z^{-1})W(z^{-1})$$

$$= \left(T_0(z^{-1}, \Lambda_0) + A^{\Lambda_0}(I_n + \Phi(z^{-1})KC)^{-1}z^{-\Lambda_0}\Phi(z^{-1})\right)B$$

where the latter term must be realised in its minimal form. Thus, from equation (7.46) and this result obtain (after some manipulation):

$$u^0(z^{-1}) = -K_1 T_0(z^{-1}, \Lambda_0)Bu^0(z^{-1})$$

$$+A^{\Lambda_0}\Phi(z^{-1})KF(z^{-1})^{-1}z(z^{-1}) + A^{\Lambda_0}(I_n + \Phi(z^{-1})KC)^{-1}z^{-\Lambda_0}\Phi(z^{-1})Bu^0(z^{-1})$$

$$= -K_1 T_0(z^{-1}, \Lambda_0)Bu^0(z^{-1})$$

$$+A^{\Lambda_0}(I_n + \Phi(z^{-1})KC)^{-1}\Phi(z^{-1})[Kz(z^{-1}) + z^{-\Lambda_0}Bu^0(z^{-1})] \qquad (7.53)$$

**Implementation of the Control Law**

The finite impulse-response term $T_0(z^{-1}, \Lambda_0)$ in the expression for the control signal (7.53) can be considered separately for implementation. Using (7.52) obtain:

$$u_f(t) = \mathcal{Z}_1^{-1}\left(T_0(z^{-1}, \Lambda_0)Bu^0(z^{-1})\right) = \sum_{\tau=-\infty}^{t} T_0(t-\tau, \Lambda_0)Bu^0(\tau)$$

$$= \sum_{\tau=-\infty}^{t} A^{t-\tau-1}diag\{(U(t-\tau-1) - U(t-\tau-1-\tau_{01}))I,$$

$$..., (U(t-\tau-1) - U(t-\tau-1-\tau_{0p}))I\}Bu^0(\tau)$$

$$u_f(t) = \begin{bmatrix} \sum_{\tau=t-\tau_{01}}^{t-1} A_{11}^{t-\tau-1}B_1 u^0(\tau) \\ \vdots \\ \sum_{\tau=t-\tau_{op}}^{t-1} A_{pp}^{t-\tau-1}B_p u^0(\tau) \end{bmatrix}$$

$$= \begin{bmatrix} B_1 u^0(t-1) + A_{11} B_1 u^0(t-2) & +...+ & A_{11}^{\tau_{01}-1} B_1 u^0(t-\tau_{01}) \\ B_2 u^0(t-1) + A_{22} B_2 u^0(t-2) & +...+ & A_{22}^{\tau_{02}-1} B_2 u^0(t-\tau_{02}) \\ \vdots & \vdots & \vdots \\ B_p u^0(t-1) + A_{pp} B_p u^0(t-2) & +...+ & A_{pp}^{\tau_{0p}-1} B_p u^0(t-\tau_{0p}) \end{bmatrix} \quad (7.54)$$

This finite impulse response term is simple to implement. The remaining terms in (7.53) denote a Kalman predictor, driven by the delayed control signal. The controller may therefore be implemented as shown in Fig. 7.4. The order of the controller is the same as the nominal plant model.

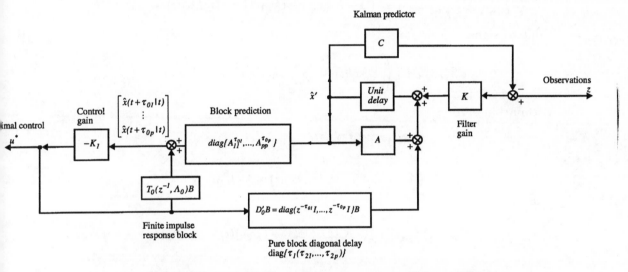

**Fig.7.4** : LQG Optimal Controller in State-space Kalman Filtering
Form for Systems with Input and Output Transport
Delay Elements ($\hat{x}'$ state estimate neglecting delays)

**Summary of the Solution**

(a) Compute the control and predictor gains $K_1$ and $K$ using the algebraic Riccati equations.

(b) Compute the finite impulse response signal $\{u_f(t)\}$, using (7.54) as:

$$u_f(t) = \sum_{\tau=-\infty}^{t} T_0(t-\tau, \Lambda_0) B u^0(\tau)$$

(c) Implement the controller, as illustrated in Fig. 7.4, which involves the Kalman predictor and a constant LQ control gain $K_1$.

### 7.3.3   Stability of the Closed Loop System

Consider the stability of the closed loop system with transport delays on inputs and outputs. The system equations may be listed as follows.

**Plant equation**

$$x(t+1) = Ax(t) + B(D_1u)(t) + D\omega(t)$$

$$z(t) = (D_2Cx)(t) + (D_2v)(t)$$

**Estimator**

$$\hat{x}'(t+1) = A\hat{x}'(t) + K\left(z(t) - C\hat{x}'(t)\right) + (D_0'Bu)(t)$$

**Control signal**

$$u^0(t) = -K_1\left(A^{\Lambda_0}\hat{x}'(t) + \sum_{\tau=-\infty}^{t} T_0(t-\tau,\Lambda_0)Bu^0(\tau)\right)$$

**Estimation error**

$$e(t+1) = (D_2'x)(t+1) - \hat{x}'(t+1)$$

$$= (A-KC)e(t) + (D_2'D\omega)(t) - K(D_2v)(t)$$

but

$$\hat{x}'(t+1) = A\hat{x}'(t) + K\left(D_2Cx(t) + (D_2v)(t) - C\hat{x}'(t)\right)$$

$$+D_0'B\left(-K_1A^{\Lambda_0}\hat{x}'(t) - K_1\sum_{\tau=-\infty}^{t} T_0(t-\tau,\Lambda_0)Bu^0(\tau)\right)$$

$$= (A - D_0'BK_1A^{\Lambda_0})\hat{x}'(t) + KCe(t) + K(D_2v)(t)$$

$$-D_0'BK_1\sum_{\tau=-\infty}^{t} T_0(t-\tau,\Lambda_0)Bu^0(\tau)$$

**Combined system equations**

$$\begin{bmatrix} e(t+1) \\ \hat{x}'(t+1) \end{bmatrix} = \begin{bmatrix} (A-KC) & 0 \\ KC & (A-D_0'BK_1A^{\Lambda_0}) \end{bmatrix} \begin{bmatrix} e(t) \\ \hat{x}'(t) \end{bmatrix}$$

$$+ \begin{bmatrix} (D_2'D\omega)(t) - K(D_2v)(t) \\ K(D_2v)(t) - D_0'BK_1 \sum_{\tau=-\infty}^{t} T_0(t-\tau,\Lambda_0)Bu^0(t) \end{bmatrix} \qquad (7.55)$$

The solution illustrated in Fig. 7.5, is optimal and involves a minimal realisation of a Kalman filtering optimal control solution. The estimation error, defined as $e(t) = (D'_2 x)(t) - \hat{x}'(t)$ is asymptotically stable ($A - KC$ is stable). The full order LQG solution, where the delays are absorbed in the state-space model, is of course stable from the separation principle.

The control law proposed is also an optimal solution and is in a minimal form. It has the control sensitivity function, defined by (7.45), where the stability is determined by the delay free return-difference matrices $F(z^{-1})$ and $F_1(z^{-1})$. It may be concluded that the system proposed has the same stability properties as that of the usual optimal full-order solution.

Note that this does not apply to the Smith predictor form of implementation. In this latter case the implementation involves additional dynamics which represent the plant. The controlled plant model output is subtracted from the observations signal. The result is that for an unstable open-loop system, unstable hidden modes are introduced so that the total closed loop system is unstable. This problem does not arise in the finite impulse response state-space solution (7.54) proposed.

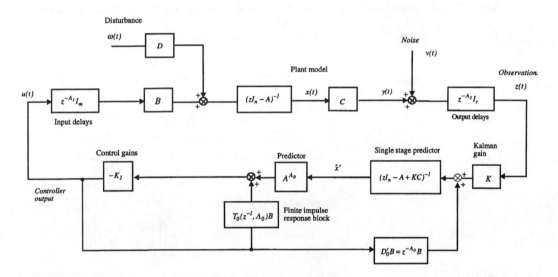

**Fig. 7.5** : Discrete LQG Optimal Regulator for a System with Control Chanel and Output Channel Delays

## 7.3.4 Unstable Open Loop Plant Example

The example which follows is for a second-order open loop unstable system. It illustrates that the time delay compensation method is effective, even for the open loop unstable case.

**Example 7.2** : *Transport Delay Compensation for Open Loop Unstable Plant*

The time lag and delay plant model will initially be presented in continuous-time form. This will be sampled and a zero order hold (ZOH) added.

**Continuous-time plant** : $W = k/(s\tau - 1)$, where $k = 1$ and $\tau = 2$. The sample time $T_s = 0.1$ seconds. The discrete-time state equation for the plant may be expressed as:

$$x_p(t+1) = A_p x_p(t) + B_p(u(t) + \xi(t))$$

$$y_p(t) = C_p x_p(t)$$

$$z_p(t) = y_p(t) + v(t)$$

where the process and measurement noise covariances $Q = 0.1$ and $R = 0.001$, respectively and the initial condition for the plant $x_p(0) = -2$.

**Cost-function Weightings**

*Error weight* $= Q_1^{1/2}/(10s + 1)$ (sampled with a ZOH)

*Control weight* $= R_1^{1/2}$

where $Q_1 = 0.0316$ and $R_1 = 0.0001$. The state equation for the error weighting:

$$x_w(t+1) = A_w x_w(t) + B_w \omega'(t)$$

$$y_w(t) = C_w x_w(t)$$

and the augmented total system, in terms of $[\; x_1^T \quad x_2^T \;]^T = [\; x_p^T \quad x_w^T \;]^T$, has the form:

$$x(t+1) = Ax(t) + B_1 u(t) + B_2 \xi(t)$$

$$y(t) = Cx(t)$$

where $A = \begin{bmatrix} A_p & 0 \\ B_w C_p & A_w \end{bmatrix}$, $B_1 = B_2 = \begin{bmatrix} B_p \\ 0 \end{bmatrix}$, $C = \begin{bmatrix} 0 & C_w \\ C_p & 0 \end{bmatrix}$.

**Transport Delay**

The explicit delay will be chosen as 100 times the sample interval, or

$$Delay = 100 \times T_s = 10 \; seconds$$

This is a difficult case since the delay is much greater than both the plant time-constant and the sample time.

**Discussion of the Simulation Results**

The system output for the delayed output case is shown in Fig. 7.6, where a 2 seconds transport delay is clearly evident. The corresponding control signal is shown in Fig. 7.7. This open-loop unstable system has clearly been stabilised.

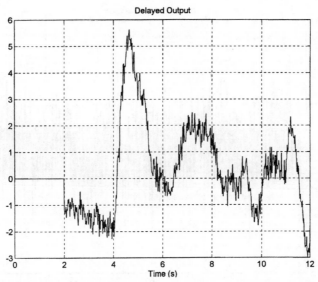

**Fig.7.6**: System Output for System with Two Seconds
Transport Delay

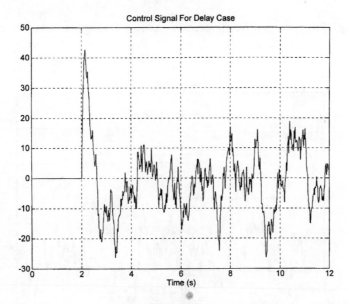

**Fig. 7.7** : Control Signal Response for System with Two Seconds Transport Delay

The effect of the transport delay does of course deteriorate the responses as a comparison with the Figs. 7.8 and 7.9 reveals. In both cases the signals are of course significantly smaller for the system without transport delay. Note that the optimal gains for both the filter and controller are the same in the two problems. The only difference lies in the prediction elements which are set to zero for the delay free case.

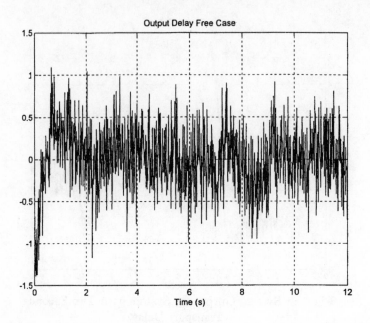

**Fig. 7.8** : Output for System Without a Transport Delay Term

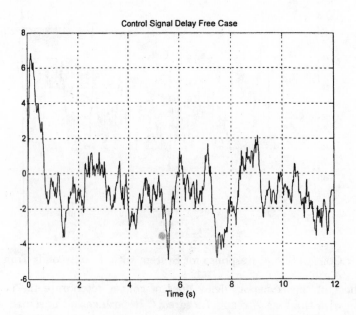

**Fig. 7.9** : Control Signal Response for a System Without
a Transport Delay Term

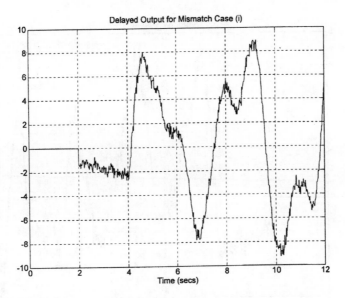

Fig. **7.10** : System Output When Gain and Time Constant are
Mismatched for Transport Delay Case (i)

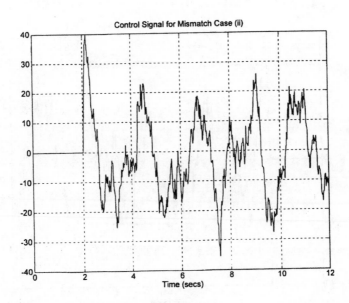

Fig. **7.11** : Control Signal Response when Time Constant and
Gain are Mismatched for Transport Delay Case (i)

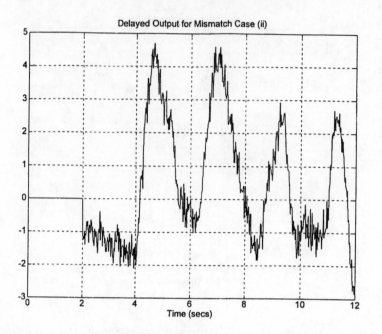

Fig. 7.12 : System Output when Gain and Time Constant are
Mismatched for Transport Delay Case (ii)

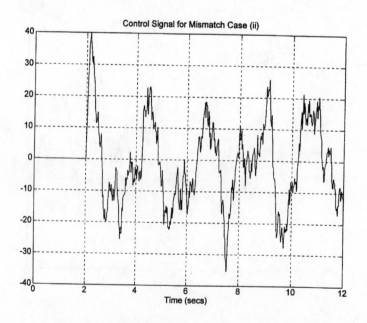

Fig. 7.13 : Control Signal Response when Time Constant Gain are
Mismatched for Transport  Delay Case (ii)

**Mismatched Case**

There are two cases of mismatch that will be considered.

*Case (i)* : gain k = 1.1, time constant = 1.9

*Case (ii)* : gain k = 0.95, time constant = 2.05

The system remains stable for these relatively small variations for both case (i) and case (ii), shown in Fig. 7.10 and 7.11, and 7.12 and 7.13, respectively. With such a large transport delay it is not too surprising that this system has poor robustness. For the case of open-loop stable systems, the robustness is of course much improved, even when the transport delay is very large.

### 7.3.5   Relationship to the Smith Predictor

It follows from equation (7.46) that the optimal control can also be implemented in the state equation form shown in Fig. 7.14. This realisation of the optimal controller has a very similar structure to a Smith predictor. To establish the link between the two, consider the following:

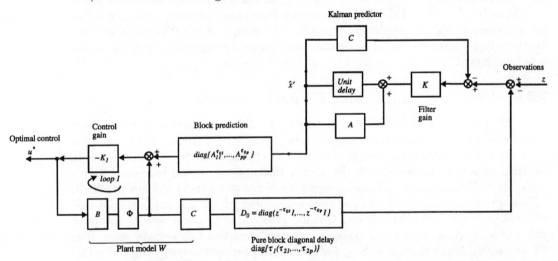

**Fig. 7.14** : LQG Optimal Controller in Smith Predictor Related Kalman Filtering Form

*($\hat{x}'$ = State estimate neglecting delays ; $D_0$ = Combined delay term)*

1. The Smith predictor has a control loop involving a model of the plant but without the transport delay terms. This loop generates the ideal delay free system response. The equivalent loop in Fig. 7.14 is shown as loop 1. This loop generates an optimal control signal for a state feedback LQ regulator, without any transport delay terms.

2. In a Smith predictor structure a signal is subtracted from the measured output of the plant (which includes the disturbance signal) that removes the component of the output signal due to the control signal. The parallel situation is illustrated on the right of Fig. 7.14, which shows the difference between two signals being

fed into the gain matrix K. That is, the bottom channel in Fig. 7.14 which represents the delayed plant output, due only to control action, removes the control component from the observations signal $z(t)$.

3. In a Smith predictor the signal which remains (minus the control action) enters the controller transfer function (often a PID controller) and produces the control action. The parallel in Fig. 7.14 is the signal amplified by the Kalman filter, which only depends upon the disturbance signal. This is predicted forwards and enters the control gain matrix $K_1$. The output of this gain matrix therefore regulates the disturbance.

Having shown the optimal solution is consistent with the Smith predictor it may also be observed that the solution is physically reasonable. Assuming the plant is open-loop stable the loop 1 in Fig. 7.14 generates an ideal control signal, not including the disturbance effect. The Kalman filter on the other hand generates a state estimate, which is only dependent upon the disturbance effects. This state estimate is predicted forward in time (using $A^{\Lambda_0}$) to compensate for the transport delays within the plant.

The main advantage of the LQG solution over the classical Smith predictor is that the disturbances will be regulated optimally. The design of the PID controller in a Smith predictor structure is not straightforward, particularly if good regulating action is to be achieved.

## 7.3.6  Hot Strip Mill Profile Control Example

The hot rolling process is described in Chapter 11 and it is of interest in this section, since there are many transport-delay elements. A hot strip finishing mill takes a bar of metal and by passing it through a number of stands reduces the thickness until the strip exits the final stand with the desired thickness. A typical bar enters with a thickness of 30 mm at 1000 °C and leaves with a reduced thickness of 2 mm at 800 °C

This applies to both aluminium and steel mills, although for this application a steel mill will be considered (Grimble and Hearns, 1998). Due to the severe environment and a lack of space, sensors to measure the strip thickness can only be used after the last stand, hence there is a delay between the measured thickness and the actual thickness exiting the last stand. The last stand of a 4-high mill is shown in Fig. 7.15 with the thickness sensors positioned some distance from the mill exit. The strip does not have a uniform thickness across the width but has a desired profile which is described by $h_e, h_m$ and $h_c$ in Fig. 7.15. These thickness values cover the strip edge ($h_e$), at a given distance close to the strip edge ($h_m$) and the centre-line thickness ($h_c$).

The actuators used to control the profile are the hydraulic capsules which adjust the vertical position of the rolls and the bending jacks. These apply a force between the work rolls in order to give the roll gap a concave form. The capsules have maximum effect on the mean thickness and minimal effect on the profile, whereas the reverse is true for the bending jacks. There are, however, significant interactions between the effect of the capsules and jacks. It is assumed that the strip has a symmetrical profile on each side and that each half has an identical profile. Since there are only two actuators $S$ and $J$ then only two distinct points on the exit profile can be controlled independently. Both actuators do of course influence the whole profile of the strip but with two actuators it is probable that it will only be possible to regulate the thickness errors to zero at two particular points. The feedback control system will therefore control the actuators $S$ and $J$, based upon the measured values $h_e$ and $h_c$ , so that zero steady-state thickness error can be achieved.

The controller will therefore be designed to regulate $h_e$ and $h_c$, while $h_m$ will be free to move naturally. There are separate X-ray thickness sensors to measure $h_e$ and $h_c$, each being a different distance from the stand exit, hence the control problem incorporates different measurement system time-delays. On each stand of the mill there are local force feedback regulators to control the average thickness but to achieve the desired product tolerance feedback from an X-ray measurement system is also required.

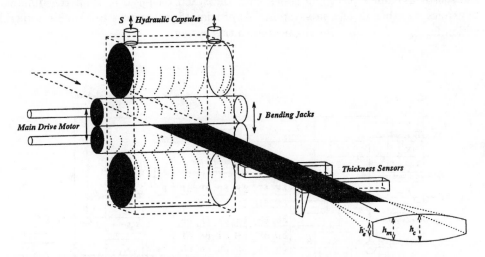

**Fig. 7.15** : Hot Strip Mill Rolling Stand and Sensors for Profile Control

## Mill Stand Modelling

The modelling of the mill stand and the plastic deformation of the strip relies on both theoretical and experimental results. For the control loop design a linear model is required for the stand and the strip which is algebraic, since all the dynamics of the stand can be considered to be in the actuators. The local thickness regulation uses the feedback of the measured rolling force $P$ through a gain $M$ to the capsule reference $S_r$ , where $M$ is given as:

$$M = \alpha \frac{\frac{\delta h_c}{\delta H_c} + \frac{\delta h_c}{\delta H_m} + \frac{\delta h_c}{\delta H_e}}{\frac{\delta P}{\delta H_c} + \frac{\delta P}{\delta H_m} + \frac{\delta P}{\delta H_e}}, \quad \alpha < 1$$

and $H_c, H_m$ and $H_e$ are the entry thicknesses to the stand in the same positions (relative to the width) as the exit thicknesses $h_c, h_m$ and $h_e$, respectively. This loop regulates the average or centre-line thickness. The inferential control for the gaugemeter loop involves positive feedback at low frequencies and to ensure stability it must have a loop gain of less than unity. If all the force and thickness sensitivities are known exactly then this loop can achieve perfect control for $h_c$. Since these sensitivities are subject to uncertainty, to ensure the loop gain is always less than unity and hence provide robustness, a detuning gain $\alpha$ is used. However, this method of providing robustness will mean that the thickness $h_c$ will not be regulated to achieve a zero steady-state error. It is the function of the X-ray thickness feedback control to remove this error and also control $h_e$.

Figure 7.16 shows the algebraic model of the stand with the force feedback and first-order dynamics for the capsule and jacks. Augmented to this model are first-order low-pass colouring filters for the entry thickness and temperature disturbances. The dynamic models for the X-ray thickness sensors should be after the transport delays in the physical representation of the system. Since each of the measurement channels are scalar systems, the order of the sensor and delay can be changed, and the sensor dynamics placed in series with the stand outputs, to keep the different dynamics together in one sub-system. Approximate integrators to provide integral action for $h_c$ and $h_e$ are also augmented to this system.

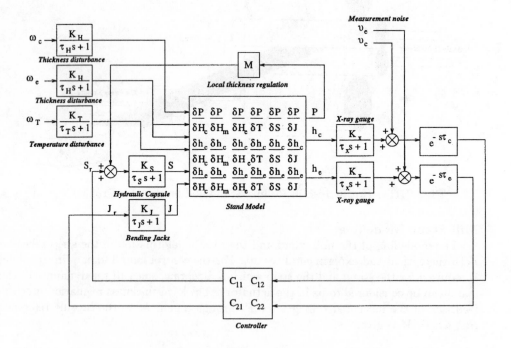

**Fig. 7.16** : Profile Control System and Plant Represented by Sensitivity Coefficients

## System Model

The system is shown in Fig. 7.16 and the plant models may now be defined in continuous-time form. Note that the form of the compensator for continuous-time systems is almost the same as in Fig. 7.14 (see Grimble and Hearns 1998).

**Stand Model:**

$$\begin{bmatrix} P \\ h_c \\ h_e \end{bmatrix} = G \begin{bmatrix} H_c \\ H_m \\ H_e \\ T \\ s \\ J \end{bmatrix}$$

where

$$G = \begin{bmatrix} 7.05 \times 10^7 & 1.35 \times 10^9 & 6.76 \times 10^3 & -3.15 \times 10^3 & -2.34 \times 10^9 & 0.15 \\ 0.640 & 0.062 & 0.030 & -8.41 \times 10^{-7} & 0.374 & 2.63 \times 10^{-11} \\ 0.031 & 0.072 & 0.645 & -8.66 \times 10^{-7} & 0.356 & 2.17 \times 10^{-11} \end{bmatrix}$$

$$\tau_c = 0.2s, \quad \tau_e = 0.5s$$

**Hydraulic Capsule:**

$$S(s) = \frac{1}{0.03s + 1} S_r(s)$$

**Bending Jacks:**

$$J(s) = \frac{10^{10}}{0.03s + 1} J_r(s)$$

**X-ray Gauge:**

$$h_{e,c}^m(s) = \frac{1}{0.01s + 1} h_{e,c}(s)$$

**Thickness disturbance:**

$$H(s) = \frac{10^{-3}}{0.1s + 1} W_{e,c}(s)$$

**Temperature Disturbance:**

$$T(s) = \frac{10}{0.1s + 1} W_T(s)$$

**Control Design Data:**

$$Q = \begin{bmatrix} 1 & 0 & 0 \\ 0 & 1 & 0 \\ 0 & 0 & 1 \end{bmatrix}, \quad R = \begin{bmatrix} 1 \times 10^{-11} & 0 \\ 0 & 1 \times 10^{-11} \end{bmatrix},$$

$$Q_1 = \begin{bmatrix} 10 & 0 \\ 0 & 10 \end{bmatrix}, \quad R_1 = \begin{bmatrix} 0.001 & 0 \\ 0 & 0.001 \end{bmatrix}$$

## Controller and Sensitivity Function Frequency Responses

The system shown in Fig. 7.16 can be expressed in state-space form but cannot be used directly as the model for design since it does not satisfy Assumption 3a. That is, the disturbances do not independently affect each of the measured delayed outputs. To overcome this problem an approximation may be made and the dynamics can be repeated twice in a block diagonal structure. The disturbances $\omega_c$ , $\omega_e$ and $\omega_T$ will therefore be duplicated to make six independent disturbances. This will mean that each output is affected by a different set of disturbances. This does not change the plant transfer-function between control signal input and plant output and this is still a correct representation of the multivariable system. The penalty to pay for satisfying this assumption is that the controller will have a larger order and stochastic properties will be sub-optimal.

Fig. 7.17 shows the frequency responses of the individual elements of the controller. Figures. 7.18 and 7.19 show the frequency plots for the closed loop sensitivity and control sensitivity functions, respectively. Although the controller was not designed to regulate $h_m$ , the sensitivity of $h_m$ to the temperature disturbance $T$ is low due to actions of the force feedback loop. The sensitivity of $h_m$ to the entry thickness disturbances is not low, because the controller can only independently control two points on the strip profile.

## Closed-loop Time Response Simulation

The controller design is assessed using input step temperature and thickness disturbances. Figures 7.20 and 7.21 show the output profile and actuator responses for a step input of 20°C which occurs at 2 seconds. The actuators first respond to the centre line thickness measurement and then to the edge measurement which has a longer delay. The Fig. 7.22 takes the three points on the profile for one side of the strip and reconstructs the full profile for the top of the strip. Figures 7.23 and 7.24 show the response of the profile and actuators to a change of 0.5 mm at the centre of the input thickness profile and a change of -0.3mm at the edge of the input thickness profile. The error for the edge and centre thickness both go to zero while the middle thickness $h_m$ includes a steady-state error which is also shown in the reconstruction of the profile in Fig. 7.25.

These results reveal that for this application the controller is effective at rejecting the input strip temperature and thickness disturbances. The design allows for the different transport delay elements. The speed of the rolls and hence the strip speed and time-delays can change but it is easy to adjust the controller since the filter $K$ and control $K_1$ gains are independent of the time-delays. These gains do not need to be re-calculated as the line speed varies. This does not of course mean that the controller is totally independent of line speed, since both the prediction and delay elements in Fig. 7.14 vary with speed.

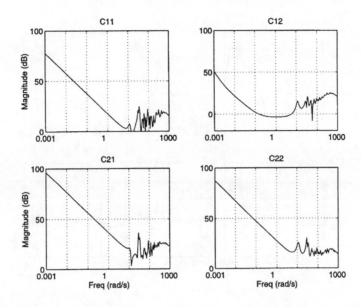

**Fig. 7.17** : Magnitude Frequency Responses for Controller Elements

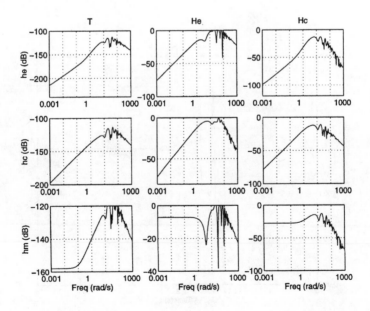

**Fig. 7.18** : Loop Sensitivity Function Frequency Responses between

Disturbance Inputs and Thickness Profile Outputs

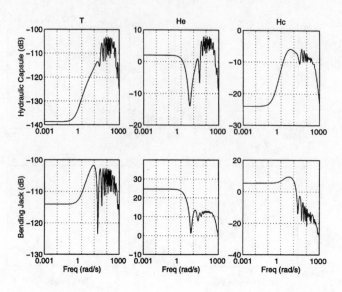

**Fig. 7.19** : Control Sensitivity Function Frequency Responses between
Disturbance Inputs and Actuator Control Signals

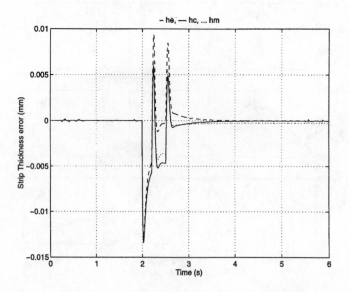

**Fig. 7.20** : Time Responses at Three Points on the Profile for
Temperature Disturbance at t = 2 Seconds

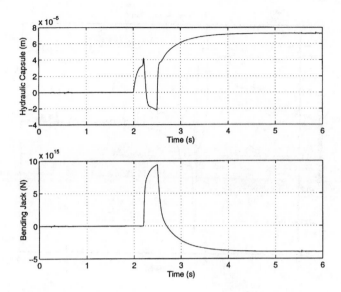

Fig. 7.21 : Actuator Time Responses for Temperature Disturbance at t = 2 Seconds

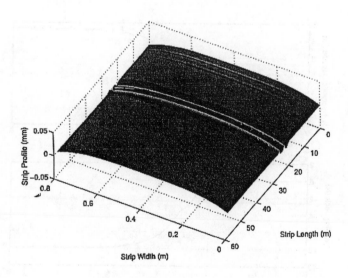

Fig. 7.22 : Strip Surface Thickness Profile for a Temperature Disturbance t=2
Seconds

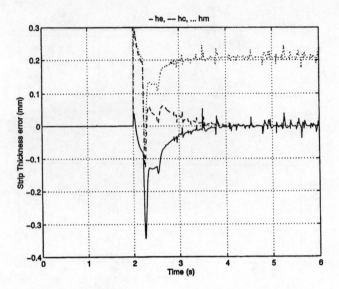

**Fig. 7.23** : Thickness Time Response at Three Points on the Profile
Input Thickness Disturbance at t=2 Seconds

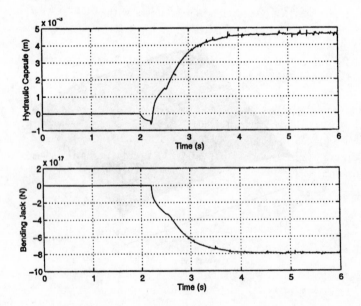

**Fig. 7.24** : Actuator time Responses for Input Thickness Disturbance at t=2
Seconds

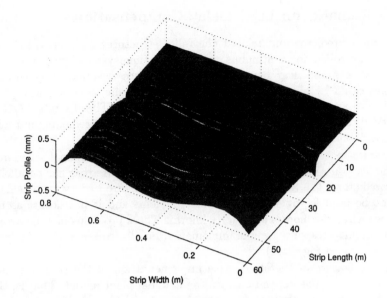

**Fig. 7.25** : Strip Surface Thickness Profile for Input Thickness Disturbance
at t=2 Seconds

## Measurement System

The determination of the strip profile is assumed to be achieved by a very particular arrangement of only two X-ray thickness sensors with different distances to the roll gap of the last stand of the finishing mill. There are several other solutions to this problem: X-ray sensors can be moved continuously lateral to the strip speed direction (i.e. Voest-Alpine in Austria) and scanning isotope sensors give 22 thickness values across the width in one line (i.e. Thyssen Steel in Germany).

## Comparison with PID Control

The controller frequency responses shown in Fig. 7.17 clearly reveal integral action at low frequency and the effect of a lead term in mid-frequency region. A Smith predictor, in conjunction with a PID controller, can therefore provide similar performance. However, there are major differences in the simplicity of the design processes for the LQG solution.

- The noise filtering and disturbance estimation is achieved naturally by the Kalman filter, whereas in the PID design it is a time consuming trade-off between performance and stochastic properties.

- The design of a multivariable PID/Smith solution is not straightforward, whereas the LQG solution is not significantly more difficult to tune (via the weighting) than in the scalar case.

- The design of the PID controller in a Smith predictor system requires that closed-loop stability be achieved for the delay free loop. This is achieved naturally in the LQG solution (leaving aside the problems of model mismatch).

### 7.3.7  Remarks on LQG Delay Compensation

The Smith predictor and related controller structures have provided a means of designing controllers for continuous and discrete-time systems with transport delays. However, it was not clear how the best disturbance rejection or noise attenuation properties could be obtained. In fact the Smith predictor only really provides the controller structure, rather than a design procedure. The discrete LQG approach presented deals naturally with such stochastic design requirements, and also applies to multivariable systems.

The state-space solution obtained involves a finite dimensional standard Kalman filter and the controller may be a represented in a structure that has parallels with the Smith predictor. The Smith predictor type of structure has two problems. Namely it cannot be used for open-loop unstable problems and it has notorious robustness difficulties when the models are poorly known. These problems may be avoided using the finite impulse response form of implementation, illustrated in Fig. 7.5 (rather than the Smith form in Fig. 7.14).

A useful feature of the state-space implementation of the controller is that the control and filter gains depend only on the nominal plant model. That is, the control and predictor gain calculations do not depend upon the length of the transport delays which is valuable in systems where the delay varies (as occurs in rolling mills when the linear speed changes). The state-space results also confirm that a novel type of *separation theorem* applies.

## 7.4  $H_\infty$ State-space Control and Estimation

The state-space approach to $H_\infty$ design has become very popular. The accepted solution followed the work of Peterson (1987) and was reported in the seminal paper by Doyle et al. (1989). The solution of the $H_\infty$ problem using the game theory approach (Basar, 1989 and Kwon et al. 1990) also provides a useful physical explanation for the results.

The main advantage of the state-space solution is the separation structure that is obtained (Hvovstov and Luksic, 1990). This has some advantages over the polynomial approach, since state-estimates can be valuable (even if they do not have the usual minimum variance interpretation). However, there are interesting relationships between the $H_\infty$ state-space and polynomial system equations (Grimble 1988, 1991a,b).

The following results are based on those presented in Yaesh and Shaked (1992). The state feedback $H_\infty$ control problem is considered first and the $H_\infty$ filtering problem is then discussed (Stoorvogel, 1992). Finally, the standard four block, output feedback $H_\infty$ optimal control problem is described and the main results are presented. Commercial toolboxes can of course be used to compute state-space $H_\infty$ solutions (Chiang and Safonov, 1988).

The solution of the output feedback synthesis problem is obtained by transforming it into a $H_\infty$ optimal estimation problem whose solution has been derived. The optimal estimates for time $t$ are assumed to be found using observations up to time $(t-1)$. The solution approach taken by Yaesh and Shaked (1992), that is followed below, is constructive and provides valuable insights into the results obtained.

**Problem Formulation**

Consider the following linear discrete time-invariant system:

$$x(t+1) = Ax(t) + B_1w(t) + B_2u(t) \qquad (7.56)$$

where $x(t) \; \epsilon R^n$, $w(t) \; \epsilon R^p$, $u(t) \; \epsilon R^m$ and $A, B_1$ and $B_2$ are constant matrices of the appropriate dimensions. The measured outputs and the controlled outputs, for the system are denoted, respectively, as:

$$z(t) = C_1 x(t) + D_{12}u(t) \qquad \text{and} \qquad y(t) = C_2 x(t) + D_{21}w(t) \qquad (7.57)$$

where $y(t) \epsilon R^\ell$ and $z(t) \epsilon R^r$. Introduce the following assumptions:

3(i) $(A, B_2)$ stabilisable, $(A, C_1)$ detectable

3(ii) $(A, B_1)$ stabilisable, $(A, C_2)$ detectable

3(iii) $\begin{bmatrix} D_{21}^T & B_1^T \end{bmatrix}^T D_{21}^T = \begin{bmatrix} I & 0 \end{bmatrix}^T$

3(iv) $D_{21}^T \begin{bmatrix} D_{12} & C_1 \end{bmatrix} = \begin{bmatrix} I & 0 \end{bmatrix}$

3(v) $B_1$ is left invertible

The first two assumptions are standard and Assumptions 3(iii) and 3(iv) can be replaced by the nonsingularity of $D_{21}D_{21}^T$ and $D_{12}^T D_{12}$, respectively. Assumption 3(v) can be always achieved by suitable definition of the disturbance signal $\{w(t)\}$.

## 7.4.1 $H_\infty$ State Regulation Problem

In this section it is assumed that the state variables are available for feedback control that are free from noise. This is not, of course, a very practical problem, although there are a few applications where critical and representative plant states are measurable. The aim is to find an optimal control strategy $\{u(t)\}$ in $\ell_2[0, \infty)$ that leads to a bounded $H_\infty$-norm of the transfer function matrix $T_{zw}$ from $\omega(t)$ to $z(t)$. By rescaling this problem, it can be brought to the form:

$$\|T_{zw}\|_\infty < 1 \qquad (7.58)$$

The results of Yaesh and Shaked (1992, 1991) for the problems of state regulation and a priori state estimation, are summarised in the following. The Assumptions 3(i), 3(iv), 3(v) are first invoked and it will also be assumed that the system is scaled, so that $C_2 = I$ and $D_{21} = 0$.

**Theorem 7.1** : $H_\infty$ *State Regulation Problem*

There exists a state feedback stabilising solution if and only if there exists $P \geq 0$ to the following discrete-time algebraic Riccati equation:

$$P = A^T P A - A^T P B_2 (I + B_2^T P B_2)^{-1} B_2^T P A$$

$$+ P B_1 (I + B_1^T P B_1)^{-1} B_1^T P + C_1^T C_1 \qquad (7.59)$$

If such a $P$ exists, then the $H_\infty$ norm relationship (7.58) is satisfied by the following stabilising constant-gain state feedback control law:

$$u(t) = -K_c x(t) = -(I + B_2^T P B_2)^{-1} B_2^T P A x(t) \qquad (7.60)$$

**Control Problem Results**

The following result relates the above solution to the state regulation problem with a different type of Riccati equation that appears in Limbeer et al. (1989). This may be used to obtain an alternative solution involving decoupled Riccati equations.

**Corollary 7.1 :**

Consider the following matrix equation,

$$X_\infty = A^T X_\infty (I - BJB^T X_\infty)^{-1} A + C_1^T C_1 \qquad (7.61)$$

where

$$B = [B_1 \quad B_2] \quad \text{and} \quad J = diag\{I, -I\} \qquad (7.62)$$

The following statements can be shown to be equivalent:

(i)

$$P = (I - X_\infty B_1 B_1^T)^{-1} X_\infty, \qquad X_\infty \geq 0 \ \text{ and } \ I - B_1^T X_\infty B_1 > 0 \qquad (7.63)$$

(ii)

$$X_\infty = (I + PB_1 B_1^T)^{-1} P \quad \text{and} \ \ P \geq 0 \qquad (7.64)$$

Moreover, if either (7.63) or $P \geq 0$ are satisfied, the state feedback gain $K_c$ of (7.60) is given by:

$$K_c = B_2^T X_\infty (I - BJB^T X_\infty)^{-1} A = B_2^T X_\infty (I - (B_1 B_1^T - B_2 B_2^T) X_\infty)^{-1} A \qquad (7.65)$$

## 7.4.2 $H_\infty$ State Estimation Problem

The problems of discrete-time $H_\infty$ state regulation and estimation have been considered in Yaesh and Shaked (1990). Two kinds of estimation problem were considered. In the first the estimation of the current state can use only previous measurements (a priori estimate), whereas in the second kind it is assumed that the current measurement is available (a posteriori estimate). In practice the processing of the current measurement will often require one time step, which is the case considered below.

The solution of the $H_\infty$ state estimation problem may now be considered. The full potential of the $H_\infty$ filter has probably still to be determined, even though it was developed back in 1987 (Grimble, 1987a). The state space form, described below, is the best known form of the filter. However, if the $H_\infty$ filter is used in signal processing problems, the effect of a dynamic weighting function can be critical. This frequency weighting is easier to understand in the frequency-domain polynomial form. Nevertheless, the state-space solution is particularly valuable in control problems where uncertainty and robustness issues are important (Yaesh, 1992).

In this state-space problem structure, assume, in addition to assumptions 3(ii), 3(iii) and 3(v), that $y(t-1)$ is available to produce the estimate $\hat{x}(t)$ of $x(t)$, and estimate $\hat{z}_1(t)$ of $z_1(t) = C_1 x(t)$. Then the following theorem may be obtained.

**Theorem 7.2 :** *$H_\infty$ State Estimation Problem*
The output estimation error can be written as:

$$\epsilon(t) = z_1(t) - \hat{z}_1(t) = C_1(x(t) - \hat{x}(t))$$

and let $T_{\epsilon w}$ denote the transfer function matrix which relates $\epsilon(t)$ to the external signal $w(t)$. Then the $H_\infty$ norm inequality:

$$\|T_{\epsilon w}\|_\infty < 1 \tag{7.66}$$

can be achieved if and only if there exists a solution $M \geq 0$ to the following discrete-time algebraic Riccati equation:

$$M = AMA^T - AMC_2^T(I + C_2MC_2^T)^{-1}C_2MA^T$$

$$+MC_1^T(I + C_1MC_1^T)^{-1}C_1M + B_1B_1^T \tag{7.67}$$

If such an $M$ exists, the norm (7.66) is satisfied by the gain matrix:

$$K_f = AMC_2^T(I + C_2MC_2^T)^{-1} \tag{7.68}$$

and the state estimation equation:

$$\hat{x}(t+1) = (A - K_fC_2)\hat{x}(t) + K_fy(t) + B_2u(t) \tag{7.69}$$

**$H_\infty$ Estimation Problem Results**
The estimate for $z_1(t) = C_1x(t)$ is given by $\hat{z}_1(t) = C_1\hat{x}(t)$. The following corollary is the dual of Corollary 7.1 and can be used to obtain the solution of the filtering problem in terms of two decoupled Riccati equations.

**Corollary 7.2 :**
Consider the following matrix equation,

$$Y_\infty = AY_\infty(I - C^TJCY_\infty)^{-1}A^T + B_1B_1^T \quad \text{where} \quad C = [C_1^T \ C_2^T]^T \tag{7.70}$$

The following statements are equivalent:

(i)

$$M = Y_\infty(I - C_1^TC_1Y_\infty)^{-1}, \quad Y_\infty \geq 0 \quad \text{and} \quad I - C_1Y_\infty C_1^T > 0 \tag{7.71}$$

(ii)

$$Y_\infty = M(I + C_1^TC_1M)^{-1} \quad \text{and} \quad M \geq 0 \tag{7.72}$$

Moreover, if either (7.71), or $M \geq 0$ are satisfied, the filter gain $K_f$ of (7.68) is given by

$$K_f = AY_\infty(I - Y_\infty C^TJC)^{-1}Y_\infty C_2^T = AY_\infty(I - Y_\infty(C_1^TC_1 - C_2^TC_2))^{-1}Y_\infty C_2^T \tag{7.73}$$

### 7.4.3  Output Feedback Control Problem

The output-feedback control problem is the most practical and valuable solution. It is shown below that the results are analogous to those in LQG control and in fact the order of the controller is equivalent. That is, the output-feedback $H_\infty$ controller has an order equal to that of the plant and disturbance models, plus the order of any dynamic cost-function weighting terms. The technique is to transform the output feedback problem to a state-estimation problem, (solved in Limbeer et al. (1989)). The main step in the solution is to transform $\{\omega(t)\}$ and $\{z(t)\}$ to new variables $\{r(t)\}$ and $\{v(t)\}$, where

$$\|T_{zw}\|_\infty = \|T_{vr}\|_\infty \quad \text{for all } u(t)\epsilon\ell_2[0,\infty) \tag{7.74}$$

From this result the problem of making the norm:

$$\|T_{vr}\|_\infty < 1 \tag{7.75}$$

is equivalent to the problem of satisfying the norm inequality (7.58).

**Cost Equivalence**

The following equivalence has been established in Limebeer et al. (1989):

$$J_1 = \|z\|_2^2 - \|w\|_2^2 = -\sum_{t=0}^{\infty}(\phi(t)-\phi^0(t))^T(J-B^TX_\infty B)(\phi(t)-\phi^0(t)) \tag{7.76}$$

where $\|.\|_2$ denotes the $\ell_2$ norm, and B is defined in (7.62). From Corollary 7.1 the matrix $X_\infty$ is the solution of (7.61). Also introduce the following signals:

$$\phi(t) = [w^T(t) \quad u^T(t)]^T, \qquad \phi^0(t) = [w^0(t)^T \quad u^0(t)^T]^T \tag{7.77}$$

The vectors $(u^0(t), w^0(t))$ are the saddle point pair (Yaesh and Shaked, 1989) that brings $J_1$ of (7.76) to an equilibrium, determined as:

$$u^0(t) = -B_2^T(I+PB_2B_2^T)^{-1}PAx(t) \quad \text{and} \quad w^0(t) = B_1^T(I+PB_2B_2^T)^{-1}PAx(t) \tag{7.78}$$

where $P$ is the solution of (7.59) and $\phi^0(t)$ can thus be written as :

$$\phi^0(t) = (J-B^TX_\infty B)^{-1}B^TX_\infty Ax(t)$$

**Factorisation Simplification**

The following useful factorisation was introduce in Limbeer et al. (1989):

$$J-B^TX_\infty B = W^TJW \quad \text{where} \quad W=[W_{ij}], \quad i,j=1,2 \tag{7.79}$$

and hence the norm expression (7.76) may be written as

$$J_1 = \|z\|_2^2 - \|\omega\|_2^2 = -\sum_{t=0}^{\infty}(\phi(t)-\phi^0(t))^TW^TJW(\phi(t)-\phi^0(t)) \tag{7.80}$$

The matrix W in (7.79) is assumed to be partitioned to be compatible with the matrix $B$ in (7.62). Expanding the partitioned equations for each term in (7.79) and recalling the relationship between $X_\infty$ and $P$ in Corollary 7.1, equation (7.64), it follows that the submatrices $W = [W_{ij}]$ satisfy the following equations:

$$W_{11}^T W_{11} = (I + B_1^T P B_1)^{-1} \quad \text{and} \quad W_{22}^T W_{22} = I + B_2^T P B_2 \qquad (7.81)$$

and

$$W_{12}^T W_{11} = -B_2^T P B_1 (I + B_1^T P B_1)^{-1} \quad \text{and} \quad W_{21} = 0 \qquad (7.82)$$

These results may be confirmed by multiplying out the terms in (7.79), using (7.81) and (7.82). If $P \geq 0$ the matrices on the right of (7.81) are positive definite and $W_{11}$ and $W_{22}$ may be obtained as the square-roots of these matrices. It then follows that the matrix $W_{12}$ may be computed from equation (7.82).

Now introduce the vector of signals:

$$\begin{bmatrix} r(t) \\ v(t) \end{bmatrix} = \begin{bmatrix} W_{11} & W_{12} \\ 0 & W_{22} \end{bmatrix} \begin{bmatrix} \omega(t) - \omega^0(t) \\ u(t) - u^0(t) \end{bmatrix} = W(\phi(t) - \phi^0(t)) \qquad (7.83)$$

and note that

$$(\phi(t) - \phi^0(t))^T W^T J W(\phi(t) - \phi^0(t)) = \|r\|_2^2 - \|v\|_2^2$$

Then, substituting in equation (7.80):

$$\|z\|_2^2 - \|w\|_2^2 = \|v\|_2^2 - \|r\|_2^2$$

From the relationship between the $H_\infty$ and $\ell_2$ norms the relationship (7.74) is obtained and the following lemma results.

**Lemma 7.1** : *Equivalent System Results*

Given $P \geq 0$ the following two statements are equivalent:

(i) $\|T_{zw}\|_\infty < 1$, where $u(t)\epsilon\ell_2[0, \infty)$ uses the measurement $y(t-1)$ of (7.57)

(ii) $\|T_{vr}\|_\infty < 1$, where the state estimate feedback control $\{u(t)\}$ is given by:

$$u(t) = -(I + B_2^T P B_2)^{-1} B_2^T P A \hat{x}(t) \qquad (7.84)$$

and $\hat{x}(t)$ solves the a priori state estimation problem of Theorem 7.2, since it satisfies (7.66), where $A, B_1, B_2, C_1, C_2$ and $D_{21}$ are replaced, respectively, by:

$$\tilde{A} = (I + B_1 B_1^T P)A, \qquad \tilde{B}_1 = (I + B_1 B_1^T P)B_1 W_{11}^T,$$

$$\tilde{B}_2 = (I + B_1 B_1^T P)B_2, \qquad \tilde{C}_1 = W_{22}(I + B_2^T P B_2)^{-1} B_2^T P A$$

$$\tilde{C}_2 = C_2 \quad \text{and} \quad \tilde{D}_{21} = D_{21} W_{11}^T \qquad (7.85)$$

**Proof** : It follows from (7.83) that $r(t) = W_{11}(w(t) - w^0(t)) + W_{12}(u(t) - u^0(t))$. Substituting the saddle point pair (7.78) in the last equation and multiplying from the left the equation by $W_{11}^T$, obtain using (7.81):

$$W_{11}^T r(t) = (I + B_1^T P B_1)^{-1}(w(t) - B_1^T P(Ax(t) + B_2 u(t))$$

Also using the state equation (7.56) obtain:

$$w(t) = W_{11}^T r(t) + (I + B_1^T P B_1)^{-1} B_1^T P x(t+1)$$

Substitute this last equation into the state and output equations (7.56) and (7.57), and use Assumption 3(iii) to obtain:

$$x(t+1) = ((I + B_1 B_1^T P)A)x(t) + ((I + B_1 B_1^T P)B_1 W_{11}^T)r(t)$$

$$+((I + B_1 B_1^T P)B_2)u(t)$$

$$\text{and} \quad y(t) = C_2 x(t) + D_{21} W_{11}^T r(t)$$

Substituting equation (7.78) in (7.83) obtain:

$$v(t) = W_{22}(u(t) + (I + B_2^T P B_2)^{-1} B_2^T P A x(t))$$

Recall from (7.85) : $\tilde{C}_1 = W_{22}(I + B_2^T P B_2)^{-1} B_2^T P A$ and hence:

$$v(t) = W_{22}u(t) + \tilde{C}_1 x(t) = W_{22}u(t) + z_1(t) \tag{7.86}$$

Noting this expression for the signal $\{v(t)\}$ the Theorem 7.2 may be invoked to obtain an estimator for $z_1(t) = \tilde{C}_1 x(t)$ where $\hat{z}_1(t) = -W_{22}u(t)$ and $v(t) = z_1(t) - \hat{z}_1(t)$. Thence, obtain: $\hat{z}_1(t) = \tilde{C}_1 \hat{x}(t)$ and (7.84) follows:

$$u(t) = -W_{22}^{-1} \tilde{C}_1 \hat{x}(t) = -(I + B_2^T P B_2)^{-1} B_2^T P A \hat{x}(t)$$

where the norm of $T_{vr}$ satisfies (7.74).

**Conditions of the Theorem**

To use the result of Theorem 7.2 it must be shown that the conditions of the theorem hold for the new system matrices of (7.85). Expanding the system matrices, recalling Assumption 3(iii), obtain:

$$\tilde{B}_1 \tilde{D}_{21}^T = (I + B_1 B_1^T P)B_1 W_{11}^T W_{11} D_{21}^T \tag{7.87}$$

$$= (I + B_1 B_1^T P)B_1(1 + B_1^T P B_1)^{-1} D_{21}^T = (I + B_1 B_1^T P)(1 + B_1 B_1^T P)^{-1} B_1 D_{21}^T = 0$$

and

$$\tilde{D}_{21} \tilde{D}_{21}^T = D_{21} W_{11}^T W_{11} D_{21}^T = D_{21}(I + B_1^T P B_1)^{-1} D_{21}^T$$

$$= D_{21}[I - (I + B_1^T P B_1)^{-1} B_1^T P B_1] D_{21}^T = I$$

The following Lemma is required (Yaesh and Shaked, 1992) to verify the validity of the rest of the conditions of Theorem 7.2.

**Lemma 7.2** : *Stabilisability and Detectability Results*

(i) $(\tilde{A}, \tilde{B}_1)$ is stabilisable

(ii) $(\tilde{A}, C_2)$ is detectable if $\|T_{vr}\|_\infty < 1$

(iii) Denote $\tilde{M}$ as the solution of the following equation:

$$\tilde{M} = \tilde{A}\tilde{M}\tilde{A}^T - \tilde{A}\tilde{M}C_2^T(I + C_2\tilde{M}C_2^T)^{-1}C_2\tilde{M}\tilde{A}^T$$

$$+\tilde{M}\tilde{C}_1^T(I + \tilde{C}_1\tilde{M}\tilde{C}_1^T)^{-1}\tilde{C}_1\tilde{M} + \tilde{B}_1\tilde{B}_1^T \qquad (7.88)$$

obtained from (7.67) by replacing the system matrices with those of (7.85). Then $(\tilde{A}, C_2)$ is detectable if $\tilde{M} \geq 0$.
The main *output feedback* theorem may now be presented.

**Theorem 7.3** : $H_\infty$ *Output Feedback Control Problem*
The output feedback control problem has a solution that satisfies the norm relationship (7.58), if there exist positive semi-definite solutions $P$ and $\tilde{M}$ of (7.59) and (7.88), respectively. If such solutions exist, one state estimate feedback controller that satisfies (7.58) is given as:

$$u(t) = -(I + B_2^T P B_2)^{-1} B_2^T P A \hat{x}(t) \qquad (7.89)$$

$$\text{where} \quad \hat{x}(t+1) = (\tilde{A} - \tilde{K}_f C_2)\hat{x}(t) + \tilde{K}_f y(t) + \tilde{B}_2 u(t) \qquad (7.90)$$

$$\text{and} \quad \tilde{K}_f = \tilde{A}\tilde{M}C_2^T(I + C_2\tilde{M}C_2^T)^{-1} \qquad (7.91)$$

**Proof** : From Yaesh and Shaked (1992).

**Sufficiency** : If $P \geq 0$ and $\tilde{M} \geq 0$ then by part (iii) of Lemma 7.2 $(\tilde{A}, C_2)$ is detectable. The sufficiency then follows from the results of Lemma 7.1 and Theorem 7.2 , noting that the rest of the conditions for Theorem 7.2 are satisfied.
**Necessity** : Suppose that $\|T_{zw}\|_\infty < 1$ with a controller $K(z^{-1})$ operating on $y$ of (7.57). The plant input signal then becomes :

$$u(t) = -(KC_2 x(t) + KD_{21}w(t))$$

Substituting the last equation in equations (7.56) and (7.57) obtain:

$$T_{zw} = T_1 + T_2 = (C_1 - D_{12}\tilde{K}_c)(zI - A + B_2\tilde{K}_c)^{-1}B_1$$

$$-[(C_1 - D_{12}\tilde{K}_c)(zI - A + B_2\tilde{K}_c)^{-1}B_2 + D_{12}]KD_{21} \qquad (7.92)$$

where the $T_1$ and $T_2$ are defined by the two terms in (7.92) and the gain:

$$\tilde{K}_c = K(z^{-1})C_2 \qquad (7.93)$$

Since by Assumption 3(iii) of Section 7.4, the $T_1$ and $T_2$ are orthogonal the function $T_1$ must have an $H_\infty$ norm which is less than one. Note that $T_1$ of (7.92) can easily be shown to be the same as the closed-loop transfer function matrix of a dynamic state feedback control problem. The fact that $P \geq 0$, follows from Theorem 7.1 and from the fact that any $H_\infty$ norm which is achievable by a dynamic state-feedback controller can also be achieved by a constant state feedback gain (see Yaesh and Shaked, (1991) for a proof). Given $\|T_{zw}\|_\infty < 1$ it follows from Lemma 7.1, that $\|T_{vr}\|_\infty < 1$ and thus from Lemma 7.2 part (ii), $(\tilde{A}, C_2)$ is detectable. The rest of the assumptions of Theorem 7.2 for the system matrices of (7.85) are also satisfied. Thus, from Theorem 7.2 $\tilde{M} \geq 0$. Given the proof of the necessary and sufficient conditions the main result, (7.89) follows from Lemma 7.1 and the expression for the estimator equations (7.68) and (7.69).

**Remarks**

(i) The $H_\infty$ solution in Theorem 7.3 requires the solution of two Riccati equations but note these are not independent, since $\tilde{M}$ in (7.88) depends upon the system matrices (7.85) which involve $P$.

(ii) The Corollaries 7.1 and 7.2 can be used to obtain two decoupled Riccati equations (Limebeer et al. 1989), using a change of variables : $Z_\infty = Y_\infty(I - X_\infty Y_\infty)^{-1}$. However, this requires the additional assumption that $A$ is nonsingular.

(iii) The $H_\infty$ output feedback controller, defined in Theorem 7.3, is clearly very similar structurally to the LQG Kalman filtering controller. It is therefore likely that the simplifications achieved in the first part of the chapter for explicit time delays, will also apply to the implementation of the $H_\infty$ observer based controllers.

## 7.5  Concluding Remarks

In the first part of the chapter the LQG synthesis problem was considered using a Wiener-Hopf approach. This enabled the results to be related to those of OJM Smith for time delay compensation. This approach also served the purpose of introducing the *separation principle* and showing how it may be applied for systems with explicit input and output channel delays.

The state-space LQG output feedback control problem results were found to have a parallel in $H_\infty$ synthesis problems in the later parts of the chapter. The solution of the output feedback $H_\infty$ optimal control problem was shown to be the solution of separate state-feedback optimal control and estimation problems. This is of course very similar to the solution of the LQG optimal control problem, via the separation principle. This result can be be helpful when explaining $H_\infty$ design to those that already have experience in using LQG controllers.

One benefit of the separation principle type of structure is that access to the estimated state variables can be valuable. The physical significance of the estimates is different in the two problem structures but estimates can be useful in applications such as condition monitoring and fault detection. There are of course benefits to be gained from generating controllers that provide a combination of $H_2$ and $H_\infty$ properties but this leads to further complication in the design and numerical algorithms (Schaper et al. 1990; Steinbuch and Bosgra, 1990; Sparks et al. 1990; Byrns et al. 1992; Furuta and Phoojaruenchanachai, 1990; Klifton and Black, 1991; Mustafa and Bernstein, 1991; Bambang et al. 1990; Bernstein et al. 1989).

## 7.6 References

Anderson, B.D.O and Moore, J.B., 1979, *Optimal filtering*, Prentice Hall Inc., New Jersey.

Åström, K.J., 1977, Frequency domain properties of Otto Smith regulators, *International Journal of Control*, Vol **26**, No 2, 307-314.

Bambang, R., Shimemura, E. and Uchida, K., 1990, The discrete time $H_2/H_\infty$ control synthesis: State feedback case, In Proceedings Korean Automatic Control Conference, pages 858-863, Seoul, Korea.

Basar, T., 1989, A dynamic games approach to controller design : disturbance rejection in discrete-time, *Proc. of the CDC Conf.* Tampa, Florida, pp. 407-414.

Bernstein, D.S., Haddad, W.M. and Nett, C.N., 1989, *Minimal complexity control law synthesis, part 2: Problem via $H_2/H_\infty$ optimal static output feedback.* Proceedings American Control Conference, pages 2506-2510.

Byrns, E., Gregory, V., Sweriduk, D. and Calise, A.J., 1992, Optimal $H_2$ and $H_\infty$ fixed-order dynamic compensation using canonical forms. *International Journal of Robust and Nonlinear Control*, **2**, pp.194-211.

Chiang, R.Y. and Safonov, M.G., 1988, *User's guide for robust control toolbox in MATLAB*, The MathWorks Inc.

Davis, M.H.A., 1977, *Linear estimation and stochastic control*, Chapman & Hall.

Doyle, J., Glover, K., Khargonekar, P.P. and Francis, B., 1989, State-space solutions to standard $H_2$ and $H_\infty$ control problems, *IEEE Trans. on Automat. Contr.*, Vol. 34, pp. 831-847.

Fuller, A.T., 1968, Optimal nonlinear control of systems with pure delay, *Int J Control*, Vol **8**, No 2, pp 145-168.

Furuta, K. and Phoojaruenchanachai, S., 1990, An algebraic approach to discrete time $H_\infty$ control problems, Proceedings American Control Conference, pages 3067-3072, San Diego, California.

Grimble, M J., 1978a, The design of stochastic optimal feedback control systems, *IEE Proceedings*, Vol. **125**, No. 11, pp. 1275-1284.

Grimble, M J., 1978b, Definition of a complex domain adjoint for use in optimal control problems, *Journal of the Institute of Measurement and Control*, pp 33-35.

Grimble, M.J., 1979a, Solution of the stochastic optimal control problem in the s-domain for systems with time delay, *IEE Proceedings*, Vol. **126**, No. 7, pp. 697-704.

Grimble, M.J., 1979b, Solution of the discrete-time stochastic optimal control problem in the z-domain, *Int. J. Systems Science*, Vol. **10**, No. 12, pp. 1369-1390.

Grimble, M.J., 1987a, $H_\infty$ design of optimal linear filters, *MTNS Conference, Phoenix, Arizona, Editors : C.I. Byrnes, C.F. Martin and R.E. Saeks, Elseiver Science Publishers, BV*, North Holland.

Grimble, M.J., 1987b, Relationship between polynomial and state-space solutions of the optimal regulator problem, *Systems and Control Letters*, **8**, pp. 411-416.

Grimble, M.J., 1988, State-space approach to $H_\infty$ optimal control, *Int. J. Systems Sci.*, Vol. **19**, No. 8, pp. 1451-1468.

Grimble, M.J., 1991a, $H_\infty$ filtering problem and the relationship between polynomial and state-space results, *IEEE CDC Conference*, Brighton.

Grimble, M.J., 1991b, Solution of discrete $H_\infty$ optimal control problems using a state-space approach, *IEEE CDC Conference*, Honolulu.

Grimble, M.J., 1998, LQG controllers for discrete-time multivariable systems with different transport delays in signal channels, *IEE Proc Control Theory Appl.* Vol. **145**, No. 5, pp. 449-462.

Grimble, M.J. and Johnson, M A., 1988, *Optimal multivariable control and estimation theory, Parts 1 and 2*, John Wiley, London.

Grimble, M.J. and Hearns, G., 1998, LQG controllers for state-space systems with pure transport delays, *Automatica*, Vol. **34**, No. 10, pp. 1169-1184.

Hvovstov, H.S. and Luksic, M., 1990, On the structure of $H_\infty$ controllers. *In Proceedings American Control Conference*, pages 2484-2489, San Diego, California.

Kalman, R.E., 1961, New methods in Wiener filtering theory, *Proc. of the Symposium on Engineering Applications of Random Function Theory and Probability*, pp. 270-388.

Kalman, R.E. and Bucy, R.S., 1961, New results in linear filtering and prediction theory, *Journal of Basic Eng.*, Vol. **83**, pp. 95-108.

Kleinman, D.L., 1969, Optimal control of linear systems with time-delay and observation noise, *IEEE Trans. on Automatic Control*, pp. 524-527.

Kliffton M. Black, 1991, A new algorithm for solving a mixed $H_2/H_\infty$ optimization problem, Proceedings American Control Conference, pages 1157-1158, Boston, Massachusetts.

Kwakernaak, H. and Sivan, R.,1972, *Linear optimal control systems*, Wiley, New York.

Kwon, W.H., Lee, J.H. and Kim, W.C., 1990, $H_\infty$ controller design via LQG game problem for discrete time systems, *Proc of Korean Automatic Control Conf.*, **19**, pp. 864-867.

Kučera, V., 1980, Stochastic multivariable control : A polynomial equation approach, *IEEE Trans. on Auto. Contr.*, Vol. AC-**25**, 5, pp. 913-919.

Lepage, W R., 1961, *Complex variables and the Laplace transform for engineers*, (McGraw Hill), p 338.

Limebeer, D.J., Green, M. and Walker, D., 1989, Discrete-time $H_\infty$ control, *Proc. 28th CDC*, Tampa, Florida, pp. 392-396.

MacFarlane, A.G.J., 1970, Return difference and return ratio matrices and their use in analysis and design of multivariable feedback control systems, *Proc. IEE*, Vol. **117**, No. 10, pp. 2037-2049.

Malek-Zavaris M. and Jamshidi, M., 1987 *Time-delay systems analysis, optimization and applications*, North Holland.

Marinescu, B. and Bourles, H., 1997, Robust predictive control for multi-input/output control systems with non-equal time delays, *European Control Conference*, Brussels.

Marshall, J.E., 1974, Extension of 0.J. Smith's method to digital and other systems, *Int J Control*, Vol **19**, No 5, 933-939.

Mustafa, D. and D.S. Bernstein, 1991, *LQG cost-bounds in discrete time $H_2/H_\infty$ control*, In Proceedings Symposium on Robust Control System Design Using $H_\infty$ and Related Methods, University of Cambridge.

Normey-Rico, J.E., Bordons, C. and Camacho, E.F., 1997, Improving the robustenss of dead-time compensating PI controllers, *Control Engineering Practice*, Vol. **5**, No. 6, pp. 801-810.

Peterson, I.R., 1987, Disturbance attenuation and $H_\infty$ optimizations: A design method based on the algebraic Riccati equation, *IEEE Trans. Auto. Contr.*, AC-**32**, pp. 427-429.

Schaper, C.D., Mellichamp. D.A. and Seborg, D.E., 1990, Stability robustness relations for combined $H_\infty$/LQG control, In Proceedings American Control Conference, pp. 3050-3055, San Diego, California.

Schmotzer, R.E. and Blankenship, G.L., 1978, A simple proof of the separation theorem for linear stochastic systems with time delays, *IEEE Transactions on Auto Control*, Vol AC-**23**, No 4, pp. 734-735.

Shaked, U., 1976a, A general transfer-function approach to linear stationary filtering and steady-state optimal control problems, *Int J Control*, Vol **24**, No 6, pp. 741-770.

Shaked, U., 1976b, A general transfer function approach to the steady-state linear quadratic Gaussian stochastic control problem, *Int J Control,* Vol **24**, No 6, pp. 771-800.

Smith, O.J.M., 1959, A controller to overcome dead time, *ISA Journal,* Vol. **6**, No. 2, pp. 28-33.

Sparks, A.G., Yeh, H.H. and Banda, S.S., 1990, Mixed $H_2$ and $H_\infty$ optimal robust control design, *Optimal Control Applications and Methods*, Vol. **11**, pp. 307-325.

Steinbuch, M., and Bosgra, O.H., 1990, Necessary conditions for static and fixed order dynamic mixed $H_2/H_\infty$ optimal control, Selected topics in *Identification Modelling and Control*, Vol. **2**, pp. 17-23.

Stoorvogel, A., 1992, *The $H_\infty$ control problem*, Prentice Hall International, London.

Yaesh, I. and Shaked, U., 1992, $H_\infty$ optimal one-step-ahead output feedback control of discrete-time systems, *IEEE Trans. on Automatic Control,* Vol. **37**, No. 8, pp. 1245-1250.

Yaesh, I. and Shaked, U., 1990, Minimum $H_\infty$ norm regulation of linear discrete-time systems and its relation to linear quadratic discrete games, *IEEE Trans. on Automatic Control*, Vol. **35**, No. 9, pp. 1061-1064.

Yaesh, I. and Shaked, U., 1991, A transfer function approach to the problems of discrete time systems : $H_\infty$ optimal linear control and filtering, *IEEE Trans. Automat. Contr.*, Vol. **36**, pp. 1272-1276.

Yaesh, I., 1992, *Linear optimal control and estimation in the minimum $H_\infty$ norm sense*, PhD thesis, Tel Aviv University, Israel.

# 8

# State-space Predictive
# Optimal Control

## 8.1  Introduction

This chapter focuses on the state-space approach to predictive optimal control and it parallels to some extent the polynomial system results described in Chapter 3. However, there is a more general view of how future setpoint or reference information can be used. This information may be incorporated into an optimal control law to provide improved tracking characteristics, together with smaller actuator changes. In some applications, such as robot control, the device is required to follow the same trajectory repeatedly and the future reference values are therefore known. Similarly, in many chemical plants the temperature of vessels must follow a desired set-point trajectory, and once again advantage may be taken of information about the future. The predictive control algorithms were applied in the early years on relatively slow processes (such as thermal processes) for the chemical, petrochemical, food and cement industries. More recently they have been applied on faster systems, such as servo drives, hydraulic systems, robotics and aerospace applications.

Although it is less common there are also applications where future disturbance variations are also known. For example in a multi-stand hot rolling mill (Chapter 11) downstream load changes on a stand change the interstand tension and therefore affect the tension and thickness between earlier stands. The main disturbance is therefore known, well before it affects earlier stands and such information is akin to future disturbance knowledge. It may therefore be used in the optimal solutions in a similar manner to the known future setpoint information.

Two alternative methods of using future set-point information are considered. In the early part of this chapter on state space systems the predictive control action is introduced, by use of a multi-step criterion. In the later part of the chapter a two/three degrees of freedom controller enables future reference signals to be fed into the tracking controller, but a single-step criterion is employed.

### 8.1.1    Model Based Predictive Control

The model based predictive control (MBPC) approach has been applied very successfully in the process industries, since it can make a substantial contribution to the profitability and competitiveness of a production plant (Konno et al. 1992). It has been used to improve the performance of control in difficult systems which contain long dead times, time varying system parameters, nonlinearities and multivariable interactions (Pike et al. 1995). Of particular importance is the ability to handle constraints on both the control and output variables. The approach is often thought to be easy to understand relative to other control design methods, and process operator acceptance of this advanced control approach seems reasonably high.

The best known MBPC method in the process industries is probably *Dynamic Matrix Control*, due to Cutler and Ramaker (1980). The most common form of the general theory is the *Generalized Predictive Control* (GPC) due to Clarke et al. (1987, 1989). Peterka[1] (1984) recognised the importance of prediction for adaptive control applications and future reference information is also used in the Multistep Multivariable Adaptive Regulator, due to Mosca (Greco et al. 1984). The term *preview action* is used by Tomizuka and Rosenthal (1979) to indicate the use of future set point knowledge.

The solution of the Linear Quadratic Gaussian Predictive Control (LQGPC) problem, when future set point information is available, has been considered by Grimble (1986, 1988), Hunt (1988) and Sebek et al. (1988), when the plant is represented in polynomial matrix form. There are a number of model based predictive control philosophies, which employ state equation models, that are related to the results presented (Li et al. 1989; Marquis et al. 1988 and Ricker, 1990). A useful summary of such techniques is included in Prett and Garcia, (1988).

### 8.1.2    Set-point Optimisation

To optimise the output of a production process or to minimise the energy usage the set-points for controllers must be optimised. In the majority of industrial processes set-points are established by operators but there is now a trend to the automatic adaptation of set-points to improve the quality of the product, or to optimise the total operation of the plant. There has recently been an attempt to develop the theoretical basis for such systems (Casavola and Mosca, 1998). New dynamic set point optimisation algorithms can be produced which enable dynamic scheduling to be undertaken. As a process moves through non-linear operating conditions it is often essential to schedule the controller with set-point variations, so that the processes are not unduly disturbed and stability is maintained.

Advances have also been made on adaptive control systems that include set-point optimisation. The SCAP adaptive predictive controller, due to Juan Martin Sanchez (1996), includes a set-point optimisation facility. Moreover, the toolbox has several other features which enable it to cope with variations in the dynamics of the plant and provide optimisation facilities.

---

[1] The contributions of Peterka were recognised in the special issue of the International Journal of Adaptive Control and Signal Processing entitled Adaptive and Self-Tuning Control: A Peterka Perspective, September 1999.

Bitmead et al. (1989) explored the relationship between LQ optimal and predictive control laws using state-space system descriptions, and also generated considerable interest in the topic of identification for control law design. The potential for predictive control methods was realised well over a decade ago by researchers such as Richalet et al. (1978, 1993), who applied the technique very successfully to a wide range of applications.

## 8.2 State-space Form of GPC Controller

The GPC controller was first used in polynomial system form (Mohtadi and Clarke, 1986) and was later derived in state-space form, by Ordys and Clarke, (1993). Only the main points in the state-space GPC solution are summarised below. Consider the system model in the state-space form:

$$x(t+1) = Ax(t) + Bu(t) + Dw(t) \tag{8.1}$$

$$y(t) = Cx(t) + \xi(t) \tag{8.2}$$

where

$x(t)$    is a vector of $n$ system states,

$u(t)$    is a vector of $m$ control signals,

$y(t)$    is a vector of $r$ measurable output signals,

and $\{w(t)\}$ and $\{\xi(t)\}$ are vector disturbances, assumed to be zero-mean, independent, Gaussian white noise signals and $A, B, G, D$, are constant matrices. The aim in this problem is to control the output signal $\{y(t)\}$, given past and some future values of the reference signal. The stochastic system description is applicable in industrial processes such as basis weight control in paper machines (Åström, 1967, 1970).

**Future Plant Outputs and States**
The future values of the plant states and outputs may be obtained as:

$$x(t+1) = Ax(t) + Bu(t) + Dw(t)$$

$$x(t+2) = A\left(Ax(t) + Bu(t) + Dw(t)\right) + Bu(t+1) + Dw(t+1)$$

$$= A^2x(t) + ABu(t) + Bu(t+1) + ADw(t) + Dw(t+1)$$

$$x(t+3) = Ax(t+2) + Bu(t+2) + Dw(t+2)$$

$$= A^3x(t) + A^2Bu(t) + ABu(t+1) + Bu(t+2)$$

$$+A^2Dw(t) + ADw(t+1) + Dw(t+2)$$

Generalising this result, the evolution of the state follows as:

$$x(t+k) = A^k x(t) + \sum_{j=1}^{k} A^{k-j} \left( Bu(t+j-1) + Dw(t+j-1) \right) \tag{8.3}$$

and the output equation has the form :

$$y(t+k) = CA^k x(t) + \sum_{j=1}^{k} CA^{k-j} \left( Bu(t+j-1) + Dw(t+j-1) \right) + \xi(t+k) \tag{8.4}$$

These output signals may be collected in the convenient vector form:

$$\begin{bmatrix} y(t+1) \\ y(t+2) \\ \vdots \\ y(t+N+1) \end{bmatrix} = \begin{bmatrix} CA \\ CA^2 \\ \vdots \\ CA^{N+1} \end{bmatrix} x(t)$$

$$+ \begin{bmatrix} CB & 0 & \cdots & 0 \\ CAB & CB & & 0 \\ \vdots & \vdots & \ddots & \\ CA^N B & CA^{N-1} & \cdots & CB \end{bmatrix} \begin{bmatrix} u(t) \\ u(t+1) \\ \vdots \\ u(t+N) \end{bmatrix}$$

$$+ \begin{bmatrix} CD & 0 & \cdots & 0 \\ CAD & CD & & 0 \\ \vdots & \vdots & \ddots & \\ CA^N D & CA^{N-1}D & \cdots & CD \end{bmatrix} \begin{bmatrix} w(t) \\ w(t+1) \\ \vdots \\ w(t+N) \end{bmatrix} + \begin{bmatrix} \xi(t+1) \\ \xi(t+2) \\ \vdots \\ \xi(t+N+1) \end{bmatrix} \tag{8.5}$$

With an obvious definition of terms the previous equation may be written as:

$$Y_{t,N} = C_N A_N x(t) + C_N B_N U_{t,N} + C_N D_N W_{t,N} + W_{t,N}^0 \tag{8.6}$$

where the following system matrices may be defined:

$$A_N = \begin{bmatrix} A \\ A^2 \\ \cdot \\ \cdot \\ A^{N+1} \end{bmatrix}, \quad B_N = \begin{bmatrix} B & 0 & \cdots & 0 \\ AB & B & & \vdots \\ \vdots & \vdots & \ddots & 0 \\ A^N B & A^{N-1}B & \cdots & B \end{bmatrix}$$

$$C_N = diag\{C,C,...,C\}, \quad D_N = \begin{bmatrix} D & 0 & \cdots & 0 \\ AD & D & & \vdots \\ \vdots & \vdots & \ddots & 0 \\ A^N D & A^{N-1}D & \cdots & D \end{bmatrix}$$

and the vectors of signals:

$$U_{t,N} = \begin{bmatrix} u(t) \\ u(t+1) \\ \vdots \\ u(t+N) \end{bmatrix}, \quad W_{t,N} = \begin{bmatrix} w(t) \\ w(t+1) \\ \vdots \\ w(t+N) \end{bmatrix}, \quad W_{t,N}^0 = \begin{bmatrix} \xi(t+1) \\ \xi(t+2) \\ \vdots \\ \xi(t+N+1) \end{bmatrix}$$

$$(8.7)$$

It is important to observe for later reference that the future calculated states depend upon future inputs and the state-vector, at time $t$. The state-vector at time $t$ does not depend upon the future values of the calculated states and it may also be noted that only the block for generating the actual state vector $x(t)$ has dynamics. That is, the system has $n$-states and only the first block includes dynamics and hence only the state vector $x(t)$ (or estimated state) will enter the optimal feedback solution.

## 8.2.1 Prediction Modelling

Noting the relationship between the stochastic signals in (8.4), the $k$-steps prediction of the output signal may be calculated from the relationship:

$$\hat{y}(t+k|t) = E\{y(t+k)|t\} = CA^k\hat{x}(t|t) + \sum_{j=1}^{k} CA^{k-j}Bu(t+j-1) \qquad (8.8)$$

Collect together the results for $\hat{y}(t+k|t)$, when $k$ changes from 1 to $N+1$, and obtain the block matrix form of the equations:

$$\hat{Y}_{t,N} = \begin{bmatrix} \hat{y}(t+1|t) \\ \hat{y}(t+2|t) \\ \vdots \\ \hat{y}(t+N+1|t) \end{bmatrix} = \underbrace{\begin{bmatrix} CA \\ CA^2 \\ \vdots \\ CA^{N+1} \end{bmatrix}}_{C_N A_N} \hat{x}(t|t)$$

$$+ \underbrace{\begin{bmatrix} CB & 0 & \cdots & 0 \\ CAB & CB & & 0 \\ \vdots & \vdots & \ddots & \\ CA^N B & CA^{N-1}D & \cdots & CB \end{bmatrix}}_{E_N} \underbrace{\begin{bmatrix} u(t) \\ u(t+1) \\ \vdots \\ u(t+N) \end{bmatrix}}_{U_{t,N}} \qquad (8.9)$$

This set of equations can clearly be written in the vector of outputs form:

$$\hat{Y}_{t,N} = F_{t,N} + C_N B_N U_{t,N} \qquad (8.10)$$

where the signal $F_{t,N}$ is defined as:

$$F_{t,N} = \begin{bmatrix} CA \\ CA^2 \\ \vdots \\ CA^{N+1} \end{bmatrix} \hat{x}(t|t) = C_N A_N \hat{x}(t|t)$$

where the vector of signals $\hat{Y}_{t,N}$ and the prediction error $\tilde{Y}_{t,N}$ are orthogonal.

## 8.2.2  GPC Performance Index and Optimal Solution

The GPC performance index to be minimised (Clarke et al. 1987, 1989) may be defined, in terms of the finite sum of future cost terms, as follows:

$$J = E\{J_t|t\} \tag{8.11}$$

where

$$J_t = \sum_{j=0}^{N}\{((y(t+j+1) - r(t+j+1))^T (y(t+j+1) - r(t+j+1))$$

$$+\lambda u(t+j)^T u(t+j)\}$$

and

$E\{.|t\}$ denotes the conditional expectation, conditioned on measurements up to time $t$.

$r(t+j+1)$ represents a vector of reference (set point) signals.

$\lambda$ is a control weighting factor $\geq 0$ .

The cost-function may be written in a more concise vector form by introducing the following setpoint and control signal vectors:

$$R_{t,N} = \begin{bmatrix} r(t+1) \\ r(t+2) \\ \vdots \\ r(t+N+1) \end{bmatrix} \quad \text{and} \quad U_{t,N} = \begin{bmatrix} u(t) \\ u(t+1) \\ \vdots \\ u(t+N) \end{bmatrix}$$

where $R_{t,N}$ is a block vector of $N+1$ future reference signals, $U_{t,N}$ is a block vector of $N+1$ future control signals. The terms in the performance index can then be expressed, noting the orthogonality of $\hat{Y}_{t,N}$ and the signals $\tilde{Y}_{t,N}$ and $R_{t,N}$, in the vector-matrix form:

$$J = E\{J_t|t\} = E\{(\hat{Y}_{t,N} - R_{t,N})^T(\hat{Y}_{t,N} - R_{t,N}) + \lambda U_{t,N}^T U_{t,N}|t\} + J_0 \tag{8.12}$$

where

$$J_0 = E\{\tilde{Y}_{t,N}^T \tilde{Y}_{t,N}|t\}$$

Substituting for the output signal vector : $\hat{Y}_{t,N}$ , from (8.10), obtain:

$$J = E\{(F_{t,N} + C_N B_N U_{t,N} - R_{t,N})^T(F_{t,N} + C_N B_N U_{t,N} - R_{t,N})$$

$$+\lambda U_{t,N}^T U_{t,N}|t\} + J_0$$

$$= E\{(F_{t,N} - R_{t,N})^T(F_{t,N} - R_{t,N}) + U_{t,N}^T B_N^T C_N^T (F_{t,N} - R_{t,N})$$

$$+ (F_{t,N} - R_{t,N})^T C_N B_N U_{t,N} + U_{t,N}^T (C_N^T B_N^T B_N C_N + \lambda I) U_{t,N} | t\} + J_0$$

The procedure for minimising this cost term, if the signals are deterministic, is almost identical to that when the conditional cost function term (8.12) is considered. That is, the gradient of the cost-function must be set to zero, to obtain the optimal control. Thus, from a rather obvious perturbation and gradient calculation (Grimble and Johnson, 1988), noting that the $J_0$ term is independent of the control signal, the vector of future GPC optimal control signals becomes:

$$U_{t,N} = (B_N^T C_N^T C_N B_N + \lambda I)^{-1} B_N^T C_N^T (R_{t,N} - F_{t,N}) \qquad (8.13)$$

The first element of this vector gives the control at time $t$ in the spirit of receding horizon optimal control. One of the main advantages of the GPC algorithm is the ease with which constraints may be introduced using quadratic optimisation. The properties of such a system were explored in Ordys and Grimble (1994).

# 8.3 State-space Form of LQGPC Controller

The GPC controller has some undesirable stability and robustness characteristics which makes it less suitable for high performance applications, where the plant characteristics may include non-minimum phase or open-loop unstable behaviour. There are of course a number of methods of improving the stability properties of the basic GPC algorithms, such as Kouvaritakis et al. (1992) and Rossiter et al. (1988). The Linear Quadratic Gaussian Predictive Controller (LQGPC) also provides improved properties in its basic form, for such systems. The LQGPC problem was introduced in Chapter 3.4, using a polynomial system description. The analysis in the following involves essentially the same cost index but the system model and assumptions are somewhat different. The results are not therefore directly equivalent to those in Chapter 3, although there are similarities.

## 8.3.1 LQGPC Dynamic Performance Index

If now an unconditional expectation operator is used the LQGPC type criterion, referred to as a *dynamic performance index* (Ordys and Grimble, 1996; Hangstrup et al. 1997; Grimble and Ordys, 2000), is obtained as:

$$J = E\left\{ \lim_{T \to \infty} \frac{1}{2T} \sum_{t=-T}^{T} J_t \right\} \qquad (8.14)$$

where $J_t$ was defined by (8.11). The use of the unconditional expectation in the criterion results in the GPC and LQGPC designs being different, even if the system and criteria have the same apparent form. This is similar to the situation which arises when considering the difference between Generalised Minimum Variance (GMV) and LQG solutions (MacGregor, 1997). The latter has much improved stability characteristics and this also applies to the difference between GPC and LQGPC solutions.

It may be noted that when the number of steps in the cost-index $N = 0$ the criterion reduces to the usual type of LQG cost-function. Letting $N > 0$ in (8.11) introduces future error and control terms, which relate to those in the GPC criterion, and these provide the novel additional features of the LQGPC cost-index.

If the vector of $N + 1$ future control signals (8.7) is considered as an input to the system and the vector of $N + 1$ future outputs (8.6) is treated as an output, then the state-space equations (8.1) can be rewritten in the form:

$$x(t + 1) = Ax(t) + \beta U_{t,N} + \Gamma W_{t,N} \tag{8.15}$$

$$Y_{t,N} = C_N(A_N x(t) + B_N U_{t,N} + D_N W_{t,N}) + W_{t,n}^0 \tag{8.16}$$

where :

$$\beta = [\ B\quad 0\quad \cdots \quad 0\ ] \quad \text{and} \quad \Gamma = [\ D\quad 0\quad \cdots \quad 0\ ]$$

Notice from equation (8.15) that the state equation is only influenced by the first (vector) element from the vector $U_{t,N}$ namely, $u(t)$. The remaining elements are needed only for the output equation and for the prediction equation. Thus, the remaining elements $(u(t + 1), u(t + 2), ..., u(t + N + 1))$ of the vector $U_{t,N}$ can be treated as predicted values of future controls, rather than the actual control signals applied to the system. Substituting for the output equation (8.6), using the performance index (8.14), obtain:

$$J = E\left\{ \lim_{T \to \infty} \frac{1}{2T} \sum_{t=-T}^{T} \left[ (Y_{t,N} - R_{t,N})^T (Y_{t,N} - R_{t,N}) + \lambda U_{t,N}^T U_{t,N} \right] \right\}$$

$$= E\left\{ \lim_{T \to \infty} \frac{1}{2T} \sum_{t=-T}^{T} \left[ \left( C_N(A_N x(t) + B_N U_{t,N} + D_N W_{t,N}) + W_{t,N}^0 - R_{t,N} \right)^T \right. \right.$$

$$\left. \left. \times \left( C_N(A_N x(t) + B_N U_{t,N} + D_N W_{t,N}) + W_{t,N}^0 - R_{t,N} \right) + \lambda U_{t,N}^T U_{t,N} \right] \right\} \tag{8.17}$$

The value of the control signal to be calculated at the time instant $t$, consists of the actual control $u(t)$ applied in the system, and the predicted values of the future controls $u(t + 1), ..., u(t + N)$ . Also note that there is an unavoidable approximation in the formulation of the above problem. That is, the future controls and outputs will be in error. This error occurs because the vector of future controls, calculated at time $t$, will not normally correspond, in this stochastic problem, with the actual control signal applied at the future times. This does not affect the plant to be controlled, that at time $t$ has input $u(t)$, and hence the stability of the closed-loop system is not affected by any error in the future predicted controls.

There are two consequences of this observation. The first is the expected result that the predicted values will not be the same as the true values but they will be optimal estimates given the assumed models. The second consequence is more subtle, since the control at time $t$ will try to minimise the errors at future times and will therefore be different to the usual control for a single-step $(N = 0)$ criterion, which attempts to minimize the traditional weighted error and control signal variances.

## 8.3.2 The Deterministic Problem

The criterion (8.14) represents the dynamic performance index for the stochastic system, however, the derivation of the output feedback stochastic controller will start from the simpler case of zero reference and disturbances (deterministic problem). The performance index may therefore be defined to have the deterministic form:

$$J = \lim_{T \to \infty} \sum_{t=0}^{T} [(C_N(A_N x(t) + B_N U_{t,N}))^T (C_N(A_N x(t) + B_N U_{t,N}))$$

$$+ \lambda U_{t,N}^T U_{t,N}] \tag{8.18}$$

and the state equation, with given initial condition, simplifies as:

$$x(t+1) = Ax(t) + \beta U_{t,N} \tag{8.19}$$

This is a state-feedback LQ problem which may be solved using Dynamic Programming or other optimisation methods. The Riccati equation solution may be obtained from, for example, Ogata (1987). The vector of current and predicted optimal control signals can then be expressed as:

$$U_{t,N} = -(\lambda I + B_N^T C_N^T C_N B_N + \beta^T P_\infty \beta)^{-1} (B_N^T C_N^T C_N A_N + \beta^T P_\infty A) x(t) \tag{8.20}$$

where $P_\infty$ is the *steady-state*, or *algebraic solution*, of the time-dependent Riccati equation (Anderson and Moore, 1971):

$$P_j = (A_N^T C_N^T C_N A_N + A^T P_{j+1} A)$$

$$- (A_N^T C_N^T C_N B_N + A^T P_{j+1} \beta)(\lambda I + B_N^T C_N^T C_N B_N + \beta^T P_{j+1} \beta)^{-1}$$

$$\times (B_N^T C_N^T C_N A_N + \beta^T P_{j+1} A) \tag{8.21}$$

**Reference Tracking**

Assume a deterministic problem with a non-zero known reference signal. The performance index (8.18) is modified to include the refernce signal as:

$$J = \lim_{T \to \infty} \sum_{t=0}^{T} \{(C_N(A_N x(t) + B_N U_{t,N}) - R_{t,N})^T (C_N(A_N x(t) + B_N U_{t,N}) - R_{t,N})$$

$$+ \lambda U_{t,N}^T U_{t,N}\} \tag{8.22}$$

As the prediction horizon is limited to $N$, only $N$ future values of the reference are assumed known at any particular time $t$. Assume that the future output reference values may in a special case, be identified with reference model states and be evaluated from the $N$ first reference signal values, as follows:

$$R_{t+1,N} = \Theta_N R_{t,N} \tag{8.23}$$

The matrix $\Theta_N$ may be selected so that the best approximation is given to the expected future behaviour of the reference signals. If for example the reference is modelled by a discrete integrator:

$$\Theta_N = \begin{bmatrix} 0 & I & \cdots & 0 \\ \vdots & \ddots & \ddots & \vdots \\ & & 0 & I \\ 0 & \cdots & 0 & I \end{bmatrix} \in R^{(N+1)\times(N+1)}$$

the value of the reference signal will remain constant after $N$ steps, or if:

$$\Theta_N = \begin{bmatrix} 0 & I & \cdots & 0 \\ \vdots & \ddots & \ddots & \vdots \\ & & 0 & I \\ 0 & \cdots & 0 & 0 \end{bmatrix}$$

the reference will be set to zero after $N$ steps.

The system equations in this deterministic problem may be listed, using (8.15) and (8.16), as:

$$x(t+1) = Ax(t) + \beta U_{t,N}$$

$$\hat{Y}_{t,N} = C_N A_N x(t) + C_N B_N U_{t,N}$$

and the predicted tracking error:

$$\hat{E}_{t,N} = R_{t,N} - \hat{Y}_{t,N}$$

The block diagram for this system is shown in Fig. 8.1.

Introduce the extended state vector, comprising the state of the system and the reference signal states. Then the extended state-space equation in the case of a simple reference model (reference signal ouputs = reference states), may be written as follows:

$$\chi_{t+1} = \begin{bmatrix} x(t+1) \\ R_{t+1,N} \end{bmatrix} = \underbrace{\begin{bmatrix} A & 0 \\ 0 & \Theta_N \end{bmatrix}}_{\Lambda} \chi_t + \underbrace{\begin{bmatrix} \beta \\ 0 \end{bmatrix}}_{\Psi} U_{t,N} = \Lambda \chi_t + \Psi U_{t,N} \qquad (8.24)$$

and the predicted output vector:

$$\hat{Y}_{t,N} = C_N A_N x(t) + B_N U_{t,N} = [\; C_N A_N \quad 0\;]\chi_t + C_N B_N U_{t,N}$$

and hence the tracking error equation gives:

$$-\hat{E}_{t,N} = \hat{Y}_{t,N} - R_{t,N} = \underbrace{[\; C_N A_N \quad -1\;]}_{L_N} \chi_t + C_N B_N U_{t,N}$$

$$= L_N \chi_t + C_N B_N U_{t,N} \qquad (8.25)$$

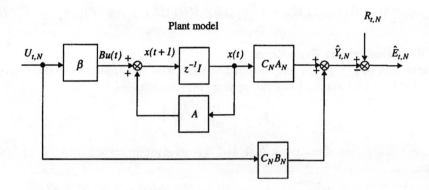

**Fig. 8.1** : State-space Model for the Predictive Output and Error Signals

The block diagram corresponding to the error equation (8.25) can be shown, as in Fig. 8.2. Substituting this equation into the performance index (8.22) the problem converts to a standard LQ optimal control problem, where the criterion:

$$J = \lim_{T \to \infty} \sum_{t=0}^{T} [(L_N \chi_T + C_N B_N U_{t,N})^T (L_N \chi_T + C_N B_N U_{t,N})$$

$$+ \lambda U_{t,N}^T U_{t,N}] \tag{8.26}$$

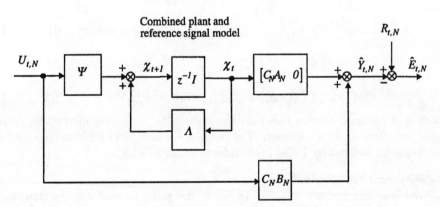

**Fig. 8.2**: Extended State Vector Model for the Predicted Output and Error Signals

The vector of current and predicted optimal control values is then given by:

$$U_{t,N} = -(\lambda I + B_N^T C_N^T C_N B_N + \Psi^T \tilde{P}_\infty \Psi)^{-1} (B_N^T C_N^T L_N + \Psi^T \tilde{P}_\infty \Lambda)\chi_t \tag{8.27}$$

where $\tilde{P}_\infty$ is the algebraic solution of the Ricatti equation:

$$\tilde{P}_j = L_N^T L_N + \Lambda^T \tilde{P}_{j+1} \Lambda - (L_N^T C_N B_N + \Lambda^T \tilde{P}_{j+1} \Psi)$$

$$\times (\lambda I + B_N^T C_N^T C_N B_N + \Psi^T \tilde{P}_{j+1} \Psi)^{-1} (B_N^T C_N^T L_N + \Psi^T \tilde{P}_{j+1} \Lambda) \qquad (8.28)$$

**Simplification of the Equations**

Equations (8.27), (8.28) may be further simplified. Assuming that the matrix $\tilde{P}_j$ is divided into four matrix blocks of appropriate dimensions:

$$\tilde{P}_j = \begin{bmatrix} \tilde{P}_j^1 & \tilde{P}_j^2 \\ \tilde{P}_j^{2T} & \tilde{P}_j^3 \end{bmatrix}$$

and using definitions of matrices, $\Lambda$, $\Psi$ and $L_N$, as given in equation (8.24) and (8.25), obtain:

$$B_N^T C_N^T L_N = [B_N^T C_N^T C_N A_N - B_N^T C_N^T], \quad \Psi^T \tilde{P}_j = [\ \beta^T \tilde{P}_j^1 \quad \beta^T \tilde{P}_j^2]$$

$$\Psi^T \tilde{P}_j \Psi = \beta^T \tilde{P}_j^1 \beta, \quad \Psi^T \tilde{P}_j \Lambda = [\ \beta^T \tilde{P}_j^1 A \quad \beta^T \tilde{P}_j^2 \Theta_N]$$

$$L_N^T L_N = \begin{bmatrix} A_N^T C_N^T C_N A_N & -A_N^T C_N^T \\ -C_N A_N & I \end{bmatrix}, \quad \Lambda^T \tilde{P}_j \Lambda = \begin{bmatrix} A^T \tilde{P}_j^1 A & A^T \tilde{P}_j^2 \Theta_N \\ \Theta_N^T \tilde{P}_j^{2T} A & \Theta_N^T \tilde{P}_j^3 \Theta_N \end{bmatrix}$$

The vector of current and predicted optimal controls then follows from (8.27) as:

$$U_{t,N} = -(\lambda I + B_N^T C_N^T C_N B_N + \beta^T \tilde{P}_\infty^1 \beta)^{-1} [(B_N^T C_N^T C_N A_N + \beta^T \tilde{P}_\infty^1 A) x(t)$$

$$+ (\beta^T \tilde{P}_\infty^2 \Theta_N - B_N^T C_N^T) R_{t,N}] \qquad (8.29)$$

The Riccati equation (8.28) may be split into two equations:

$$\tilde{P}_j^1 = (A_N^T C_N^T C_N A_N + A^T \tilde{P}_{j+1}^1 A) - (A_N^T C_N^T C_N B_N + A^T \tilde{P}_{j+1}^1 \beta)$$

$$\times (\lambda I + B_N^T C_N^T C_N B_N + \beta^T \tilde{P}_{j+1}^1 \beta)^{-1} (B_N^T C_N^T C_N A_N + \beta^T \tilde{P}_{j+1}^1 A) \qquad (8.30)$$

$$\tilde{P}_j^2 = -A_N^T C_N^T + A^T \tilde{P}_{j+1}^2 \Theta_N - (A_N^T C_N^T C_N B_N + A^T \tilde{P}_{j+1}^1 \beta)$$

$$\times (\lambda I + B_N^T C_N^T C_N B_N + \beta^T \tilde{P}_{j+1}^1 \beta)^{-1} (\beta^T \tilde{P}_{j+1}^2 \Theta_N - B_N^T C_N^T) \qquad (8.31)$$

where it is assumed that in this case the state $x(t)$ can be reconstructed, since the model and plant input are known. The use of a Kalman filter for state reconstruction was discussed in Chapter 7 and is considered again in §8.5.5.

**Stability and Robustness**

Note that the Riccati equation (8.30) is the same as that for the deterministic regulator problem (see (8.21)). If the reference is null the expressions for the optimal control signals (8.20) and (8.29) are the same. This confirms the expected result that the addition of the reference signal does not change the stability or robustness properties of the optimal control. In the GPC literature it is sometimes inferred that greater robustness is achieved through the use of future set point information but, as in this case, the robustness properties are actually influenced by the use of the multi-step criterion. It also follows that the reference or set-point models can be assumed to be null for the robustness analysis, since stability robustness depend only on the properties of the feedback loop.

### 8.3.3 Stochastic Disturbance Case

The full performance index, given in equation (8.14), that includes the reference signal, will be now considered. Using the model including the extended state, given in equation (8.24), the performance index, for the stochastic case, can be expressed as follows:

$$J = E\left\{ \lim_{T\to\infty} \frac{1}{2T} \sum_{t=-T}^{T} \left[ (Y_{t,N} - R_{t,N})^T (Y_{t,N} - R_{t,N}) + \lambda U_{t,N}^T U_{t,N} \right] \right\}$$

$$= E\left\{ \lim_{T\to\infty} \frac{1}{2T} \sum_{t=-T}^{T} \left[ (C_N A_N x(t) - R_{t,N} + C_N B_N U_{t,N} + C_N D_N W_{t,N} + W_{t,N}^0)^T \right. \right.$$

$$\times (C_N A_N x(t) - R_{t,N} + C_N B_N U_{t,N} + C_N D_N W_{t,N} + W_{t,N}^0) + \lambda (U_{t,N}^T U_{t,N})] \}$$

$$= E\{ \lim_{T\to\infty} \frac{1}{2T} \sum_{t=-T}^{T} \left[ (L_N \chi_t + C_N B_N U_{t,N} + C_N D_N W_{t,N} + W_{t,N}^0)^T \right.$$

$$\times (L_N \chi_t + C_N B_N U_{t,N} + C_N D_N W_{t,N} + W_{t,N}^0) + \lambda U_{t,N}^T U_{t,N}] \} \tag{8.32}$$

To obtain the optimal solution in this stochastic case the covariance matrices are required, that result from the multiplication of terms within the performance index. However, recalling the definitions (8.7), and the independence of the noise sources, it can be shown that:

$$E\left\{ W_{t,N}^0 W_{t,N} \right\} = 0 \,, \quad E\left\{ \chi_t^T W_{t,N} \right\} = 0 \,, \quad E\left\{ \chi_t^T W_{t,N}^0 \right\} = 0 \,,$$

The vector $U_{t,N}$ is calculated at time instant $t$ and includes predictions of future controls which are based on information available at time $t$. It follows that:

$$E\left\{ U_{t,N}^T W_{t,N}^0 \right\} = 0 \quad \text{and} \quad E\left\{ U_{t,N}^T W_{t,N} \right\} = 0$$

Thus, in equation (8.32), all cross-covariances associated with the disturbances $W_{t,N}^0$ and $W_{t,N}$ can be set to zero. The optimal control may therefore be obtained from the *Separation Principle*, by substituting for the state estimate, in place of the state, in equation (8.29):

$$U_{t,N} = -(\lambda I + B_N^T C_N^T C_N B_N + \beta^T \tilde{P}_\infty^1 \beta)^{-1}$$

$$\times [(B_N^T C_N^T C_N A_N + \beta^T \tilde{P}_\infty^1 A)\hat{x}(t|t) + (\beta^T \tilde{P}_\infty^2 \Theta_N - B_N^T C_N^T)R_{t,N}] \tag{8.33}$$

where $\hat{x}(t|t)$ can be obtained from a standard Kalman filtering state estimation equation (Chapter 7). There is no need to provide an estimator for the reference model states $R_{t,N}$ since they are known and are independent of other state models.

**Remarks**

As noted the LQGPC controller defined here is not the direct state-space equivalent to the polynomial results in Chapter 3. The design philosophy used here is slightly different, particularly with regard to the predicted signals.

The optimal control problem to be solved will compute the vector of controls $U_{t,N}$ for the output equation (8.6). Note that the predicted output signals for $t + 2, t + 3, .., t + N + 1$ are not related here to the predicted control signals only, but to the control signal at time t, as well.

In the present analysis the future controls are computed to minimise the criterion, but they do not affect the next future predicted controls (unlike the results in Chapter 3). It is the assumption regarding the modelling of the predicted signals which is different in the two cases. The state model employed here enables optimal predicted signal estimates to be obtained, based upon the actual control $u(t)$ and the future predicted controls. This is a self-correcting mechanism, since even if models are inaccurate the predicted output will be dependent upon $u(t)$ and a build up of modelling errors will not arise, since the actual input to the plant is utilised in the prediction. Moreover, at time $t$ the control is chosen, like in GPC designs, to minimise not only the errors at time $t$ but also the predicted errors. The GPC and LQGPC controllers are very suitable for supervisory control in power and other large systems, as described by Pike et al. (1995).

**Relationship to GPC Algorithms**

Consider the case of an open loop stable and minimum-phase system where the Riccati solution $\tilde{P}'_\infty = 0$ then (8.33) gives:

$$U_{t,N} = -(\lambda I + B_N^T C_N^T C_N B_N)^{-1} \left[ B_N^T C_N^T C_N A_N \hat{x}(t|t) - B_N^T C_N^T R_{t,N} \right]$$

and note that the above expression may be written as:

$$U_{t,N} = (\lambda I + B_N^T C_N^T C_N B_N)^{-1} B_N^T C_N^T [R_{t,N} - F_{t,N}]$$

This expression is clearly identical to (8.13).

The physical explanation for this result, that followed from the assumption $\tilde{P}'_\infty = 0$, is not quite so straightforward but a related justification can be obtained in terms of the finite interval criterion. If say a single-step criterion is considered:

$$J_1 = E \left\{ \sum_{t=0}^{1} \left[ (Y_{t,N} - R_{t,N})^T (Y_{t,N} - R_{t,N}) + \lambda U_{t,N}^T U_{t,N} \right] \right\}$$

and if the end state weighting is null, then from (8.28):

$$\tilde{P}_0 = L_N^T L_N - (L_N^T C_N B_N)(\lambda I + B_N^T C_N^T C_N B_N)^{-1}(B_N^T C_N^T L_N)$$

Now consider the case where $C_N B_N$ is square and full rank and is large compared with $\lambda I$. Clearly $\tilde{P}_0$ will tend to the null matrix and the above expression for the future controls $U_{t,N}$ is again obtained.

## 8.4 State-space GPC with Through Terms

The linear, time-invariant, discrete-time, finite-dimensional, $r \times m$ multivariable system of interest in this section, includes through terms, as illustrated in Fig. 8.3 (Ordys and Pike, 1998). These represent very fast dynamics which are sometimes approximated by non-dynamic through terms between system inputs (control or disturbance) and outputs. The subsystem $S_1$ denotes the plant model and $S_2$ represents the coloured measurement noise model. Note that the subsystem $S_2$ model output may also represent an uncontrolled disturbance. This is a disturbance that the controller should not attempt to reject, like the wave motion forces in a ship positioning system (Chapter 12). The subsystem $S_3$ represents an unobservable subsystem which can also be used as a dynamic control weighting model. The reference signal is generated by the subsystem $S_0$.

The plant model which is represented by subsystem $S_1$ involves a number of generalisations to cover the types of industrial process of interest. If the predictive control system is to be used for a supervisory, or upper level, control system then it will be operated at a lower sample rate than the regulating loops. It will therefore be slow relative to the behaviour of the regulating loops. These fast dynamics can be approximated by through terms in the state equation model for the supervisory system.

The driving zero-mean, white noise signals $\{v(t)\}$ and $\{\xi_i(t)\}$, where $\xi_i(t)\epsilon R^{q_i}$, are assumed to have the following covariance matrices :

$$cov[v(t), v(\tau)] = R_f \delta_{t\tau} > 0 \quad \text{and} \quad cov[\xi_i(t), \xi_i(\tau)] = I_{q_i}\delta_{t\tau}$$

where $i = \{0, ..., 3\}$, and the cross-covariance matrices are assumed to be null.

The state equation model for the plant is therefore assumed to be of the following form:

**State :**

$$x_1(t+1) = A_1 x_1(t) + B_1 u(t) + D_1 \xi_1(t) + P_1 \xi_1^0(t)$$

**Output :**

$$y(t) = C_1 x_1(t) + E_1 u(t) + F_1 \xi_1^0(t)$$

**Observations :**

$$z(t) = y(t) + v(t) + n(t)$$

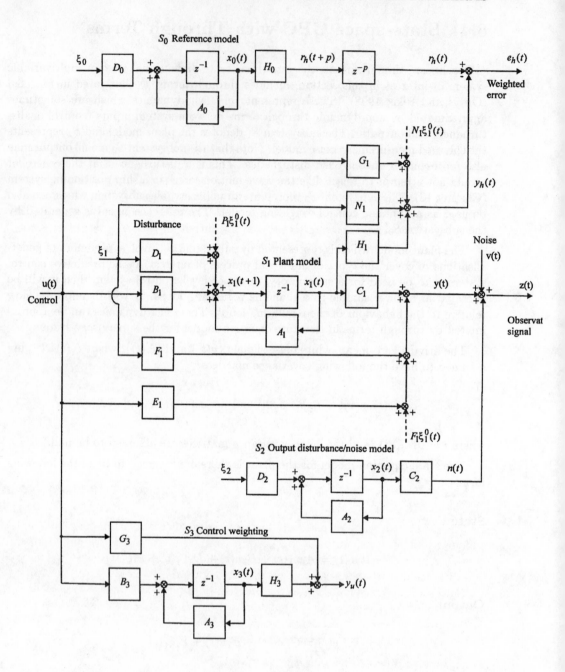

**Fig. 8.3:** State Space System Model with Through Terms Disturbance and Reference Models

**Inferred output :**

$$y_h(t) = H_1 x_1(t) + G_1 u(t) + N_1 \xi_1^0(t)$$

and the state $x_1(t) \ \varepsilon \ R^{n_1}$, output $y(t) \ \varepsilon \ R^r$ , control $u(t) \ \varepsilon \ R^m$, inferred output $y_h(t)$ $\varepsilon \ R^{n_h}$, disturbances $\xi_1(t) \ \varepsilon \ R^q$, $\xi_1^0(t) \ \varepsilon \ R^{q_0}$, and noise $v(t) \ \varepsilon \ R^r$ .

**Effect of the Through Terms**

The above equations that involve $\xi_1^0(t)$ entering the output and inferred output equations complicate the predictor (Anderson and Moore, 1979) and the separation principle results. However, a simple modification to the problem avoids such difficulties. This is explained below, where it is shown that $P_1, F_1$ and $N_1$ can be set to zero, if the state model is redefined. The plant equations can be modified by removing the white noise disturbance through terms from the output and inferred output equations. Assume that the disturbance driving noise $\xi_1^0(t) = \xi(t-1)$ where $\xi(t)$ is white noise of covariance matrix $I$. Clearly $\xi_1^0(t)$ is also white noise with the same covariance matrix. Let a state $x_\xi(t)$ be defined as: $x_\xi(t) = \xi(t-1)$ then the plant equation above may be written as:

$$\begin{bmatrix} x_1(t+1) \\ x_\xi(t+1) \end{bmatrix} = \begin{bmatrix} A_1 & P_1 \\ 0 & 0 \end{bmatrix} \begin{bmatrix} x_1(t) \\ x_\xi(t) \end{bmatrix} + \begin{bmatrix} B_1 \\ 0 \end{bmatrix} u(t) + \begin{bmatrix} D_1 \xi_1(t) \\ \xi(t) \end{bmatrix}$$

$$y(t) = \begin{bmatrix} C_1 & F_1 \end{bmatrix} \begin{bmatrix} x_1(t) \\ x_\xi(t) \end{bmatrix} + E_1 u(t)$$

$$z(t) = y(t) + v(t) + n(t)$$

$$y_h(t) = \begin{bmatrix} H_1 & N_1 \end{bmatrix} \begin{bmatrix} x_1(t) \\ x_\xi(t) \end{bmatrix} + G_1 u(t)$$

Clearly the same white noise driving term adds to the input of the plant model and contributes to each output equation. These equations are therefore able to model the same physical situation and have the same form as the original state equations but with the disturbance term $P_1$, and the through terms $N_1$ and $F_1$, set to zero.

Note that this reformulation of the problem does not add a significant number of states, since they depend upon the number of elements in the disturbance vector $\xi_1^0(t)$. This number is usually small, since it only represents the components that contribute to the output equations via the through terms.

**Summary of System Model**

It follows that the system model for analysis may be summarised as follows:

$$x_1(t+1) = A_1 x_1(t) + B_1 u(t) + D_1 \xi_1(t) \tag{8.34}$$

$$y(t) = C_1 x_1(t) + E_1 u(t) \tag{8.35}$$

$$z(t) = y(t) + v(t) + n(t) \tag{8.36}$$

$$y_h(t) = H_1 x_1(t) + G_1 u(t) \tag{8.37}$$

**Inferred Future Plant Outputs**

For later use a model is required to predict the future values of the inferred output signal. From equation (8.37) for the inferred output obtain:

$$y_h(t+1) = H_1(A_1 x_1(t) + B_1 u(t) + D_1 \xi_1(t)) + G_1 u(t+1)$$

$$= H_1 A_1 x_1(t) + H_1 B_1 u(t) + G_1 u(t+1) + H_1 D_1 \xi_1(t)$$

$$y_h(t+2) = H_1 A_1^2 x_1(t) + H_1 A_1 B_1 u(t) + H_1 B_1 u(t+1) + G_1 u(t+2)$$

$$+ H_1 A_1 D_1 \xi_1(t) + H_1 D_1 \xi_1(t+1)$$

$$y_h(t+3) = H_1 A_1^3 x_1(t) + H_1 A_1^2 B_1 u(t) + H_1 A_1 B_1 u(t+1) + H_1 B_1 u(t+2)$$

$$+ G_1 u(t+3) + H_1 A_1^2 D_1 \xi_1(t) + H_1 A_1 D_1 \xi_1(t+1) + H_1 D_1 \xi_1(t+2)$$

The above results may be collected and generalised as:

$$y_h(t+p) = H_1 A_1^p x_1(t) + \sum_{j=1}^{p} H_1 A_1^{p-j}(B_1 u(t+j-1) + D_1 \xi_1(t+j-1))$$

$$+ G_1 u(t+p) \tag{8.38}$$

These output signals may be collected in an $N$ block output vector form:

$$
\begin{bmatrix} y_h(t+1) \\ y_h(t+2) \\ \vdots \\ y_h(t+N) \end{bmatrix}
=
\begin{bmatrix} H_1 A_1 \\ H_1 A_1^2 \\ \vdots \\ H_1 A_1^N \end{bmatrix} x_1(t)
$$

$$
+
\begin{bmatrix}
H_1 B_1 & G_1 & \cdots & 0 & 0 \\
H_1 A_1 B_1 & H_1 B_1 & \ddots & \vdots & \vdots \\
\vdots & \vdots & \ddots & G_1 & 0 \\
H_1 A_1^{N-1} B_1 & H_1 A_1^{N-2} B_1 & \cdots & H_1 B_1 & G_1
\end{bmatrix}
\begin{bmatrix} u(t) \\ u(t+1) \\ \vdots \\ u(t+N) \end{bmatrix}
$$

$$+ \begin{bmatrix} H_1 D_1 & 0 & \cdots & 0 \\ H_1 A_1 D_1 & H_1 D_1 & \ddots & \vdots \\ \vdots & \vdots & \ddots & 0 \\ H_1 A_1^{N-1} D_1 & H_1 A_1^{N-2} D_1 & \cdots & H_1 D_1 \end{bmatrix} \begin{bmatrix} \xi_1(t) \\ \xi_1(t+1) \\ \vdots \\ \xi_1(t+N-1) \end{bmatrix} \qquad (8.39)$$

The model for the predicted outputs, as represented by this equation, is as shown in Fig. 8.4. With an obvious definition of terms this equation for the inferred outputs and the observations equation, may be written in the more concise form:[2]

$$Y_{t+1,N}^h = H_N A_N x_1(t) + G_N U_{t,N} + N_N W_{t,N} \qquad (8.40)$$

where $H_N = diag\{ H_1 , \cdots , H_1 \}$,

$$A_N = \begin{bmatrix} A_1 \\ A_1^2 \\ \vdots \\ A_1^N \end{bmatrix}, \quad G_N = \begin{bmatrix} H_1 B_1 & G_1 & 0 & \cdots & 0 & 0 \\ H_1 A_1 B_1 & H_1 B_1 & G_1 & \ddots & \vdots & \vdots \\ \vdots & \vdots & \ddots & \ddots & & 0 \\ H_1 A_1^{N-1} B_1 & H_1 A_1^{N-2} B_1 & \cdots & & H_1 B_1 & G_1 \end{bmatrix}$$

$$N_N = \begin{bmatrix} H_1 D_1 & 0 & 0 & \cdots & 0 \\ H_1 A_1 D_1 & H_1 D_1 & 0 & \ddots & \vdots \\ \vdots & \vdots & \ddots & \ddots & 0 \\ H_1 A_1^{N-1} D_1 & H_1 A_1^{N-2} D_1 & \cdots & & H_1 D_1 \end{bmatrix}$$

and the vector of control signals, disturbances and reference signals:

$$U_{t,N} = \begin{bmatrix} u(t) \\ u(t+1) \\ \vdots \\ u(t+N) \end{bmatrix}, \quad W_{t,N} = \begin{bmatrix} \xi_1(t) \\ \xi_1(t+1) \\ \vdots \\ \xi_1(t+N-1) \end{bmatrix}, \quad R_{t,N}^h = \begin{bmatrix} r_h(t) \\ r_h(t+1) \\ \vdots \\ r_h(t+N-1) \end{bmatrix}$$

$$(8.41)$$

---

[2] Note there is a slight change in notation in the vector definitions relative to the previous sections

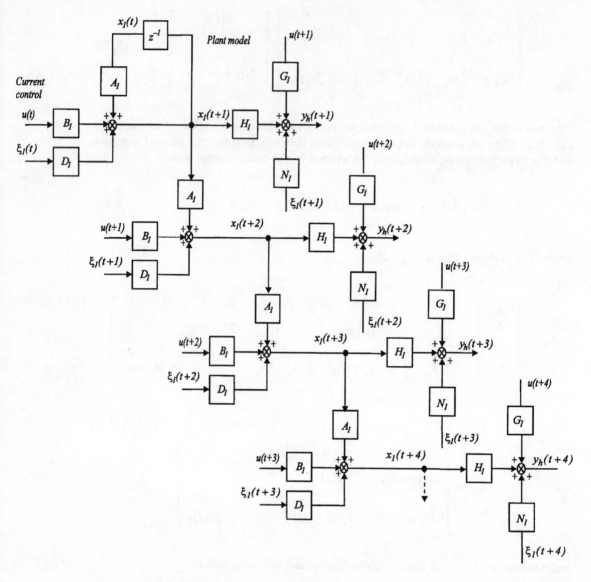

**Fig. 8.4** : Future Modelled Inferred Plant Outputs for a System
with Through Terms

**Prediction Model**

The $k$-steps prediction of the output signal may therefore be calculated (Grimble and Johnson, 1988) from the relationship:

$$\hat{y}_h(t+p|t) = E\{y_h(t+p)|t\}$$

$$= H_1 A_1^p x_1(t|t) + \sum_{j=1}^{p} H_1 A_1^{p-j} B_1 u(t+j-1) + G_1 u(t+p)$$

Collecting together the results $\hat{y}_h(t+p|t)$ for $p$ changing from 1 to $N$, obtain the block matrix form of the equations:

$$\hat{Y}^h_{t+1,N} = \begin{bmatrix} \hat{y}_h(t+1|t) \\ \hat{y}_h(t+2|t) \\ \vdots \\ \hat{y}_h(t+N|t) \end{bmatrix} = \underbrace{\begin{bmatrix} H_1 A_1 \\ H_1 A_1^2 \\ \vdots \\ H_1 A_1^N \end{bmatrix}}_{H_N A_N} \hat{x}_1(t|t)$$

$$+ \underbrace{\begin{bmatrix} H_1 B_1 & G_1 & 0 & \cdots & 0 \\ H_1 A_1 B_1 & H_1 B_1 & G_1 & \vdots & \vdots \\ \vdots & \vdots & \ddots & G_1 & 0 \\ H_1 A_1^{N-1} B_1 & H_1 A_1^{N-2} B_1 & \cdots & H_1 B_1 & G_1 \end{bmatrix}}_{G_N} \underbrace{\begin{bmatrix} u(t) \\ u(t+1) \\ \vdots \\ u(t+N) \end{bmatrix}}_{U_{t,N}} \qquad (8.42)$$

This equation can clearly be written in the more concise form:

$$\hat{Y}^h_{t+1,N} = H_N A_N \hat{x}_1(t|t) + G_N U_{t,N} \qquad (8.43)$$

**Prediction Error**

Let the vector for the inferred output estimation error be denoted by $\tilde{Y}^h_{t,N} = Y^h_{t,N} - \hat{Y}^h_{t,N}$ then,

$$Y^h_{t,N} = \hat{Y}^h_{t,N} + \tilde{Y}^h_{t,N}$$

and the vector of tracking error signals:

$$R^h_{t,N} - Y^h_{t,N} = (R^h_{t,N} - \hat{Y}^h_{t,N}) - \tilde{Y}^h_{t,N}$$

For later use note that $\tilde{Y}^h_{t,N}$ is orthogonal to $\hat{Y}^h_{t,N}$ and the reference signals $R^h_{t,N}$. Substituting from the prediction equation (8.43), the tracking error, becomes:

$$R^h_{t+1,N} - Y^h_{t+1,N} = (R^h_{t+1,N} - H_N A_N \hat{x}_1(t|t) - G_N U_{t,N}) - \tilde{Y}^h_{t+1,N} \qquad (8.44)$$

## 8.4.1 GPC Performance Index and Optimal Solution

The GPC criterion will be chosen to be similar to that previously discussed in equations (8.11) and (8.12), where the criterion $J = E\{J_t|t\}$ and

$$J_t = (R^h_{t+1,N} - Y^h_{t+1,N})^T (R^h_{t+1,N} - Y^h_{t+1,N}) + \lambda U^T_{t,N} U_{t,N}$$

Substituting from (8.44) obtain:

$$J_t = (R^h_{t+1,N} - H_N A_N \hat{x}_1(t|t) - G_N U_{t,N})^T (R^h_{t+1,N} - H_N A_N \hat{x}_1(t|t) - G_N U_{t,N})$$

$$+\lambda U_{t,N}^T U_{t,N} + J_0$$

$$= (R_{t+1,N}^h - H_N A_N \hat{x}_1(t|t))^T (R_{t+1,N}^h - H_N A_N \hat{x}_1(t|t))$$

$$-U_{t,N}^T G_N^T (R_{t+1,N}^h - H_N A_N \hat{x}_1(t|t)) - (R_{t+1,N}^h - H_N A_N \hat{x}_1(t|t))^T G_N U_{t,N}$$

$$+U_{t,N}^T (G_N^T G_N + \lambda I) U_{t,N} + J_0 \tag{8.45}$$

The GPC optimal control law, for the through term case, may then be obtained as in the previous problem:

$$U_{t,N} = (G_N^T G_N + \lambda I)^{-1} G_N^T (R_{t+1,N}^h - H_N A_N \hat{x}_1(t|t)) \tag{8.46}$$

This result may be compared with the GPC solution in equation (8.13). The results are essentially the same, noting that this result is for the through term case and there is a slight notational change in the vector definitions.

## 8.5   State-space LQGPC with Through Terms

The LQGPC problem will now be solved for the more general system description that includes through terms. However, first it will be useful to introduce a more sophisticated reference model. The reference signal $\{r_h(t)\}$ will be assumed to be generated by the asymptotically stable linear stochastic state equation system model:

$$x_{r0}(t+1) = A_r x_{r0}(t) + D_r \xi_0(t)$$

$$z_0(t) = C_r x_{r0}(t) + v_0(t)$$

The zero-mean white-noise sources $\{\xi_0(t)\}$ and $\{v_0(t)\}$ are assumed to be statistically independent. The noise free output $\{y_0(t)\}$ represents the desired future values of the reference signal $p$ steps-ahead. That is,

$$r_h(t+p) = y_0(t) = C_r x_{r0}(t)$$

where $p \geq N \geq 1$ is greater than or equal to the number of steps in the criterion. This is, of course, assuming that the reference might be available for the same, or more steps into the future, than is used in GPC type algorithms.

New state equation variables may be defined, when $p > 1$, as delayed versions of the reference signal:

$$x_{r1}(t) = r_h(t+p-1) = C_r x_{r0}(t-1)$$

$$x_{r2}(t) = r_h(t+p-2) = x_{r1}(t-1)$$

$$\vdots \qquad\qquad \vdots \qquad\qquad \vdots$$

$$x_{r(p-1)}(t) = r_h(t+1) = x_{r(p-2)}(t-1)$$

The evolution of the reference signal is illustrated in Fig. 8.5. The model represents the reference at time $t+1$ and at all future times up to time $t+p$. These future reference values can be collected in a vector and a state-equation obtained as:

$$
\begin{bmatrix}
x_{r0}(t+1) \\
x_{r1}(t+1) \\
x_{r2}(t+1) \\
\vdots \\
x_{r(p-1)}(t+1)
\end{bmatrix}
=
\begin{bmatrix}
A_r & 0 & \cdots & & 0 \\
C_r & 0 & \cdots & & \\
0 & I & & & \\
\vdots & & \ddots & & 0 \\
0 & 0 & \cdots & I & 0
\end{bmatrix}
\begin{bmatrix}
x_{r0}(t) \\
x_{r1}(t) \\
x_{r2}(t) \\
\vdots \\
x_{r(p-1)}(t)
\end{bmatrix}
+
\begin{bmatrix}
D_r \\
0 \\
\vdots \\
0
\end{bmatrix}
\xi_0(t)
$$

(8.47)

which may be written in the vector form:

$$x_0(t+1) = A_0 x_0(t) + D_0 \xi_0(t) \tag{8.48}$$

where the state-vector for the reference signal is denoted as:

$$x_0(t) = \begin{bmatrix} x_{r0}^T(t) & x_{r1}^T(t) & \cdots & x_{r(p-1)}^T(t) \end{bmatrix}^T$$

The next and future reference values can then be obtained as:

$$r_h(t+1) = \begin{bmatrix} 0 & 0 & \cdots & 0 & I \end{bmatrix} x_0(t) \quad \text{(Next reference)}$$

$$
\left.
\begin{aligned}
r_h(t+2) &= \begin{bmatrix} 0 & 0 & \cdots & I & 0 \end{bmatrix} x_0(t) \\
&\ \vdots \\
r_h(t+p-1) &= \begin{bmatrix} 0 & I & 0 & \cdots & 0 \end{bmatrix} x_0(t) \\
r_h(t+p) = H_0 x_0(t) &= \begin{bmatrix} C_r & 0 & \cdots & 0 \end{bmatrix} x_0(t)
\end{aligned}
\right\}
\quad \text{(Future reference values)}
$$

These equations can be collected and written as:

$$
\begin{bmatrix}
r_h(t+1) \\
r_h(t+2) \\
\vdots \\
r_h(t+p-1) \\
r_h(t+p)
\end{bmatrix}
=
\begin{bmatrix}
0 & 0 & \cdots & 0 & I \\
0 & 0 & \cdots & I & 0 \\
\vdots & \vdots & \ddots & & \\
0 & I & 0 & \cdots & 0 & 0 \\
C_r & 0 & 0 & \cdots & & 0
\end{bmatrix}
\begin{bmatrix}
x_{r0}(t) \\
x_{r1}(t) \\
x_{r2}(t) \\
\vdots \\
x_{r(p-1)}(t)
\end{bmatrix}
$$

which can be written in the concise form:

$$R_p(t+1) = C_p x_0(t) \tag{8.49}$$

The $N$ future setpoint or reference values in the cost-function can be denoted as:

$$
R_{t+1,N}^h = C_{RN} R_p(t+1) =
\begin{bmatrix}
I & 0 & & 0 & & 0 \\
0 & I & & & & \vdots \\
\vdots & & \ddots & & | & \\
0 & \cdots & & I & 0 & \cdots & 0
\end{bmatrix}
R_p(t+1)
$$

$$\underbrace{\hphantom{I \quad 0 \quad \quad I}}_{N} \quad \underbrace{\hphantom{0 \quad \cdots \quad 0}}_{p-N+1}$$

$$= C_{RN}C_p x_0(t) \qquad (8.50)$$

In the Generalised Predictive Control problems $C_{RN}$ will be a square matrix, since there will only be as many future reference values as steps in the criterion ($N$ steps where $N \geq 1$). In the analysis here the future set point knowledge can be of number $p > N$ and this can be an advantage. For example if $N$ is chosen as unity the multi-step criterion reduces to a single step and yet future reference information may still be utilised in this LQGPC approach.

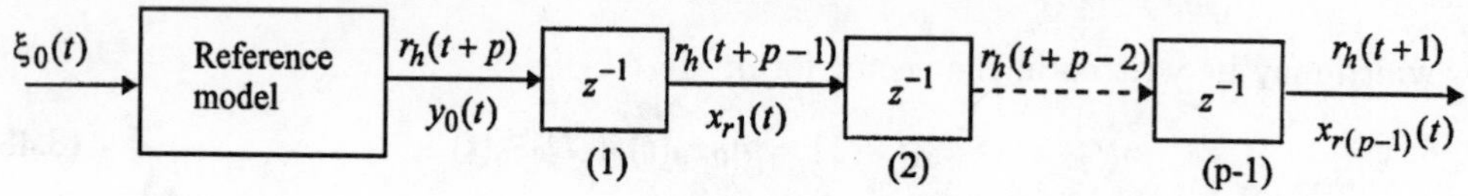

**Fig. 8.5** : Reference Signal Generating Mechanism for the Future and Current Inferred Outputs

## 8.5.1 Augmented System Model Including Through Terms

The equations for the total system will now be obtained and these will determine the size of the Riccati equations in the control and estimation problems. Combining the state equations for the reference, plant, noise and control weighting obtain the total augmented system as:

$$
\begin{bmatrix} x_0(t+1) \\ x_1(t+1) \\ x_2(t+1) \\ x_3(t+1) \end{bmatrix} = \begin{bmatrix} A_0 & 0 & 0 & 0 \\ 0 & A_1 & 0 & 0 \\ 0 & 0 & A_2 & 0 \\ 0 & 0 & 0 & A_3 \end{bmatrix} \begin{bmatrix} x_0(t) \\ x_1(t) \\ x_2(t) \\ x_3(t) \end{bmatrix} + \begin{bmatrix} 0 & 0 \\ B_1 & 0 \\ 0 & 0 \\ B_3 & 0 \end{bmatrix} U_{t,N}
$$

$$
+ \begin{bmatrix} D_0\xi_0(t) \\ D_1\xi_1(t) \\ D_2\xi_2(t) \\ 0 \end{bmatrix} \qquad (8.51)
$$

which may be written in the more concise form:

$$X(t+1) = AX(t) + BU_{t,N} + D\xi(t) \qquad (8.52)$$

where the following block matrices and partitioned vectors may be defined:

$$A = diag\{A_0, A_1, A_2, A_3\}, \qquad D = diag\{D_0, D_1, D_2, 0\}$$

$$
B = \begin{bmatrix} 0 & 0 \\ B_1 & 0 \\ 0 & 0 \\ B_3 & 0 \end{bmatrix}, \quad U_{t,N} = \begin{bmatrix} u(t) \\ \hline U'_{t,N} \end{bmatrix} = \begin{bmatrix} u(t) \\ \hline u(t+1) \\ \vdots \\ u(t+N) \end{bmatrix}, \quad \xi(t) = \begin{bmatrix} \xi_0(t) \\ \xi_1(t) \\ \xi_2(t) \\ 0 \end{bmatrix}
$$

The first component of the observations input to the filter may now be written in the form:

$$z(t) = CX(t) + v(t) = \begin{bmatrix} 0 & C_1 & C_2 & 0 \end{bmatrix} X(t) + v(t) \qquad (8.53)$$

where the output map $C = \begin{bmatrix} 0 & C_1 & C_2 & 0 \end{bmatrix}$. The second component of the observations signal, due to the future reference signal, becomes:

$$z_0(t) = HX(t) + v_0(t) = \begin{bmatrix} H_0 & 0 & 0 & 0 \end{bmatrix} X(t) + v_0(t)$$

For some problems the white noise signal $\{v_0(t)\}$ may be added to this measurement of the reference signal. However, when a simple model for the reference signal is used and reference states equal outputs, then $C_r = I$ and $x_{r0}(t) = r_h(t + p)$ and there is no need to estimate the vector:

$$x_0(t) = [r_h^T(t + p), \quad r_h^T(t + p - 1), ..., \quad r_h^T(t + 1)]^T$$

The corresponding element of the optimal control signal vector will then be found to involve a control gain fed by these stored values of the future reference signal components.

## 8.5.2   Current and Future Tracking Error Terms

The equation for the future outputs to be costed in the criterion may be written in terms of the state vector $X(t)$, using (8.40), as:

$$Y_{t+1,N}^h = \begin{bmatrix} 0 & H_N A_N & 0 & 0 \end{bmatrix} X(t) + G_N U_{t,N} + N_N W_{t,N} \qquad (8.54)$$

and from the reference vector equation (8.50):

$$R_{t+1,N}^h = \begin{bmatrix} C_{RN} C_p & 0 & 0 & 0 \end{bmatrix} X(t)$$

The error signal may now be written, using these two equations, as:

$$E_{t+1,N} = R_{t+1,N}^h - Y_{t+1,N}^h$$

$$= \begin{bmatrix} C_{RN} C_p & -H_N A_N & 0 & 0 \end{bmatrix} X(t) - G_N U_{t,N} - N_N W_{t,N} \qquad (8.55)$$

**Remarks**

- A dynamic weighting on the future error terms could be introduced in the cost-function by simply redefining the reference $S_0$ and plant $S_1$ models.

- The subsystem $S_2$ does not really enter the control problem calculation, since the gain from these states will be null (Grimble and Johnson, 1988). Thus, calculations can be simplified by reordering the state equations and omitting these subsystem elements in the calculation.

**Subsystem $S_3$**

The subsystem $S_3$ can be used for control signal costing and from Fig. 8.3 the subsystem equations become:

$$x_3(t+1) = A_3 x_3(t) + B_3 u(t)$$

$$y_u(t) = H_3 x_3(t) + G_3 u(t) \tag{8.56}$$

The output of the subsystem $S_3$ can be included in the cost-function. Combining the error and $y_u(t)$ signal terms, equations (8.55) and (8.56) obtain:

$$\tilde{E}_{t+1,N} = \tilde{H} X(t) + \tilde{G} U_{t,N} + \tilde{W}_{t,N} \tag{8.57}$$

where

$$\tilde{E}_{t+1,N} = \begin{bmatrix} E_{t+1,N} \\ -y_u(t) \end{bmatrix} = \begin{bmatrix} R^h_{t+1,N} - Y^h_{t+1,N} \\ 0 - y_u(t) \end{bmatrix} \tag{8.58}$$

and $\tilde{H}, \tilde{G}$ and $\tilde{W}_{t,N}$ are defined from an obvious correspondence of terms with the augmented equation:

$$\tilde{E}_{t+1,N} = \underbrace{\begin{bmatrix} C_{RN} C_P & -H_N A_N & 0 & 0 \\ 0 & 0 & 0 & -H_3 \end{bmatrix}}_{\tilde{H}} X(t)$$

$$+ \underbrace{\begin{bmatrix} -G_N U_{t,N} \\ -G_3 u(t) \end{bmatrix}}_{\tilde{G} U_{t,N}} + \underbrace{\begin{bmatrix} -N_N W_{t,N} \\ 0 \end{bmatrix}}_{\widetilde{W}_{t,N}} \tag{8.59}$$

For later use note that the error term can be separated into prediction and prediction error terms where,

$$\tilde{E}_{t+1,N} = \begin{bmatrix} R^h_{t+1,N} - Y^h_{t+1,N} \\ -y_u(t) \end{bmatrix} - \begin{bmatrix} \tilde{Y}^h_{t+1,N} \\ 0 \end{bmatrix} = \tilde{H} \hat{X}(t|t) + \tilde{G} U_{t,N} - \begin{bmatrix} \tilde{Y}^h_{t+1,N} \\ 0 \end{bmatrix} \tag{8.60}$$

The augmented system model is shown in Fig. 8.6.

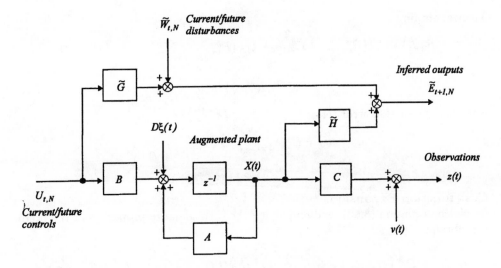

**Fig 8.6** : Augmented Plant Model Including Direct Feed-through Term

### 8.5.3 LQGPC Dynamic Performance Criterion

The dynamic performance index to be minimised in the following is defined like (8.14), to be time-averaged, as:

$$J = E\left\{\lim_{T \to \infty} \frac{1}{2T} \sum_{t=-T}^{T} J_t\right\} \tag{8.61}$$

where $J_t$ is defined, similar to the previous results, as:

$$J_t = \sum_{j=1}^{N}(r_h(t+j) - y_h(t+j))^T Q_j(r_h(t+j) - y_h(t+j)) + y_u^T(t)y_u(t)$$

$$+ \sum_{j=0}^{N} u^T(t+j)R_j u(t+j) \tag{8.62}$$

where $E\{.\}$ denotes the unconditional expectation operator, and the error $Q_j$ and control $R_j$ weightings can be time dependent. Observe that there is one more control signal weighting term than error term. This arises because of the presence of the plant through terms which implies the weighted output depends at time $t$ on both $u(t)$ and $u(t-1)$.

The cost-function can be simplified by introducing the block diagonal weighting matrices $\tilde{Q}_n = diag\{Q_1, ..., Q_N\}$ and $\tilde{R} = diag\{R_0, ..., R_N\}$ and by recalling the orthogonality of the prediction, reference and prediction error terms. The $J_t$ term can therefore be written in the vector/matrix form:

$$J_t = (R_{t+1,N}^h - Y_{t+1,N}^h)^T \tilde{Q}_N(R_{t+1,N}^h - Y_{t+1,N}^h) + y_u^T(t)y_u(t) + U_{t,N}^T \tilde{R}U_{t,N}$$

Thence, obtain :

$$E\{J_t\} = E\{(R_{t+1,N}^h - \hat{Y}_{t+1,N}^h)^T \tilde{Q}_N (R_{t+1,N}^h - \hat{Y}_{t+1,N}^h) + y_u^T(t) y_u(t)$$

$$+ U_{t,N}^T \tilde{R} U_{t,N}\} + J_0 \qquad (8.63)$$

where the prediction error term:

$$J_0 = E\{(\tilde{Y}_{t+1,N}^h)^T \tilde{Q}_N \tilde{Y}_{t+1,N}^h\}$$

## Cost-function Expansion

Noting equation (8.63), and employing the orthogonality properties of the predictor, obtain:

$$E\{J_t\} = E\left\{ \left( \tilde{H}\hat{X}(t|t) + \tilde{G}U_{t,N} \right)^T \tilde{Q} \left( \tilde{H}\hat{X}(t|t) + \tilde{G}U_{t,N} \right) + U_{t,N}^T \tilde{R} U_{t,N} \right\} + J_0$$

$$= E\{\hat{X}(t|t)^T \tilde{H}^T \tilde{Q} \hat{X}(t|t) + U_{t,N}^T \tilde{G}^T \tilde{Q} \tilde{H} \hat{X}(t|t) + \hat{X}(t|t)^T \tilde{H}^T \tilde{Q} \tilde{G} U_{t,N}$$

$$+ U_{t,N}^T (\tilde{G}^T \tilde{Q} \tilde{G} + \tilde{R}) U_{t,N}\} + J_0 \qquad (8.64)$$

where $\tilde{Q} = diag\{Q_1, ..., Q_N, I\}$ .

Note that the estimator states are present in the cost-function. Following the same argument as in the traditional proof of the separation principle, it is noted that the estimator states $\hat{X}(t|t)$ (based on the system in (8.53)), are measurable. The solution therefore follows from the state-feedback LQ control problem : $U_{t,N} = -K_c \hat{X}(t|t)$. The solution for the LQ state feedback optimal control problem gain $K_c$ was discussed briefly in the previous Chapter 7.1. However, the cost function must include cross terms for present purposes and this result is summarised below.

**Theorem 8.1** : *LQG State Feedback Optimal Control Problem Solution*

Consider a discrete-time system with measurable states and the following state equation:

$$x(t+1) = Ax(t) + Bu(t) + D\xi(t)$$

where $\{\xi(t)\}$ is a white noise sequence. Assume the performance index is given by:

$$J = \lim_{T \to \infty} \frac{1}{2T} E\left\{ \sum_{t=-T}^{T} [x^T(t) \ u^T(t)] \begin{bmatrix} Q & M \\ M^T & R \end{bmatrix} \begin{bmatrix} x(t) \\ u(t) \end{bmatrix} \right\} \qquad (8.65)$$

where $Q$ is a positive semi-definite Hermitian matrix and $R$ is a positive-definite Hermitian matrix, such that,

$$\begin{bmatrix} Q & M \\ M^T & R \end{bmatrix} > 0$$

The optimal control gain and control signal can be computed as:

$$u(t) = -K_c x(t) = -[R + B^T P B]^{-1}[B^T P A + M^T]x(t) \qquad (8.66)$$

where $P$ satisfies the discrete steady-state Riccati equation:

$$P = (Q - MR^{-1}M^T) + (A - BR^{-1}M^T)^T P[I + BR^{-1}B^T P]^{-1}(A - BR^{-1}M^T)$$

or

$$P = Q + A^T P A - (M + A^T P B)(R + B^T P B)^{-1}(M^T + B^T P A) \qquad (8.67)$$

**Proof** : See Ogata (p.831, 1987) and Kwakernaak and Sivan (p.547, 1972).

### 8.5.4 Separation Principle Proof

A review of the previous steps reveals that a special LQG problem has been established. That is, the system model was first extended to provide future inferred outputs. This model not only involves the $n$ state variables for time $t$ but provides the vector of future outputs, using future disturbance information. The sytem to be controlled can therefore be redrawn, as shown in Fig. 8.6, where only the output and input vectors include future information. The state model in this case is for the augmented system in Equation (8.52), which includes reference, plant, noise and control weightings.

The multi-step criterion (8.61) was then written in vector-matrix form and the future outputs were substituted from (8.10) using: $Y_{t+N}^h = \hat{Y}_{t+N}^h + \tilde{Y}_{t+N}^h$. After using the orthogonality of predictions and prediction error, the cost-function has the form given in equation (8.65). This includes the state estimate for the augmented system in Fig. 8.6 The resulting criterion involves cross-product cost terms but is otherwise the same as in conventional LQG control, where states are available (the states in this case being $\hat{X}(t|t)$). Since the state estimates are known and are driven by white noise (the innovations) the solution follows from the state-feedback Theorem 8.1 above. This is the usual argument employed in the proof of the *Separation Principle*, by Kwakernaak and Sivan (1972). The optimal control follows as:

$$U_{t,N} = -K_c \hat{X}(t|t) \qquad (8.68)$$

The equations to be solved, to compute the output feedback optimal controller, may now be derived. Assume that state estimates are available for the system of equations (8.52) and shown in Fig. 8.6. Then the control law can be computed from an obvious identification of cost function terms, in (8.64) and (8.65):

$$Q = \tilde{H}^T \tilde{Q} \tilde{H}, \quad R = \tilde{R} + \tilde{G}^T \tilde{Q} \tilde{G}, \quad M = \tilde{H}^T \tilde{Q} \tilde{G} \qquad (8.69)$$

and the optimal control with state estimate feedback:

$$U_{t,N} = -K_c \hat{X}(t|t) = -[\tilde{R} + \tilde{G}^T \tilde{Q} \tilde{G} + B^T P B]^{-1}(B^T P A + \tilde{G}^T \tilde{Q} \tilde{H})\hat{X}(t|t) \qquad (8.70)$$

where $P$ satisfies the following matrix Riccati equation:

$$P = \tilde{H}^T \tilde{Q} \tilde{H} + A^T P A$$

$$-(\tilde{H}^T \tilde{Q} \tilde{G} + A^T P B)(\tilde{R} + \tilde{G}^T \tilde{Q} \tilde{G} + B^T P B)^{-1}(\tilde{G}^T \tilde{Q} \tilde{H} + B^T P A) \qquad (8.71)$$

**Remarks**

(i) The order of the compensator which results is the same as the Kalman filter for the system in equation (8.52). The controller will therefore have the same order as an LQG controller but may be very different because of the changes to the cost weightings.

(ii) The difference between an LQGPC and an LQG controller is not in the state estimation problem but lies in the control gain computation.

The Kalman filter to generate the state estimates is to be described but first robustness properties, when states can actually be measured, will be discussed.

### Robustness of the State Feedback Solution

The above LQGPC optimal control, when state feedback is available, has the same robustness properties as LQ optimal control. To prove the validity of this assertion consider the structure of the model in Fig. 8.4 The vector of future controls $U_{t,N}$ includes the current control $u(t)$ and only the actual control signal $u(t)$ affects the states of the plant. The future controls clearly do not affect $x_1(t+1)$. Thus, stability is therefore determined by only the computed signal $u(t)$ and not the whole vector of future controls $U_{t,N}$. The robustness properties, established for LQ systems (Anderson and Moore, 1971) will therefore hold for the system of (8.52) with input $U_{t,N}$. However, from the above argument stability is determined only by the signal $u(t)$ which feeds into the real plant states. Robustness properties, for the real plant, should therefore be the same, even if the controls applied in the future are not what are predicted in the future controls vector $U_{t,N}$. It follows that errors in computing the future predicted control increments will not affect the stability characterictics. Such errors will of course affect the predicted outputs which may be important in constrained optimisation versions (where quadratic programming is used).

## 8.5.5   The Discrete-time Kalman Filter

The Kalman filtering results are well known (Kalman, 1961) and the single-stage Kalman predictor was introduced in Chapter 7. However, the system of interest includes through terms of a particular structure and cross covariance noise terms. The following theorem summarises the required Kalman filtering results.

**Theorem 8.2** : *Discrete-time Kalman Filtering Problem and Solution*
Consider a system with state and observations equation of the form:

$$x(t+1) = Ax(t) + Bu(t) + D\xi(t)$$

$$z(t) = Cx(t) + Eu(t) + v(t) \tag{8.72}$$

where the sequence of zero mean, disturbance and noise signals, $\begin{bmatrix} \xi^T(t) & v^T(t) \end{bmatrix}^T$ are uncorrelated random vectors with the variance matrix:

$$\begin{bmatrix} Q_0 & M_0 \\ M_0^T & R_0 \end{bmatrix}$$

Then the optimal state estimates can be computed to minimise the variance of the estimation error:

$$J = E\{\tilde{x}^T(t|t)\tilde{x}(t|t)\}$$

where the estimation error is defined as:

$$\tilde{x}(t|t) = x(t) - \hat{x}(t|t)$$

and the initial time $t_0 \to -\infty$. The state estimates can be computed from the following recursive relationship:

$$\hat{x}(t+1|t+1) = A\hat{x}(t|t) + Bu(t)$$

$$+K_f(z(t+1) - CA\hat{x}(t|t) - CBu(t) - Eu(t+1)) \tag{8.73}$$

where the Kalman filter gain:

$$K_f = \left[(APA^T + Q)C^T + M\right]R_f^{-1} \tag{8.74}$$

and $P$ satisfies the steady-state matrix Riccati equation:

$$P = APA^T + Q - \left[(APA^T + Q)C^T + M\right]R_f^{-1}\left[C(APA^T + Q) + M^T\right] \tag{8.75}$$

where

$$R_f = R + C(APA^T + Q)C^T + CM + M^T C^T$$

$$Q = DQ_0D^T, \quad M = DM_0, \quad R = R_0 \tag{8.76}$$

**Proof :** The variance matrix for the process and noise vectors:

$$\begin{bmatrix} D\xi(t) \\ v(t) \end{bmatrix}$$

can be obtained, using the definitions of the signal $\xi(t), v(t)$ variances as:

$$\begin{bmatrix} Q & M \\ M^T & R \end{bmatrix} = \begin{bmatrix} DQ_0D^T & DM_0 \\ M_0^T D^T & R_0 \end{bmatrix}$$

The optimal linear filter then follows from the results given in Kwakernaak and Sivan (p. 500, 1972).

**Remarks:**

(i) The Kalman filter may easily be generated from the above results, for the system shown in Fig. 8.3, and defined in equations (8.52), (8.53).

(ii) Given the state estimates the output feedback LQGPC optimal control follows from (8.68) for the augmented system model.

# 8.6 LQ State-space Predictive Optimal Control

The use of future set-point information is usually associated with GPC or LQGPC type multi-step control laws. However, future set-point information may easily be incorporated into the usual LQ criterion optimal control solution (Grimble 1990, 1994). The state-space approach to this class of problems, which does not require a multi-step criterion, is illustrated below.

## 8.6.1 System Description

The discrete-time system will be assumed to be linear and time-invariant and to be driven by independent white noise signals. The reference subsystem driving noise is also independent of the measurement noise and plant disturbance signals.

**Plant Model:**

$$x_g(t + 1) = A_g x_g(t) + B_g u(t) + \xi_g(t) \tag{8.77}$$

$$y(t) = C_g x_g(t) \tag{8.78}$$

$$z_g(t) = C_g x_g(t) + v_g(t) \tag{8.79}$$

where $\{\xi_g(t)\}$ and $\{v_g(t)\}$ are zero-mean independent white noise signals representing the disturbance and measurement noise, respectively. The plant is assumed to be stabilisable and detectable.

**Reference Model**

The reference signal $\{r(t)\}$ will be represented by the asymptotically stable linear stochastic system model :

$$x_r(t + 1) = A_r x_r(t) + \xi_r(t) \tag{8.80}$$

$$y_r(t) = C_r x_r(t) \tag{8.81}$$

$$z_r(t) = y_r(t) + v_r(t) \tag{8.82}$$

The zero-mean white-noise sources $\{\xi_r(t)\}$ and $\{v_r(t)\}$ are assumed to be independent. The noise-free output $\{y_r(t)\}$ represents the desired future values of the reference signal $p$ steps-ahead. That is,

$$r(t + p) = C_r x_r(t) \tag{8.83}$$

New state variables may be defined as delayed versions of the reference state:

$$x_{r1}(t) = r(t + p - 1) = C_r x_r(t - 1) \tag{8.84}$$

$$x_{r2}(t) = r(t + p - 2) = x_{r1}(t - 1) = C_r x_r(t - 2)$$
$$\vdots \qquad\qquad \vdots \qquad\qquad \vdots \tag{8.85}$$
$$x_{rp}(t) = r(t) = x_{r(p-1)}(t - 1) = C_r x_r(t - p)$$

This model represents the reference at time $t$ and at all future times up to time $t + p$. The assumption made is of course that the reference trajectory is known $p$ steps ahead. These future reference values can be collected in the vector $\tilde{x}_r(t)$ where

$$\tilde{x}_r(t) = \left[ x_{r1}^T(t), ..., x_{rp}^T(t) \right]^T$$

The combined state-equation model, including the plant, becomes:

$$
\begin{bmatrix}
x_g(t+1) \\
x_r(t+1) \\
x_{r1}(t+1) \\
x_{r2}(t+1) \\
\vdots \\
x_{rp}(t+1)
\end{bmatrix}
=
\begin{bmatrix}
A_g & 0 & 0 & \cdots & 0 \\
0 & A_r & 0 & & \\
0 & C_r & 0 & & \vdots \\
0 & 0 & I & & 0 \\
\vdots & & & \ddots & \\
0 & 0 & \cdots & I & 0
\end{bmatrix}
\begin{bmatrix}
x_g(t) \\
x_r(t) \\
x_{r1}(t) \\
x_{r2}(t) \\
\vdots \\
x_{rp}(t)
\end{bmatrix}
$$

$$
+
\begin{bmatrix}
B_g \\
0 \\
0 \\
0 \\
\vdots \\
0
\end{bmatrix}
u(t)
+
\begin{bmatrix}
\xi_g(t) \\
\xi_r(t) \\
0 \\
0 \\
\vdots \\
0
\end{bmatrix}
\tag{8.86}
$$

$$
\begin{bmatrix}
z_g(t) \\
z_r(t)
\end{bmatrix}
=
\begin{bmatrix}
C_g & 0 & 0 & \cdots & 0 \\
0 & C_r & 0 & \cdots & 0
\end{bmatrix}
\begin{bmatrix}
x_g(t) \\
x_r(t) \\
\tilde{x}_r(t)
\end{bmatrix}
+
\begin{bmatrix}
v_g(t) \\
v_r(t)
\end{bmatrix}
\tag{8.87}
$$

These equations may be written in the usual augmented state-space form:

$$x(t+1) = Ax(t) + Bu(t) + \xi(t)$$

$$z(t) = Cx(t) + v(t)$$

## 8.6.2 LQ State Feedback Predictive Control Problem

In this section it is assumed that the system states are available for feedback. This is of course unrealistic since states of the reference model will not be accessible. However, this is a useful starting point, since the separation principle can be invoked later to derive the two-degrees of freedom LQG controller.

**Steady-state Cost-function**

$$J = E\{x^T(t)Q_c x(t) + u^T(t)R_c u(t)\} \tag{8.88}$$

where the initial time $t_o \rightarrow \infty$.

The expression for the tracking-error becomes:

$$e(t) = r(t) - y(t) = x_{rp}(t) - C_g x_g(t) = H_c x(t) \quad (8.89)$$

where

$$H_c = [-C_g \quad 0 \quad \cdots \quad 0 \quad I] \quad (8.90)$$

The cost-function should include the error term $e^T(t)Q_h e(t)$ which can be arranged by defining $Q_c$ in the regulating cost (8.88) as:

$$Q_c = H_c^T Q_h H_c \quad (8.91)$$

where $Q_h$ can be selected to cost the error terms. Cross terms in the cost-function are null in this case.

**Control Algebraic Riccati Equation**

The solution to the steady-state LQG optimal control problem, when the system is assumed stabilisable and detectable and the states are available for feedback, was defined in Theorem 8.1, as:

$$u(t) = -K_c x(t) \quad \text{where} \quad K_c = (B^T P_c B + R_c)^{-1} B^T P_c A \quad (8.92)$$

and $P_c$ denotes the solution to the **Control Algebraic Riccati Equation** (CARE):

$$P_c = Q_c + A^T P_c A - A^T P_c B(B^T P_c B + R_c)^{-1} B^T P_c A \quad (8.93)$$

**Gain Expression**

The feedback control gain has an interesting polynomial form. Let the optimal feedback control law (8.92) be written as:

$$u(t) = -K_c x(t) = -K_{cg} x_g(t) + K_{cr} x_r(t) + K_{c1} x_{r1}(t) + ... + K_{cp} x_{rp}(t)$$

$$= -K_{cg} x_g(t) + K_{cr} x_r(t) + (K_{c1} z^{-1} + ... + K_{cp} z^{-p}) r(t+p) \quad (8.94)$$

where $r(t+p) = C_r x_r(t)$.

**Example 8.1** : $H_2$ *Scalar Prediction Problem : State Feedback*

The following example will illustrate the mechanism for introducing prediction and will indicate more clearly the structure of the controller derived. The plant is open-loop unstable and the noise sources are zero-mean and independent. The noise variances are all assumed to be unity with the exception of the reference measurement noise signal which has a variance = 0.1.

**Plant:**

$$x_g(t+1) = 2x_g(t) + u(t) + \xi_g(t), \quad y_g(t) = x_g(t)$$

**Reference:**

$$x_r(t+1) = x_r(t) + \xi_r(t), \quad y_r(t) = x_r(t)$$

Assume that the reference is known for $p$ steps-ahead, then the state-equation becomes:

$$
\begin{bmatrix} x_g(t+1) \\ x_r(t+1) \\ x_{r1}(t+1) \\ x_{r2}(t+1) \\ \vdots \\ x_{rp}(t+1) \end{bmatrix} = \begin{bmatrix} 2 & 0 & 0 & \cdots & 0 \\ 0 & I & 0 & & 0 \\ 0 & I & 0 & & 0 \\ 0 & 0 & I & & 0 \\ \vdots & & \ddots & & \vdots \\ 0 & 0 & 0 \cdots & 0\,I & 0 \end{bmatrix} \begin{bmatrix} x_g(t) \\ x_r(t) \\ x_{r1}(t) \\ x_{r2}(t) \\ \vdots \\ x_{rp}(t) \end{bmatrix} + \begin{bmatrix} 1 \\ 0 \\ 0 \\ 0 \\ \vdots \\ 0 \end{bmatrix} u(t) + \begin{bmatrix} \xi_g(t) \\ \xi_r(t) \\ 0 \\ \\ \vdots \\ 0 \end{bmatrix}
$$

$$
\begin{bmatrix} z_g(t) \\ z_r(t) \end{bmatrix} = \begin{bmatrix} 1 & 0 & 0 & \cdots & 0 \\ 0 & 1 & 0 & \cdots & 0 \end{bmatrix} \begin{bmatrix} x_g(t) \\ x_r(t) \\ x_{r1}(t) \\ \vdots \\ x_{rp}(t) \end{bmatrix} + \begin{bmatrix} v_g(t) \\ v_r(t) \end{bmatrix}
$$

The reference $r(t) = x_{rp}(t)$ or $r(t+p) = x_r(t)$. The error weighting term in the cost-function can be defined using $H_c$ where

$$H_c = [-1\ 0\ 0\ ..\ 0\ 1] \text{ and } Q_c = H_c^T H_c.$$

For example, if the reference is known for *3 steps-ahead*, the state equation and the weightings become:

$$
\begin{bmatrix} x_g(t+1) \\ x_r(t+1) \\ x_{r1}(t+1) \\ x_{r2}(t+1) \\ x_{r3}(t+1) \end{bmatrix} = \begin{bmatrix} 2 & 0 & 0 & 0 & 0 \\ 0 & I & 0 & 0 & 0 \\ 0 & I & 0 & 0 & 0 \\ 0 & 0 & I & 0 & 0 \\ 0 & 0 & 0 & I & 0 \end{bmatrix} \begin{bmatrix} x_g(t) \\ x_r(t) \\ x_{r1}(t) \\ x_{r2}(t) \\ x_{r3}(t) \end{bmatrix} + \begin{bmatrix} I \\ 0 \\ 0 \\ 0 \\ 0 \end{bmatrix} u(t) + \begin{bmatrix} \xi_g(t) \\ \xi_r(t) \\ 0 \\ 0 \\ 0 \end{bmatrix}
$$

$$
\begin{bmatrix} z_g(t) \\ z_r(t) \end{bmatrix} = \begin{bmatrix} 1 & 0 & 0 & 0 & 0 \\ 0 & 1 & 0 & 0 & 0 \end{bmatrix} x(t) + \begin{bmatrix} v_g(t) \\ v_r(t) \end{bmatrix}
$$

where $r(t) = x_{r3}(t)$ or $r(t+3) = x_r(t)$, $H_c = [-1\ 0\ 0\ 0\ 1]$ and

$$
Q_c = H_c^T H_c = \begin{bmatrix} 1 & 0 & 0 & 0 & -1 \\ 0 & 0 & 0 & 0 & 0 \\ 0 & 0 & 0 & 0 & 0 \\ 0 & 0 & 0 & 0 & 0 \\ -1 & 0 & 0 & 0 & 1 \end{bmatrix}, \quad R_c = r_c = 0.1.
$$

The computed state-feedback gains for different prediction intervals, are shown in Table 8.1.

**Table 8.1** : State-feedback Gain Matrices ($K_c$)

| p = 0 | | | | |
|---|---|---|---|---|
| 1.8642 | -0.7856 | | | |
| p = 1 | | | | |
| 1.8642 | -0.7856 | 0 | | |
| p = 2 | | | | |
| 1.8642 | -0.10668 | -0.67896 | 0 | |
| p = 3 | | | | |
| 1.8642 | -0.01449 | -0.09219 | -0.67896 | 0 |
| p = 4 | | | | |
| 1.8642 | -0.00197 | -0.01252 | -0.0922 | -0.67896 | 0 |
| *(Plant) state* | *(r(t+p))* | *(r(t+p+1))* | ⋯ | *(r(t+1))* | *(r(t))* |

# 8.7  LQG Predictive State Estimate Feedback

The solution of the LQG optimal control problem where only outputs, rather than states, are available for feedback, follows from the separation principle of stochastic optimal control theory (Kwakernaak and Sivan, 1972). The resulting 2 DOF system structure (Grimble, 1994, 1995) is illustrated in Fig. 8.7. To minimise the cost-function a steady-state Kalman filter must first be calculated, based on the plant and reference state-equations (8.77) and (8.80). Assume that the disturbance covariance matrix $Q_f$ is factorised as: $Q_f = H_f H_f^T$ and that the system $[A, H_f]$ is detectable. The Kalman estimator may then be computed from the state equations for the total plant and reference models (8.86), (8.87), using either the Kalman predictor results outlined in Chapter 7 or the Kalman filtering solution given in Theorem 8.2. If the estimates are to be computed using the previous values of the observations signals then the following results are required.

**Kalman :** *Single-Stage Predictor : $\hat{x}(t|t-1)$*

The solution to the steady-state Kalman filtering problem, at time $t$, using observations up to time $t$ - 1, gives an estimator equation:

$$\hat{x}(t+1|t) = A\hat{x}(t|t-1) + Bu(t) + K_f(z(t) - C\hat{x}(t|t-1)) \tag{8.95}$$

where the Kalman filter gain matrix:

$$K_f = AP_f C^T [CP_f C^T + R_r]^{-1} \tag{8.96}$$

and $P_f$ satisfies the **Filter Algebraic Riccati Equation (FARE)**:

$$P_f = AP_f A^T - AP_f C^T [CP_f C^T + R_f]^{-1} CP_f A^T + Q_f \tag{8.97}$$

The estimator transfer-function matrix, when $u(t) = -K_c\hat{x}(t|t-1)$ is given as:

$$\hat{x}(t|t-1) = (zI - A + K_f C + BK_c)^{-1} K_f z(t)$$

**Kalman filter : $\hat{x}(t|t)$**

The Kalman filter, when observations are available up to time $t$, has the estimator equation:

$$\hat{x}(t|t) = \hat{x}(t|t-1) + K_f[z(t) - C\hat{x}(t|t-1)] \qquad (8.98)$$

where

$$\hat{x}(t|t-1) = A\hat{x}(t-1|t-1) + Bu(t-1)$$

$$K_f = P_f C^T [CP_f C^T + R_f]^{-1} \qquad (8.99)$$

The estimator transfer-function matrix, when $u(t) = -K_c\hat{x}(t|t)$, is given as:

$$\hat{x}(t|t) = (zI - (I - K_f C)(A - BK_c))^{-1} K_f z(t+1)$$

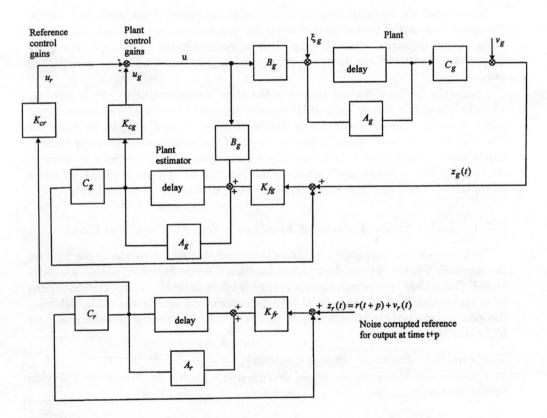

**Fig. 8.7** : Structure of Two Degrees of Freedom LQG Controller

**Example 8.2** : *Scalar Prediction Problem : State Estimate Feedback*

Consider again Example 8.1 and the computation of the Kalman filters for different prediction intervals. Partitioning the (1,1) block of the system $A$, output $C$ and noise covariance $Q_f, R_f$ matrices reveals that the plant is decoupled from the prediction/reference subsystem and its gain may be computed independently. Assuming measurements are available up to time $t-1$ the computed Kalman filter gains, for different prediction intervals are shown in Table 8.2; where $K_f = AP_fC^T(CP_fC^T + R_f)^{-1}$.

**Table 8.2 : Kalman Filter Gain Matrices $K_f$**

| $p=0$ | | $p=1$ | | $p=2$ | | $p=3$ | | $p=4$ | |
|---|---|---|---|---|---|---|---|---|---|
| 1.618 | 0 | 1.618 | 0 | 1.618 | 0 | 1.618 | 0 | 1.618 | 0 |
| 0 | 0.9161 | 0 | 0.9161 | 0 | 0.9161 | 0 | 0.9161 | 0 | 0.9161 |
| | | 0 | 0.9161 | 0 | 0.9161 | 0 | 0.9161 | 0 | 0.9161 |
| | | | | 0 | 0.0769 | 0 | 0.0769 | 0 | 0.0769 |
| | | | | | | 0 | 0.0065 | 0 | 0.0065 |
| | | | | | | | | 0 | 0.00054 |

Notice that the Kalman filter gains are small for some of the prediction states. These states do not of course contribute to the innovations since they are not coupled to the output through the output map. This suggests that the filter might be slightly simplified by setting these predictor gain elements (elements 4,2 up to $p+2,2$) to zero.

Since the feedback control gain from the state corresponding to $r(t)$ is zero, for all cases where $p > 0$ (see Table 8.1), there is no need to estimate this state. That is, the delayed reference states in the Kalman filter and the final gains in Table 8.2 (for cases $p = 1$ to 4) need not be implemented. If the plant has a unit-delay there is only a need to implement a delay line of $p-1$ elements. This provides a measure of the increase in complexity and dynamic order, caused by the use of future reference information. The closed loop controller may be implemented as shown in Fig. 8.7.

## 8.7.1   LQG State Estimate Feedback for Finite-time Cost

The finite-time optimal control problem is now considered, where the upper limit on the summation in the optimal control cost-function is finite. However, so that a steady-state Kalman filter is obtained, the initial time is taken to be $t_o \rightarrow \infty$. The finite-time behaviour is therefore only associated with the control gain calculation. The following theorem summarises the state feedback results that will be needed (Kwakernaak and Sivan, 1972).

**Theorem 8.3** : *Stochastic Optimal Controller*

Consider the state equation model described in equation (8.72) and the following finite-time cost-function:

$$J = E\{\sum_{j=t_0}^{t_f-1} (x^T(j)Q_c(j)x(j) + u^T(j)R_c(j)u(j)) + x^T(t_f)Q_c(t_f)x(t_f)\} \qquad (8.100)$$

The optimal control signal can be calculated as:

$$u^o(t) = -K_c(t)\hat{x}(t) \qquad (8.101)$$

where $\hat{x}(t) = \hat{x}(t|t)$ *or* $\hat{x}(t|t-1)$ . If $t_0$ tends to $-\infty$ then these estimates can be obtained from the steady-state Kalman filter, or the single-stage predictor described at the start of this section. The optimal control gain matrix:

$$K_c(j) = (R_c(j) + B^T P_c(j+1)B)^{-1}B^T P_c(j+1)A \qquad (8.102)$$

is obtained using the time-varying Riccati difference equation:

$$P_c(j) = A^T P_c(j+1)A$$

$$-A^T P_c(j+1)B(R_c(j) + B^T P_c(j+1)B)^{-1}B^T P_c(j+1)A + Q_c(j) \qquad (8.103)$$

where

$$-\infty < j \le t_f - 1 \text{ and } P_c(t_f) = Q_c(t_f).$$

### 8.7.2  Receding-horizon Optimal Control

The receding-horizon optimal control law (Kwon and Pearson, 1977) for stochastic systems is motivated by the preceding results and employs the steady-state Kalman filter and a separation principle structure. However, the control law is changed, since the cost-function costs error and control terms only within a sliding time frame, from $t$ to $t+T$.

Let the cost-function (8.100) be modified by replacing $t_f$ by $t_f = t+T$ and let the weightings be fixed, according to the time-difference from $t_f$

$$
\begin{aligned}
Q_c(t+T) &= Q_{co}, & \\
Q_c(t+T-1) &= Q_{c1}, & R_c(t+T-1) &= R_{c1}, \\
Q_c(t+T-2) &= Q_{c2}, & R_c(t+T-2) &= R_{c2}, \\
&\ \ \vdots & &\ \ \vdots \\
Q_c(t) &= Q_{cT}, & R_c(t) &= R_{cT}.
\end{aligned}
$$

In the receding horizon control problem the weightings $Q_c(j)$ and $R_c(j)$ are assumed to be zero for $j < t$ (or equivalently $Q_{ci}$ and $R_{ci}$ are zero for $i > T$). The criterion may now be written as:

$$J = E\{ \sum_{j=t}^{t+T-1} (x^T(j)Q_c(t+T-j)x(j) + u^T(j)R_c(t+T-j)u(j))$$

$$+ x^T(t+T)Q_{co}x(t+T)\} \qquad (8.104)$$

The constant optimal-control gain matrix in the receding horizon optimal control problem now becomes:

$$K_c(t) = (R_{cT} + B^T P_c(t+1)B)^{-1}B^T P_c(t+1)A \qquad (8.105)$$

where

$$P_c(j) = A^T P_c(j+1)A - A^T P_c(j+1)B(R_c(t+T-j)$$

$$+B^T P_c(j+1)B)^{-1}B^T P_c(j+1)A + Q_c(t+T-j) \qquad (8.106)$$

and $P_c(j)$ must be solved backwards in time, using $P_c(t+T) = Q_{co}$.

If the initial time is again assumed to be $t_o \rightarrow \infty$ the steady-state Kalman filter may be used to provide the state estimates. The only difference with the finite-time LQG optimal control law is that, when the cost weights are constant, the receding horizon gain matrix $K_c(t)$ will be constant, for all time $t$, rather than being a time-varying function. This receding horizon control law may be compared with the Generalised Predictive Controller (GPC) of Clarke et al. (1987), or with the other predictive controllers such as MUSMAR and DMC.

Having established the link with other predictive approaches it is worth questioning whether the receding horizon law (Kwon and Pearson, 1977) is better than the infinite-time control law discussed in previous sections. It is of course not a true finite-time control law, since a receding horizon controller and filter has an finite-time impulse response. The stability properties are not so well defined and indeed receding-horizon controllers can give unstable closed-loop designs in some cases. Moreover, the infinite-time LQG control law, with constant weights, has effectively one parameter to choose in the scalar case and the system will be closed loop stable for all chosen values. This may be compared with some of the other predictive control laws, where a number of parameters must be carefully selected, if stability is to be achieved.

Bitmead, Gevers and Wertz (1989) have considered the relationship between predictive and LQG control law and have provided significant insights into the stability properties of the two philosophies. They broadly conclude that the improved stability properties of LQG design are well worth the numerical overhead. The good stability properties of LQG deigns also offer advantages over the DMC algorithms. The only feature in the DMC algorithms, which is not available in LQG strategies, is the method of handling process constraints (Garcia and Morshedi, 1986).

## 8.8  Concluding Remarks

The state-space approach to solving the family of GPC or LQGPC control laws was first considered that included very general plant descriptions. The multi-step cost-function employed provides one method of introducing future set-point (or disturbance) information. This approach has the merit that the equations may be written in a form where standard constrained optimisation algorithms may be applied. The optimal solutions can then be obtained when there are hard limits on actuator movement, or rates of change, and on output variables. Such techniques have been very successful in applications (Morningred et al. 1991; Morari, 1989; Ordys and Grimble, 1996; Pottman and Seborg, 1997).

The second part of the chapter dealt with a different method of utilising future setpoint knowledge. It was first shown that future reference information can be introduced and that the resulting LQ or LQG controllers are very easy to calculate using a standard Riccati equation procedure. The increase in the order of the reference controller depended upon the number $p$ future reference values. However, there was little increase in the complexity of the solution. The first change to the standard LQG solution involved inputting the reference for time $t + p$ into the controller, rather than the reference for time $t$. The Kalman filter included additional delay elements and a set of corresponding filter gains. There was also a set of control gains feeding forward from the delayed future reference state estimates. These changes did not add to the structural complexity of the solution and are very straightforward to implement.

The advantage of using future reference information is clear. Tracking errors are much reduced and control signal variations are also reduced. For suitable applications these advantages considerably outweigh the small increase in complexity. In fact most of the improvement is often obtained when the prediction interval $p \times \Delta T$ is approximately twice the desired closed-loop time constant, where $\Delta T$ denotes the sample interval.

The usual methods of constrained optimisation may be applied to the LQGPC cost index optimisation (Kock et al. 1998) in much the same way as for GPC designs. Upper level *static* set point optimisation may also be introduced (Hansen et al. 1998). The most serious problem in the design of predictive control laws lies in the difficulties due to modelling uncertainties (Ordys, 1993). The design of truly robust predictive controllers remains a challenging problem.

## 8.9 References

Anderson, B.D.O. and Moore, J.B., 1979, *Optimal filtering*, Information and System Science Series, Ed. T. Kailath, Prentice Hall, New Jersey.

Anderson, B.D.O. and Moore, J.B., 1971, *Linear optimal control*, Ed. Newcomb, R.W., Prentice-Hall, Elect. Eng. Series.

Åström, K.J., 1967, Computer control of a paper machine - an application of linear stochastic control theory, *IBM Journal of Research and Development*. No. **11**.

Åström, K.J., 1970, *Introduction to stochastic control theory*, Academic Press, New York.

Bitmead R.R., Gevers M. and Wertz, V., 1989, Optimal control redesign of generalized predictive control, *IFAC Symposium on Adaptive Systems in Control and Signal Processing*, Glasgow, Scotland, April.

Casavola, A. and Mosca, E., 1998, A predictive reference governor with computational delay, *European Journal of Control*, Vol. **4**, pp. 241-248.

Clarke, D.W. and Mohtadi, C., 1989, Properties of generalised predictive control, *Automatica*, Vol. **25**, No. 6, pp. 859-875.

Clarke, D.W., Mohtadi, C. and Tuffs, P.S., 1987, Generalized predictive control, Part 1, The basic algorithm, Part 2, Extensions and interpretations, *Automatica*, Vol. **23**, No.2, pp. 137-148.

Cutler, C.R. and Ramaker, B.L., 1980, Dynamic matrix control - A computer control algorithm, *Proceedings JACC*, San Francisco.

Garcia, C.E. and Morshedi, A.M. 1986, Quadratic programming solution of dynamic matrix control (QDMC), *Chemical Eng., Commun.* Vol. **46**, pp. 73-87.

Grimble, M.J., 1986, Feedback and feedforward LQG controller design, *American Control Conference, Seattle,* Washington, June.

Greco, C., Menga, G., Mosca, E. and Zappa, G., 1984, Performance improvement of self-tuning controllers by multistep horizons : The MUSMAR approach, *Automatica*, Vol. **20**, No. 5, pp. 681-699.

Grimble, M.J., 1988, Two-degrees of freedom feedback and feedforward optimal control of multivariable stochastic systems, *Automatica*, Vol. **34**, No. 6, pp. 809-817.

Grimble, M.J., 1990, LQG predictive optimal control for adaptive applications, *Automatica*, Vol. **26**, No. 6, pp. 949-961.

Grimble, M.J., 1994, State-space approach to LQG multivariable predictive and feedforward optimal control, *Trans. of the ASME, Journal of Dynamic Systems, Measurement and Control.* Vol. **116**, pp. 610-615.

Grimble, M.J., 1995, Two DOF LQG predictive control, *IEE Proc.* Vol. **142**, No. 4, July, pp. 295-306.

Grimble, M.J. and Johnson, M.A., 1988, *Optimal control and stochastic estimation, Vols I and II,* John Wiley, Chichester.

Grimble, M.J. and Ordys, A., 2000, Predictive control for industrial applications, Plenary at IFAC Conference Control Systems Design, Bratislava.

Hangstrup, M., Ordys, A.W. and Grimble, M.J., 1997, Dynamic algorithm of LQGPC predictive control, *4th International Symposium on Methods and Models in Automation and Robotics,* Miedzyzdroje.

Hansen, J.F., Ordys, A.W. and Grimble, M.J., 1998, A toolbox for simulation of multilayer optimisation system with static and dynamic load distribution, IFAC Symposium on Large Scale Systems: Theory and Applications, Patras, Greece.

Hunt, J.J., 1988, A single-degree of freedom polynomial solution to the optimal feedback/feedforward stochastic tracking problem, *Kybernetika*, Vol. **24**, No. 2, pp. 81-97.

Kalman, R.E., 1961, Proc. of the Symp. on Eng. Applications of Random Function Theory and Probability, pp. 270-388.

Kock, P. Ordys, A.W. and Grimble, M.J., 1998, Constrained predictive control design for multivariable systems, *SIAM Conference on Control and its Applications,* Jacksonville, Florida.

Konno, Y., Hashimotot, H. and Tomizuka, M., 1992, On prediction and preview in control systems, *Preprints of the 21st Symposium on Control Theory*, Kariya, Japan.

Kouvaritakis, B., Rossiter, J.A. and Chang, A.O.T., 1992, Stable generalised predictive control : an algorithm with guaranteed stability, *IEE Proceedings-D*, Vol. **139**, No. 4.

Kwakernaak, H. and Sivan, R., 1972, *Linear optimal control systems*, John Wiley.

Kwon, W.H. and Pearson, A.E., 1977, A modified quadratic cost problem and feedback stabilization of a linear system, *IEEE Trans on Auto. Cont.* Vol. AC-**22**, No. 5, pp. 838-842.

Li, S., Lim, K.Y. and Fisher, D.G., 1989, A state space formulation for model predictive control, *AIChE, J.*, **35**, 241-249.

Martin Sanchez, J.M. and Rodellar, J., 1996, *Adaptive predictive control : from the concepts to plant optimization*, Prentice Hall International, Hemel Hempstead.

Marquis, P. and Broustail, J.P., 1988, SMOC, A bridge between state space and model predictive controllers - application to the automation of a hydrotreating unit, *IFAC Workshop on Model Based Control*, Atlanta, GA.

MacGregor, J.F., 1977, Discrete stochastic control with input constraints, *Proc. IEE*, Vol. **124**, No. 8, pp. 742-744.

Mohtadi, C. and Clarke, DW., 1986, Generalized predictive control, *Proc. of 25th Conf. on Decision and Control*, Athens, Greece, pp. 1546-1541.

Morningred, J. Duane, Mellichamp, D.A. and Seborg, D, 1991, A multivariable adaptive nonlinear predictive controller, *ACC*, Boston, Massacheusetts, pp. 2776-2781.

Morari, M. and Zafiriou, E., 1989, *Robust process control*, Prentice-Hall, Englewood Cliffs, N.J.

Ogata, K., 1987, *Discrete-time control systems*, Prentice Hall International Series, New Jersey.

Ordys, A.W., 1993, Model-system parameter mismatch in GPC control, *International Journal of Adaptive Control and Signal Processing*, Vol. **7**, pp. 239-253.

Ordys, A.W. and Grimble, M.J., 1994, Evaluation of stochastic characteristics for a constrained GPC algorithm, In : *Advances in Model-Based Predictive Control.*, edited by David Clarke, Oxford University Press, Oxford.

Ordys, A.W. and Grimble, M.J., 1996, A multivariable dynamic performance predictive control with application to power generation plants, *IFAC World Congress*, San Francisco.

Ordys, A.W. and Clarke, D.W.,1993, A state-space description for GPC controllers, *Int. J. Systems Sci.*, Vol. **23**, No. 2.

Ordys, A.W. and Pike, A.W., 1998, State-space generalized predictive control incorporating direct through terms, 37th IEEE Control and Decision Conference, Tampa, Florida.

Peterka, V., 1984 Predictor based self-tuning control, *Automatica*, Vol. **20**, No. 1, pp. 39-50.

Pike, A.W., Grimble, M.J., Shakhour, S. and Ordys, A.W., 1995 *Predictive control*, in the Control Handbook, Editor W.S. Levine, CRC Press Inc.

Pottmann, M. and Seborg, D.E., 1997, Nonlinear predictive control strategy based on radial basis function models, *Computers Chemical Engineering*, Vol. **21**, No. 9, pp. 9665-980.

Prett, D.M. and Garcia, C.E., 1988, *Fundamental process control*, Butterworth, Stoneham, M.A.

Richalet, J., Rault, A., Testud, J.L. and Papon, J., 1978, Model predictive heuristic control applications to industrial processes, *Automatica*, **14**, pp. 414-428.

Richalet, J., 1993, Industrial applications of model based predictive control, *Automatica*, Vol. **29**, No. 8, pp. 1251-1274.

Ricker, N.K., 1990, Model predictive control with state estimation., *Ind. Eng. Chem. Res*, **29**, pp. 374-382.

Rossiter, J.A., Kouvaritakis, B. and Rice, M.J., 1998, A numerically robust state-space approach to stable-predictive control strategies, *Automatica,* Vol. **34**, No. 1, pp. 65-73.

Sebek, M., Hunt, K.J. and Grimble, M.J., 1988, LQG regulation with disturbance measurement feedforward, *Int. J. Control*, Vol. **47**, No. 5, pp. 1497-1505.

Tomizuka, M. and Rosenthal, D.E., 1979, On the optimal digital state vector feedback controller with integral and preview actions. Transactions of the ASME, Vol. 101, pp. 172-178.

# 9

# QFT and Frequency Domain Design

## 9.1 Introduction

This chapter is concerned with various aspects of control systems design. Most of the previous material dealt with synthesis problems in control and estimation theory. That is, the emphasis was on the solution of the optimal control and filtering problems. This is not exactly what might be considered a design procedure, since algorithm development only provides a computational method. Such techniques might be considered design procedures, when it is clear how to satisfy an industrial performance specification. In this chapter, attention will concentrate on the design issues which in optimal systems reduces to the problem of cost function weighting selection and system disturbance specification. However, attention will focus here on a frequency domain approach to design, referred to as Quantitative Feedback Theory (QFT), and developed by Isaac Horowitz (1963).[1]

The QFT design approach is from a very different perspective to optimal control, since it involves direct manipulation of frequency responses. However, it throws light on the real robust design problem and is therefore valuable, both in its own right, and for the insights it provides for $H_2/H_\infty$ solutions.

The chapter begins with an introduction to Quantitative Feedback Theory (Azvine and Wynne, 1992). This frequency domain robust control design approach is almost the opposite, in strategy, to that of the optimal solutions previously discussed. In fact the approach focuses on the design problems and tries to ensure the controllers obtained are robust. There is therefore an argument for developing procedures that combine the best features of these very different philosophies (Breslin, 1996). Recent work has considered the complementary features of these methods. The stability and robustness QFT provides has given valuable insights into optimal control applications.

The following discussion of the QFT design method is in terms of continuous-time plant models, since this simplifies the analysis, and is very similar to the approach

---

[1] See the special issue of the International Journal of Robust and Nonlinear Control, edited by Osita Nwokah, on Horowitz and QFT Design Methods, Vol. 4, No. 1, January-Feburary 1994.

for discrete-time systems. This was also the first case considered by Horowitz. The results apply to the discrete-time case in a very similar manner. Let the plant transfer-function be represented by $W(s)$, where $s$ is the Laplace-transform variable. The plant frequency response can be denoted by $W(j\omega)$.

## 9.2 Quantitative Feedback Theory

In his seminal contribution Isaac Horowitz (1963) introduced a frequency domain design technique which he later refined into what is now known as quantitative feedback theory (QFT). His philosophy was straightforward: *feedback is introduced into systems to reduce the effects of uncertainties and nonlinearities which cannot be measured.* The amount of feedback which is needed is a function of this uncertainty, and the required performance of the closed-loop system. The cost of this feedback can be assessed in terms of bandwidth which should be minimised if problems with noise amplification, unmodelled high frequency dynamics and resonances are to be avoided. The plant uncertainty and the closed loop requirements should therefore be specified quantitatively so that at each stage of the design process the cost of feedback can be assessed.

The QFT method is a frequency domain based design technique Horowitz (1963, 1991, 1992), which approaches the robust control synthesis problem directly. The QFT approach is probably the only method which enables a controller to be designed to satisfy a given specification in a transparent and quantitative manner. The major advantage of QFT is that the trade-offs between the design requirements are clearly evident at all stages of the design process, rather than after the controller has been calculated, (as with $H_\infty$ or LQG optimal control designs). The QFT method extends highly intuitive classical frequency domain loop shaping concepts to cope with uncertainties and simultaneous requirements on performance specifications. It is extremely appealing to practical engineers trained in classical concepts (Doyle, 1986; D'Azzo and Houpis, 1988).

The main disadvantage of QFT is that it is difficult to develop a pro-forma design procedure, where relatively unskilled engineers can follow a well defined route with limited tuning variables. In fact, engineers using QFT must have a very good understanding of classical design and the likely impact of frequency response changes on the time-domain response. This is the opposite of $H_\infty$ and LQG design where a form filling type of design procedure can be developed and where formalised design procedures will always give a stable, but not necessarily good, multivariable design.

The *QFT* method was one of the first approaches to exploit feedback to achieve desired performance in the presence of plant uncertainty. Results have been published for single-input, single-output plants (Horowitz and Sidi, 1972), for multi-input, multi-output linear plants (Horowitz and Loeder, 1981; Yaniv and Horowitz, 1986) and for various classes of nonlinear and time-varying systems (Horowitz, 1976). The design of multivariable controllers by QFT methods is not so straightforward and this is discussed later in the chapter. The popularity of the method has not increased so quickly as might have been expected, partly because of the lack of good computer tools during its early development. In addition step by step routines have often been presented by means of worked examples so that it is difficult to identify the generic applicability of the method. Not all problems have the standard features of these examples and without precise guidelines it is not clear how to translate the experience of these examples into new situations, and to understand the limitations of the method. However, Houpis and Rasmussen (1999) provide a clear exposition of the various stages in QFT design.

The QFT approach has the obvious advantage that it is close to engineers' existing experience on classical deign methods. It also provides facilities to deal with uncertainty which are not available in traditional methods. More recent tools, such as $H_\infty$ design, also show promise but are very different to the existing frequency domain procedures used in parts of the Aerospace industry. The QFT approach therefore has the attractive feature of providing a link with existing classical techniques, and at the same time providing the robustness features needed in today's high performance systems. Ideally tools of the future should combine the attractive features of the QFT and $H_\infty$ design approaches. For example, a feature that optimal methods do not normally provide is that controller complexity and system requirements cannot be traded in a iterative solution, whereas this is a central feature of the QFT design process.

One of the main objectives of the QFT design procedure is to design a low order controller with a minimum bandwidth. This minimises problems due to high frequency noise, resonances or high frequency unmodelled dynamics. It also reduces costs, since in general, increasing bandwidth is accompanied by greater actuation costs. It is, of course, often necessary that the closed-loop bandwidth for a system be made reasonably large. This ensures a fast step response for a single degree of freedom control system. However, it is desirable to have a small bandwidth for the controller, since noise is then attenuated, limiting wear and tear on actuators and there is less likelihood that high frequency resonances in the system will be excited. In fact, a second objective of any design is to deal with uncertainty by ensuring a rapid controller roll off rate so that there is less dependence of the design on possible unmodelled high frequency dynamics (Wu et al. 1998). If the system has hardware constraints, the number of poles in the controller may also have to be limited. The QFT design method can easily accommodate this type of requirement.

The QFT design procedure converts closed-loop specifications into constraints on a nominal open-loop transfer function, and these are called QFT bounds. In fact, if $H_\infty$ design has been computed prior to the QFT design, the sensitivity weightings can be utilised to provide starting values for the QFT solution.

**QFT Design Features**

Some of the features of the QFT approach may be summarised as follows:

1. Since QFT is similar to classical design the trade-offs between stability, disturbance level, uncertainty, performance and controller characteristics are relatively easy to understand and achieve.

2. The gain of the system, or amount of feedback, can be related directly to the plant uncertainty, disturbance level and performance requirements. The amount of feedback gain may be limited, reducing the cost of feedback.

3. Low loop gains are achieved which avoid problems due to sensor noise amplification, saturation and high frequency uncertainties due to unmodelled dynamics.

4. Controllers can be designed to cope with a family of plant models (nonlinear systems can be approximated by a number of linearised models at different operating points). It is possible to obtain one feedback controller which will cover the full envelope of responses. Note that there is no need to verify stability conditions for plants inside the templates.

5. The amount of feedback is chosen depending upon the amount of plant and disturbance uncertainty and according to the specified performance requirements. A controller designed using QFT is robust to a wide range of plant variations and can also be chosen to satisfy performance specifications simultaneously.

6. The designer can specify the desired structure and order of the controller at the beginning of the design process. The structure can be based upon classical schemes such as PID or lead lag compensators, up to advanced LQG or $H_\infty$ type structures.

7. The specifications on noise and disturbance models do not have to be complete in the sense that they cover the full frequency spectrum. It is more realistic to assume these are defined only in certain frequency ranges. The QFT approach can cope with this type of incomplete information.

8. It is possible to determine what specifications are achievable early in the design process and this is not normally the case with optimal control synthesis methods.

## Plant Model Uncertainties and Example

The plant characteristics are represented by a set of linear and time-invariant system equations with transfer functions over a range covering the operational envelope of the system. Feedback control systems are of course only needed because of the uncertainty in the system models. If the plant dynamics are known exactly and the plant is open-loop stable then an open-loop controller should be adequate. However, uncertainty in the plant description and the disturbances normally make an open-loop solution unacceptable and a closed loop system is needed. It is therefore not surprising that the treatment of uncertainty is central to the QFT design procedure. In fact QFT design allows for plant uncertainty, performance specifications and design limitations, as an integral part of the design process.

A typical industrial example, where the plant model changes significantly, is illustrated in Figures 9.1 to 9.3. This shows, the Bode diagrams for a set of plant characteristics (calculated at different nonlinear operating points) for the power generation example in Chapter 10. The plant has three outputs and a single input and the figures show the frequency responses for each channel. There are clearly substantial gain and phase variations and in fact the models change from being minimum phase to non-minimum phase. By the use of QFT a single compensator and prefilter may be obtained which stabilises the full range of plant parameter variations. The fact that such a controller will stabilise all the given plant models does not of course guarantee that in the actual non-linear system the closed-loop will be stable, but experience suggests that this is often the case.

Examples of uncertain systems in the aerospace industry are numerous including Phillips (1994), Reynolds et al. (1996), Snell and Stout (1996), Wu and Reichert (1996) and Wu (1991).

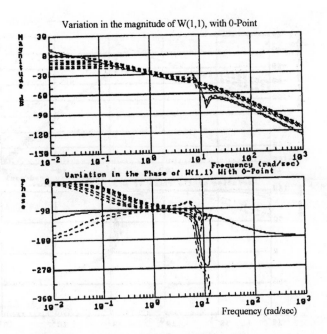

**Fig. 9.1**: Magnitude and Phase of Plant Frequency Response to Output 1 Over Range of Operating Conditions

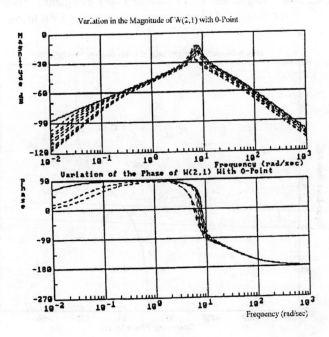

**Fig. 9.2** : Magnitude and Phase of Plant Frequency Response to Output 2 Over Range of Operating Conditions

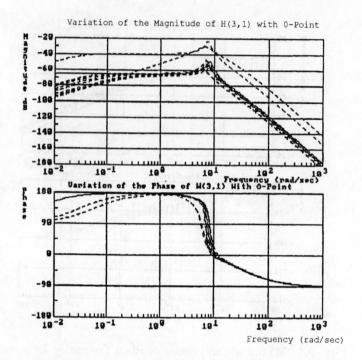

**Fig. 9.3** : Magnitude and Phase of Plant Frequency Response
to Output 3 Over Range of Operating Conditions

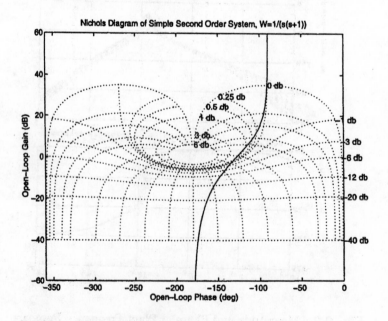

**Fig. 9.4** : Nichols Chart for a Simple Second-Order System

The M-circles in fig. 9.4 were drawn for $M_m$ 0 0.1, 0.2, 0.3, 0.4, 0.5, 0.6, 0.7, 0.8, 0.9, 1.0, 1.1, 1.2, 1.3, 1.4, 1.5, 2.0, 3.0, 4.0, 5.0.

## 9.2.1 Frequency Responses and the Nichols Chart

The basic tool of QFT is the Nichols chart (NC) shown in Fig. 9.4. The Nichols chart, developed by Professor Nathaniel Nichols, involves a frequency response plot of the open loop transfer function with axes of phase and magnitude. The NC has vertical and horizontal axes involving the log magnitude of the open-loop transfer-function $G(j\omega)$ in dB and the phase (arg $(G(j\omega))$, respectively. For the present let $M(j\omega)$ represent the closed-loop transfer-function, or control ratio, of a unity feedback continuous-time system:

$$M(s) = G(s)/(1 + G(s)) \tag{9.1}$$

The NC includes contours of constant log magnitude of $M$ $(Lm(M))$ and constant phase $\alpha = arg(M(j\omega))$. From a plot of the open-loop frequency response $G(j\omega)$ the values of $Lm(M(j\omega))$ and phase $\alpha(j\omega)$ can be found. That is, the amplitude and phase of the closed-loop frequency response can be determined for each frequency and the maximum value of the closed-loop frequency response amplitude $M_m$ can therefore be determined. The Nichols chart is a useful tool for relating the open-loop frequency response, which needs shaping, to the desired closed loop response.

To achieve a given peak magnitude $M_m$ the open-loop frequency response plot should be moved vertically up and down by changing a scalar gain multiplier $(kG(j\omega))$, until the contour of constant $M_m$ is touched. At a fixed frequency the variation in the plants frequency response describes a set complex number points referred to as a *template*. Once the templates have been defined, for each frequency of interest, the plants frequency response is known to lie in a band, which covers any of the points within the templates or on their borders.

It is necessary to introduce *bounds* on the allowable range of the gain variation of the nominal open loop transfer function. At selected frequencies, this enables magnitude constraints on the specific closed loop transfer function of interest to be satisfied.

**QFT Design**

The design specifications for a system normally involve a combination of time-domain and frequency domain requirements. The QFT approach is a frequency do-main based method, which accommodates time-domain requirements indirectly. If time-domain specifications are given they must be transformed into the frequency-domain. Time-domain criteria like overshoot and settling time are then related to the frequency-domain requirements. This does not guarantee that time-domain specifica-tions are met but experience reveals that this approach is often adequate. The QFT specifications normally involve the maximum amplitude of the closed-loop transfer function frequency response over a given frequency range and also a maximum ampli-tude limit for the sensitivity function which can be frequency dependent. The robust stability specification has the form:

$$\left| \frac{W(j\omega)C_0(j\omega)}{1 + W(j\omega)C_0(j\omega)} \right| \leq \mu \ for \ all \ W \in P, \ \omega \geq 0 \tag{9.2}$$

The performance specifications consist of constraints on the magnitude of the closed-loop frequency responses. A *robust performance* problem occurs when the performance specifications must be met for all transfer-functions which can occur in an uncertain system. In this case the performance specification must be satisfied for all possible variations of the uncertain system. Structured uncertainty models are normally used to represent uncertainty in the low to medium frequency range and unstructured uncertainty in the high frequency range. The uncertainty can be represented by templates on the Nichols diagram, showing the variations of the system frequency response, over the entire range of parameters, for a given frequency. The objective is to synthesise a controller to meet all of the specifications, including the robust performance problem, simultaneously.

**Open-loop Characteristics** : For the nominal plant $W_0(j\omega)$ the nominal loop transmission :

$$Lm\{L_0\} = Lm\{W_0 C_0\} = Lm\{W_0\} + Lm\{C_0\}$$

whereas for all other plants $W(j\omega)$:

$$Lm\{L\} = Lm\{W C_0\} = Lm\{W\} + Lm\{C_0\} \tag{9.3}$$

Thus, for a particular frequency, $\omega = \omega_i$ the variation $\delta_p(j\omega_i)$ in $Lm\{L(j\omega_i)\}$ is given by:

$$\delta_p(j\omega_i) = Lm\{L(j\omega_i)\} - Lm\{L_0(j\omega_i)\} = Lm\{W(j\omega_i)\} - Lm\{W_0(j\omega_i)\}$$

and

$$\angle \Delta W(j\omega_i) = \angle L - \angle L_0 = (\angle W + \angle C_0) - (\angle W_0 + \angle C_o) = \angle W - \angle W_0$$

## 9.2.2  Stability Criterion

Define the loop transmission of the feedback system shown in Fig. 9.5 as: $L(s) = W(s)C_0(s)H(s)$, where $C_0(s)$ and $H(s)$ are fixed. All transfer-functions in the loop are assumed rational and at least proper, although the loop transmission $L(s)$ is assumed to be strictly proper. Define the standard Nyquist contour, $\Gamma$, as in Fig. 9.6 where j-axis indentations are added to account for any imaginary poles of $L(s)$. Assume that $\Gamma$ is chosen to be large enough to include all the unstable poles of $L(s)$. Let $n$ denote the total multiplicity of these poles. An implicit assumption is that no unstable cancellations take place in forming the product $W(s)C_0(s)H(s)$. The Nyquist plot is the image of $L(s)$ under $\Gamma$. The Nyquist stability criterion for the feedback system shown in Fig. 9.5, with $W(s) \in P$ , may be stated as follows.

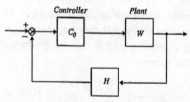

**Fig. 9.5** : Continuous-time Feedback System

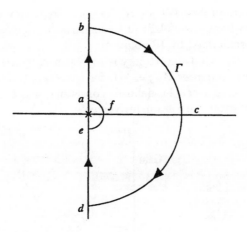

Fig. 9.6 : Continuous-time System Nyquist Contour

**Theorem 1** : The feedback system in Fig. 9.5 is stable if and only if the Nyquist plot of $L(s)$ does not intersect the point $(-1, 0)$, and encircles it $n$ times in the counterclockwise direction.

The gain and phase margins are used in classical frequency domain design to give a measure of the degree of stability of the system. These do of course relate to the distance between the open loop frequency responses and the critical point (-1, 0). In QFT an alternative approach is taken, using the maximum amplitude of particular closed-loop transfer functions. For the range of predicted parameter variations the peak magnitude of the sensitivity function, or complementary sensitivity function, must be limited. Clearly the closer the open-loop response can approach the critical point the larger will be the peak of these closed-loop frequency responses. Note that similar remarks can be made regarding $H_\infty$ design since it enables the maximum value of certain frequency responses to be limited. However, this involves the nominal system description, whereas in QFT design the results apply for the range of system models at a particular frequency. Moreover, the distance from the critical point is clearly evident during the QFT design process (rather than at the end of the $H_\infty$ design process).

The stability criterion used for Nyquist plots can be translated onto the Nichols chart. The point $(-1, 0)$ on the Nyquist plot is equivalent to the critical point (-180°,0 dB) on the Nichols chart. This implies that the winding number, of passing above the critical point on the Nichols chart, must be equal to the number of unstable open-loop poles, to guarantee the stability of the closed-loop system.

### 9.2.3  Uncertainty and Plant Templates

The plant uncertainty can be shown on a Nichols chart by a region around the plant frequency response that was referred to earlier as a template. The plant templates are calculated at particular frequencies and represent the variations in the structured plant uncertainty. These templates are employed to define the bounds within which the open-loop frequency response characteristic must lie. A template encloses all possible frequency responses $W(j\omega_k, \alpha)$ for some frequency $\omega_k$. Fig. 9.7 shows uncertainty templates for a given plant at frequencies $\{\omega_1, ..., \omega_5\}$, with the boundary of the templates approximated by 10 points.

In QFT design the plant uncertainty is represented by a set of templates $\wp$ within which all frequency responses $W(j\omega_k, \alpha)$, for some $\omega_k$, are included. For each frequency point the set $\wp_{\omega k}$ can be defined, consisting of a finite number of elements, determined by the uncertain parameter values.

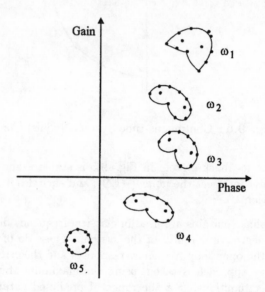

**Fig. 9.7** : Typical Bounds with a Nominal Loop Frequency
Response L(jω) that Satisfies the Bounds

**Bounds**

The Fig. 9.8 shows a typical open-loop frequency response, which satisfies the bounds for $\omega_1$ to $\omega_5$ with $\omega_5 > \omega_4 > ... > \omega_1$. The bounds for $\omega_1$ to $\omega_4$ are robust performance bounds, that require the nominal loop $|L_0(j\omega)|$ gain to exceed a minimum gain at each frequency point. The high frequency band (at $\omega_5$) is a robust stability bound. This requires the nominal loop gain to maintain a desired distance from the critical (-1) point for stability. This limits the maximum loop gain at high frequency.

The *stability bounds* are calculated using the phase-margin specification and the uncertainty templates. The *performance bounds* are obtained using the upper and lower limits on the desired frequency domain response and the uncertainty templates. The disturbance bounds are determined by both the templates and the disturbance model upper limit. Good disturbance rejection and tracking action is assured if the compensator provides sufficient gain of the open-loop frequency response above the disturbance and tracking bounds, as displayed on the Nichols chart. The variation of the plant due to parametric changes can be expressed as:

$$\delta_W(s) = \max_{\alpha \in \Omega} |W(j\omega, \alpha)|_{dB} - \min_{\alpha \in \Omega} |W(j\omega, \alpha)|_{dB}$$

where $\Omega$ denotes the parameter space of the plant $W$.

A bound is obtained by determining all possible positions on the Nichols chart for which the uncertainty template of $W(j\omega, \alpha)$ can be translated (without rotation) where the performance specification (in terms of sensitivities) satisfies the magnitude requirement. This was a very laborious manual task using the Nichols chart but is now straightforward via standard computer packages, (Borghesani et al. 1994).

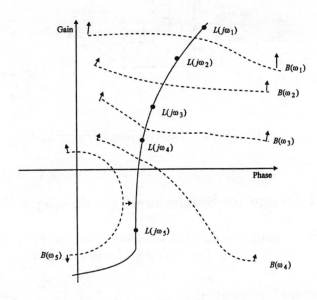

**Fig. 9.8** : Uncertainty Templates for a Plant at

Discrete Frequency Points

**Example 9.1** : *Parametric Uncertainty*

Consider the model with uncertain gain $k$ and poles $\alpha, \beta$ :

$$W(s) = \frac{k\alpha\beta}{s(s+\alpha)(s+\beta)} \tag{9.4}$$

where $k \in [0, \ 10]$, $\alpha \in [1, \ 5]$ and $\beta \in [1, \ 4]$ . The parameter variations lie within the region indicated in Fig. 9.9. The region of uncertainty may therefore be represented by the 8 LTI plants which lie on the boundary of the parameter space.

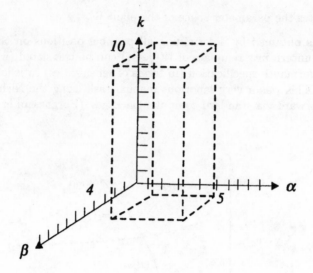

**Fig. 9.9** : Parameter Space Diagram

## 9.2.4  QFT Design for Single-input Single-output Systems

It is useful to identify the essential features of the SISO QFT design procedure for linear time invariant plants. The main steps are now summarised below.

### System Structure Specification

The system is assumed to be two degrees of freedom with reference input $r(s)$, noise $n(s)$, input and output disturbances $d_1(s)$ and $d_2(s)$. (The disturbances are not assumed to be applied simultaneously). The system is shown in Fig. 9.10.

### Selection of the Nominal Plant Model

The QFT design approach is based upon pointwise design at the trial frequencies. The *nominal* plant model must be defined to be a representative function in the uncertainty set. This is needed to generate the bounds and to synthesise the controller. Any plant model within the region of uncertainty may be selected as the nominal model. However, a plant model, whose NC template point is at the lower left hand corner of the template, is normally recommend (Houpis 1996). Any of the plant models may be chosen to be the nominal model for the bound calculation, so long as the same plant model is used for each frequency. Numerical difficulties may also arise if a nominal plant model is chosen to be too close to a resonant frequency.

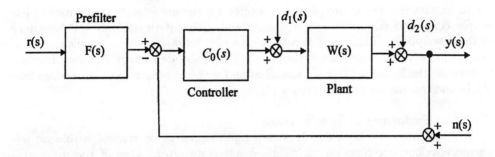

**Fig. 9.10** : Two Degrees of Freedom Control System

## Uncertainty Description for the Plant Model W(s)

Plant uncertainty is described by the variations of $W(j\omega)$ over the set of uncertain plant parameters. More generally a nonlinear plant is represented by a series of linear time invariant (LTI) transfer-functions, over a desired operating range. There are several aspects to consider:

(i) The selection of a point set of frequencies $\{\omega_1, \omega_2, ..., \omega_N\}$ over which to evaluate the plant uncertainty. These frequency points are known as the trial frequencies.

(ii) The variations of $W(j\omega_i)$ over the uncertain parameters for given $\omega_i$ translates into a template Tem$\{W(j\omega_i)\}$ on the Nichols chart. The plant templates are evaluated for each frequency point.

(iii) The relationship between the template and the loop transfer $L(s) = C_0(s)W(s)$ gives rise to rules for manipulating the template. For example by measuring the amplitude in decibels $L(j\omega_i)\ dB = C_0(j\omega_i)\ dB + W(j\omega_i)\ dB$ is equivalent to a vertical translation of the $W(j\omega_i)\ dB$ point.

(iv) In many of the examples, the plant uncertainty is described by interval plants, where the transfer function polynomials have coefficients varying over prespecified real parameter intervals. Thus, the uncertainty is explicitly defined by the templates, $Tem\{W(j\omega_i)\}$.

(v) The basic QFT design approach is not changed whether the uncertainty in the plant models is described by non-parametric or parametric models.

In QFT design a set of plants, rather than a single plant model, is considered. Thus, at each frequency point the variation in the magnitude and phase of the plants yields a set of points on the Nichols chart, rather than a single point. For a particularly frequency point a connected region, referred to earlier as a *template,* must be constructed which covers the range of uncertainty. To avoid undue conservatism the template must be chosen to be the smallest convex polygram that encloses all of the points.

The templates obtained are then used to define regions, or the so called *bounds* in the complex frequency domain. The open-loop frequency responses must lie within the bounds if the performance, stability specifications and disturbance requirements are to be met (for the entire plant set covered by the uncertainty description). For simply connected templates it is necessary and sufficient to use only the boundary of these templates. This result is related to a well known result in complex variable theory, namely the *maximum principle*.   The main step in QFT design involves translating the frequency domain specification for the uncertain feedback system into the bounds on the complex frequency plane.

### Tracking Performance Specification

A major objective is to enable a step input signal to be tracked within an adequate envelope specification with zero steady-state error.  One of the innovative features of the Horowitz method was the translation of a range of desired step-responses into a frequency domain performance requirement. Thus, by ad-hoc transfer function fitting, a Bode plot specification for an acceptable closed-loop transfer $|T(j\omega)| = |y(j\omega))|/|r(j\omega)|$ is created and used. The tracking specification defines the acceptable range of variations of the closed-loop tracking response of the system, due to uncertainty and disturbance inputs. It is generally defined in the time-domain as:

$$y(t)_L \leq y(t) \leq y(t)_U \tag{9.5}$$

where $y(t)$ denotes the system tracking response to a step input and $y(t)_L$ and $y(t)_U$ denote the lower (positive)and upper tracking bounds in the time domain, respectively. It may be transformed into the frequency domain, as:

$$B_{RL}(\omega) = |T_{RL}(j\omega)| \leq |T_R(j\omega)| \leq |T_{RU}(j\omega)| = B_{RU}(\omega)$$

where $T_R(s)$ denotes the filtered closed-loop tracking transfer function and $T_{RL}(s)$ and $T_{RU}(s)$ are the transfer functions of the lower and upper tracking bounds. The real values of the bounds are denoted by the functions $B_{RL}(\omega)$ and $B_{RU}(\omega)$, respectively. The performance bounds are derived using the uncertainty templates and the upper and lower limits for the frequency-domain responses. Typical bounds are illustrated in Fig. 9.11.

The time-domain tracking performance can be defined as follows:

**Overshoot :**

$$M_p \leq x_1\%$$

**Settling Time :**

$$T_s \leq x_2 \ seconds$$

The initial form of the upper tracking bound transfer-function model can be defined as a typical second-order model:

$$T_{RU}(s) = \frac{\omega_n^2}{s^2 + 2\xi\omega_n s + \omega_n^2}$$

Based on the desired peak magnitude $M_p$ and settling time $T_s$, the damping $\xi$ and $\omega_n$ can be determined. For non-minimum phase processes the tolerances on $T_{RU}(j\omega)$ must also be specified and satisfied in the design. An additional zero can be added to the model to flare the tracking bounds in the high frequency range. The QFT design objective is to reduce the effect of plant parameter variations, by shaping the open loop frequency responses, so that the Bode plots of the closed loop system, fall between the bounds $B_{RU}(\omega)$ and $B_{RL}(\omega)$. At the same time stability margins, noise rejection requirements and external disturbance rejection specifications, must be met.

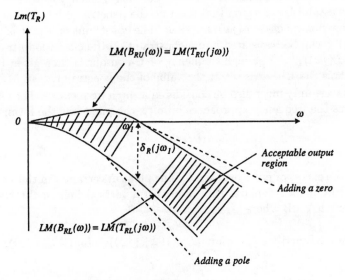

**Fig. 9.11** : Performance Bounds on the Reference to Output
Frequency Response

### Disturbance Rejection Specification

Two separate cases are considered corresponding to the presence of either $d_1(s)$ or $d_2(s)$. Two disturbance transfer-functions are important :

$$T_{d_1}(s) = \frac{1}{1 + C_o(s)W(s)} \quad and \quad T_{d_2}(s) = \frac{W(s)}{1 + C_o(s)W(s)}$$

In both cases, they are rearranged to exploit the geometry of the Nichols chart so that disturbance rejection bounds can be readily constructed and used. The disturbance rejection properties of the compensator are based upon keeping the loop transfer function gain above the disturbance bound on the Nichols chart.

The frequency domain bounds on the disturbance models have the form:

$$|T_{d1}(j\omega)| = \left| \frac{1}{1 + W(j\omega)C_0(j\omega)} \right| \leq B_{d1}(\omega), \quad \omega \in [\omega_1, \, \omega_2]$$

$$|T_{d2}(j\omega)| = \left| \frac{W(j\omega)}{1 + W(j\omega)C_0(j\omega)} \right| \leq B_{d2}(j\omega), \quad \omega \in [\omega_3, \, \omega_4]$$

The disturbance rejection models require an upper bound only, unlike the tracking specification that involves both upper and lower bounds.

## The Universal High Frequency Contour

The range of permitted step responses implies a minimum damping ratio $\xi$ for the dominant poles of the closed-loop transfer $T(s) = W(s)C_0(s)/(1+W(s)C_0(s))$. Thus, in the frequency-domain, this specifies a bound $M_m$ on the Nichols chart which must not be exceeded, constraining the range of acceptable open-loop frequency response plots. This boundary is known as the *universal high-frequency boundary,* or U-contour. This bound establishes a region that must not be penetrated by the templates and the loop transmission function $L(j\omega)$, for all $\omega$. The constraint on the sensitivity function $(1+L(j\omega))^{-1}$ can be translated into a constraint on the closed-loop transfer-function $L(j\omega)(1+L(j\omega))^{-1}$. The high frequency or U-contour is then given by an M-circle on the Nichols chart, that is set at the value of the constraint, (say $M_\ell$).

Plant uncertainty must also be considered at high frequencies. For many industrial processes, as the frequency approaches infinity, the plant has the form:

$$\lim_{\omega\to\infty} |W(j\omega)| = |K/(j\omega)^n| \qquad (9.6)$$

where $n$ is the excess of poles over zeros. The plant uncertainty in this case is therefore represented by plant templates that approach a vertical line on the Nichols chart of constant length V dB where,

$$\Delta Lm(W(j\omega)) = \lim_{\omega\to\infty} (Lm(W_{\max}(j\omega)) - Lm(W_{\min}(j\omega)))$$

$$= Lm(K_{\max}) - Lm(K_{\min}) = V \ dB$$

Thus, to ensure the high frequency boundary is satisfied, for all variations of the plant at high frequencies, the forbidden region must be stretched vertically by V dB, as shown in Fig. 9.12.

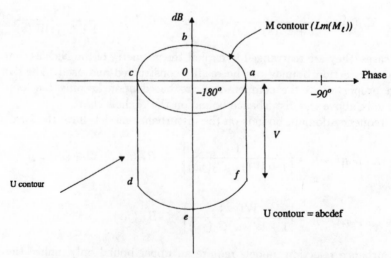

**Fig. 9.12** : The Universal High Frequency Boundary

## Composite Bound

If both tracking and disturbance rejection specifications are to be satisfied a composite bound must be constructed. To satisfy both of these specifications the point on the loop transmission $L(j\omega)$, corresponding to a particular frequency, must lie on or above the boundary for this frequency point. This is illustrated in Fig. 9.13, which reveals the combined boundary consists of points that at any phase angle on the Nichols diagram have the largest gain (Azvine and Wynne, 1992). The limiting boundary is determined by examining the nature of the inequalities, that are used to obtain each set of frequency boundaries.

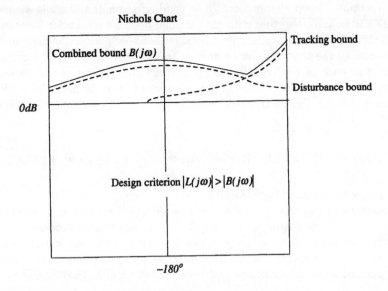

Fig. 9.13 : A Combined Tracking and Disturbance Bound

The QFT method requires the construction of the composite bound and the selection of a nominal plant model. The nominal plant frequency response and the boundaries may be listed as:

(i)  $W_0(j\omega)$ ~Nominal plant

(ii)  $B_R(\omega)$ ~Tracking boundary

(iii)  $\begin{cases} B_{d_1}(\omega) \\ \quad or \\ B_{d_2}(\omega) \end{cases}$  ~Input disturbance bounds

(iv)  $U(\omega)$ ~Universal high-frequency boundary or contour

The various bounds include a robust stability bound, tracking bound, disturbance rejection and high frequency bounds. The most restrictive bound is taken as the *composite bound*. The composite bound is determined from a combination of the robust stability bound, the tracking bound, the ultra high frequency bound and the disturbance bound. After integrating these bounds together to obtain the composite bound, the controller can be designed during the *loop shaping* stage. If there is no interaction between the bounds then the most restricted bound can be taken to be equal to the composite bound. If the bounds are intersecting, then the composite bound must be constructed from the most restricted regions of each bound. Once the appropriate combinations are available, a *composite boundary*, $B_0(\omega)$ is created. The desired loop transfer function $L_0(j\omega)$, is then chosen to satisfy some simple rules. For example, $L_0(j\omega)$ cannot penetrate the *U-contour*.

### Loop Shaping and Controller Generation

The *loop shaping* stage involves the process of designing the nominal open loop transfer function. Loop shaping design is conducted on the Nichols diagram using the composite bounds, together with the nominal open loop transfer function plot. In the construction of the controller, dynamical elements are added to the nominal plant model to change the shape of the open loop transfer-function. The boundaries must be satisfied at each of the trial frequency points. At the end of the design process the controller is obtained by forming the product of these terms. Given the desired loop transfer $L_0(j\omega)$, the controller is synthesised using the nominal plant $W_0(j\omega)$, by construction, using:

$$L_0(j\omega) \ = \ W_0(j\omega)C_0(j\omega), \text{ where the controller } C_0(j\omega) = \prod_{k=o}^{\tilde{k}} C_{ok}(j\omega)$$

where $C_0$ may be found as $C_0 = L_0/W_0$.

The loop shaping process therefore involves a curve fitting problem where the poles and zeros of the compensator are chosen. This is a very visible process and the trade offs between compensator complexity and system performance are clearly evident. For the controller to be realisable the nominal loop frequency response $L_0(j\omega)$ must satisfy Bode's gain and phase relationships and hence the locus of $L_0(j\omega)$ cannot be arbitary.

### Pre-filter design

The pre-filter must be designed to ensure the output of the system satisfies the tracking specification. This design is also carried out in the frequency domain but using a Bode diagram. The frequency response of the closed loop transfer function for the plant containing the uncertainties is moved into the specification envelope by using the pre-filter. This then ensures the required tracking performance is met. A further constructive procedure is given to synthesise the prefilter $F(s)$ (D'Azzo and Houpis, 1988). This uses the relationships:

$$T(j\omega) = L(j\omega)/(1 + L(j\omega)) \qquad \text{(Closed-loop frequency response)}$$

and

$$T_R(j\omega) = F(j\omega)L(j\omega)/(1 + L(j\omega)) = F(j\omega)T(j\omega) \qquad \text{(Output/reference)}$$

Then, for $\omega = \omega_i$ the magnitude of the gain from the reference to the output:

$$Lm\{T_R(j\omega_i))\} = Lm\{F(j\omega_i)\} + Lm\{T(j\omega_i)\}$$

This relationship can be used in reverse to determine a suitable prefilter $F(s)$ from the frequency-domain data for $T_R(s)$ and $T(s)$.

### 9.2.5   The QFT Design Procedure Example

An extended design example will now be presented which illustrates the steps involved in the design process.

**Example 9.2** : *QFT Design Process*

If structured uncertainty is present in the design problem, a subset of plants must be chosen from all possible plant cases. The plants in this subset should describe the full range of uncertainty present in order that the resulting design will be sufficiently robust. In this example the set of possible plants is described by the following transfer function:

$$\frac{a}{(s+2)(s+b)} \text{ where } a \in [1, 10] \text{ and } b \in [1, 3] \tag{9.7}$$

From the above set a subset of plants is chosen to represent the full range of uncertainty. Plants were chosen with the plant parameters having the following values shown in Table 9.1.

**Table 9.1** : Variation of Plant Parameters

|         | a  | b |
|---------|----|---|
| Plant 1 | 1  | 1 |
| Plant 2 | 1  | 3 |
| Plant 3 | 10 | 1 |
| Plant 4 | 10 | 3 |

The bode plots for these plants are shown in Fig. 9.14. It can be seen that the varying parameters of the plant transfer function change both the magnitude and phase responses.

**Generation of Templates**

The uncertainty varies across the frequency spectrum and this variation can be represented on the Nichols chart by the set of templates. As noted above a template is a graphical representation of the plant uncertainty generated at a specific frequency. A template is formed by calculating the gain and phase, at the selected frequency, for each plant in the set. The range of gain and phase values forms a region of uncertainty, at each frequency point, on the Nichols chart.

The designer must choose a frequency array, that is a set of frequency points spaced over the range of interest. The assumption is that if the design specifications are met, at each of these discrete frequency points, then the specifications are also met, at all points between them. As the frequency at which the plant responses are calculated is gradually increased, the shape of the template changes. This behaviour can be seen in the Figure 9.15. In general, the template shape will become a vertical line at high frequencies. Above some frequency the template will no longer change shape, thus no additional frequencies need be considered.

If, over a particular range of frequency, the template rapidly varies in shape then this variation must be captured within the QFT design. A concentration of frequency points can be inserted into the frequency array over this range. Consequently all variations in the uncertainty should be represented within the design. In this example the frequency array is chosen as $\{0.5, 1, 5, 10, 50, 200\}$. The templates resulting from evaluating the plant cases at these frequencies are shown on Fig. 9.15.

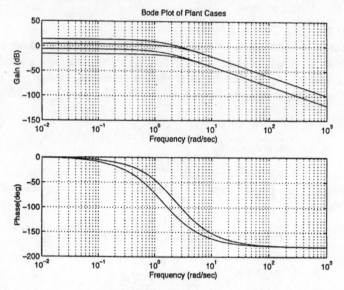

Fig. 9.14 : Bode Frequency Responses for Plant with Varying Parameters

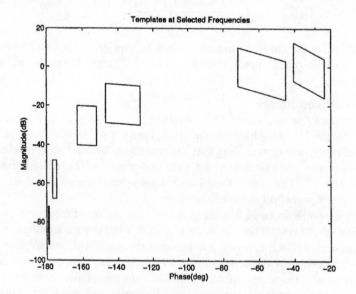

Fig. 9.15 : Template Generation for the Example

## Selection of the Nominal Model

One plant, normally a member of the uncertainty set, must be chosen as the nominal plant for the design. The control system design will be performed for this plant. The QFT design process attempts to ensure that the system designed for the nominal plant will also maintain robust performance and stability for all other plants of the uncertainty set. The normal convention in QFT design is to choose the lower left plant of the template as the nominal. However, in principle any plant model can be selected as the nominal model. In this example the plant $1/((s + 1)(s + 2))$ was selected as the nominal model for the design.

### Definition of Performance and Stability Requirements

Depending on the application, there are numerous measures to specify the desired behaviour of the control system. In QFT design the most common requirements involve stability, tracking performance and disturbance rejection. The stability requirement can be specified in QFT by defining a minimum permissible gain margin or phase margin. This effectively limits the magnitude of the complementary sensitivity function to ensure an adequate stability margin.

The performance requirement can be specified in either the time-domain or the frequency-domain. In the time-domain a minimum rise time and a maximum settling time can be specified. These specifications are translated into the frequency domain to give upper and lower limits on the closed-loop performance. Alternatively, specifications can be stated directly in the frequency domain.

### Stability Bound Generation

The Nichols chart has lines known as M-contours, which denote the points where there is constant closed loop magnitude. When limiting the closed-loop magnitude response of a system, a region around the (0 dB,-180 degree) point of the Nichols chart becomes forbidden. An $M$-contour with a 3 dB bounded contour is shown in Fig. 9.16. If the area within this contour is violated then the closed-loop magnitude response will exceed the desired maximum value. Shaping the open loop response to avoid this region is a key part of the QFT design process.

When there is no uncertainty present in the open-loop response then the stability requirement is to avoid the open-loop response passing within the specified $M$-contour of the Nichols chart. However, in the presence of uncertainty, the parameters of the system, for which the design is being performed, can vary. The forbidden region, must therefore be enlarged in order to account for the resulting variations in open-loop response. The enlarged region will be shaped such that, if the nominal plant avoids this region, then all plants of the plant set will not violate the forbidden M-contour.

This forbidden region, otherwise known as the stability bound, is calculated using the $M$-contour, the possible plant variation (template) and the frequency response of the nominal plant. Fig. 9.17 shows the stability bound generated using the 0.5 rad/s template. Since the template shape varies with frequency the stability bound will also vary with frequency. For each template a corresponding stability bound is generated. Hence, if a design is to be robustly stable the open-loop response, at each particular frequency, cannot violate the corresponding stability bound. The full set of stability bounds for all templates is shown in Fig. 9.18.

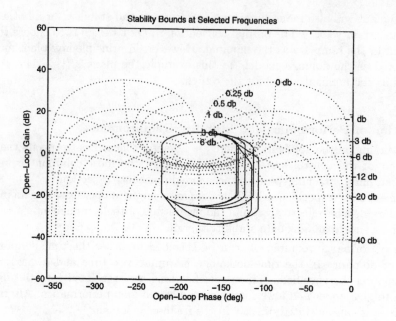

**Fig. 9.16** : Nichols Diagram with 3 dB Contour Bold

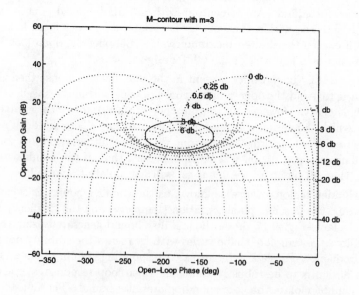

**Fig. 9.17** : Stability Bound for 0.5 Rads/sec Template

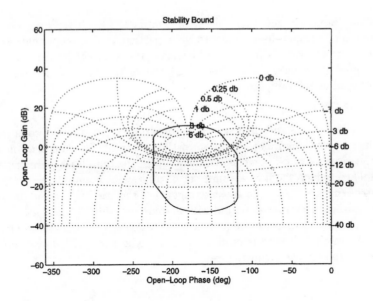

**Fig. 9.18** : Full Set of Stability Bounds for all Templates

### Tracking Performance Generation

Tracking specifications for the closed-loop performance of the control system can be included in the QFT design process. These specifications are translated into upper and lower bounds on the closed loop magnitude response. For these specifications to be met, the uncertainty of the closed-loop response must be smaller than the magnitude difference between the upper and lower bound.

The templates define the magnitude of the uncertainty of the open-loop response at the selected frequencies. On the Nichols chart the $M$-contours define lines of constant closed-loop magnitude. The uncertainty of the closed-loop response can be determined by the difference between the minimum and maximum $M$-contour crossings. As the template is shifted vertically upwards on the Nichols chart (equivalent to increasing the open-loop gain), the difference between the value of the minimum and maximum $M$-contour crossings decreases, that is, the closed-loop uncertainty decreases. The template must be shifted vertically until the closed-loop uncertainty is equal to the maximum allowed uncertainty, defined by the tracking specifications.

This process is carried out at every phase value on the Nichols chart. The result is a set of minimum gain values which are joined to form a tracking bound. The tracking bound represents a minimum value of gain for the open-loop response. If the open-loop response falls below this bound then the tracking performance requirements of the system cannot be met.

In this example, for a step response, the maximum settling time was 5 seconds and the minimum rise time was 2 seconds. These specifications are transformed into the frequency domain resulting in the limits shown in Fig. 9.19. All of the closed loop responses must lie within these upper and lower limits. The tracking bound, generated using these closed loop tracking specifications and the template at 0.5 rad/s, is shown on Fig. 9.20.

A tracking bound is generated for each template.  Each tracking bound applies a constraint on the open-loop response at a particular frequency, corresponding to a template.  In general, as frequency increases, the gain constraint imposed by the tracking bound decreases.  The Fig. 9.21 shows the full set of tracking bounds generated by all of the templates.

### Grouping of QFT bounds

The stability bounds and the tracking bounds produce two sets of constraints which must be met by the open-loop response of the system.  The two sets of constraints apply at the same set of frequencies (those at which the templates were generated) and it is possible to reduce them to a single set of composite bounds.  A *composite bound* is made up of the *worst case* (or most stringent) of the stability bound or the tracking bound.  In general a composite bound will be a combination of the tracking bound and stability bound, since at some frequencies the tracking bound is more stringent and at others the stability bound is more stringent.  The Fig. 9.22 shows the composite bounds generated for the frequency set.

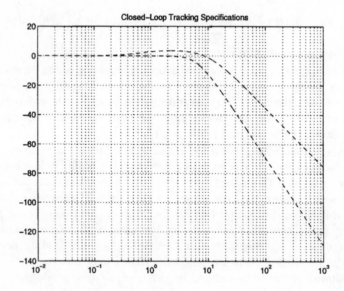

**Fig. 9.19** : Frequency Domain Tracking Performance Specification

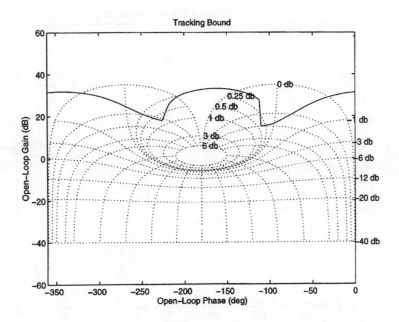

Fig. 9.20 : Tracking Bound and Template for 0.5 rad/sec

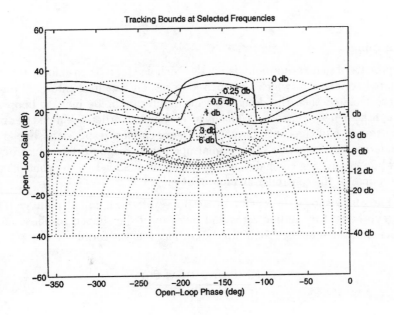

Fig. 9.21 : Tracking Bounds Computer at Chosen Frequency Points

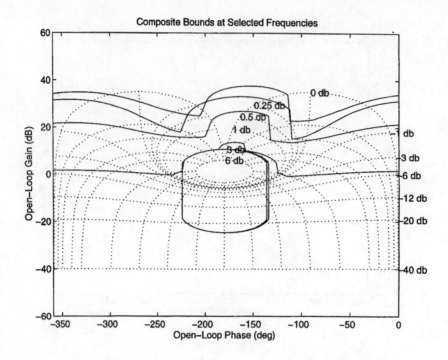

**Fig.9.22** : Composite Bounds Evaluated at all Selected Frequencies

## Loop Shaping

The process of generating bounds on the Nichols Chart is to inform the designer of prohibited regions into which the open-loop response cannot enter. If any of the bounds are violated then the design specifications cannot be met. Loop shaping involves synthesising a controller, element by element, in order to shape the open-loop frequency response of the control system. The open-loop response is shaped so that the bounds are not violated. An optimal open loop response lies on or above the bounds while minimising the gain. Elements are also added to the controller to roll off the response of the controller at high frequencies.

The Fig. 9.23 shows the open-loop response of the nominal plant. The controller is synthesised by adding zeros and poles until the shaped response meets the requirements of the bounds. A suitable controller has the form:

$$\frac{3200(s + 0.5)(s + 20)}{s(s + 200)}$$

The shaped response is shown in Fig. 9.24. It can be seen that the bounds are satisfied at each frequency.

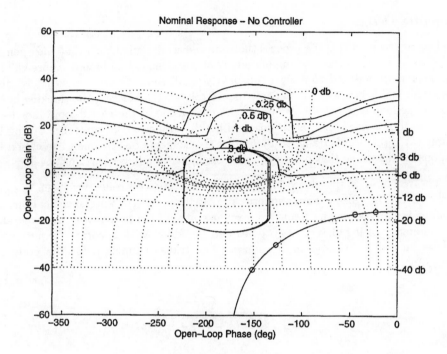

Fig. 9.23 : Open-loop Frequency Response of Plant

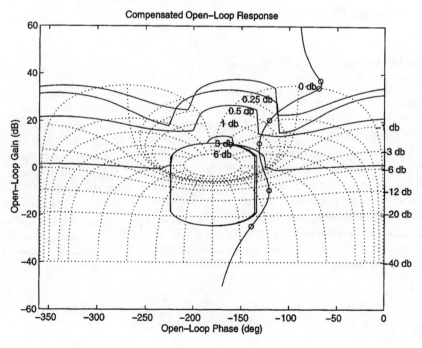

Fig. 9.24 : Compensated Open-loop Frequency Response
and Composite Bounds

**Pre-filter Design**

During the loop shaping process the controller is designed such that the open loop response meets all of the tracking bounds. The resulting closed-loop response will have its uncertainty reduced to a level sufficiently small that it can lie completely within the upper and lower closed-loop performance tolerances. Meeting the tracking bounds during the loop shaping process limits the magnitude of the closed-loop uncertainty. The pre-filter is used for this purpose. Like the design of the controller during loop shaping, the prefilter is designed by finding the elements one by one. The closed loop response is shaped so that it falls between the upper and lower tracking specifications.

The response of the closed-loop system, after the controller has been designed, is shown in Fig. 9.25. The response is not within the tracking specifications so a prefilter is necessary. The frequency response magnitude needs to be reduced above the frequency of 9 rad/sec. Two poles can be added with a corner frequency of 9 rad/sec in order to shape the response. The transfer function for the prefilter is defined as:

$$F(s) = \frac{81}{(s+9)(s+9)} \tag{9.8}$$

The pre-filtered response is shown in Fig. 9.26. It can be seen that the response now meets the design specifications.

**Validation of Open-loop Responses**

Once the controller and prefilter have been designed for the nominal plant it is necessary to check that the design is valid for all of the possible plant cases. Although the design is valid for the nominal plant there is no guarantee that the design will be valid for all cases. When the frequency array for the design has been poorly selected it is possible that key variations in uncertainty have not been captured in the design. If, upon checking all open loop responses against the stability specification, any of the plant responses violate the forbidden region, then the frequency at which this violation takes place must be noted. The design can then be repeated with the offending frequency added to the frequency array for the design. This revision guarantees that the uncertainty, at that frequency, will not be ignored in the design. Fig. 9.27 shows all of the plant cases plotted against the 3 dB M-contour. None of the responses pass through the forbidden region.

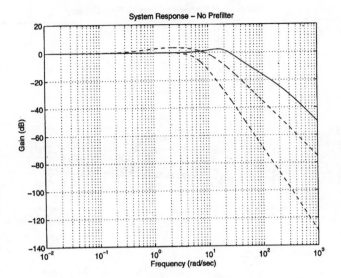

**Fig. 9.25** : Frequency Response of Closed-loop System Without Prefilter

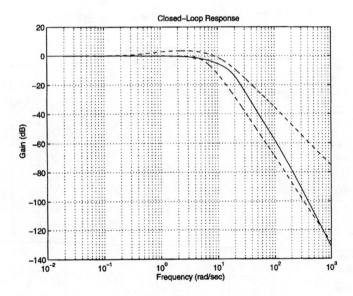

**Fig. 9.26** : Frequency Response of Closed-loop System
with Prefilter

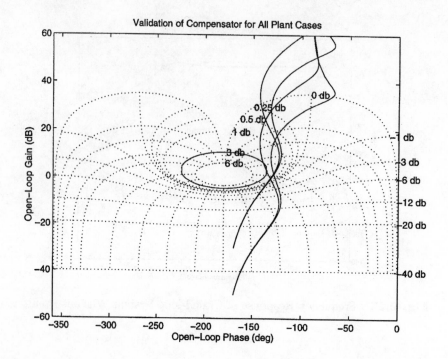

Fig. 9.27 : Validation of Closed-loop Stability

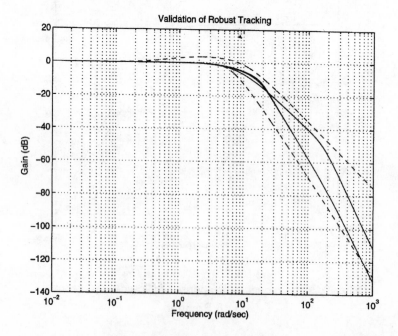

Fig. 9.28 : Closed-loop Frequency Responses and Tracking
Specifications for all Plant Cases

**Validation of Closed-loop Tracking Responses**

Once the stability of all plant cases has been verified it is necessary to check their tracking performance. All closed-loop responses are plotted against the closed-loop tracking specifications in Fig. 9.28 and these lie within the upper and lower tracking bounds.

The closed-loop frequency responses, at the frequencies contained in the frequency array, should be within the bounds but at intermediate frequencies it is possible that the bounds are breached. If the uncertainty is too large for the responses to fit between the upper and lower tracking specifications, then the frequency at which the breach occurs must also be added to the frequency array for the design.

## 9.2.6 Summary of Design Steps

1. Specify the desired closed-loop tracking model.

2. Define the desired closed-loop disturbance model.

3. Specify the $j$ linear time invariant plant models, which will define the boundary of the plant parameter uncertainty. Note that these models represent both the nominal plant and the perturbed plant for different choices of parameter values. In the case of non-linear systems the models may also have different orders, in addition to the parametric variations.

4. Calculate plant templates at specific frequencies $\{\omega_i\}$ which describe pictorially the region of plant parameter uncertainty on the Nichols diagram.

5. Define the nominal plant transfer function model $W_0$, using one of the representative functions in the uncertainty set.

6. Calculate the stability, or U-contour, on the Nichols diagram.

7. Determine the bounds on the Nichols chart for the disturbance and tracking requirements, based on the specifications in Steps 1 and 2 and using the templates in Step 4. The closed-loop magnitude specifications are thereby transformed into magnitude constraints on the nominal open-loop response (defining the bounds).

8. Synthesise the nominal loop transfer-function $L_0 = W_0 C_0$ that satisfies the stability contour requirement and has a gain on or above the bounds at each trial frequency. By satisfying the bounds for the nominal plant, the specifications for all plants, described by the uncertainty, are satisfied.

9. The composite bounds and the nominal open-loop transfer-function are plotted at the trial frequencies and the controller is designed by adding dynamic elements to the nominal plant, changing the shape of the open loop transfer function so that the boundaries are satisfied. The controller is obtained from a combination of these elements.

10. Determine the tracking prefilter to satisfy the tracking specifications.

11. Compute the time response of the system for the different plant models, using the computed controller, to validate the solution.

## 9.3  QFT for Multi-input Multi-output Systems

The MIMO QFT procedure has a general framework, which seeks to produce a sequence of SISO designs that can be solved by the SISO QFT procedure (Yaniv and Horowitz, 1986; Houpis, 1987; Yaniv, 1995). The two degrees of freedom system structure of Fig. 9.10 is used but controller $C_0(s)$, prefilter $F(s)$ and plant $W(s)$ are now taken to be multivariable systems. An $m \times m$ feedback control system can be represented by $m^2$ equivalent multi-input single-output (MISO) feedback control systems.

The system configuration is assumed to be multivariable and $n$-square so that $C_0(s), F(s)$ and $W(s) \epsilon R^{n \times n}(s)$. The plant transfer-function has elements with positive pole-zero excess. The plant elements are taken to be uncertain within a class of all possible plant cases $\wp$. The method described below applies to plants $W(s)$ for which the inverse $W^{-1}(s)$ has stable elements.

Let the prefilter $F(s) = [f_{ij}(s)]_{i=1}^n {}_{j=1}^n$ and $C_0(s) = diag\{c_1(s), ..., c_n(s)\}$. Each diagonal controller element is assumed to have a positive pole-zero excess and to be selected so that the elements of the closed-loop transfer-function matrix $T_R(s) \epsilon R^{n \times n}(s)$ are stable and satisfy:

$$0 \leq \alpha_{uv}(\omega) \leq |t_{uv}(j\omega)| \leq b_{uv}(\omega) \quad \text{for the set of all } W \epsilon \wp$$

where $T_R(s) = [I_n + W(s)C_0(s)]^{-1}W(s)C_0(s)F(s) = [t_{ij}(s)]_{i=1}^n {}_{j=1}^n$.

### MIMO QFT Analysis

The method exploits three structural features:

(i) *Structure of the SISO QFT problem :* When noise is assumed null the output (from Fig. 9.10):

$$y = \left[\frac{WC_oF}{1+WC_o}\right] r + \left[\frac{1}{1+WC_o}\right] d \tag{9.9}$$

(ii) *Sequential SISO Design Problem :* Write the tracking transfer:

$$T_R(s) = [I_n + W(s)C_0(s)]^{-1}W(s)C_0(s)F(s) = [I_n + C_0^{-1}(s)W^{-1}(s)]^{-1}F(s)$$

and rearranging gives:

$$T_R(s) + C_o^{-1}(s)W^{-1}(s)T_R(s) = F(s) \tag{9.10}$$

Write the inverse of the plant model:

$$W^{-1}(s) = \left[\frac{1}{Q_{ij}(s)}\right]_{i=1 \ j=1}^{n \quad n}$$

then

$$C_0^{-1}(s)W^{-1}(s) = \left[\frac{1}{L_{ij}(s)}\right]_{i=1 \ j=1}^{n \quad n} = \left[\frac{1}{c_i(s)Q_{ij}(s)}\right]_{i=1 \ j=1}^{n \quad n}$$

and the equation for $T_R(s)$ yields:

$$[t_{ij}(s)]_{i=1j=1}^{n\ n} + \left[\frac{1}{L_{ij}(s)}\right]_{i=1j=1}^{n\ n} [t_{ij}(s)]_{i=1j=1}^{n\ n} = [f_{ij}(s)]_{i=1j=1}^{n\ n}$$

or considering the individual elements of this matrix relationship:

$$t_{k\ell}(s) + \sum_{j=1}^{n} L_{kj}^{-1}(s)t_{j\ell}(s) = f_{k\ell}(s) \qquad (9.11)$$

and

$$L_{kk}(s)t_{k\ell}(s) + t_{k\ell}(s) + \sum_{j=1,j\neq k}^{n} L_{kk}(s)L_{kj}^{-1}(s)t_{j\ell}(s) = L_{kk}(s)f_{k\ell}(s)$$

Thence, there are $n^2$ transmission elements in this matrix equality of the form:

$$t_{k\ell}(s) = \frac{1}{(1+L_{kk}(s))}\left(L_{kk}(s)f_{k\ell}(s) - \sum_{j=1,j\neq k}^{n} L_{kk}(s)L_{kj}^{-1}(s)t_{j\ell}(s)\right) \qquad (9.12)$$

$$t_{kl}(s) = \left[\frac{L_{kk}(s)}{1+L_{kk}(s)}\right]f_{kl}(s) - \left[\frac{1}{1+L_{kk}(s)}\right]\sum_{j=1,j\neq k}^{n}\left(L_{kk}(s)L_{kj}^{-1}t_{jl}(s)\right)$$

for $l = 1, ..., n$ and $k = 1, ..., n$. In a quantitative statement of the problem there are tolerance bounds on each $t_{ij}$ leading to a set of $n^2$ acceptable regions which must be specified in the design. This expression has the format of the SISO QFT problem and by moving forward from $k = 1$ through to $k = n$ (solving $n$ SISO QFT problems per step) the complete MIMO problem can be solved.

(iii) *Efficient Recursive Procedure.* An efficient recursive formula may be derived based on both (i) and (ii) above. The details can be found in Yaniv and Horowitz (1986).

The proof for the MIMO QFT procedure is constructive and inductive, the procedure is proven to lead to a stable design. A $3 \times 3$ multivariable example is given by Yaniv and Horowitz (1986), extending results obtained previously by Horowitz and Loeder (1981).

**Remarks**

1. The methods of Horowitz depend on careful construction and good design skills. Horowitz emphasises these requirements in the polemic preface of Horowitz (1992).

2. The uncertainty description based on interval plants would appear to have its origin in the aerospace industry, where linear plant gains range over several orders of magnitude and the gain scheduling technique is commonly used.

3. Some fundamental issues which have been investigated include the question of whether a QFT design is robustly stable and the identification of the existence conditions for the QFT design (Jayasuriya and Zhao, 1994). A stability theorem for Nichols charts has been determined by Cohen et al. (1994). Further related work on the robust stability of interval plants has been published by Keel and Bhattacharyya (1994).

## 9.3.1   Nonlinear QFT

Nonlinear plant characteristics can be represented by a set of linear time-invariant transfer-functions which cover a prescribed range of parametric uncertainty. These linear models are obtained by linearising the full nonlinear plant dynamics. This set of transfer functions do not necessarily have the same denominator orders. The specifications can be represented in terms of linear time invariant (LTI) transfer-functions which represent lower $B_\ell(s)$ and upper $B_u(s)$ boundaries, that can be portrayed on a Bode magnitude diagram.

The nonlinear plant is therefore described by a set of $j$ minimum-phase LTI plants of the form $W = W_i(s)$ for $i = \{1, 2, ..., j\}$ and these define the structured plant uncertainty. Applying QFT to the plant, with structured uncertainty, ensures that the nonlinear plant will satisfy the desired performance specifications over the prescribed region. A single controller can be designed using QFT for the whole set of plant models, or a scheduled controller can be obtained dealing with distinct ranges of operation.

## 9.3.2   QFT and its Relationship to H∞ Control Theory

There are several fundamental differences between $H_\infty$ control theory and QFT. The most obvious is the way in which the uncertainty is represented. In $H_\infty$ control theory the uncertainty is represented by magnitude bounded sets and the lack of phase information introduces conservatism into the solution. The QFT methodology addresses the robust stability and the robust performance problems simultaneously. The proof of Nordgren et al. (1994) is given below in a form which shows how the QFT problem can be formulated as an $H_\infty$ problem.

Consider the scalar plant set described by

$$\mathcal{P} = \{W(\alpha, s)(1 + \Delta_n) : \alpha \epsilon \Omega, \; \Delta_n \in \Delta\} \tag{9.13}$$

$$\Delta = \{\Delta_n \in RH_\infty : |\Delta_n(j\omega)| < |m(\omega)|\}$$

where $\alpha$ represents the structured plant uncertainty and $m(\omega)$ is a known function.

Recall that the tracking response requirements are represented as upper and lower bounds on the closed-loop transfer function. In the time-domain, these specifications are given as:

$$y_L(t) \le y(t) \le y_U(t) \tag{9.14}$$

A mean response can be defined as:

$$m_R(t) = \frac{y_U(t) + y_L(t)}{2}$$

and a maximum allowable variation from the mean can be defined as

$$v(t) = \frac{y_U(t) - y_L(t)}{2}$$

Thus, the specification may be written, for a scalar system, as:

$$(y(t) - m_R(t))^2 \le (v(t))^2$$

To facilitate translation of this result into the frequency domain, where Parseval's theorem will be used, this condition is relaxed to the weaker requirement:

$$\int_0^\infty (y(t)) - m_R(t))^2 dt \le \int_0^\infty (v(t))^2 dt \tag{9.15}$$

By Plancherel's theorem, this condition is satisfied if

$$|T_R(\alpha, j\omega) - M_R(j\omega)| < |V(j\omega)| \quad \forall \; \omega \in [0, \infty) \tag{9.16}$$

where the output/reference transfer function $T_R(\alpha, j\omega)$ is given as:

$$T_R(\alpha, j\omega) = \frac{L(\alpha, j\omega)}{1 + L(\alpha, j\omega)} R(j\omega)$$

and the open-loop transfer function is given by

$$L(\alpha, j\omega) = W(\alpha, j\omega) C_0(j\omega)$$

Assume that the mean tracking response has been obtained for some nominal plant. That is

$$M_R(j\omega) = \frac{L_0(j\omega)}{1 + L_0(j\omega)} R(j\omega)$$

Then, equation (9.16) can be written as:

$$\left| \frac{1}{(1 + L(\alpha, j\omega))} \frac{(L(\alpha, j\omega) - L_0(j\omega))}{L_0(j\omega)} \right| \le \left| \frac{V(j\omega)}{M_R(j\omega)} \right| \tag{9.17}$$

If the *sensitivity function* is defined as:

$$S(\alpha, j\omega) = \frac{1}{1 + L(\alpha, j\omega)}$$

then equation (9.17) can be written as,

$$\left| S(\alpha, j\omega) \left[ \frac{W(\alpha, j\omega) - W_0(j\omega)}{W_0(j\omega)} \right] \right| \le \left| \frac{V(j\omega)}{M_R(j\omega)} \right| \tag{9.18}$$

where the following result was utilised:

$$\frac{L(\alpha, j\omega) - L_0(j\omega)}{L_0(j\omega)} = \frac{(W(\alpha, j\omega) - W_0(j\omega))}{W_0(j\omega)}$$

(this last step assumes there is negligible uncertainty associated with the implementation of the controller, $C_0(s)$).

If the bound on the relative uncertainty in the plant is defined as,

$$\delta_w(\omega) = \max_{W \epsilon \mathcal{P}} \left| \frac{(W(\alpha, j\omega) - W_0(j\omega))}{W_0(j\omega)} \right|$$

and a bounding function $H_w(\omega)$, is defined as:

$$H_w(\omega) > \left| \frac{M_R(j\omega)}{V(j\omega)} \right|$$

then equation (9.18) can be re-written as :

$$\max_{\omega} |H_w(\omega) S(\alpha, j\omega) \delta_w(\omega)| \leq 1$$

which is clearly the usual *weighted sensitivity minimisation problem*. The only difference between this formulation and the $H_\infty$ problem is that in this case the uncertainty is considered to be structured, as opposed to the norm-bounded descriptions normally used in $H_\infty$ design.

The tracking constraint can be specified, for all $\alpha \epsilon \Omega$ and all $\omega$, as:

$$|S(j\omega)| \leq \frac{1}{(H_w(\omega)\delta_w(\omega))} = M_T(j\omega) \tag{9.19}$$

A similar specification can be included for disturbance rejection as:

$$|S(j\omega)| \leq M_D(\omega)$$

The tracking and disturbance specifications are now satisfied if,

$$|S(j\omega)| \leq M_T(\omega)$$

where

$$M_T(\omega) = \min_{\omega}\{M_T(\omega), M_D(\omega)\}$$

Equation (9.19) can be rewritten to include the unstructured uncertainty as,

$$\left| \frac{1}{1 + L(\alpha, j\omega)[1 + \Delta(j\omega)]} \right| \leq M_T(\omega)$$

and if $T(\alpha, j\omega) = (1 + L(\alpha, j\omega))^{-1}L(\alpha, j\omega)$ then this is equivalent to,

$$\left| \frac{1}{(1 + L(\alpha, j\omega))[1 + T(\alpha, j\omega)\Delta(j\omega)]} \right| \leq M_T(\omega)$$

This can be manipulated into the equivalent inequality,

$$|M^{-1}(\omega)S(\alpha, j\omega)| + |T(\alpha, j\omega)m(\omega)| \leq 1 \tag{9.20}$$

which is the usual *mixed sensitivity* $H_\infty$ robust control problem, if the sensitivities are defined as: $S(\alpha, j\omega) = S(j\omega)$ and $T(\alpha, j\omega) = T(j\omega)$.

### 9.3.3    Disadvantages of the QFT Design Approach

 (i) The technique does require a deep understanding of classical control concepts.

 (ii) The approach does not directly specify the structure of the controller which must be chosen by the designer.

 (iii) If the closed loop frequency response specification are tight then a high order loop compensator may result, even though QFT design is able to achieve reasonably low order solutions in general.

 (iv) In the multivariable problem QFT normally utilises diagonal controllers and this is a highly restrictive condition that may limit performance in many important practical problems. With diagonal feedback it is sometimes the case that excessive bandwidth results, and this is the very quantity that single input, single output QFT designs seek to limit.

 (v) Manual tuning is required to achieve the desired frequency response and it may not be clear what should be considered the best solution.

### 9.3.4    Advantages of the QFT Approach

 (i) The trade-off between achieving the specifications and controller complexity can be made relatively easy and in fact the normal procedure is to start with a low order compensator and increase the order while trying to achieve the desired frequency response.

 (ii) The robustness margins are very apparent from the design process.

 (iii) The design method provides one of the few solutions which deals with parametric uncertainty directly.

## 9.4    Concluding Remarks

Most of the emphasis in previous chapters has been on the optimal control of systems and this involves using a synthesis technique. In some sense optimal control involves using a black box approach with cost function tuning. Optimal methods therefore have the advantage of simplicity but the disadvantage that the effects of the tuning variables (cost weightings), on the important stability and robustness margins, are not so clear.

The QFT approach on the other hand provides great insights into the real robustness issues but it has the disadvantage that classical design expertise is required. This can be difficult for sophisticated systems which include open-loop unstable, non-minimum phase or multivariable dynamics. The obvious conclusion is that a design philosophy is required, which in some senses merges the best features of these techniques, to gain the simplicity of use of optimal designs and the real insights and robustness of QFT solutions.

There is the possibility of using the $H_\infty$ design approach for the initial stage of a multivariable system and to then utilise QFT to obtain improved robustness properties of the principal loops. This seems a promising approach and has been demonstrated by Breslin (1996) on a flight control system example (Breslin and Grimble, 1997). In fact the development of fault tolerant flight control systems is also a valuable applications areas for this technique (Keating et al. 1995; Fontenrose and Hall, 1996; Bossert, 1994; Wu et al. 1999).

## 9.5   References

Azvine, B. and Wynne, R.J., 1992, A review of quantitative feedback theory as robust control system design techniques, Transactions of the *Institute of Measurement and Control*, Vol. **14**, No. 5, pp. 265-279.

Breslin, S.G., 1996, *On aircraft flight control*, Thesis for PhD, Industrial Control Centre, Department Electronic and Electrical Engineering, University of Strathclyde.

Breslin, S.G. and Grimble, M.J., 1997, Longitudinal control of an advanced combat aircraft using quantitative feedback theory, American Control Conference, New Mexico.

Borghesani, C., Chait, Y. and Yaniv, O., 1994, Quantitative feedback theory toolbox : for use with Matlab (User's Guide) Math-Works.

Bossert, D.E., 1994, Design of robust quantitative feedback theory controllers for pitch attitude hold systems, *J. Guidance, Control and Dynamics*, Vol. **17**, No. 1, pp. 217-219.

Cohen, N., Chait, Y., Yaniv, O. and Borghesani, C., 1994, Stability using Nichols charts, *Int. J. Robust and Nonlinear Control.*

D'Azzo, J.J. and Houpis, C.H., 1988, *Linear control system analysis and design: conventional and modern*, Third Edition, McGraw-Hill, New York, pp. 686-742.

Doyle, J.C., 1986, QFT and robust control, *Proceedings ACC*, Seattle, pp. 1691-8.

Fontenrose, P.L. and Hall, C.E., 1996, Development and flight testing of quantitative feedback theory pitch rate stability augmentation systems, *J. Guidance, Control and Dynamics*, Vol. **19**, No. 5, pp. 1109-1115.

Horowitz, I.M., 1963, *Synthesis of feedback systems*, Academic Press, New York..

Horowitz, I.M., 1976, Synthesis of feedback systems with nonlinear time uncertainty plants to satisfy quantitative performance specifications, *IEEE Transactions on Automatic Control*, **64**, pp. 123-130.

Horowitz, I.M., 1992, *Quantitative feedback design theory*, Vol. 1. QFT Publications, 4470 Grinnell Ave., Boulder, Colorado.

Horowitz, I.M, 1991, Survey of quantitative feedback theory, *Int. J. Control*, Vol. **53**, No. 2, pp. 255-291.

Horowitz, I.M. and Sidi, M., 1972, Synthesis of feedback systems with large plant ignorance for prescribed time domain tolerance, *International Journal of Control*, Vol. **16**, No. 2, pp. 287-309

Horowitz, I.M. and Leoder, C., 1981, Design of a 3x3 multivariable feedback systems with large plant uncertainty, Int. J. Control, Vol. 33, pp. 677-699.

Houpis, C.H., 1996, *Quantitative feedback theory technique*, published in the Control Handbook, Edited by William S. Levine, CRC Press.

Houpis, C.H., 1987, Quantitative feedback theory (QFT): technique for designing multivariable control systems, Air Force Wright Aeronautical Laboratories, FWAL; TR;86-3107, Wright-Patterson Air Force Base, Dayton, Ohio.

Houpis, C.H. and Rasmussen, S.J., 1999, *Quantitative feedback theory, fundamentals and applications,* Ed. N. Munro, Marcell Dekker, New York.

Jayasuriya, S. and Zhao, Y., 1994, Stability of quantitative feedback designs and the existance of robust QFT controllers, *Int. J. of Robust and Nonlinear Control.*

Keating, M.S., Pachter, M. and Houpis, C.H., 1995, QFT applied to fault tolerant flight control systems design, Proceedings of *ACC*, pp. 184-199.

Keel, L.H. and Bhattacharyya, S.P., 1994, Control system design for parametric uncertainty, Int. J. Robust and Nonlinear Control (In Press).

Nordgren, R.E., Nwokah, O.D.I. and Franchek, M.A., 1994, New formulations for quantaitive feedback theory, *International Journal of Robust and Nonlinear Control*, Vol. **4**, No. 1, pp. 47-64.

Phillips, S., 1994, A quantitative feedback theory FCS design for the subsonic envelope of the VISTA F-16 including configuration variation, Masters Thesis, AFIT/GE/ENG/94D-24 Air Force Institute of Technology, Wright-Patterson Air Force Base, Ohio. pp. 255-291.

Reynolds, O.R., Pachter, M. and Houpis, C.H., 1996, Full envelope flight control system design using quantitative feedback theory, *J. Guidance, Control and Dynamics*, Vol. **19**, No. 1, pp. 23-29.

Snell, S.A. and Stout, P.W., 1996, Quantitative feedback theory with a scheduled gain for full envelope longitudinal control, *J. Guidance, Control and Dynamics*, Vol. **19**, No. 5, pp. 1095-1101.

Wu, S.F. and Reichert, G., 1996, Design and analysis of a lateral flight control and path guidance systems for commercial aircraft, DGLR-JT96-094, pp. 493-502, Deutcher Luft und Raumfahrtkongreb DGLR-Jahrestagung, Dresden.

Wu, S.F., Grimble, M.J. and Breslin, S.G., 1998, Introduction to quantaitive feedback theory for lateral robust flight control systems design, *Control Engineering Practice,* **6**, pp. 805-828.

Wu, S.F., Grimble, M.J. and Wei, W., 1999, QFT based robust/fault tolerant flight controller design for a remote pilotless vehicle, *IEEE CCA Conference*, Kona, Hawaii.

Wu, D.P., 1991, *Studies on integration techniques in flight management systems*, PhD Dissertation, Nanjing University of Aeronautics and Astronuatics, Nanjing, P.R. China.

Yaniv, O. and Horowitz, I.M., 1986, A quantitative design method for MIMO linear feedback systems having uncertain plant, *Int. J. Control.* Vol. **43**, No. 2, pp. 401-421.

Yaniv, O., 1995, *MIMO QFT using non-diagonal controllers*, Int. Journal Control, Vol. 61, No. 1, pp. 245-253.

# Part III

# Industrial Applications

# 10

# Power Generation and Transmission

## 10.1 Introduction

The highly competitive market in electrical power generation and the need to improve both the quality and security of electrical supply has required improvements in control at all level of the power generation, transmission and distribution network. There is a need to meet tight voltage regulation requirements and to cope with rapidly changing load conditions. The electrical trading system is also changing which imposes new demands on the management of the network.

Multi-level control systems are needed for the control of energy production and transmission/distribution in power networks. The control system must be able to supervise and control voltages and reactive power for the total power supply system, including the generators, power stations, transmission networks and regional/national dispatchers. The lower level of control is referred to as the primary control level, where adaptive or robust regulating controllers are used. The later sections of the chapter include examples of both $H_2$ and $H_\infty$ voltage regulation design examples.

The higher level involves the regional voltage regulator and the reactive power regulators for the control of power stations, providing the secondary voltage regulation. There is a role for predictive control methods at the higher levels of the hierarchy. By applying an integrated multi-level control, philosophy, the stability and the security of the network voltage should be improved. The chances of a possible collapse of the system can also be reduced through the use of the co-ordinated control of reactive power. The aim of coordinated control is to provide an improvement in the overall power system performance, in terms of safety, continuity of operation, optimisation of resources, operational costs and the quality of control.

The voltage regulators for the transmission system determine the quality and security of supply and the economy of operation. Voltage control requires co-ordination of the reactive power sources. The desired voltage profile in the high voltage network must be maintained, while there are continuously changing loads and even network perturbations (infrequently).

The voltage control and the primary turbine speed regulation control are the most important regulation loops (Pike et al. 1995). The voltage control of the generators is accomplished by the Automatic Voltage Regulators (AVR's). The speed control system adjusts the steam flow into the turbines in response to changes in shaft speed. The necessary mechanical power is provided to meet the demanded electrical load. The block diagram of a steam turbine and a generator system is shown in Fig. 10.1. A power-station normally involves a number of alternators connected to a busbar and the wider electrical transmission and distribution system, as shown in Fig. 10.2.

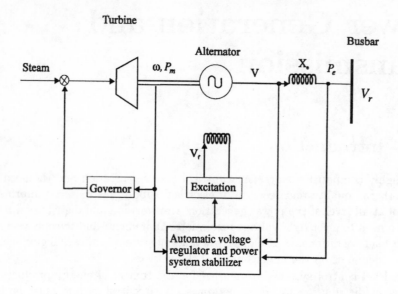

**Fig. 10.1** : Power System Control Problem Showing Steam Turbine and Alternator
($\omega$ = rotor speed, $P_m$ = mechanical power, $V_f$ = excitation voltage
$V$ = alternator output voltage, $X_e$ = external reactance,
$P_e$ = electrical power and $V_r$ = busbar voltage)

**Power System Control Difficulties**
There are various difficulties in designing a power control system :

1. The presence of nonlinear elements in the system.

2. Various disturbance inputs to the system.

3. Variations in the network structure which result in dynamic order changes.

4. The stochastic nature of disturbances and noise.

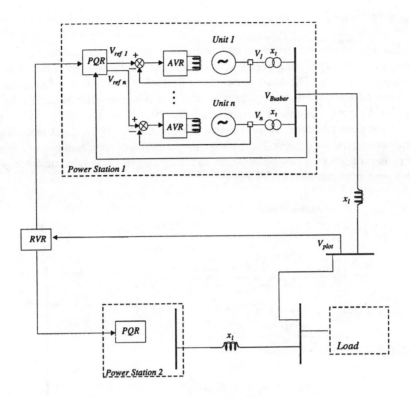

**Fig. 10.2** : Network Hierarchical Voltage Control Structure
(RVR = Regional Voltage Regulator and PQR = Reactive Power Regulator)

## 10.1.1  Automatic Voltage Regulator Design

There are a number of control problems in the generation and distribution of electrical power but attention will concentrate here on the Automatic Voltage Regulator (AVR) and the stability of the resulting multi-loop system. The frequency and load control involves the Automatic Generator Control (AGC) system which is introduced first, (Sharaf et al. 1986; Wu et al. 1988, 1992; Xia and Heydt, 1983; Law et al. 1994).

The main components in a steam turbine driven alternator, feeding a main busbar, are shown in Fig. 10.1. A power station will typically have four alternator/turbine sets connected to a national grid, as shown in Fig. 10.3. The generator automatic control system has three major objectives: (i) To hold the system frequency at, or very close, to a specified nominal value (50 Hz in the UK, 60 Hz in the US), (ii) To determine the correct levels of the interchange of power between control sectors, (iii) To control the generating capacity on each unit to maintain the capacity at the most economic level.

A load change on the system will produce a frequency change that has a magnitude depending upon the droop characteristics of the governor, and the frequency characteristics of the load. After a load change has occurred, control action must start to restore the frequency to the nominal value. This speed correction is normally accomplished by adding integral action in the controller for the control of generator speed.

Sudden and severe disturbances can result in significant variations in the system parameters and the operating points and linear controllers may not be able to maintain transient stability. The local system may lose its stability in the first load swing, unless severe countermeasures are put into effect, such as dynamic resistance breaking, loading shedding etc. Feedback linearisation has been used to cope with the nonlinearities in the power system. However, this approach relies open exact cancellation of the nonlinear terms to achieve the desired linear input-output behaviour. Since the system configuration may not be constant in fault conditions, it is virtually impossible to model such interconnections precisely.

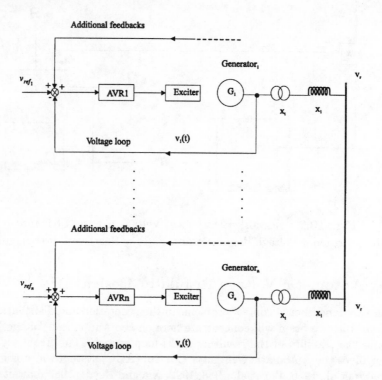

**Fig. 10.3** : Power Station Connected to an External Network

## 10.1.2  Synchronous Machine Stability

DeMello and Concordia (1969) provided an early seminal paper on the stability of synchronous machines under small perturbations. They noted that the trend was towards larger power generation units with higher performance, and the use of unit reactive power generating and transmission equipment. More emphasis was subsequently placed on control equipment, to provide compensation that could offset the reduction in stability margins, stemming from these trends in equipment design. The same trends have continued and computing capacity has significantly increased, making it easier to provide effective dynamic compensation control equipment. The problem they considered was the case of a single synchronous machine connected to an infinite bus through an external reactance. In multi-machine situations, there can be a wide band of the modes of oscillation and adequate damping must be introduced into the system (Arcidiacono et al. 1980).

## 10.2 Automatic Voltage Regulator Design

An Automatic Voltage Regulator (AVR) maintains the voltage output of a generator within strict limits, usually set by a regulating authority. The generator must remain synchronised to the grid that provides the interconnection to other power stations and distribution centres. The synchronisation torques also result from the magnetic fields, which provide damping torques (Brasca, 1993; Ledwich, 1979).

The voltage, power and frequency control loops of the generator are very non-linear and if the gain is increased in an AVR designed by classical methods, then under certain operating conditions the system can become unstable. To stabilise the system, electrical power and frequency (or rotational speed) measurements can be used for feedback. The resulting multiloop cascaded control system is known as a *Power System Stabiliser* (PSS). Since generators must operate over a wide range of conditions which moves them into different non-linear operating regions, the system can be considered to move through a set of linear operating points. To achieve adequate performance under all operating conditions requires a robust scheduled fixed controller (Shen et al. 1997), or an adaptive controller (Pierre, 1987; Chen et al. 1994). Recently fixed restricted structure optimal controllers have been developed that can optimise over a set of linearised operating point dependent models (Grimble, 2000).

The mechanical power input (from the steam turbine) and the field excitation voltage can be considered to be the system inputs. A two-input two-output problem arises if mechanical power and excitation inputs are taken as the inputs and output voltage and frequency are controlled. These multivariable problems involve the control of fast dynamics and lightly damped modes. Thyristor based excitation systems can be assumed to have a small time constant of between 0.03 and 0.05 seconds.

Most large power stations have four or more alternators connected to a common busbar and each has its own individual AVR. The primary control function for each AVR is to control the terminal voltage of that particular alternator. A major design objective is to accommodate system variations and this is often achieved by increasing the electro-mechanical damping, using the feedback of electrical power and angular speed (Brascia and Johnson 1994).

The presence of the AVR in a classically designed system often degrades the stability of the so called electro mechanical modes which are due to the combined turbine and generator system. The power systems stabiliser (PSS), uses the aforementioned additional feedbacks, to compensate for this effect. Unfortunately, the addition of these power and speed feedback terms introduces interaction between the loops, that can cause a deterioration in the voltage control performance. As the operating point moves the linearised transfer functions for the open-loop output voltage can change from an unstable to a stable model. There is also movement in the zeros of this transfer function. Either a robust or an adaptive controller can be used to cope with this type of variation (Cheng et al. 1986, Cheetah and Walker, 1982; Fork et al. 1988;

Forrest, 1992; Forrest et al. 1992).

The transfer function to voltage contains a complex conjugate pair of transmission zeros and the position of these zeros is determined by the operating point. The transfer functions to active electrical power $P_e$ and to angular speed $\omega$ do not contain this complex conjugate zero pair. If additional feedbacks are not used, the complex conjugate transmission zeros restrict the achievable performance.

### AVR System Requirements

- Zero steady state voltage error to a step change in the voltage reference.

- A terminal voltage rise time ranging from 0.2 to 2 secs depending upon the operating point.

- The desired closed-loop bandwidth for the voltage loop can be chosen to be about 10 radians/second.

- A 5% settling time from 1.5 to 10 secs for the output voltage over all operating conditions.

- An overshoot of less than thirty percent on the voltage signal output for a step change in reference.

- The electrical power output should have a 5% settling time of less than 10 seconds when a step disturbance arises due to a voltage reference change.

- The controller should ensure the dominant complex modes included in the transfer from excitation to electrical power and rotor speed have a minimum damping factor of about 0.2.

Brasca and Johnson (1994) have considered how the above requirements can be met for an LQG AVR design. They applied single input LQG design to a power generation example but also considered the wider implications of multivariable design.

## 10.2.1   Electro-mechanical Equations

The dynamics of a synchronous generator, connected to an electrical network, can be represented in the form shown in Fig. 10.4. The electro-mechanical loop contains a double integrator and is therefore almost critically stable (Arcidiacono et al. 1980). The torque generation and load angle loop is a pure oscillator with zero damping, if the electrical load damping is neglected. The natural frequency increases with the synchronising power coefficient and with the inverse of the size of the machine. This loop is illustrated in Fig. 10.4.

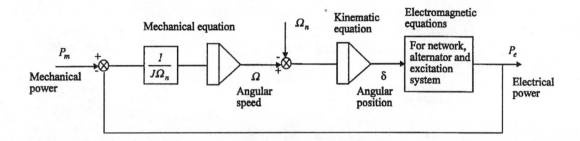

**Fig. 10.4** : Network, Alternator and Excitation System Electro-mechanical Loop

The mechanical equations represented in the first two blocks in Fig. 10.4 have the form:

$$P_m - P_e = J\Omega_n \frac{d\Omega}{dt} \qquad (10.1)$$

where $P_m$ = mechanical power; $P_e$ = electrical power; $J$ = moment of inertia of the turbine and alternator; $\Omega$ = Angular speed of the rotor; $\Omega_n$ = Synchronous speed.

The third block in this diagram results from the kinetic equation:

$$\Omega - \Omega_n = \frac{d\delta}{dt} \qquad (10.2)$$

where $\delta$ = angular position of the rotor (from a nominal point moving at synchronous angular speed). The fourth block represents the electro-mechanical equations of the alternator and network including the excitation control system. It relates the electrical power generated $P_e$ and the load angle $\delta$.

At a particular oscillation frequency, braking torques are developed in phase with the machine rotor angle and in phase with the machine rotor speed. The first type of these torques are termed synchronising torques and the second are referred to as damping torques. The stability of the system can be affected by a reduction in either of these torques. The damping torques are particularly important to the stability of machine operation, as systems are becoming more dependent on the automatic control of excitation voltage (Chow et al. 1992 and Doraiswami et al. 1984).

## 10.2.2 Electro-mechanical Oscillations

The dynamics of a single synchronous generator connected to an electrical network can be simplified to the form shown in Fig. 10.5. The electro-mechanical loop contains a double integrator and is therefore almost critically stable. The block diagram illustrating the interactions which occur within a voltage control and power generation system is shown in Fig. 10.5. The electro-mechanical loop is at the top of this diagram and the voltage control loop is at the bottom. The interaction between the voltage control loop and the electro-mechanical loop is characterised by:

1. A set of poles which may be real or complex and have a high damping factor.

2. A couple of complex poles with a low damping factor.

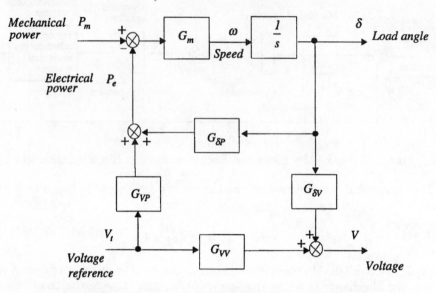

**Fig. 10.5:** Linearised Automatic Voltage Regulator

The first set of modes which are heavily damped have a dominant influence on the response of the electro-magnetic variables such as voltage, reactive power and excitation current. However, they have a rather low visibility in the response of the electro-mechanical variables, such as active electrical power and angular speed. These modes are referred to as the electro-magnetic modes (Brasca et al. 1993) and they have a natural frequency of $0.5 \sim 2$ Hz.

The oscillatory modes associated with the low damping factor have a dominant effect on the electro-mechanical variables but rather a low visibility in the electro-magnetic variables. The oscillations associated with this pair of highly damped modes are referred to as the electro-mechanical oscillations. The electro-mechanical oscillations are of course related to the double integrator in the electro-mechanical loop. Damping in the electro-mechanical loop is a function of the third block $G_{\delta p}$, shown in Fig. 10.5, which is dependent upon the electro-magnetic equations for the turbine/alternator system. The damping depends upon a positive phase contribution being made in the electro-mechanical loop by the transfer-function:

$$G_{\delta p}(s) = \Delta P_e(s)/\Delta \delta(s) \qquad (10.3)$$

The phase advance introduced by the transfer-function $G_{\delta p}(s)$ depends upon the design of the automatic voltage control loop and this is necessary for the stability of the electro-mechanical loop. The transfer-function depends upon the following: alternator parameters; network parameters; the AVR parameters and in particular the dynamic gain; the generator operating point; the possible presence of feedback from electric power and angular speed which can be added to the input of the voltage regulator.

### 10.2.3  Nonlinear Operation

An automatic voltage regulating system is very nonlinear but linearised models can be obtained for particular operating conditions. The nonlinear operating conditions are determined by a number of parameters including:

(a) the external reactance $x_e$ between the alternator and the external network.

(b) the amount of reactive power generated by the alternator.

(c) the amount of active power generated.

(d) the voltage generated by the alternator.

The reactive power generated can be positive or negative, depending upon the operating conditions, between the limits of 0.3 and 0.4 per unit. The active power during normal operation will lie between 0.3 and 0.9 per unit. Voltage variations are typically regulated between 0.95 per-unit in the under-excited situation to 1.05 per-unit in the over-excited condition. The external reactance of the system is dependent upon the network structure and the location of the power generating elements. Values of 0.1 to 0.4 per unit can be assumed typical. For the operating conditions of interest the natural frequency changes between 1.1 to 1.6 Hz and the actual damping lies between 0.08 and 0.37 (with the voltage loop open). Closing the voltage loop can destabilise the system but the use of additional feedbacks can enable both stability and performance to be enhanced.

### 10.2.4  Additional Feedback Control Signals

As noted the damping of the electro-mechanical oscillations can be improved by using additional feedback signals from power and speed to the voltage control loop. Unfortunately this feedback increases the interaction between the voltage and the electro-mechanical loops. The improvement of the damping of the lightly damped modes, in the electro-mechanical loop, is balanced by the deterioration in the performance of the voltage regulator. This arises because the electro-magnetic poles which are dominant in the voltage loop are moved by the additional feedback signals. The movement is of course related to the gain in these loops and if this gain is limited in magnitude, the deterioration in the voltage regulating performance can also be limited.

A design compromise therefore exists where the gain of the additional feedback loops should be set to a value to provide a reasonable damping for the electro-mechanical oscillations and yet at the same time not move the electro-magnetic poles too dramatically. An alternator/network system is of course strongly dependent upon the non-linear operating points and hence a primary aim must be to ensure stability for all possible operating conditions.

There was a move to larger alternator units in the 1960's and there were also demands for higher performance and both of these trends resulted in systems with dreceasing stability margins. This led DeMello and Concordia (1969) to first recommend the use of the additional stabilising feedbacks $K_p$ and $K_\omega$, to improve the stability of the Automatic Voltage Regulation system. The structure of such a system,

using a simple PI controller for voltage set-point tracking, is illustrated in Fig. 10.6,
The benefits of using adaptive systems for load and frequency control were considered
by Pierre (1987). Unfortunately, up to the present time there have only been a few
applications of adaptive control (Irving et al. 1979; Ibrahim et al. 1989; Kaliah et
al. 1984) in power generation systems, mainly because of the difficulty of guarantee-
ing reliable operation. A limited authority adaptive controller, which only allows the
power and speed gains to be adapted, should be easier to certify and should reduce
the possibility of unpredictable behaviour. Such a system was proposed by Brasca
et al. (1993) and is illustrated in Fig. 10.7. This approach is one example of what
has been termed limited authority adaptive control (Waal et al. 1996). This was a
term introduced by Grimble (1997), to describe adaptive systems where part of the
controller is fixed (non-adaptive).

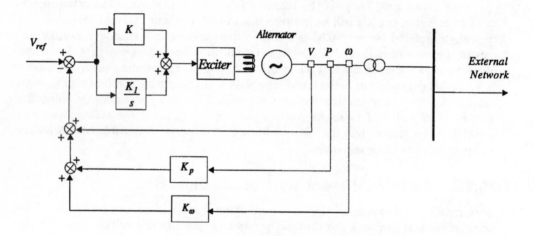

**Fig. 10.6**: Automatic Voltage Regulator Structure

There are many examples of *limited adaptive authority control*. For example a
feedback control system may require part of the controller structure to be fixed because
of the need for filtering action. That is, in certain applications a notch filter is needed
to remove unwanted disturbance signals from the control activity. If this disturbance is
not changing then the filter can be fixed and the remainder of the controller adapted.
Clearly, this is a particular example of limited authority adaptive control, where a
section of the controller is fixed. In some applications another situation occurs where
the gain of the plant varies slowly with operating condition. The major time constants
in the system may remain approximately constant and there is only a need to adapt to
the gain variations. An adaptive controller which essentially compensates for a static
gain variation is a further example of limited authority adaptation. Some authors have
also proposed using the Youla parameterisation for implementing adaptive control
structures. In this parameterization there is a Youla frequency response term which
can be adapted to improve robustness properties. Such a scheme would also fall within
the limited authority adaptive control concept. The main objective of this technique
is to simplify the numerical algorithms and to improve the reliability of operation. It

is hoped that more predictable performance will be obtained by limiting the variations within the controller. In such cases it is clearly easier to determine the likely impact of such changes by relatively simple techniques. For example, if simple gain variations are involved root loci techniques may be employed to analysis possible behaviour. The power system application described here is an excellent example of the benefits limited authority adaptation might offer.

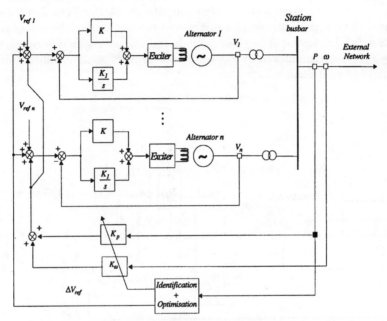

**Fig. 10.7** : Structure of Limited-authority Adaptive AVR

## 10.3   $H_2$ Automatic Voltage Regulator Design

The $H_2$ design of an automatic voltage regulator (AVR) and power system stabiliser (PSS) is now considered. Both 1 and 2.5 DOF structures will be considered (Grimble, 1992a, 1994a). The system is highly nonlinear, but a linearised model is used for a particular operating point. The performance of the AVR can be improved by the use of power and speed feedback signals, providing a power system stabiliser capability (Brasca, 1993). In classical design the individual loop controllers would be designed independently but the solution here takes into account the interaction naturally (Grimble, 1992b). It is a special case of a multivariable LQG solution. A single machine connected to an infinite bus system was shown in Fig. 10.1, and will now be discussed. The operating points and linearised models are dependent on the different reactive powers being generated by the machine.

The single input to the system corresponds with the excitation voltage for the alternator. The disturbances to the system arise through variations in mechanical power, changes in the load and the voltage of the network to which the machine is connected.

The primary voltage control system is of interest which controls the terminal voltage of the generator. Stabilising feedbacks are needed to reduce oscillations between the power station and the network. Oscillations between groups of generators can occur at frequencies as low as 0.5 Hz. The AVR tends to reduce the stability of the electro-mechanical modes and the additional output feedbacks improve stability. As noted, the frequency of the electro-mechanical modes is about 1 to 2 Hz. The damping of such a mode (without additional feedbacks) can be less than 0.05 and may even become negative in some operational situations. A damping factor of 0.2 can be assumed to be acceptable. In classical design the addition of the power and speed feedbacks which improve stability can degrade the voltage regulation properties. The SIMO optimal solution, shown in Fig. 10.8, can provide good stability margins and regulating loop performance.

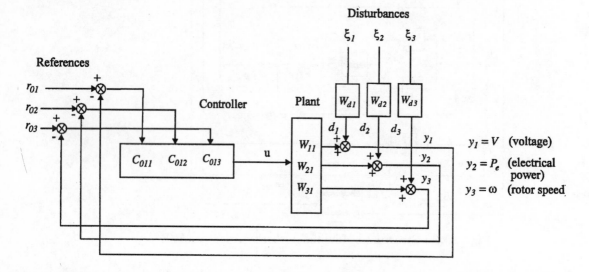

**Fig. 10.8**: Single-input Three-output Power System Model Structure

The plant model which follows is for an underexcited condition, where the reactive power is negative and the system is open loop unstable. If the external reactance of the system is increased, the open loop plant becomes more unstable.

Note that the plant models were obtained from a continuous-time nonlinear simulation, validated against real plant data, and the results are therefore computed in this continuous-time form. It may easily be shown that the previous polynomial results apply to the continuous time problem where $z^{-1}$ is replaced by the Laplace operator $s$ (excepting the $z^{-g}$ terms in Theorem 5.1 which must be replaced by unity).

**Plant Model :**

$$W_{11} = \frac{0.55[(s - 0.2545666)^2 + 12.78804^2](s + 17.78186)}{(s - 2.848749 \times 10^{-2})[(s + 0.4176358)^2 + 7.647355^2]}$$
$$\times (s + 16.1598)(s + 37.03344)$$

$$W_{21} = \frac{9[(s+0.0001)^2 + (4.98678x10^{-7})^2](s+17.77778)}{(s-2.848749 \times 10^{-2})[s+.4176258)^2 + 7.647355^2]}$$
$$\times (s+16.1598)(s+37.03344)$$

$$W_{31} = \frac{-20(s+0.0001)(s+18)}{(s-2.848749 \times 10^{-2})[(s+0.4176258)^2 + 7.647355^2]}$$
$$\times (s+16.1598)(s+37.03344)$$

**Disturbance Models :**

$$W_{d1} = \frac{0.0001}{(s^5 + 54s^4 + 700s^3 + 3600s^2 + 35000s - 1000)}$$

$$W_{d2} = \frac{0.000001(9s^3 + 160.0018s^2 + 0.032s + 1.6 \times 10^{-6})}{(s^5 + 54s^4 + 700s^3 + 3600s^2 + 35000s - 1000)}$$

$$W_{d3} = \frac{0.000001(-20s^2 - 360.002s - 0.036)}{(s^5 + 54s^4 + 700s^3 + 3600s^2 + 35000s - 1000)}$$

**Reference and Noise Models :**

$$W_{r1} = \frac{0.000001}{(s^5 + 54s^4 + 700s^3 + 3600s^2 + 35000s - 1000)}$$

$$W_{r2} = W_{r3} = 10^{-10} \times W_{r1}$$

The measurement noise terms are assumed to be set to zero.

The three input and single ouput controller that minimises the $H_2$ cost index (5.20) can be computed using Theorem 5.1 and the following weightings.

**Feedback Controller Cost Weightings:**

$$Q_{c0} = H_{q0}^* H_{q0} \quad , \quad R_{c0} = H_{r0}^* H_{r0}$$

where $H_{q0} = diag\{4000/s, \quad 600, \quad 10\}$ and $H_{r0} = 0.002 \ (s+20)^4$

**Tracking Controller Cost Weightings :**

$$Q_{c1} = H_{q1}^* H_{q1} \quad , \quad R_{c1} = H_{r1}^* H_{r1}$$

where $H_{q1} = diag\{4000/s, \quad 100, \quad 1\}$ and $H_{r1} = 2(1 + s/20)(1 + s/5)^3$ and ideal response model :

$$W_i = diag\{1/(1+s), 1, 1\}$$

**Computed Feedback Controller** (after cancellation of near common factors):

$$C_0(1,1) = \frac{\begin{array}{c} 443.9(s+1.315244)(s^2 + 0.7454086s + 57.2394) \\ \times(s+16.15969)(s+37.03344) \end{array}}{\begin{array}{c} (s+0.0004)(s^2 + 21.88548s + 314.806) \\ \times(s+17.88275)(s+43.80819) \end{array}}$$

$$C_0(1,2) = \frac{850.6237((s - 2.848749 \times 10^{-2})(s - 11.46512)}{(s^2 + 0.0002s + 10^{-8})(s^2 + 21.88548s + 314.806)} \\ \times (s + 16.17921)(s + 37.05376)) \\ \overline{\phantom{(s^2 + 0.0002s + 10^{-8})(s^2 + 21.88548s + 314.806)}} \\ \times (s + 17.77778)(s + 17.88275)(s + 43.80819)$$

$$C_0(1,3) = \frac{-1.066711 \times 10^{-2}(s - 2.848958 \times 10^{-2})}{(s + 0.0004)(s^2 + 21.88548s + 314.806)} \\ \times (s - 9.000166)(s + 16.1552)(s + 37.03344) \\ \overline{\phantom{(s + 0.0004)(s^2 + 21.88548s + 314.806)}} \\ \times (s + 17.88275)(s + 18)(s + 43.80819)$$

## 10.3.1   Frequency and Time-domain Results

The frequency and time response for the three output and single input system will now be presented. The frequency responses of the plant model are considered first. The plant frequency responses are shown in Fig. 10.9. The resonance frequency is at 7.7 rad/sec. The voltage feedback loop has a high gain at low frequency and the power and speed feedback loops have respectively, derivative or double derivative action at low frequencies, respectively. The frequency responses of the disturbance models are shown in Fig. 10.10. These models were deliberately chosen to have similar characteristics as those of the plant itself.

The step response of the voltage control loop must be reasonable with small overshoot and low interaction to the other system outputs. The reference model for the voltage control loop is for simplicity chosen to be similar to that for the low frequency disturbance inputs. This avoids increasing the degree of the controller. Measurement noise was set at a zero value for the design, since the cost function weighting on the control signal includes lead terms to ensure the controller rolls off at high frequencies.

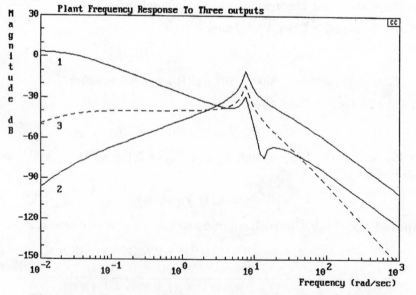

Fig. 10.9 : Frequency Responses of the Three Plant Outputs

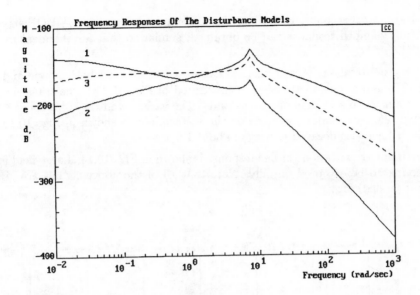

**Fig. 10.10** : Frequency Responses of the Three Disturbance Models

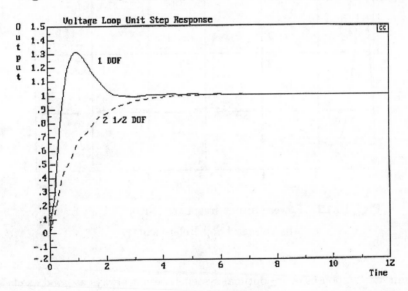

**Fig. 10.11** : Unit Step Response of the Voltage Regulator Loop
for 1 and 2.5 DOF Designs

## Time Responses

The unit step response of the voltage control loop is shown in Fig. 10.11. This shows both the 1 and $2\frac{1}{2}$ degree of freedom designs. The system has small non-minimum phase zeros and is open loop unstable. It is not therefore surprising that the step response for the 1 DOF design has some overshoot and there is a limit to how fast this response could be made without getting unacceptable overshoots in either the voltage or other loops. By adding the tracking controller in a $2\frac{1}{2}$ DOF structure, a critically damped response can be obtained, similar to that for the ideal response model $(W_{di})$

A step in voltage reference demand causes a transient on both power and rotor speed. The power output transient is illustrated in Fig. 10.12. The transients have decayed almost to zero in about 1.5 seconds. The variation in the rotor speed when a unit step reference demand is made on the voltage loop is shown in Fig. 10.13. The transients have also decayed to zero in about 1.5 seconds.

The input or excitation voltage response is shown in Fig. 10.14. Large field forcing is of course to be excepted initially. Note that all of the above signals are given in per-unit (scaled) form.

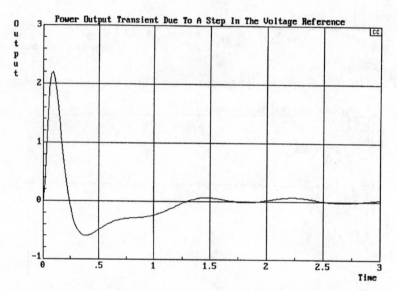

Fig. 10.12 : Power Output Response Due to a Unit Step
in Voltage Loop Reference

The unit step responses of $H_2$ optimal systems are not always as good as classical designs.    The least squares approach involves the integral of error squared in deterministic problems and the minimization of error variance in stochastic problems. This does not necessarily correspond to improving unit step time responses. In fact the $H_\infty$ design method often provides improved unit-step responses and if $H_2$ properties are also important this suggests the use of combined $H_2$ and $H_\infty$ algorithms.

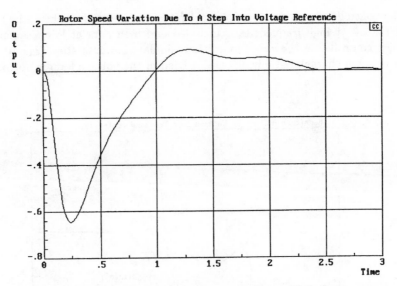

**Fig. 10.13**: Rotor Speed Response Due to a Unit Step
in Voltage Reference

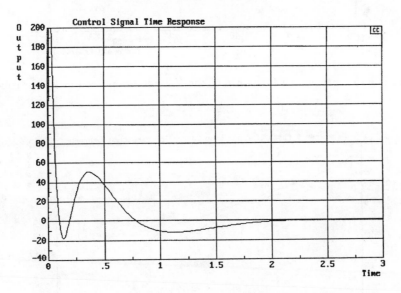

**Fig. 10.14** : The Excitation System Input for a Unit Step
in Control Reference Demand

**Controller Frequency Responses**

Not surprisingly the controller frequency response (Fig. 10.15) has a large gain
and phase shift in the same frequency range where the plant resonance resides. The
controller has limited gain at high frequency and integral action at low frequencies.
Notice that it could be approximated by an integral and proportional controller with
a notch filter characteristic.

The controller frequency response for the power feedback is shown in Fig. 10.16. Good roll off at high frequencies is included and high gain at low frequencies. The feedback controller in the speed loop has a similar characteristic (Fig. 10.17). Note that the notch characteristic which was needed in the voltage loop is not present.

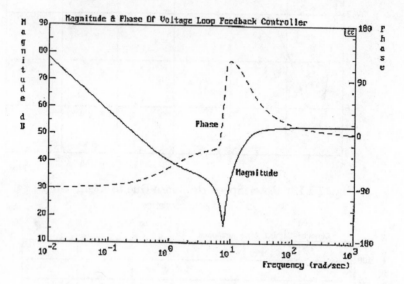

Fig. 10.15 : Frequency Response of the Voltage Loop Feedback Controller

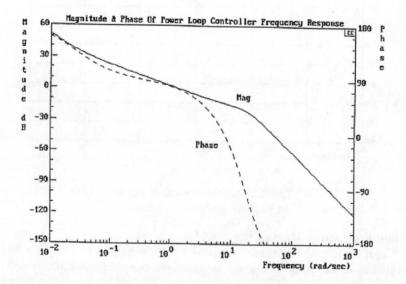

Fig. 10.16 : Frequency Response of the Power Loop Feedback Controller

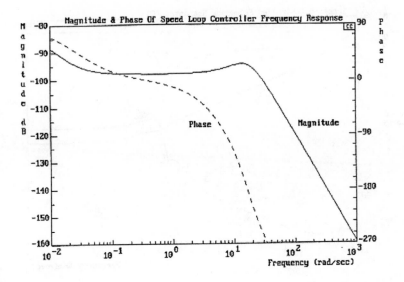

**Fig. 10.17** : Frequency Response of the Speed Loop Feedback Controller

The tracking controller frequency responses for the system in the $2\frac{1}{2}$ DOF structure are shown in Fig. 10.18. The physical justification for the particular frequency responses is not obvious. In fact there is not a very clear classical design procedure for such a controller. The solution is of course provided naturally in an optimal framework. Only the voltage loop response is important in this case and hence the weightings were small on the power and speed errors.

Recall that the combined feedback and tracking controllers are represented by the compensator $C_{01}$. The frequency responses of the elements corresponding to the voltage $C_{01}(1,1)$, power $C_{01}(1,2)$ and speed $C_{01}(1,3)$ loops are shown in Fig. 10.19.

Note that the power loop and disturbance transfer functions include almost a double differentiation and this leads to the need for integral action in the power feedback loop controller. If the power feedback includes a bias then integral action will not be acceptable but it will be assumed here that adequate sensors are available.

**Sensitivity Function**

The sensitivity function frequency response is shown in Fig. 10.20. Note that the peak magnitude of the sensitivity function is 1.351 at a frequency of 4.272 rads/sec. In classical design a rule of thumb is that the overshoot on the sensitivity will bear a close relationship to the step response overshoot, which is also the case here.

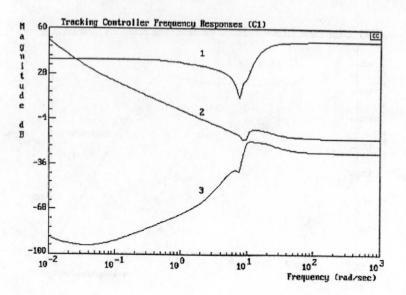

Fig.10.18: Frequency Response of the Tracking controllers
for the Voltage, Power and Speed Outputs

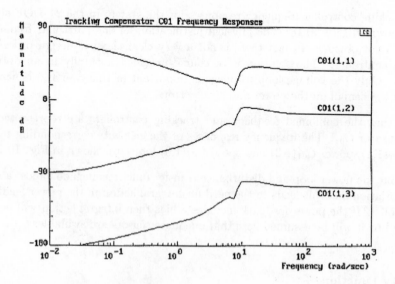

Fig. 10.19 : Tracking Compensator $C_{01}$ Frequency Responses
for the Voltage, Power and Speed Loops

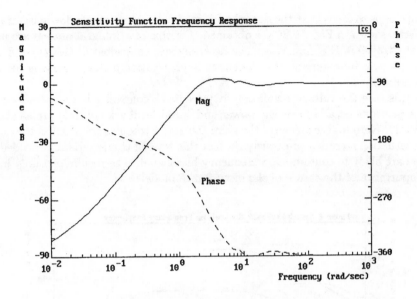

**Fig. 10.20** : Sensitivity Function Frequency Responses Measured
at the Control Signal Input

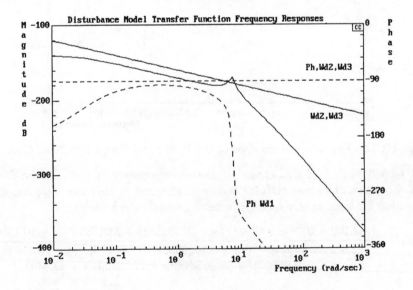

**Fig. 10.21**: Frequency Response of the Disturbance Model
for the Voltage, Power and Speed Signals

## Modified Disturbance Model

The choice of disturbance models has an important influence on the form of the
controller characteristics, particularly for the power and speed feedback loops. The
disturbance models for these loops were earlier chosen to have the same type of resonance characteristics as those of the plant transfer itself. However, if the common

assumption is made is that the disturbance models are dominantly low pass, the characteristics shown in Fig. 10.21 are obtained. For this case integral models are used on both the transfers $W_{d2}$ and $W_{d3}$. The deterministic equivalent of this type of disturbance is a step function and the stochastic interpretation a slowly varying integrated white noise signal.

In this case the voltage regulating loop feedback controller is very similar to that for the previous occasion but the power and speed feedback controls are as shown in Fig. 10.22. Note that in this case the gains fall at low frequencies which is the opposite to the situation recorded previously. In fact this type of characteristic is needed if the sensors are likely to contain a low frequency bias signal. The results certainly indicate the importance of the choice of the disturbance models.

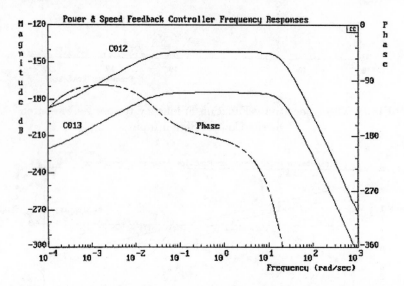

**Fig. 10.22** : Frequency Responses of the Power and Speed Feedback Controllers

A second feature of this choice of disturbance model is also evident from Fig. 10.22. Note that the phase shift for the two controllers is the same. Inspection of the expressions for these controllers, given below, reveals the reason :

$$C_0(1,2) \frac{-2.102 \times 10^{-2}(s+.0001)[(s+.4176258)^2 + 7.647355^2](s+16.1598)}{\substack{(s + 2.848749x10^{-2})[(s + 0.4459652)^2 + 7.652165^2](s + 16.15845) \\ \times [(s + 10.96063)^2 + 13.92587^2](s + 17.78503)(s + 43.81475)}}$$

$$C_0(1,3) \frac{-4.6279 \times 10^{-4}(s+1.0069x10^{-4})[(s+0.4176258)^2 + 7.647355^2]}{\substack{(s + 2.848749x10^{-2})[(s + 0.4459652)^2 + 7.652165^2](s + 16.15845) \\ \times [(s + 10.96063)^2 + 13.92587^2](s + 17.78503)(s + 43.81475)}}$$

The two controllers have the same transfer function except that the power loop has a gain which is approximately 45 times that of the speed loop transfer. For such a case, there are clear computational savings that could be made in obtaining these controllers which would be valuable in adaptive systems.

## 10.4   $H_\infty$ Automatic Voltage Regulator Design

The $H_\infty$ design of an automatic voltage regulator and power system stabiliser for both 1 and 2.5 DOF structures will now be considered. The system and problem are similar to those considered in the previous section.

Let the following system models be defined:

$$num1 = (0.55s^3 + 9.5s^2 + 85s + 1600)$$

$$num2 = ((9s + 160) \times s^2)$$

$$num3 = ((-20s - 360) \times s)$$

$$den = s^5 + 54s^4 + 700s^3 + 3600s^2 + 35000s - 1000 = den1 \times den2$$

where

$$den1 = (0.02848749 - s)((s + 0.4176258)^2 + 7.6473552))(s + 37.03344)$$

$$den2 = (s + 16.1598)$$

**Plant Model:**

$$W = \begin{bmatrix} W_{11} \\ W_{21} \\ W_{32} \end{bmatrix} = \begin{bmatrix} num1/den \\ num2/den \\ num3/den \end{bmatrix}$$

and the polynomials:

$$B = \begin{bmatrix} B_{11} \\ B_{21} \\ B_{32} \end{bmatrix} = \begin{bmatrix} num1 \\ num2 \times s \\ num3 \times s \end{bmatrix}$$

and

$$A = diag\{A_{11}, A_{22}, A_{33}\}$$

where $A_{11} = $ den, $A_{22} = $ den $\times s$, $A_{33} = $ den $\times s$.

**Disturbance Model:**

$$W_d = diag\{W_{d1}, W_{d2}, W_{d3}\} = diag\{C_{d1}/A_{11}, \;\; C_{d2}/A_{22}, \;\; C_{d3}/A_{33}\}$$

where $C_{d1} = 10^{-4}, C_{d2} = 10^{-8} \times$den, $C_{d3} = 10^{-8} \times$den.

**Reference Model:**

$$W_r = diag\{W_{r1}, W_{r2}, W_{r3}\} = diag\{E_{r1}/A_{r1}, \;\; E_{r2}/A_{r2}, \;\; E_{r3}/A_{r3}\}$$

where $E_{r1} = 10^{-6}$, $E_{r2} = 10^{-16}$, $E_{r3} = 10^{-16}$ .

$$E = diag\{E_{r1},\ E_{r2} \times den,\ E_{r3} \times den\}$$

and $A_{r1} =$ den1,    $A_{r2} = s$,    $A_{r3} = s$

The measurement noise will be assumed zero in the simulation results which follow.

**Generalised Cost:**

Given certain assumptions the functions $T_0$ and $T_1$ can be shown to be zero. Assuming, null measurement noise:

$$Q_{c0} = H_{q0}^* H_{q0}, \qquad R_{c0} = H_{r0}^* H_{r0}, \qquad G_{c0} = -H_{q0}^* H_{r0}$$

with $W_i = I_r, H_f = I_r$.

**Weighting Selection and Design Issues**

The $H_\infty$ cost-function is to be minimised for both feedback and tracking terms. The error and control weighting terms for the feedback controller ($Q_{c0}$ and $R_{c0}$) have the frequency responses shown in Fig. 10.23.

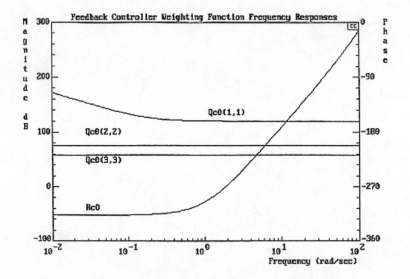

**Fig. 10.23** : Frequency Response of Feedback Controller Tracking Error and Control Weighting Functions

The shape of these frequency responses depends upon the following:

1. Integral action is required in the voltage control loop and hence the error weighting term for this loop includes an integrator.

2. The weightings on the voltage loop are much greater than on the power and speed feedbacks, since the role of these loops is to improve system stability.

3. The control weighting function includes lead terms, so that the controller does not have too high a gain at high frequency.

4. The usual rule that the bandwidth of the closed loop system is determined by the crossover point between the error and control weighting functions, does not apply exactly in this case because of the multiloop nature.

Consider now the frequency responses of the tracking controller, shown in Fig. 10.24. The selection of these weightings is based upon the following ideas.

(i) Tracking error must be penalised at low frequencies in the voltage feedback loop but there is no requirement for tracking on the other two loops, so that the corrresponding error weightings can be set to a low value.

(ii) The controller weighting function includes the lead term to ensure the controller rolls off at high frequencies.

(iii) The crossover frequency between the characteristics is at a reasonably high frequency, although there is not a direct correlation between the speed of the response of the system and this crossover point. The lower the control weighting and the higher the crossover frequency the faster will be the system response.

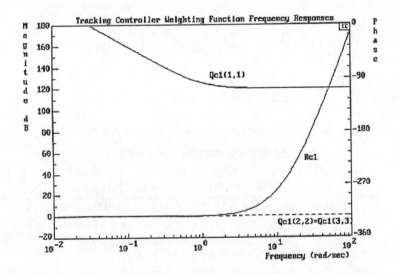

Fig. 10.24 : Frequency Response of Tracking Controller Error and Control Weighting Functions

**Weighting Function Definitions**

**Feedback control weightings:**

$$Q_{c0} = diag H_{q011}^* H_{q011}, \ H_{q022}^* H_{q022}, \ H_{q033}^* H_{q033}$$

$$R_{c0} = H_{r0}^* H_{r0} \quad and \quad G_{c0} = -[0, \ 0 \ , 0]^T$$

where

$$H_{q011} = 200(1 + s/0.1)/s, \quad H_{q022} = 60, \quad H_{q033} = 10$$

and

$$H_{r0} = 0.00625(s + 1)^4$$

**Tracking control weightings:**

$$Q_{c1} = diag\left\{H_{q111}^* H_{q111}, \ H_{q122}^* H_{q122}, \ H_{q133}^* H_{q133}\right\}$$

$$R_{c1} = H_{r1}^* H_{r1} \quad and \quad G_{c1} = [0, \ 0, \ 0]$$

where

$$H_{q111} = 100(1 + s/3)/s, \quad H_{q122} = 10, \quad H_{q133} = 2$$

and

$$H_{r1} = 0.03(1 + s/10)^4$$

Notice that this is a mixed sensitivity type of cost-function (with null cross product terms) as defined by (63).

**Computed $H_\infty$ Feedback Controller**

The algorithm used to compute the controllers involved the approximation mentioned earlier that $T$ is a constant (or represented by its maximum value). The three controller terms for the voltage, power and speed loops are respectively:

$$C_0(1,1) = \frac{\begin{array}{c} -0.5750344(s + 0.0940344)(s^2 + 3.265857s + 19.3615) \\ \times(s^2 + 7.482898s + 19.66633)(s + 4.486526)(s^2 + 0.252573s + 57.93452) \\ \times(s^2 + 1.040668s + 59.86669)(s + 16.15976)(s + 37.0.3344) \end{array}}{\begin{array}{c} s(s^2 + 7.468646s + 19.59273)(s^2 + 3.233045s + 19.65562) \\ \times(s + 4.492749)(s^2 + 6.75191s + 51.3992) \\ \times(s^2 + 1.137333s + 61.02896)(s + 8.126404) \end{array}}$$

$$C_0(1,2) = \frac{\begin{array}{c} 2.239482(s + 10^{-4})(s - 0.0284874)(s + 0.492682) \\ \times(s + 3.937327)(s^2 + 6.112064s + 15.75014)(s^2 + 0.2184153s + 65.40228 \end{array}}{\begin{array}{c} (s + 0.02848749)(s^2 + 7.468646s + 19.59273)(s^2 + 3.233045s + 19.65562) \\ \times(s + 4.392749)(s^2 + 6.75191s + 51.3992) \\ \times(s^2 + 1.137333s + 61.02896)(s + 8.126404) \end{array}}$$

$$C_0(1,3) = \frac{\begin{array}{c} -2.543113 \times 10^{-3}(s + 0.0001)(s - 0.02848749)(s + 3.979654) \\ \times(s^2 + 6.202844s + 15.99795)(s^2 + 0.2709231s + 66.52635)(s - 49.29886) \end{array}}{\begin{array}{c} (s + 0.02848749)(s^2 + 7.468646s + 19.59273)(s^2 + 3.233045s + 19.65562) \\ \times(s + 4.492749)(s^2 + 6.75191s + 51.3992) \\ \times(s^2 + 1.137333s + 61.02896)(s + 8.126404) \end{array}}$$

### 10.4.1 Frequency and Time Domain Results

The time and frequency responses for the SIMO system will now be considered. The plant frequency responses for this $H_\infty$ design are shown in Fig. 10.25. The frequency response of the disturbance models is shown in Fig. 10.26. The voltage loop model was chosen to have a similar characteristic to the plant itself. The disturbance models for the remaining loops are integrators to represent low frequency disturbances.

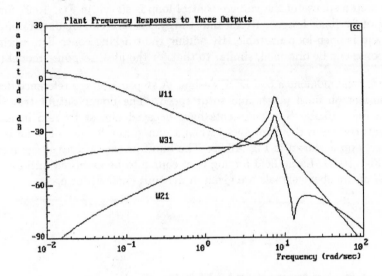

**Fig. 10.25** : Single-input Multi-output Plant Transfer Function Frequency Responses

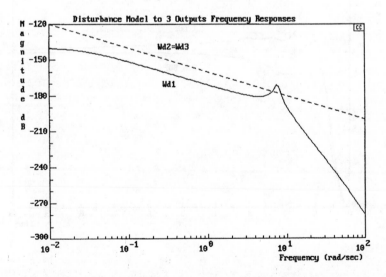

**Fig. 10.26** : Disturbance Model on the Three Plant Outputs

The step response of the voltage control loop must be reasonable with small overshoot and low interaction to the other system outputs. The reference model for the voltage control loop is chosen to be the same as a slow plant mode. This avoids some increase in the degree of the controller.

### Time Responses

The unit step response of the voltage control loop is shown in Fig. 10.27 for both the 1 and $2\frac{1}{2}$ of freedom designs. Recall that the system has two small non-minimum phase zeros and is open-loop unstable. By adding the tracking controller, a critically damped response can be obtained, similar to that for the ideal response model ($W_{di}$).

Consider for the moment a one DOF design. A step in voltage reference demand causes a transient on both power and rotor speed. The power output transient is illustrated in Fig. 10.28. The transients have decayed almost to zero in about 4 seconds. The variation in the rotor speed when a unit step reference demand is made on the voltage loop is shown in Fig. 10.29. The input or excitation voltage response is shown in Fig. 10.30. Large field forcing is of course to be excepted initially. Note again that all of the above signals are given in per-unit (scaled) form.

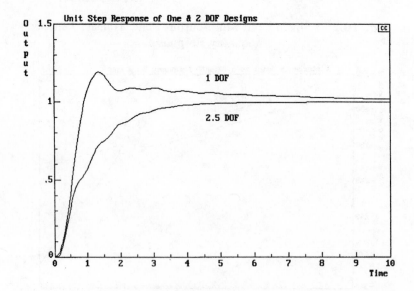

**Fig. 10.27** : Unit Step Respsone of the Voltage Control Loop

for 1 and 2.5 DOF Designs

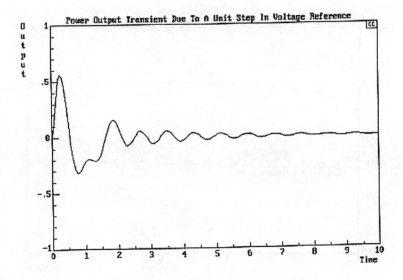

Fig. 10.28 : Transient in the Power Output due to Unit Step Demand
in the Voltage Reference

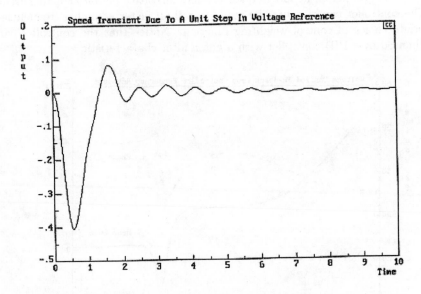

Fig. 10.29 : Transient Response in the Rotor Speed due to a
Unit Step Response in the Voltage Reference

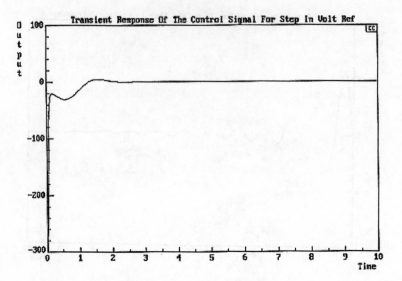

**Fig. 10.30** : Plant Input Voltage Resulting from a Step Reference
in the Voltage Loop

**Controller Frequency Responses**

The voltage loop controller frequency response (Fig. 10.31) has a large gain and phase shift in the same frequency range where the plant resonance resides. The controller has integral action at low frequencies and derivative action at higher frequencies. The controller can of course be made to roll-off faster at higher frequencies by appropriate choice of control weighting function. Notice that the controller can be approximated by a PID controller with a notch filter characteristic.

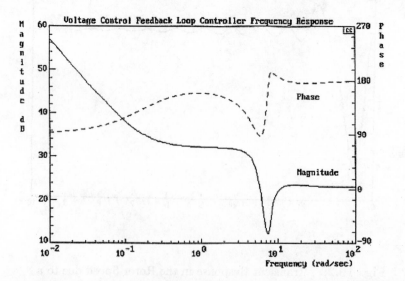

**Fig. 10.31** : Frequency Response of the Voltage Feedback Loop Controller

The controller frequency response for the power feedback loop is shown in Fig. 10.32. Good roll off at high frequencies is included and low gain at low frequencies. The feedback controller in the speed loop has a similar characteristic (Fig. 10.33). Note that the notch characteristic which was needed in the voltage loop is not present.

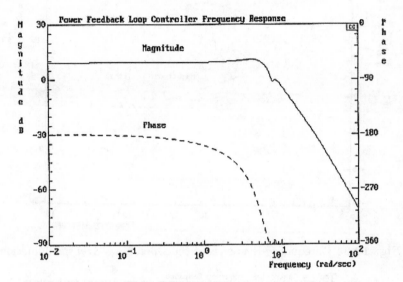

**Fig. 10.32** : Frequency Response of Power Feedback Loop Controller

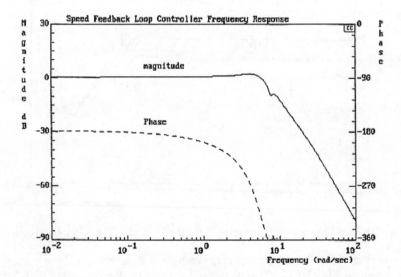

**Fig. 10.33** : Frequency Response of the Speed Feedback Loop Controller

The open loop transfer function frequency response is shown in Fig. 10.34. Note that the unity gain crossover frequency is in the region expected, according to the crossover frequency for the error and control weighting functions. The system also has integral action at low frequencies and reasonable roll off at high frequencies.

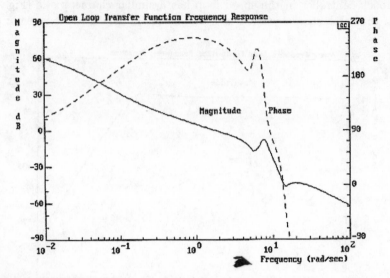

Fig. 10.34 : Frequency Response of the Open Loop Transfer Function

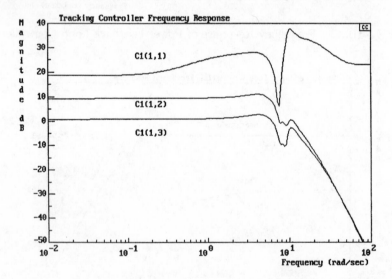

Fig.10.35 : Frequency Response Magnitude of the Tracking Controller

The tracking controller frequency responses for the system in the $2\frac{1}{2}$ DOF structure are shown in Fig. 10.35. The physical justification for the particular frequency responses is not obvious. In fact there is not a very clear classical design procedure for such a controller. The solution is of course provided naturally in an optimal framework. Note that only the tracking error for the voltage loop is to be minimised in this case and hence the weightings on the power and speed loops were set to low values which explains the low gains for the respective controllers.

Recall that in the 2.5 DOF design the combined feedback and tracking controllers are represented by the compensator $C_{01}$. The frequency responses of the elements corresponding to the voltage $C_{01}(1,1)$, power $C_{01}(1,2)$ and speed $C_{01}(1,3)$ loops are shown in Fig. 10.36. This is of course the same as the response of a 2 DOF controller.

**Sensitivity Function**

The sensitivity function frequency response is shown in Fig. 10.37. Note that the peak magnitude of the sensitivity function is 1.583 at a frequency of 4.5 rads/sec. In classical design a rule of thumb is that the overshoot on the sensitivity will bear a close relationship to the step response overshoot, as observed in this example.

The $H_\infty$ design approach is particularly relevant for minimising sensitivity functions. The $H_\infty$ criterion can be used to measure the peak value of the sensitivity function frequency response and this can then be optimised using the $H_\infty$ controller. In practice a criterion must include control sensitivity minimisation to also limit actuator variations. However, the $H_\infty$ design method provides a natural tool for the minimization of sensitivity.

Power system faults cause major difficulties and the $H_\infty$ design method also provides a mechanism of dealing with these fault conditions. That is, a possible line fault can be treated as an uncertainty. The $H_\infty$ controller can therefore be made robust to such fault conditions. This is of course the basis of so called *reliable control* which has been discussed previously.

The major problem in using $H_\infty$ design for power system transmission control lies in the high order of the controllers obtained. Traditional controllers for such systems are of relatively low order and are often of simple PID form. Users may only be convinced of the value of $H_\infty$ solutions if improved reliability and robustness can be guaranteed.

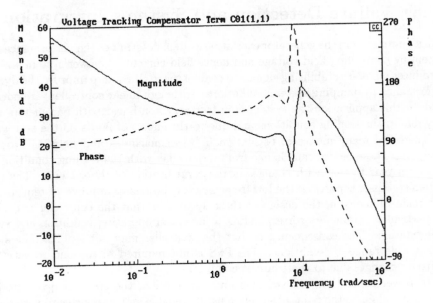

Fig. **10.36** : Frequency Response of the Compensator $C_{01}$

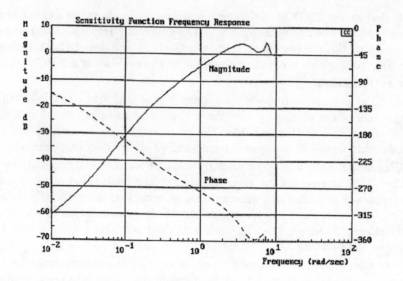

**Fig. 10.37** : Frequency Response of the Sensitivity Function

**$H_2$ and $H_\infty$ Comparison**

Clearly either the $H_2$ or the $H_\infty$ solutions would be acceptable, although the responses obtained are rather different. The design process was a little simpler for the $H_2$ solution although good step responses are more difficult to achieve. The problems of implementation would be the same for the two design approaches.

## 10.5 Failure Detection and Systems Integration

In a power system the generator excitation system determines the performance of the system by controlling field voltage and hence field current. A power system stabiliser introduces auxiliary stabilising signals to control the excitation to improve the dynamic performance by damping system oscillations. Although linear controllers are normally used for this application the system is in fact very non-linear with parameters which vary due to the loading conditions and due to the effects of faults on the system.

There are many modes of oscillation between machines and between groups of machines. These modes can change from day to day with the loading conditions and also from year to year with changes in the configuration of the system. The power system stabiliser suppresses the low frequency oscillations to improve system stability. The main difficulty in the design of these systems is that the controller is designed for particular parameters corresponding to a typical operating condition and system configuration. The consequence is that the controller may not stabilise the system over a wide range of conditions. The PSS is not required to respond to very rapid transient changes due to short circuit conditions.

In power generation, transmission and distribution, the entire network from the generating plant to the point of supply must be reliable and responsive to variations in demand. There is often the need in the generation problem for the integrated control system to be combined with intelligent condition monitoring, performance monitoring and optimisation (Uchida et al. 1981). In the case of the transmission and distribution system the emphasis is towards the automation of the distribution network to achieve optimal use of resources, while ensuring continuity of supply. This can be accomplished by integrating intelligent network management and information systems with the supervisory control and monitoring systems, for remote plant (such as protection systems, switch gear etc.). The system integration problem also involves the need to integrate products from many different suppliers. This includes, plant, sensors, condition monitoring systems, control and data acquisition systems, and process management and optimisation systems.

## 10.6 Conclusions

A polynomial solution was presented in the earlier chapters to both the $H_2$ and $H_\infty$ design problems for SIMO systems. This was very suitable for the AVR design problem discussed above. These solutions are applicable to multi-loop systems having several measurements and a single input. In fact there are subtleties in the design of SIMO systems which are not addressed in classical control design. For example, in what order should loops be tuned and how can the best stability characteristics be achieved. The optimal control approach provides a natural means of designing such systems with useful tuning variables (an error weighting for each output).

The interest in this problem was motivated by the need to develop an industrial controller for a SIMO power system application, which was described in the $H_2$ and $H_\infty$ design examples. The polynomial approach is relatively straightforward and the ARMAX model form is very suitable if an adaptive control solution is desired. Moreover, it is even possible to obtain stabilising low order optimal controllers that will be suitable for a set of the linearised operating points. The so called multiple model restricted structure optimal controllers are based on a polynomial systems framework (Grimble, 2000). Such techniques are also of value for wind-turbine control (Grimble, 1994b).

There is also considerable potential for the predictive control methods introduced in Chapters 3 and 8. Predictive control is very suitable for the supervisory control of power systems (Katebi et. al. 1995; Ordys and Grimble, 1996; Duan et al. 1997), by invoking the constrained optimisation features.

## 10.7 References

Arcidiacono, V., Ferrari, E., Marconato, R., Dos Ghali, J. and Grandez, D., 1980, Evaluation and improvement of electromechanical oscillation damping by means of eigenvalue-eigenvector analysis, *IEEE Trans. On Parallel and Distributed Systems*, Vol., **98**, pp. 771-78.

Brasca, C.L., 1993, Polynomial systems control design for synchronous generators, MPhil Thesis, Industrial Control Centre, University of Strathclyde, Glasgow.

Brasca, C.L., Arcidiacono, V. and Corsi, S., 1993, An adaptive excitation controller for synchronous generators, *Proceedings of 2nd IEEE Conference on Control Applications*, Vol. **1**, pp. 305-318, Vancouver, Canada.

Brasca, C.L. and Johnson, M.A., 1994, On automatic voltage regulator design for synchronous generators, *The 3rd IEEE Conference and Control Applications*, Vol. **1**, pp. 199-206.

Cheetah, R.G. and Walker, P.A.W., 1982, A turbo generator self-tuning regulator, *Int. J.Control.* Vol. **36**, No. 1, pp. 127-142.

Chen, G.P., Malik, O.P. and Hancock, G.C., 1994, Implementation and experimental studies of an adaptive self-optimisation power systems stabilizer, *Control Engineering Practice*, Vol. **2**(6), pp. 969-977.

Cheng, S.J., Chew, Y.S., Malik, O.P. and Hope G.S., 1986, An adaptive synchronous machine stabiliser, *IEEE Transactions on Power Systems,* Vol. **1**(3), pp. 101-9.

Chow, J.H., Harris, L.P., Kale, M.A., Othman, H., Sanchez-Gascci, J.J. and Terwilliger, G.E., 1992, Robust control design of power system stabilizers using multivariable frequency domain techniques, *Int.J. of Robust and Nonlinear Control*, Vol. **2**, No. 2, pp. 123-138.

DeMello, F.P. and Concordia, C., 1969, Concept of synchronous machine stability as affected by excitation control, *IEEE Trans. On PAS*, Vol. **87**, No., 4, pp. 324-333.

Doraiswami, R., Sharaf, A.M. and Castro, J.C., 1984, A novel excitation control design for multivariable-machine power systems, *IEEE Transactions on Power Apparatus and Systems.* Vol. **103**(5), pp. 1052-58.

Duan, J., Grimble, M.J. and Johnson, M.A., 1997, Multivariable weighted predictive control, *Journal of Process Engineering,* Vol. **7**, No. 3, pp. 219-235.

Fork, K. and Schreurs, C., 1988, An adaptive regulator of the excitation of synchronous generators, *IFAC Symposium on Power Systems, Modelling and Control Application*, pp. 141-147.

Forrest, S.W., 1992, Self-tuning LQG control: Theory and Applications, PhD thesis, Industrial Control Centre, University of Strathclyde, Glasgow.

Forrest, S.W., Grimble, M.J. and Johnson, M.A., 1992, On LQG self-tuning control: Implementation and application, *Trans. Inst. MC.* Vol. **14**, No. 5, pp. 243-253.

Grimble, M.J., 1997, New directions in adaptive control theory and applications, *IFAC-IFIP-IMACS Conference on the Control of Industrial Systems*, Belfort, France, pp.95-110.

Grimble, M.J., 1992a, Youla parameterized $2\frac{1}{2}$ degrees of freedom LQG controller and robustness improvement cost weighting, *IEE Proceedings*, Pt. D., Vol. **139**, No. 2, pp.147-160.

Grimble, M.J., 1992b, Model reference predictive LQG optimal control law for SIMO systems, *IEEE Trans. On Auto. Control*, Vol. **37**, No. 3, pp. 365-371.

Grimble, M.J., 1994a, *Robust Industrial Control*, Prentice Hall, Hemel Hempstead.

Grimble, M.J., 1994b, Two and a half degrees of freedom LQG controller and application to wind turbines, *IEEE Trans. AC*, Vol. **39**, No. 1, pp. 122-127

Grimble, M.J., 2000, Restricted structure LQG optimal control for continuous-time systems, *IEE Proceedings*, Part D, Control Theory and Applications, vol. 147, No. 2, pp. 185-195.

Ibrahim, A.S., Hogg, B.W. and Sharaf, M.M., 1989, Self-tuning automatic voltage regulators for a synchronous generator, *IEE-Proceedings*, Part D., Vol. **136** (5), pp. 252-260.

Irving, E., Barret, J.P., Charcossey, C. and Monville, J.P., 1979, Improving power network stability and unit stress with adaptive generator control, *Automatica*, Vol. **15**, No.1, pp. 31-46.

Kaliah, J., Malik, O.P. and Hope, G.S., 1984, Excitation control of synchronous generators using adaptive regulators - Parts I and 2, *IEEE Trans. On Pas.*, Vol. **103**, No. 5, pp. 887-910.

Katebi, M.R., Johnson, M.A. and Pike, A.W., 1995, Constrained control for power generation plant, *IFAC Conference on Large Scale Systems*, London.

Law, K.T., Hill, D.J. and Godfrey, N.R., 1994, Robust coordinated AVR-PSS, *IEEE Trans. On Power Systems*, Vol. **9**, pp. 1218-1225.

Ledwich, G., 1979, Adaptive excitation control, *Proc. IEE.*, Vol. **126**, pp. 249-253.

Ordys, A.W. and Grimble, M.J., 1996, A multivariable dynamic performance predictive control with application to power generation plants, *IFAC World Congress*, San Francisco.

Pierre, D.A., 1987, A perspective on adaptive control of power systems, *IEEE Trans. on Power systems*, Vol. **2**, pp. 387-396.

Pike, A.W., Katebi, M.R., Johnson, M.A. and Ordys, A.W., 1995, Hierarchical modelling and interactive simulation of large power plants, *IFAC Conference on Large Scale Systems*, London.

Sharaf, S.M.Z., Hogg, B.W., Abdalla, O.H. and El-Sayed, M.L., 1986, Multivariable adaptive controller for a turbogenerator, *IEE Proceedings*, Part D, Vol. **133** (2), pp. 83-89.

Shen, C., Malik, O.P. and Chen, T., 1997, A robust power system stabilizer design, *Optimal Control Applications and Methods*, Vol. **18**, pp. 179-193.

Uchida, M., Nakamura, N. and Kawai, K., 1981, Application of linear programming to thermal power plant control, *IFAC 8th Triennial World Congress*, Vol. XX, paper 976-912, Koyto, Japan.

Waal, van de, E., Grimble, M.J. and Brasca, C.L.,1996, Self-tuning AVR control with limited and full authority adaptation, *American Control Conference.*

Wu, Q.H. and Hogg, B.W., 1988, Adpative controller for a turbogenerator systems, *IEE Proceedings, Part D*, Vol. **135.**, pp. 35-42.

Wu, Q.H., Hogg, B.W. and Irwin, G.W., 1992, A neural network regulator for turbogenerators, *IEEE Trans. On Neural Networks*, Vol., **3** No. 1.

Xia, D. and Heydt, G.T., 1983, Self-tuning controller for generator excitation control, *IEE Trans. On PAS*, Vol. **102**, No. 6, pp. 1877-1885.

# 11

# Design of Controllers for Metal Processing

## 11.1  Introduction

Advanced multivariable robust control design methods have the most potential in strongly interacting, complex, high speed electro-mechanical systems. The multi-stand rolling processes are good examples of such a system. The aim of the chapter is to illustrate the types of control structure employed and typical simulation responses. The optimal control results were obtained using state-space based commerical tool-boxes and the skill therefore lies in setting up the models and the cost index.

A single stand rolling mill (Gunawardene, 1981) normally includes metal strip supplied in coiled form which passes through the mill and is then rewound at the exit side of the mill. Tension is maintained using both the unwind coiler at the entry to the mill and the rewind coiler at the exit from the mill stands. The gauge or strip thickness is controlled through the control of the force or load on the stand. The thickness is measured along the centre line of the strip and this is called the *centre line gauge*. A greater reduction in thickness can be achieved using a number of rolling stands in series and this is referred to as a Tandem Hot or Cold rolling mill (Bryant, 1973).

Metal strip is normally first rolled in a tandem hot rolling mill and later enters the cold mills for further reduction in the strip thickness (Grimble et al. 1978). A typical tandem hot strip mill plant layout is illustrated in Fig. 11.1. The slab is first heated to the required temperature in the reheating furnace and is then successively reduced in thickness, first in the roughing mill stands and finally in the finishing mill train. At the exit of the finishing mill the strip is cooled by the run-out table cooling sprays and is then coiled by the down coiler. The main difference between a hot and a cold strip mill lies in the use of loopers, which are placed between the hot mill stands, to control the strip tension. The structure of a hot mill showing the loopers is illustrated in Fig. 11.2 (Anbe et al. 1996; Clark et al. 1996).

459

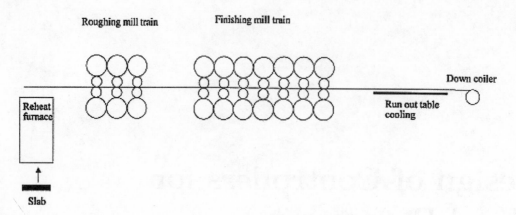

**Fig.11.1** :  Hot Strip Mill Process Line Layout

The input to a hot strip tandem finishing mill is a bar of metal at high temperature (typically 1100 degrees °C for mild steel).  The tandem mill normally reduces the thickness by a factor of 10 and since the width of the strip does not change (because of friction in the roll gap), there will also be a corresponding increase in the length of the bar by a factor of 10. A looper is placed between the stands of a tandem mill to store some strip so that disturbances do not cause the strip to become either too loose or tight. An ideal looper maintains the strip tension constant and this has the effect of decoupling the control action on the upstream and downstream stands (Duysters et al. 1994; Fukushima et al. 1998).

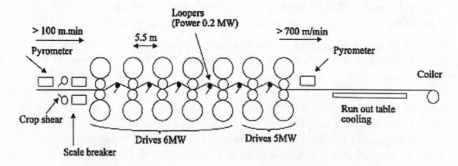

Fig.  11.2 : Example of a Hot Strip Finishing Mill Train

## 11.1.1   Hot Strip Mill Objectives

The general objective of any hot rolling mill control system design is to maximise the production of steel (or aluminium) strip at a minimum cost, while maintaining the product quality parameters of Gauge, Temperature, Profile, Shape, Width and Surface Finish. The production cost comprises the cost of equipment, energy consumption, raw material usage, human resources and the cost of maintenance facilities. The control of product quality and the minimisation of the rolling time for a coil are two of the important control problems to be considered Reeve et al. (1999). There is no direct mechanism to control the width of the strip in the finishing mill train and the width control will therefore be neglected. The main quality parameters to be considered are the centre-line gauge, thickness profile, shape profile and exit temperature, (Harakawa et al. 1995; Hayashi et al. 1987).

There are currently strong commercial pressures to improve the cross-strip thickness profile of strip produced by hot strip mills. The so called *shape* or *flatness* of the strip is discussed and explained in §11.4 which follows. Advances in the measurement of profile and shape have enabled the introduction of sophisticated profile and flatness control systems.

## 11.1.2 Background on Multivariable Control in Hot Mills

Between 1981 and 1996 there were 18 papers published documenting multivariable control design techniques applied to hot strip finishing mills, either as a feasibility exercise, or involving implementation on an actual mill. Table 11.1 lists publications on the multivariable control of mills. It is apparent that most of the published work on this subject stems from Japan and that a direct tension measurement is always used, except for the work of Hearns et al. (1996a,b). Three of the papers (Okada (1996), Duysters (1994) and Nakagawa (1990)) control the strip tension, looper angle and strip thickness, whereas the rest only control tension and angle. It would appear that the current use of multivariable control in finishing mills is generally where the instrumentation is available to measure all the variables and in some cases the plant states. The most common design techniques used are LQ optimal state feedback or output feedback $H_\infty$ optimal control (Imanari et al. 1995).

**Typical Hot Mill Data**

- Bending jacks : magnitude = 180 tonnes

- Hydraulic capsules : magnitude = 40 mm stroke

  force : 1800 tonnes per capsule

- Screw-downs : [2.839,3.66,3.66,3.66,1.1,1.1,1.1] = rate constraint in mm/second, for stands F7 to F1

- Main drives : [1120,1030,872,526,335,205] = magnitude constraint in m/min

- Actuators : all first-order low-pass models, with unity steady-state gain

  - Bending jacks: 25 rad/s bandwidth

  - Hydraulic capsules : 100 rad/s bandwidth

  - Screw-down motors : 5 rad/s bandwidth

  - Main drives : 10-15 rad/s bandwidth

- Sensors for a finishing mill :

- Shapemeter : discrete model, 0.25 second delay and four point moving average, sampling time of 0.25 seconds, 10 m/s exit speed from mill, 5-10 % noise standard deviation

- Thickness meter : 30 rad/s and standard deviation of noise 0.1%

- Pyrometer : unity gain

- Profile meter : assumed to be an array of thickness meters. Same bandwidth but much larger noise, 5-10%

• Sensors at each stand assumed unity gain (force, speed, looper angle etc.)

**Typical Quality and Dimensional Parameters**

| | |
|---|---|
| Gauge | 2 - 2.5 mm ±0.05 mm |
| Profile | Crown : 30 - 90 $\mu$m |
| Shape | $\pm$ 20 I units |
| Width | 0.7cm - 1.5m |

**Other Performance Measures**

• Surface finish, Temperature, Cost of energy consumption (80MW).

• Raw material usage, Human resources, Maintenance costs.

Note that the term crown, referred to above, is the difference in thickness between the gauge at the centre and the edge of the strip (normally taken as 0.25 mm from the true strip edge). The shape profile, referred to above, is discussed in more detail in §11.4.2.

Table **11.1** : Publications on Multivariable Hot Strip Mill Control

| Author | Year | Country | Technique | Implemented |
|--------|------|---------|-----------|-------------|
| Kotera | 1981 | Japan | Decoupler | yes |
| Hamada | 1985 | Japan | Decoupler | yes |
| Tsuji | 1987 | Japan | State-feedback | yes |
| Hayashi | 1987 | Japan | State-feedback | yes |
| Fukushima | 1988 | Japan | • Decoupler<br>• State-feedback | yes<br>yes |
| Kawaguchi | 1988 | Japan | State-feedback | yes |
| Nakagawa | 1990 | Japan | State-feedback | yes |
| Holton | 1991 | Canada | • Decoupler<br>• State-feedback | no<br>no |
| Seki | 1991 | Japan | State-feedback | yes |
| Shioya | 1993 | Japan | H-infinity | no |
| Duysters | 1994 | Holland | Characteristic-loci | no |
| Randall | 1994 | UK | • Reverse frame<br>• LQG<br>• H-infinity | no<br>no |
| Harakawa | 1995 | Japan | State-feedback | no |
| Imanari | 1995 | Japan | • State-feedback<br>• H-infinity | yes<br>yes |
| Clark | 1996 | UK | PI | yes |
| Okada | 1996 | Japan | State-feedback | yes |
| Anbe | 1996 | Japan | • Decoupler<br>• H-infinity<br>• Inverse-linear quadratic | yes<br>yes<br>yes |
| Hearns | 1996 | UK | H-infinity | no |

### 11.1.3 Finishing Mill Features

The main dynamic features of the hot finishing mill process will now be identified (Ginzburg, 1989).

**End point control:** The material specifications must be satisfied at the exit of the finishing stand. There are therefore a number of degrees of freedom for the control of the load distribution between the stands. Questions which arise include: How accurately should the thickness, profile, temperature and tension be controlled on the earlier stands?

**Sequential process:** The hot strip mill finishing train is a sequential process. That is the inputs to a stand are the outputs of the up-stream stands. The stands are linked together through the use of the loopers. If the control tasks are equally distributed between the stands and if the stands are assumed to have effective fully autonomous localised control systems, no co-ordination is needed. However, this is an unrealistic scenario because of the effects of disturbances. Future information via feedforward action from previous stands, must therefore be incorporated in the control action.

**Massflow balance:** Central to the operation of the finishing mill train is the mass flow balance between the stands. The static and dynamic mass flows into and out of a stand must be equal, to avoid a build up of material between stands. This can be achieved using the looper to adjust the tension and/or the main drive motors to vary the stand speed. The mass flow balance relationships can be considered to provide constraints on the stand control action.

**Speed of operation:** The speed of the process is high compared to a conventional process control problem. In some cases an update frequency of 50 Hz is used for accurate control. It is therefore necessary to develop simple and computationally efficient control algorithms for implementation.

**Interactions between the mill stands:** The main source of the interactions between the stands is the strip tension which is manipulated by the loopers and the main drive motors. A conventional control strategy attempts to maintain a constant tension between the stands in an effort to decouple the effect of disturbances between the stands.

**Constraints:** The performance of many control systems can be improved if the process can be safely near to the actuator and process constraints. To achieve the best performance of the hot mill control system, it is important to incorporate the bending, profile and flatness constraints into the design.

**Co-ordination:** A degree of co-ordination between the stand control systems is required to compensate for the effects of unpredictable disturbances. The co-ordination can be performed by manipulating the local feedback loops by the supervisory system adjusting the regulating loop set points.

**Model dependency:** Due to the restricted number of sensors available for gauge, profile and shape measurements, the performance of any control system for the hot strip mill will inevitably depend, to some extent, on the accuracy of the models used in the control design. The conventional procedure is to adapt the models to mill speed and type of steel. Adaptation and gain scheduling leads to problems such as switching, non-linear behaviour, and high computational burden. It is therefore, important to employ robust control strategies allowing for the modelling errors or uncertainty.

**Disturbances:** The most important disturbance in hot strip mills arises from the so called skid chill marks. Whilst the slab is in the re-heat furnace it is placed on skid rails that will be at a lower temperature and hence cause a local reduction in temperature. When the strip is being rolled there are therefore areas of strip which are harder, since they are at a lower temperature. The automatic gauge control system must therefore compensate for these yield stress variations. The looper control system normally keeps the mass flow constant, whilst the thickness control system is rejecting such disturbances. If, for example, the input thickness increases over a length of the strip, then with an ideal gauge control system, the output speed will temporarily increase and the looper arm will move up to increase the amount of strip between the stands temporarily.

**Interstand cooling systems:** The interstand cooling systems on a hot mill consists of arrays of spray nozzles which are distributed across the width of the mill. They can be used for both an average cooling effect and to distribute cooling across the roll surfaces to influence the roll profile and hence the strip profile and flatness. Any temperature control system is of course relatively slow and the dynamics of the actuators for this system can be ignored.

**Run out table cooling:** The run out tables at the output of a hot strip mill have an important influence on the metallurgical properties of the strip, such as the microstructure. It is necessary to accurately control the cooling temperature through the use of controllable water jets and water curtains. Consistent properties are required both across the width and through the thickness. A set-up model is normally used to tune the control systems with respect to variations in finishing mill temperature, speed, thickness variations, water temperature and heat transfer conditions. A feedforward temperature control strategy can be employed using an intermediate pyrometer. If the target cooling temperature cannot be achieved through the use of the run out table cooling then a modified exit speed is demanded using the finishing mill set-up model.

**Work roll shifting:** Rolls can have work roll shifting which is activated between the rolling of coils so that a more uniform wear occurs on the work rolls. Work roll shifting can only be employed during the rolling of a coil in special mills and will not be discussed further here.

**Roll bending:** Bending jacks are used on most mills to be able to vary the distribution of load across the strip at the work roll gap. The back-up-roll and work roll bending jacks enable the work roll profile and strip flatness to be controlled.

## 11.2 Hot Mill Modelling

The control system design depends upon the availability of reasonable stand and interstand models. The main components of a rolling mill stand are shown in Fig. 11.3, and the important mechanical parameters are illustrated in Fig. 11.4. The view across the mill is shown in Fig. 11.5 and the application of the total rolling load $P$ and of the roll bending jack forces are illustrated. The total load on the mill affects the average strip thickness and the roll bending forces enable the flatness and profile to be controlled.

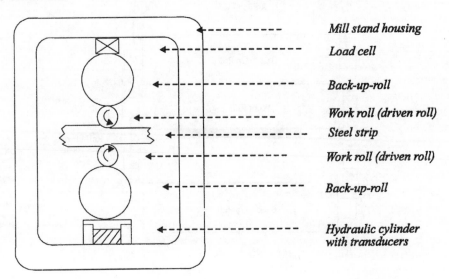

**Fig. 11.3** : Mill Terminology for a Four-high Rolling Mill Stand

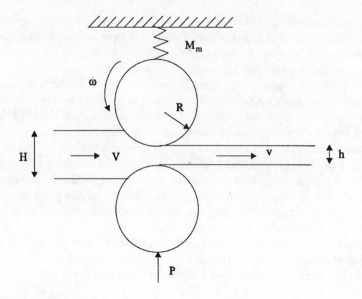

M = mill modulus
R = work roll radius                    P = rolling force
H = strip input gauge                   v = strip output velocity
h = strip output gauge                  V = strip input velocity
s = roll gap before load applied.       $\omega$ = work roll angular velocity

**Fig. 11.4** : Single Stand Rolling Mill Variables

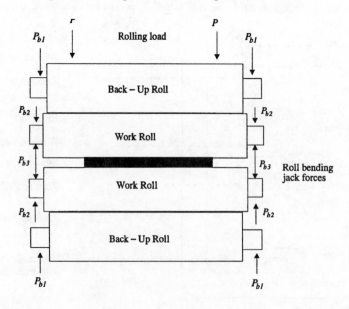

**Fig. 11.5** : Schematic Diagram Showing Rolling Load and Roll Bending Forces

### 11.2.1 Mill Equations

The load on a mill stand (Grimble, 1976b) is controlled by electric screw down motors or hydraulic capsules. The mill stand stretches elastically under load and the mill stretch relationship has the following features :

(i) Linear over a significant range of roll force $P$ values.

(ii) Form of the characteristic is independent of the initial roll gap setting $S$.

(iii) Mill modulus $M_m$ is the inverse slope of the roll force/mill stretch curve and represents the spring constant for the complete mill housing.

If it is assumed that the output gauge is equal to the roll gap, then the relationship between the output gauge, rollgap setting and the extension due to the mill stretch characteristic becomes:

$$h = S + e_x \tag{11.1}$$

where

$\qquad h$ = output gauge
$\qquad S$ = rollgap setting
$\qquad e_x$ = mill extension
The linearised form of this equation, for the first-order variable changes:

$$\delta h = \delta S + \frac{\delta P}{M_m} \tag{11.2}$$

The strip modulus $M_s$ which relates the strip thickness change to rolling load, is the inverse of the slope of the roll-force reduction curve. Linearising this characteristic about the operating point gives:

$$\delta S - \delta H = \frac{\delta P}{M_s} \tag{11.3}$$

### 11.2.2 Gaugemeter Principle in Classical Control

If the input thickness to the stand $\delta H$ increases, then the roll force increases and the stand will stretch. To return the output gauge to the setpoint, the required roll gap position change $S$, assuming the linear part of the characeristic, is given as:

$$\delta S = -\frac{\delta P}{M_m} \tag{11.4}$$

The gauge correction can be calculated given $M_m$ and the roll force change $\delta P$. The mill modulus will change as the rolls wear out and after each change of the work rolls or back up rolls. The automatic gauge control (AGC) output is the reference for the actuators controlling the work roll positions that determine the rolling force. The mill modulus can be measured from the mill stretch characteristic, shown in Fig. 11.6, and this operation is illustrated in Fig. 11.7. The Automatic Gauge Control (AGC) output is the reference for the hydraulic capsule stroke and is the original reference signal, modified by $\delta S$. This simple approach to gauge control is referred to as the *gaugemeter principle*.

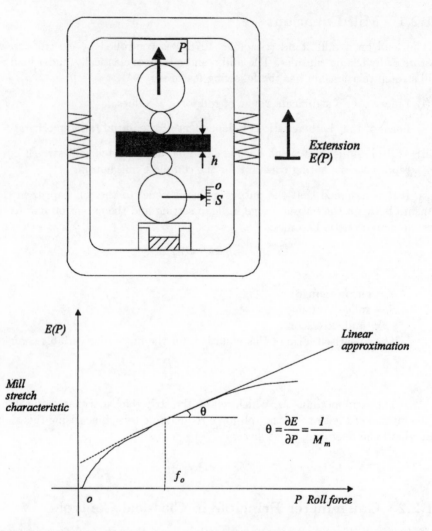

Fig. 11.6 : Mill Stand Housing and the Mill Stretch Characteristic

## 11.2.3  Measurements

For the finishing mill of interest, variables measured include: *gauge, profile, shape, temperature, speed and width.* The unloaded roll gap position can be set using either screw-down motors or hydraulic capsules. Each stand includes load cells to measure the rolling force. The X-ray thickness gauge output will include measurement noise so that the output is filtered. The filter choice is a compromise between the response time and the noise attentuation and this is arranged through adjusting the filter time constant. The continuous-time transfer-function relationship between the measured gauge $h_m$ and the actual gauge $h$, with Gaussian noise $n$, becomes:

$$h_m = \frac{K_m}{(\tau_x s + 1)}(h + n)$$

where

$K_m$ = steady state gain of the filter

$\tau_x$ = time constant of the filter

The filter time-constant is usually adjusted between 1 ms and 200 ms. A typical specification is 0.15% noise with time constant $\tau_x = 10$ ms.

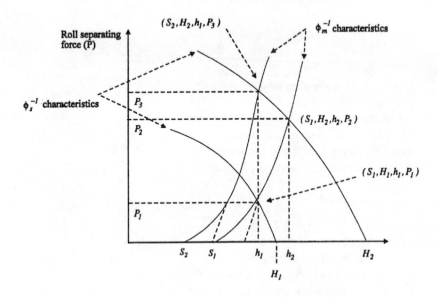

**Fig. 11.7** : Operation of the Gaugemeter Thickness Control Indicating Movement Between Operating Points

## 11.3  Control of Interstand Tension

The variation in the interstand strip tension has an important impact on the dimensional accuracy and the threading performance of hot strip finishing mills. It is therefore desirable that the interstand strip tension and the looper angle be kept constant but this is difficult because of the interaction between variables (Saito et al. 1985).

The tension between stands is controlled by a looper, as shown in Fig. 11.8. The range of looper operation is limited and the looper angle should therefore be maintained at a constant value. The interstand strip tension and the looper angle are controlled using both the looper and the mill stand motors. Neglecting the interaction with thickness, the control system has two inputs (position and speed reference) and two outputs (angle and tension). The relationship between the looper angle and the torque (or equivalently looper motor current) and the strip tension, is a nonlinear function of the looper arm angle. To keep the looper arm angle constant any changes in the looper motor current must be accompanied by changes in the mill stand drive speed.

## 11.3.1   Strip Tension Calculation

The calculation of the strip tension, noting the geometry in Fig. 11.8, is straight-forward based on fundamental relationships: derived by Johnson et al. (1999). It will be assumed that the dimensional change in the strip length is much greater than the change in the cross-sectional area. The tension model is based on the longitudinal stress and strain relationship (Grimble, 1976a):

$$\sigma = E\left(\frac{\Delta L}{L}\right) \tag{11.5}$$

where

$\sigma$ is the longitudinal tension stress
$L$ is the length
$\Delta L$ is the extension in length
$(\Delta L/L)$ is the longitudinal strain
$E$ is the Young's modulus of elasticity

The unstretched strip strip length $L_s$ at time $t$:

$$L_s(t) = L_0 + \int_{t_0}^{t} \Delta\nu(\tau)d\tau \tag{11.6}$$

where $\Delta\nu(\tau) = v_i(\tau) - V_{i+1}(\tau)$
$L_0$ = strip length at initial time $t = t_o$
$v_i$ = strip exit velocity from stand $i$
$V_{i+1}$ = strip entry velocity to stand $i + 1$

From Fig. 11.8 the strip length between the two stands is approximately:

$$L(\theta(t)) = l_1(\theta(t)) + l_2(\theta(t))$$

$$l_1(\theta(t)) = [(r + R\sin\theta - y)^2 + (a + R\cos\theta)^2]^{1/2}$$

$$l_2(\theta(t)) = [(r + R\sin\theta - y)^2 + (L - a - R\cos\theta)^2]^{1/2}$$

where $R$ is the length of the looper arm and the constant $a$ denotes the distance of the looper pivot point from the upstream stand centre line. These may be combined using the Young's modulus relationship, to obtain the tension stress:

$$\sigma(t) = E[L(\theta(t)) - L_s(t)]/L_s(t)$$

If the looper has an inertia $J_L$, with respect to the pivot, then Newton's Second Law of Motion can be applied to obtained :

$$J_L\ddot{\theta} = T_m - T_{load} \tag{11.7}$$

where $T_m$ is the applied looper torque and $\mathrm{T}_{load}$ is the load torque:

$$T_{load} = T_\sigma + T_s + T_\ell + T_b + T_d$$

Consider a strip width $w$, thickness $h$, looper-arm length $R$ and looper roll radius $r$, then the load torque due to strip tension follows as:

$$T_\sigma = \sigma wh[R\cos\theta f_1(\theta) + (R\sin\theta + r)f_2(\theta)]$$

where

$$f_1(\theta) = (r + R\sin\theta - y)(\frac{1}{l_1} + \frac{1}{l_2})$$

$$f_2(\theta) = (\frac{L - \alpha - R\cos\theta}{l_2} - \frac{\alpha + R\cos\theta}{l_1})$$

For a strip density $\rho$ and gravitational acceleration $g$, the torque to support the strip weight follows as:

$$T_s = Rg\rho wh(l_1 + l_2)\cos\theta$$

The torque required to support the looper arm of mass $M_\alpha$ and roller of mass $M_r$ follows as:

$$T_\ell = RgM_r\cos\theta + \frac{R}{2}gM_\alpha\cos\theta$$

The torque required to bend the strip with yield stress $\sigma_{ys}$ over the looper roll follows as:

$$T_b = R\cos\theta(\frac{\sigma_{ys}wh^2}{4})(\frac{1}{l_1} + \frac{1}{l_2})$$

The frictional damping torque follows as:

$$T_d = c\dot\theta$$

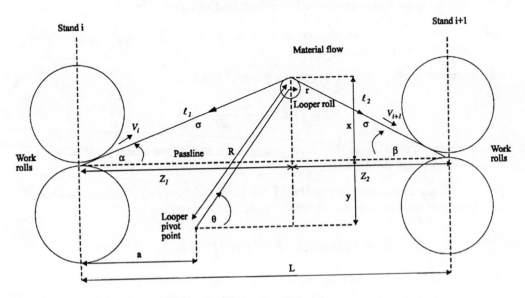

**Fig. 11.8 :** Stands and Looper Interstand Geometry

## 11.3.2   Classical Design of Looper Control Systems

The loopers must regulate the strip tension between the stands and compensate for any disturbances introduced. One of the main disturbances is the roll gap change which occurs due to the variations of the thickness control system. In addition to regulating strip tension the variations in the looper angle must be controlled. The changes in the looper angle should enable the deviations in the strip tension to be accommodated.   Loopers perform at least two important functions:

(i) Prevent changes in the width and thickness of strip by regulating the interstand tension.

(ii) Prevent the formation of loops of strip between the stands by regulating the mass flow.

The tension is controlled by controlling the looper motor torque and the looper angle or height, which is used as a measure of the mass flow. The looper angle is then maintained at a reference value by controlling the speed of the upstream stand drive motor. The looper angle affects the strip loop length and the tension changes the load torque. However, in conventional systems these functions are controlled independently.

One of the main disturbances entering the system arises from the so called skid chill marks, mentioned earlier. When the slab is in the reheating furnace it rests upon water cooled skids and this results in non-uniform heating of the slab. As the slab is rolled there is a variation in the hardness of the material, resulting in thickness variations in the strip output from the finishing mill train. In fact, both strip tension and looper angle also change as the skid chill strip zones go through the mill. The frequency of the skid chill disturbances is proportional to the rolling speed and for the present purposes it may be assumed to be in the range 0 to 2 rads/sec. The use of the loopers for mass flow control can be summarised as follows:

(i) Looper angle $\theta$ is maintained at the reference $\theta_{ref}$ by adjusting the speed of the upstream stand drive motor.

(ii) Reference angle is related to reference looper height $x_{ref}$

$$\theta_{ref} = \sin^{-1}\left(\frac{x_{ref} + y - r}{R}\right)$$

where $r$ is the radius of the looper roll and y is the height of the passline above the looper arm pivot point.

(iii) The looper angle error fed to the PI (proportional-integral) classical controller generates a speed trim signal for the stand drive motor:

$$\Delta\omega_m(t) = K_p\left(\theta_{ref}(t) - \theta(t)\right) + K_I \int_0^t \left(\theta_{ref}(\tau) - \theta(\tau)\right)\ d\tau$$

where $\Delta\omega_m$ is the speed trim reference signal, $K_p$ is the proportional gain and $K_I$ is the integral gain of the controller. A typical roll drive motor control system, for a rolling stand, is shown in Fig. 11.9.

(iv) The new drive motor speed reference :

$$\omega_r = \omega_s + \Delta\omega_m$$

where $\omega_r$ is the reference angular velocity and $\omega_s$ is the setup angular velocity, (reference speed neglecting disturbances).

A classical interstand tension and looper angle control structure is illustrated in Fig. 11.10.

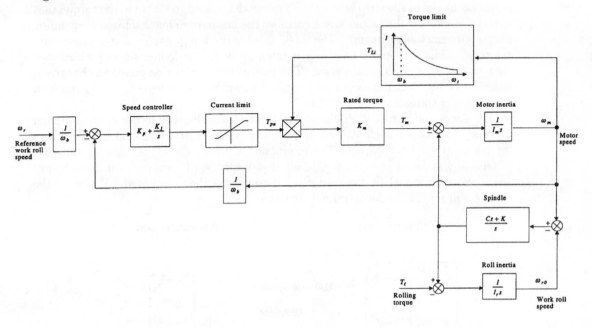

**Fig. 11.9** : Stand Motor Speed Control and Roll Dynamics

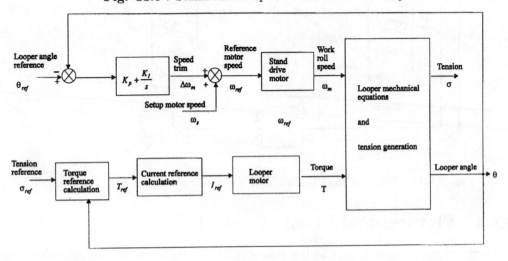

**Fig. 11.10** : Traditional Looper Massflow and Tension Control System

### 11.3.3   Classical Gauge Control System Design

The conventional solution for the regulation of thickness, angle and tension is shown in Fig. 11.11. The Automatic Gauge Control (AGC) uses force feedback via the inverse of the mill modulus (as described in §11.2.2) to change the capsule position. A PI controller is used to regulate the angle by trimming the speed of the upstream stand drive motor. The massflow control also includes a speed trim signal from changes of the downstream motor speed. The strip tension is regulated open-loop by using a static nonlinear model of the strip geometry. This model is used to estimate the torque that must be applied to the looper arm to achieve the desired reference tension (depending on the current looper angle). The AGC is effective but provides disturbances into the tension loop and more emphasis is often given to the looper angle performance than to the strip tension regulation. The gauge control often responds to changes in massflow faster than the looper can regulate massflow. This behaviour can result in large tension transients and instability.

Harakawa and Furuta (1995) proposed a two stage advanced looper controls design process. The looper angle control system was designed first, followed by the design for the strip tension controls. The interaction between the looper system and that of tension was found to be so significant that feedforward compensation from looper angle reference signal to the control of tension was then employed. However, the system still retained a decentralised structure.

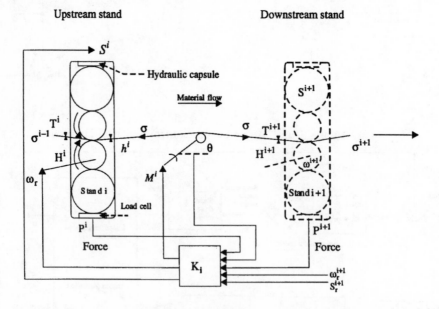

Fig. 11.11 : Conventional Tension and Thickness Regulator Structure

## 11.4   Flatness and Profile Control

There are many control loops in hot rolling mills but the main process variables to be controlled are thickness, tension, thickness profile and flatness, (Grimble and

Hearns, 1998, 1999). The control of the flatness of rolled strip is often referred to as being *shape control*. However, shape control is a misnomer and does not refer to the thickness profile but is related to the distortions which can be present in rolled strip, if the tension stress distribution is not uniform (Grimble, 1975, Grimble and Fotakis, 1982). When strip is under high tension in a rolling mill it may appear to be flat but once the strip is cut into sections it may not lie flat on a flat surface. This occurs when the strip contains internal stresses, and hence the strip is said to have poor shape.

Variations in the thickness profile across the strip can cause differential elongation of the strip, leading to a non-uniform stress distribution, which is referred to as strip flatness. If these internal stresses cause visible flatness variations (see Fig. 11.12) it is referred to as *manifest shape* but if the tension is sufficiently high these variations are hidden from view, and it is then referred to as *latent* shape. Once the strip is removed from the mill and it is not under tension, *latent shape* may become visible and *manifest* shape results.

Flatness or poor shape is one of the most difficult strip properties to control. Wavy edges to the strip, centre buckling and other strip defects can occur due to a poor stress distribution. The attainment of a desired strip thickness profile is normally considered secondary to the requirements to achieve proper flatness. However, strip is supplied to customers according to minimum thickness over the full width of the strip, rather than by the coil weight. Thus, thickness profile control is also of importance.

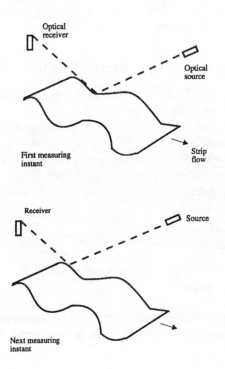

**Fig. 11.12** : Manifest Shape Measurement

## 11.4.1   Factors Affecting the Strip Profile

The strip profile that results during the rolling process depends upon the rolling force which is itself determined by the roll gap actuator positions. The profile is also determined by the bending forces produced by roll bending jacks, inserted between the roll bearing housings (see Fig. 11.5). The rolling force actuators are designed to have a large effect on the average thickness of the strip but minimal effect on the thickness profile. The bending forces introduced by the roll bending jacks are designed to have the main effect on the thickness profile. Unfortunately, there are significant interactions between both types of actuator, the mean thickness and the thickness profile.

There are other mechanisms to control the roll gap profile. The ground camber on the back-up and work rolls can be used to compensate for the tendency for the edges of the strip to be reduced more than the middle sections. This over-rolling of the strip edges is avoided by using rolls that have a barrel shape. The application of rolling load then results in a more uniform thickness profile. Although the load is applied at the bearings, located at the roll ends, the ground camber compensates for the resulting roll bending.

As the temperature builds on the rolls they also change their profile due to expansion and this is referred to as the *thermal camber* on the rolls. The ground camber will of course be affected by roll wear and the rolls have to be reground regularly. The work rolls gain heat from the mechanical work energy and from conduction and radiation from the strip itself. They lose heat through convection due to the cooling fluid, through radiation to the air, and through conduction, to the back-up rolls.

## 11.4.2   Parameterisation of Strip Profile and Flatness

The following terms may be used for the parameterisation of the flatness profile:
*Crown* : Difference between the average edge thickness and the centre-line gauge. The edge is normally defined to be 25 mm or 40 mm from the actual strip edge, leading to the terms Crown-25 and Crown-40.
*I-units for flatness* : Relative difference in length between one point on the width and the average (i.e. differential strain) multiplied by $10^5$ gives the flatness in I-units. It is common to measure flatness at the strip edges and the middle.
*Polynomial parameterisations*

- *Zero'th order component* is the average or centre-line component.

- *First order component*, often called squew, is normally controlled using an asymmetrical roll gap setting.

- *Quadratic component* is normally controlled using symmetrical roll bending jack action. The quadratic *component* is the highest order component used in standard models.

- Higher order components are sometimes used to model strip edge effects (e.g. 15th and 27th order).

## 11.4.3   Hot Mill Shape Measurement

Manifest strip shape can be measured with laser/camera sets. The flatness or shape measurements are then obtained by triangulation. This type of shape meter may only operate when the head of the strip is on the run out table and not once it becomes under the coiler tension.

A stress distribution type of shape meter can measure the stress distribution or flatness by using a looper with segmented roll including transducers. The different tensions on corresponding strip sections can be measured by load cells mounted in the measuring roll segments.

Flatness measurement rolls are normally divided along their length into segments which enable an estimate of the stresses in the strip to be obtained. There is usually a non-uniform distribution in the size of the regions at the edge of the strip in comparison to the centre region. A higher resolution at the strip edges is desirable, since the gradient of the stress distribution characteristic is greater at the edges.

## 11.4.4   Flatness Control Systems

A poor stress distribution arises when the thickness profile of the strip entering the stand is different to the thickness profile which exists from the stand (leaving aside the natural reduction in thickness which occurs). Consider, for example, a rolled strip which has a uniform thickness across the strip width that enters the stand. If the work roll profile is such that the exiting strip thickness is greater in the central regions than at the edges, then internal stresses will be introduced. Because of friction in the roll bite the strip cannot spread sideways and hence the reduction in thickness corresponds with an increase in the speed of the strip exiting the stand. In the above case if the strip were cut into longitudinal sections, this would imply the central region was moving slower than the edge regions. In practice, of course, the strip is one homogenous mass and hence internal stresses are built up which can only be relieved when the strip is out of the mill and not under high tensions. In this case, the strip can take its natural form, that in this example would give rise to a so-called long edge, visible as a wavy edge to the strip.

A flatness control system consists of actuators for flatness control which modify the roll gap profile and a measuring system to estimate the so called shape profile. A typical four high mill configuration was shown in Fig. 11.5. A stiff mill housing supports both work rolls and back-up-rolls, and the rolls are mounted in chocks within the mill housing. Both work roll bending forces and back-up-roll bending can be applied. This enables the roll gap profile to be modified and indirectly changes the strip shape profile (McNeilly et al. 1996).

**Flatness Actuators**

If different forces are applied on different sides of the mill, so called tilt action can be generated, which provides one mechanism for shape control. Symmetrical bending action may also be imposed through the work-roll bending jacks. The coolant sprays may also be controlled, providing an additional slower mechanism for flatness control. Flatness control systems are normally rather slower than the strip thickness controls. A tilt adjustment mechanism can be used to adjust the gap position, by equal amounts in opposite directions, on each side of the mill. This enables the centre line gap position to be maintained unaltered. To achieve work roll bending action the work roll jacks are adjusted, so that the work roll necks are moved closer together or further apart. For normal bending action an equal change is made on both sides of the mill.

**Coolant Flow Control**

Coolant sprays act on the work roll and back up roll surfaces. The nozzles on the sprays are arranged in banks parallel to the roll axis. A number of coolant spray bars are normally employed so that a vertical zone of nozzles can be obtained acting in a particular zone of strip. By turning nozzles on or off in a vertical cooling zone, the thermal crown on the rolls can be varied, altering the flatness in different sectors across the strip width.

## 11.4.5   Thickness Profile Control

The main emphasis is on centre-line gauge control which might be considered the representative thickness of the rolled strip (Hearns and Grimble 1998a,b, 1997). However, strip profile is also important in both steel and aluminium rolling. Crown control systems have been developed to enable transverse control of thickness to be achieved in rolling mills. The crown on the work rolls which determines profile can sometimes be controlled directly, but it will certainly be dependent upon temperature, which depends upon the mill setup. The setup refers to the setpoint settings determining the nominal stand speeds, forces and looper angles (in hot mills). Note that there is a limit to the amount of crown control which can be applied on a single stand, because the shape (flatness profile) may deteriorate when too much bending is applied. The sheet crown is defined as:

$$crown = (H_c - H_e)/H_c \qquad (11.8)$$

where

$H_c$ = thickness at the strip centre
$H_e$ = thickness at the strip edge

It is possible to grind a different crown on both the work rolls and the back up rolls of steel rolling mills. However, if the strip width is changed the ground crown may be inappropriate. For this reason work rolls have been developed where the crown can be varied mechanically. The rolls are not of course solid and the profile is changed hydraulically by internally applied forces. A typical maximum increase in the roll diameter would be 2 mm.

**Mill Setup**

To achieve the target exit crown the work roll crown, work roll force and coolant spray pattern must be set appropriately. Work Roll Bending (WRB) involves the use of roll bending jacks applied between the spindles of the different rolls. The work roll thermal camber can be modified by the spray bars, which affects the flow of coolant. There are typically four spray bars per stand and about forty nozzles on a spray bar. Sprays are typically used in patterns. The sheet crown at the exit of the stand is given by the following equation:

$$C_{ri} = \zeta_i C_i + \eta_i C_{ri-1} \qquad (11.9)$$

where

$\zeta_i$ : Imprinting ratio on stand $i$ (proportion of the roll gap profile which appears in the exit strip profile).

$\eta_i$ : Heredity crown coefficient on stand $i$ ; (proportion of the entry strip crown which appears in the exit strip crown).

$C_{ri}$ : Sheet crown at the exit of the $i$th stand

$C_i$ : Mechanical crown is obtained from the following equation:

$$C_i = \alpha_{pi}P_i + \alpha_{bi}P_{bi} + \alpha_{wi}C_{wi} \qquad (11.10)$$

where

$P_i$ : Separating force

$P_{bi}$ : Work roll bending force

$C_{wi}$ : Work roll crown

$\alpha_{pi}, \alpha_{bi}$ , $\alpha_{wi}$ : Influence coefficients.

## Profile Control

Strip profile meters for the measurement of profile are now available. It is therefore possible to develop a multivariable control system to change the various actuators to achieve the target strip profile. Typically a finishing mill will have work roll bending (Fig. 11.5) on each stand for strip profile control. It may also include special work rolls or back-up-rolls for direct control of roll profile. A much slower work roll cooling temperature control system is also used. Exit temperature from the stand may be controlled by varying both overall mill speed and the interstand cooling. The speed control has a continuous reference input and the faster the speed the higher will be the temperature in the strip generated. Coolant control is often only available in discrete levels having minimum, intermediate and maximum coolant flows. The major disturbances are due to the entry temperature variations and the incoming strip thickness profile.

## Robust Control Design in Profile Control

Since temperature changes are relatively slow, compared with the roll bending mechanisms and rolling load control, temperature and bending systems can be designed independently. Consider the design of the roll bending/load control system where thermal effects can be considered to represent very slow disturbances. The profile meter will include a lag term which normally averages the measurements obtained. Hydraulic actuators are relatively fast but the transport delays between stands are important and these vary with the speed of the line. Although the controller can be scheduled accurately with line speed, the same controller will be used for given speed zones requiring robustness to speed variations. However, the greatest uncertainty arises because of the difficulty in obtaining the constant coefficient which enables the exit crown to be calculated. In most systems, where profile control is added, there are insufficient measurements to enable the coefficients to be calculated accurately. The coefficients may be found from sophisticated computer models of mills, but these are not always available for existing mill systems.

Uncertainty in these constant coefficients and errors which arise due to the variations in the transport delays can be represented by multiplicative uncertainties. A sensitivity minimisation problem can be defined which ensures robustness is achieved and actuator demands are limited. The main advantage of using $H_\infty$ design in this application is that the system is truly multivariable. Consistency of product quality in the face of incoming profile variations and temperature changes is central to the design problem. This is therefore a natural $H_\infty$ design problem, where the disturbance models are poorly defined but some information on worst case disturbances is often available.

# 11.5   Classical Control of Hot Strip Mills

The overall total control design strategy may be summarised in terms of the regulating loop, coordinated and open-loop control policies. The classical control strategy for hot strip mills will now be summarised, where the variables to be controlled and the actuators are detailed.

**Regulator control**

1. Gauge : using roll-gap settings.

2. Looper angle : using preceding stand roll speed.

3. Strip tension : using looper torque.

4. Flatness on last stand : using roll bending force.

**Coordinated control**

1. Temperature : using roll speed of all stands.

2. Gauge : using gauge set-point or roll-gap setting of all stands.

3. Mass flow: using roll speeds of all stands preceding the interstand zone, where a looper angle error occurs.

**Open-loop control**

1. Bending force trim is changed in response to roll force changes and thermal camber model predictions.

2. Gauge disturbance information is fed forward to downstream stands.

The classical design of the regulating loops will first be discussed and attention will then turn to the upper-level supervisory control problem, namely coordinated control.

### 11.5.1   Classical Gauge and Tension Control

The conventional regulating loop strategy, illustrated in Fig. 11.13, is that separate designs should be undertaken for each stand and interstand combination. The controllers $K_i$ for each stand and interstand may be made up from smaller sub-controllers but the designs can be carried out such that the interactions are taken into consideration. The measured variables are the stand speed, stand force and looper angle, while the outputs of the controller are the looper torque reference, capsule position reference and stand speed trim. The controller also has inputs of the downstream stand speed for global massflow control. The variables to be controlled are the strip gauge, crown, tension, shape and looper angle (within limits). The references into each stand/interstand module are the tension, gauge, crown, shape and speed references. The proposed advanced control solutions will also have a decentralised structure.

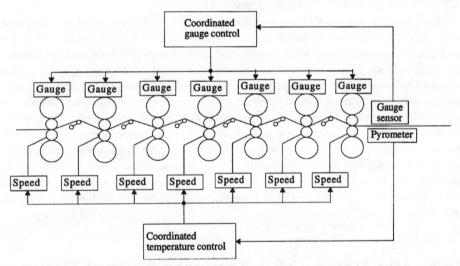

**Fig. 11.13** : Coordinated Control Scheme for Thickness and Temperature Control

### 11.5.2   Classical Cordinate Control

Coordinated control for a hot mill strip describes the control action that takes place simultaneously over several stands in response to an error at one location (normally the finishing mill exit). The first example of coordinated control is the response to a strip thickness measurement that involves an X-ray measurement taken approximately 3 metres after the last stand exit. The gauge error from this measurement is fed back to all the stands, as a trim on the gauge reference signal. The classical controller through which the error is amplified is typically of a Proportional plus Integral type (PI). Each trim signal may be weighted in line with the gauge reduction schedule, so that earlier stands receive a larger absolute trim. The second coordinated control action, which is associated with an exit measurement, is that of the control of average strip temperature. Typically each stand drive motor has its speed accelerated at a constant rate in response to temperature dropping below a given set-point (the strip does not gain energy with time so the temperature never rises when the mill speeds are constant). The global acceleration of the mill is called *zooming*. Fig. 11.13 illustrates the generalised architecture of gauge and temperature control.

The remaining coordinated control action of interest is one for maintaining constant mass flow. Traditionally a local deviation from mass flow is detected by a change in looper height, and consequently a change in looper angle. The regulator control then adjusts the preceding stand drive motor speed to bring the looper angle back to its reference position. The speeds at all other preceding stands are also adjusted. The trim that is added to the stand speeds in this case is weighted according to the set-up speed at each stand: thus, the trim is smaller at stand 1 than at stand 2 and so on. This strategy, whereby stands preceding the interstand region (where the change occurs) have their speeds adjusted, leaves the last stand with a constant speed. This stand is known as the pivot stand. Common practice to use the middle stand as the pivot stand. The mass-flow control scheme acts like a set of $n$ co-ordinated controllers, where $n$ is the number of interstand regions.

The mass-flow control schemes operate on the values of stand speed relative to each other, whereas the temperature control operates on the absolute values. This gives the degrees of freedom required to control the temperature independently of the mass-flow irregularities. The main task of the upper level control action is to co-ordinate the stand references and resolve any conflicting control actions. This control function usually involves a mix of logical expressions and numerical control algorithms. The mechanisms to control the process at this level are set point manipulation, controller adaptation and control structure reconfiguration. Predictive control has produced promising results in the process industries based on the optimisation of a cost index incorporating the future behaviour of the process using linear predictors. This technique can be applied at a level above the closed-loop regulators and the regulator set points then become the manipulated variables for the predictive supervisory controllers.


## 11.6   Hot Rolling Mill Advanced Control Design

The Fig.11.14 shows three stands of a mill connected by metal strip which is supported by the looper roll. Recall that the control objective is to regulate the strip tension stress ($\sigma$), strip thickness ($h$) and looper angle ($\theta$) by using the measurements of looper angle ($\theta$), stand force ($P$) and strip tension ($\sigma$) (if available) as inputs to a multivariable controller. This controller must generate the actuator references of capsule position ($S_r$), main drive motor speed ($\omega_r$) and looper motor torque ($M_r$).

The disturbances that must be regulated are the strip temperature and thickness entering stand $i$, the entry tension of stand $i$ and the velocity of the strip entering stand $i+1$. A controller is to be designed for each stand zone and there are six identical zones connected to each other. As well as the local performance of each zone, the global stability of the connected system must also be considered (Hearns et al. 1996b). The problems of designing a multivariable controller, achieving performance against unknown disturbances, and ensuring global stability, are all issues which can be tackled effectively by utilising either an $H_2$ or $H_\infty$ design methodology, (Grimble and Hearns, 1998).

### 11.6.1  Summary of Control Requirements

The control objectives are typically:

1. *Bandwidth* : The maximum bandwidth $\omega_b$ should not exceed 100 rad/s.

2. *Performance*: The strip thickness, strip tension and looper angle should be regulated given the external disturbances, skid chill temperature and backup roll eccentricity. The highest priority should be the strip thickness and the lowest priority should be the looper angle.

3. *Robustness* : The closed-loop system should be stable for a variation in the mill modulus $M_m$ of $\pm 20\%$ for each stand.

4. *Robustness* : The controller should operate over the full speed and temperature range experienced in the rolling process.

5. *Controller Complexity* : The controller can be a multivariable controller but with a minimum complexity and adequate roll off at high frequencies.

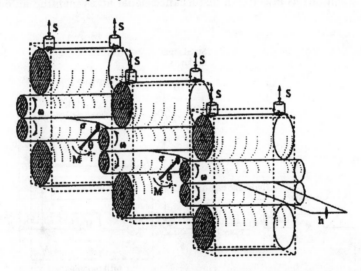

**Fig. 11.14** : Layout of Three Stands and Loopers

### 11.6.2  Mill Modulus Uncertainty

To analyse the mill modulus robustness problem (Hearns and Grimble 1997), the stand load control dynamics will be considered, without the interactions with the tension dynamics. This simplification aids understanding and enables a comparison to be made with the robustness of the conventional AGC. If $M_m$ is the mill modulus and $M_s$ is the strip gain the linear equations describing the behaviour of the stand and strip (only considering the entry thickness disturbance) are given as:

$$h = S + \frac{P}{M_m} \qquad (11.11)$$

$$P = M_s(H - h) \tag{11.12}$$

Figure 11.15 shows the thickness control loop using force ($P$) feedback from a load cell to drive the hydraulic capsule position $S$. To represent the uncertain mill modulus it is convenient to use an inverse additive uncertainty description. The actual mill modulus $M_m$ is represented by a nominal value $\bar{M}_m$ with feedback through a perturbation $\Delta$ and uncertainty weightings $W_1$ and $W_0$. The perturbation $\Delta$ is bounded, $\|\Delta\|_\infty \leq 1$ , therefore $W_1$ and $W_0$ have to be chosen such that $M_m$ can vary within the desired range:

$$M_m = \frac{\bar{M}_m}{1 - \bar{M}_m W_1 \Delta W_0} \tag{11.13}$$

If $M_m \in [0.8\bar{M}_m, 1.2\bar{M}_m]$ and $\|\Delta(j\omega)\| \leq 1$, for all $\omega$ , then letting $W_1 = 1$ and $W_0 = 0.25/\overline{M}_m$ will ensure that as $\Delta$ varies from -1 to 1, the mill modulus $M_m$ can cover the specified range. In order to design a robust feedback controller the feedback loop is broken, so that the weightings $W_1$ and $W_0$ can be considered (in the standard system description) as injecting a disturbance signal and weighting an output signal, respectively.

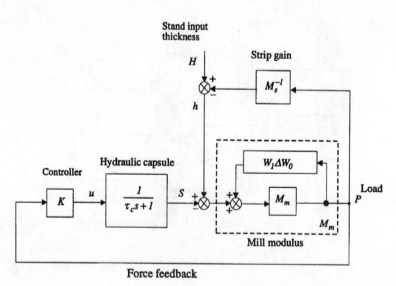

**Fig. 11.15** : Thickness Control with Inverse Additive Uncertain Mill Modulus
Description

The Fig. 11.16 shows the closed-loop system, augmented with the weightings, that both determine performance and represent uncertainty. The $W_s$ is usually a low-pass filter to weight the exit thickness. The transfer $W_c$ is a high-pass filter to weight the actuator reference, and $W_d$ is a low-pass filter to represent the input strip thickness disturbance. If the gain, as seen from $\omega_{mi}$ to $z_{mo}$, is less than unity, for the closed-loop system, then the closed-loop will be stable, as the mill modulus changes within its range. The transfer $W_0$ is for weighting the rolling force, whereas $W_1$ can be considered to be equivalent to the input thickness model $W_d$. To reduce the conservativeness $W_1$ can be removed and $W_d$ can be used both as the uncertainty input weighting and the disturbance model. This change is shown in Fig. 11.17 with $W_0 = 0.25/(W_d \bar{M}_m)$.

The inverse additive uncertainty model description was used because it enabled the input thickness disturbance model, also to be used for robust stability purposes, and the rolling force to be weighted, which is a variable that has a physical significance. The strip gain depends on the temperature and this can also be modelled by the uncertainty feedback loop, as shown in Fig. 11.18. This is an additive uncertainty description and when the loop is broken it will be equivalent to weighting the force for a system with thickness disturbance. This suggests the uncertainty for $M_m$ and $M_s^{-1}$ can be conveniently represented by one uncertainty model.

This problem reveals the importance of the dynamic cost-function weighting models in $H_\infty$ design. These capture the information on uncertainty models and they also determine the performance achieveable. However, disturbance models can also be used as design variables and in some sense they play the same role as the cost weightings. That is, the weightings and disturbance models are simply filters placed at different points in the system which determine the closed-loop properties of the resulting design.

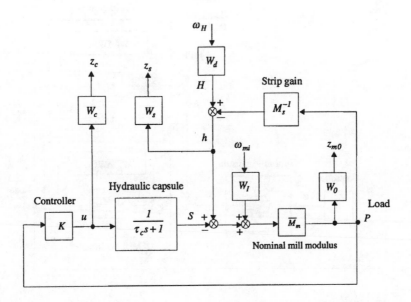

**Fig. 11.16** : Plant Augmented with Performance and Uncertainty Weightings

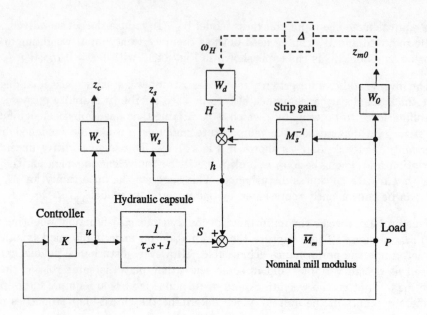

Fig. 11.17 : $H_\infty$ Thickness Control Robust Design Model

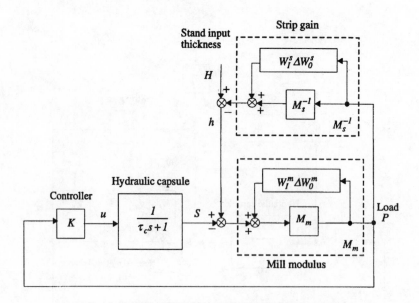

Fig. 11.18: Mill Modulus and Strip Gain Uncertainty

## $H_\infty$ Thickness Control

If $C$ denotes the transfer-function for the hydraulic capsule then the three output sensitivities that must be weighted are defined from the following relationships:

*Thickness sensitivity*:

$$h = \left( \frac{1 + M_m C K}{1 + M_m C K + M_s^{-1} M_m} \right) H = S_h H \qquad (11.14)$$

*Force sensitivity*:

$$P = \left( \frac{M_m}{1 + M_m C K + M_s^{-1} M_m} \right) H = S_p H \qquad (11.15)$$

*Control sensitivity*:

$$S_r = \left( \frac{K M_m}{1 + M_m C K + M_s^{-1} M_m} \right) H = S_u H \qquad (11.16)$$

The $H_\infty$ sensitivity minimisation problem (Skogestad and Postlethwaite, 1958) is to find a stabilising controller $K$ such that:

$$\left\| \begin{array}{c} W_s S_h W_d \\ W_0 S_p W_d \\ W_c S_u W_d \end{array} \right\|_\infty \leq 1 \qquad (11.17)$$

If (11.16) is satisfied, then the closed-loop will be robustly stable for changes in the mill modulus and will possess the desired nominal performance.

### Conventional Thickness Control System Performance

Since the conventional control is static the ideal gain is $K = -1/M_m$ since the steady-state gain of the thickness sensitivity ($S_h$) will be zero. For robust stability the closed-loop poles should be analysed. The return-difference for the nominal system in Fig. 11.15 may be written as: $F(s) = 1 + \left( \frac{K}{1 + \tau_c s} \right) \frac{M_m}{(1 + M_m/M_s)}$. If the controller $K = -1/\hat{M}_m$ the characteristic equation becomes:

$$s + \frac{1}{\tau_c} \frac{M_s}{(M_s + M_m)} \left( 1 + \frac{M_m}{M_s} - \frac{M_m}{\hat{M}_m} \right) = 0 \qquad (11.18)$$

Since $\tau > 0, M_m > 0$ and $M_s > 0$, then the closed-loop system is stable, if and only if:

$$1 + \frac{M_m}{M_s} - \frac{M_m}{\hat{M}_m} > 0 \qquad (11.19)$$

or

$$\hat{M}_m > \frac{M_m M_s}{M_s + M_m} \qquad (11.20)$$

For closed-loop stability, in the worst case mill modulus conditions:

$$\hat{M}_m > \frac{1.2 \bar{M}_m M_s}{M_s + 1.2 \bar{M}_m} \qquad (11.21)$$

Now $0 < M_m/M_s << 1$ and robust stability is achieved if:

$$\hat{M}_m \geq 1.2\bar{M}_m \tag{11.22}$$

The nominal thickness performance sensitivity with $K = -1/(1.2\bar{M}_m)$ is:

$$S_h = \frac{1 + M_m CK}{1 + M_m CK + M_s^{-1} M_m} = \frac{1 - 1.2^{-1}}{1 - 1.2^{-1} + M_s^{-1} \bar{M}_m} \tag{11.23}$$

A trade off is needed to provide robust stability ( $\hat{M}_m$ larger than the mill modulus) and achieve nominal performance (determined by $S_h$). The strip gain $M_s$ is not a constant but will decrease as the strip temperature increases. However, this uncertainty is not a problem for the AGC, since if $M_s$ decreases it will actually increase the stability margin.

### Trading-off Performance Against Robustness

The uncertainty for $M_m$ was represented within this $H_\infty$ problem structure by a norm bounded feedback loop. Since the parameter can be represented by a complex number and the uncertainty is real then this approximation of the uncertainty implies the condition for robust stability is only sufficient. The controller will satisfy the combined performance and stability specification for the range of $M_m$ if:

$$\|W_0 S_p W_I\|_\infty \leq 1 \tag{11.24}$$

The AGC controller $K = -1/(1.2\bar{M}_m)$ will be robustly stable but will not necessarily satisfy equation (11.23). The AGC controller achieves improved robustness by decreasing the nominal performance. The advantage of the $H_\infty$ design approach is that the robust stability and nominal performance objectives are stated as two separate (but connected) objectives in the cost-function. To find a controller which will satisfy the different objectives, the weights might have to be changed to tradeoff the performance against robustness.

## 11.6.3   Advanced Mill Control Structure

The choice of the control structure for the thickness and tension control is important to the design (Hearns and Grimble, 1998b, Okada et al. 1996). A single centralised controller can be ruled out because it would only enable that controller to operate when the mill was fully threaded. It would also make commissioning the controller difficult and it would be intolerant to certain types of fault. The design of a single centralised controller for a very high order plant would also present numerical difficulties. A decentralised controller does not have these limitations but the problem arises of how the control objectives are allocated to the individual controllers. This solution is aided by the fact that the process is sequential and consists of subsystems with the same structure and similar dynamics.

The structure for control design is based on a *stand zone* decentralised control philosophy, whose models are illustrated in Fig. 11.19. That is, a particular stand will be associated with the downstream interstand to make one stand/interstand unit for control design purposes. The objective for each of these units is to control the thickness $h$, tension $\sigma$ and angle $\theta$, using the control inputs $\omega_r$, $S_r$ and $M_r$. The main problem is the lack of measurements, with only force $P$ and angle $\theta$ being measurable. This means the controller will have to infer the effects of the upstream and downstream disturbances on thickness $h$ and tension $\sigma$ from these measurements. Note that none of the variables which are fedforward or fedback from one unit to the next (determining total stability) can be measured directly. The measurements of load $P$ and angle $\theta$ should give a good indication of the disturbances entering upstream. However the effect of the disturbance $V^{i+1}$, entering from downstream, is a major disturbance to the strip tension.

Figure 11.20 for the stand zone standard system model shows the four main exogenous input components $H^{i+1}$, $T_{en}^{i+1}$, $S^{i+1}$ and $\omega^{i+1}$ that influence $V^{i+1}$. The last two of these changes $S^{i+1}$ and $\omega^{i+1}$ are actuator changes from the downstream stand, which are measurable, and the effect of the others can be estimated using $P^{i+1}$. The conclusion is that the controller $K_i$ should be supplied with the measurements $P^i$, $\theta^i$, $P^{i+1}$, $S^{i+1}$ and $\omega^{i+1}$. Instead of measuring $S^{i+1}$ and $\omega^{i+1}$, the actuator references $S_r^{i+1}$ and $\omega_r^{i+1}$ can be used as feedforward signals and can be taken straight from the outputs of the controller $K^{i+1}$.

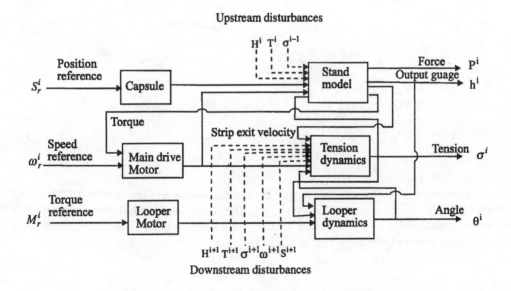

**Fig. 11.19** : Stand Zone and Related Models

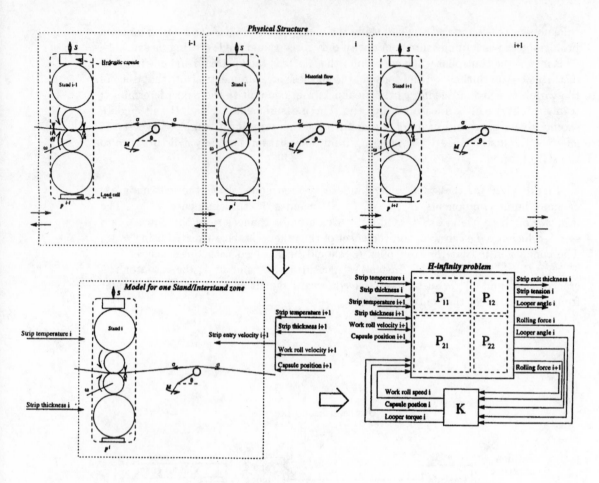

**Fig. 11.20** : Physical Motivation for Control Structure

The complete multivariable control structure is shown in Fig. 11.21 with the speed and capsule references being fed-forward to the upstream stand and the downstream force measurement also being used for the upstream controller. The overall controller structure $K_i$ can be partitioned as follows:

$$\begin{bmatrix} \omega_r^i \\ S_r^i \\ \overline{M_r^i} \end{bmatrix} = \begin{bmatrix} K_{11}^i & K_{12}^i \\ K_{21}^i & K_{22}^i \end{bmatrix} \begin{bmatrix} \theta^i \\ P^i \\ P^{i+1} \\ \overline{\omega_r^{i+1}} \\ S_r^{i+1} \end{bmatrix} \quad (11.25)$$

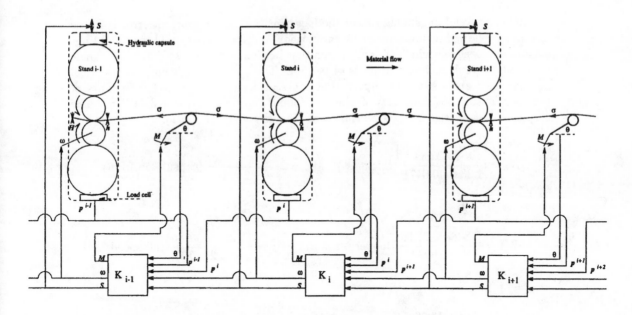

**Fig. 11.21** : Regulating Loop Control Structure with Feedforward
and Feedback Control

### 11.6.4   Performance and Disturbance Weightings

The weightings to be used in the state-space $H_\infty$ design should be proper, stable, low order and for multivariable systems they may often be taken as diagonal matrices (Zhou et al. 1996). Performance weights should, as far as possible, have a physical justification (Skogestad and Postlethwaite, 1996). Weights can be thought of as shaping closed-loop transfer functions (in size and bandwidth) in a cost-function partitioned as in (11.24), for the standard system model shown in Fig. 11.22.

The cost-function to minimise the output error ($z_s$) at low frequencies and the control signal ($z_c$) at high frequencies, for a low frequency disturbance signal ($w_d$) and high frequency measurement noise ($w_n$), is of the form:

$$\max_{\omega \neq 0} \frac{\|z\|_2}{\|\omega\|_2} = \left\| \left[ \begin{array}{cc} W_s(I - GK)^{-1}GKW_n & W_s(I - GK)^{-1}W_d \\ W_cK(I - GK)^{-1}W_n & W_cK(I - GK)^{-1}W_d \end{array} \right] \right\|_\infty \quad (11.26)$$

To achieve these objectives the matrix $W_d$ should include low pass filters with bandwidths determined by the expected disturbances. The noise models $W_n$ should be high pass filters with break frequencies chosen to be the same as the minimum significant measurement noise frequencies. For a multivariable system $W_d$ and $W_n$ may be chosen as diagonal matrices with each element having a bandwidth which reflects the frequency content of the signal. The transfer $W_s$ is a low-pass filter with a bandwidth related to the desired frequency content of the error signal. The transfer $W_c$ is a high-pass filter with a cut-off frequency equal to the desired closed-loop control bandwidth.

The error weightings must reflect the fact that the most important error is in the thickness, followed in importance by the tension and angle errors. These weightings are applied to signals which are scaled such that tension is in units $2 \times 106$ N/m$^2$, angle is in radians and thickness in units of 0.1mm. The control weightings are identical because they are applied to signals that are scaled such that speed is in units of rad/sec, capsule position units of mm and looper torque in units of 3000 Nm. The output and error weightings were chosen as (Lundstrom et al. 1991):

$$W_h = \frac{s + 42.589}{1.33s + 0.0171} \text{ (thickness)}, \quad W_s = \frac{11.95s + 77.85}{s + 9.2 \times 10^3} \text{(capsule position)}$$

$$W_\sigma = \frac{0.828s + 11.66}{s + 0.0093} \text{ (looper tension)}, \quad W_\omega = \frac{11.95s + 77.85}{s + 9.2 \times 10^3} \text{(roll speed)}$$

$$W_\theta = \frac{0.1s + 13.56}{1.29s + 0.0308} \text{ (looper angle)}, \quad W_M = \frac{11.95s + 77.85}{s + 9.2 \times 10^3} \text{(looper torque)}$$

The disturbance models involve filters on the signal inputs and can be considered as weightings that are fixed by the physics of the problem. The disturbance models used in the design are entry strip thickness ($W_H$), entry strip temperature ($W_T$), capsule position reference ($W_{S_r}$), speed reference ($W_{\omega_r}$) and strip tension ($W_{\sigma_{ex}}$). All the measurements of force, angle and the feedforward signals have independent additive noise sources with the same frequency response model ($W_n$).

$$W_H = \frac{0.1x10^{-3}}{s + 1} \quad , \qquad W_T = \frac{50}{5s + 1}$$

$$W_{S_r} = \frac{5 \times 10^{-4}}{0.1s + 1}, \qquad W_{\omega_r} = \frac{0.5}{0.1s + 1}$$

$$W_{\sigma_{ex}} = \frac{3 \times 10^6}{s + 1} \quad , \qquad W_n = \frac{0.003s + 2.79 \times 10^{-8}}{0.0597s + 99.82}$$

The stand zone controllers can be computed using the above weightings and disturbance models, and the plant models for particular stands. The feedforward controls are the subject of a separate calculation.

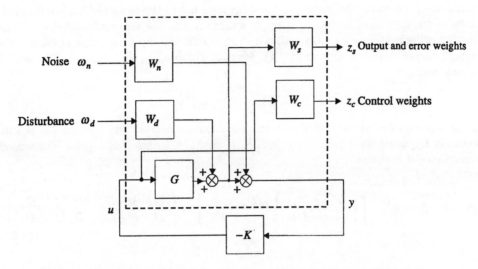

**Fig. 11.22** : Mixed Sensitivity Problem Standard System Description

## 11.6.5 Stand Zone Control Design Philosophy

The $H_\infty$ controller can be evaluated for a stand zone once the system and weightings (for a zone $i$) are collected together in the *standard system form* (Zhou et al. 1995). The main steps are summarised below.

**Unscaled system used for design**

The generalised error, measured output, control and exogenous inputs for the standard system model may be rearranged and collected as:

$$z = [\sigma^i \; \theta^i \; h^i \; P^i]^T$$

$$y = [\theta^i \; P^i \; P^{i+1} \; \omega_r^{i+1} \; S_r^{i+1}]^T$$

$$u = [\omega_r^i \; S_r^i \; M_r^i]^T$$

$$\omega = [H^i \; T_{en}^i \; H^{i+1} \; T_{en}^{i+1} \; \sigma^{i+1} \; \omega_r^{i+1} \; S_r^{i+1}]^T$$

$$\begin{bmatrix} z \\ y \end{bmatrix} = \begin{bmatrix} M_{11} & M_{12} \\ M_{21} & M_{22} \end{bmatrix} \begin{bmatrix} \omega \\ u \end{bmatrix} \quad (11.27)$$

**Scaled system for design**

Since the system inputs and outputs are in different units the sysem has to be normalised to make the error outputs of roughly the same magnitude, or importance, and to enable a disturbance vector of unity norm to be applied. Scaling the measurement outputs and control inputs provides a better conditioned system, helping to prevent numerical problems. The scaling method involves use of the expected maximum deviation of the signals to normalise them to a unity norm. For this multivariable system the following matrices are diagonal, with the maximum expected value of each component on the diagonal $D_z = z_{\max}$, $D_y = y_{\max}$, $D_u = u_{\max}$, $D_\omega = \omega_{\max}$. The scaled signals are:

$$\bar{z} = D_z^{-1} z, \quad \bar{y} = D_y^{-1} y, \quad \bar{u} = D_u^{-1} u, \quad \bar{\omega} = D_\omega^{-1} \omega$$

If the assumptions about the maximum deviation are correct and a controller is successfully obtained then $\|z\|_2 \leq 1$, $\|y\|_2 \leq 1$, $\|u\|_2 \leq 1$ and $\|\omega\|_2 \leq 1$. The scaled design model becomes:

$$S = \begin{bmatrix} D_z^{-1} & 0 \\ 0 & D_y^{-1} \end{bmatrix} \begin{bmatrix} M_{11} & M_{12} \\ M_{21} & M_{22} \end{bmatrix} \begin{bmatrix} D_\omega & 0 \\ 0 & D_u \end{bmatrix} = \begin{bmatrix} D_z^{-1} M_{11} D_\omega & D_z^{-1} M_{12} D_u \\ D_y^{-1} M_{21} D_\omega & D_y^{-1} M_{22} D_u \end{bmatrix}$$
$$(11.28)$$

**Weighted system for design**

All the weightings are diagonal with first-order elements. The feedforward signals have measurement noise added to ensure the feedforward element of the controller rolls off at high frequencies. Thence let,

$$W_s = diag\{W_\sigma, W_\theta, W_h, W_p\} , \quad W_n = diag\{W_{n_1}, ..., W_{n_s}\}$$

and

$$W_c = diag\{W_\omega, W_S, W_M\} ,$$

$$W_d = diag\{W_H, W_T, W_H, W_T, W_{\sigma_{ex}}, W_{\omega_r}, W_{s_r}\}$$

The weighted open-loop system model is obtained as:

$$G = \left[ \begin{array}{cc|c} W_s S_{11} W_d & 0 & W_s S_{12} \\ 0 & 0 & W_c \\ \hline S_{21} W_d & W_n & S_{22} \end{array} \right]$$
$$(11.29)$$

**Full-order controller calculation**

The objective is to find a stabilising controller such that:

$$\|T_{zw}\|_\infty = \|F_\ell(G, K)\|_\infty = 1$$
$$(11.30)$$

The construction of $G$ and $T_{zw}$ from the individual models and weightings is shown in Fig. 11.23. A controller can be found which minimises the $H_\infty$ norm but as the calculation nears the state-space optimal solution it becomes ill-conditioned. By solving for the sub-optimal case, with a bound of unity, the frequency weighted objectives can be satisfied approximately. If higher, or different, performance is required then the weights can be adjusted.

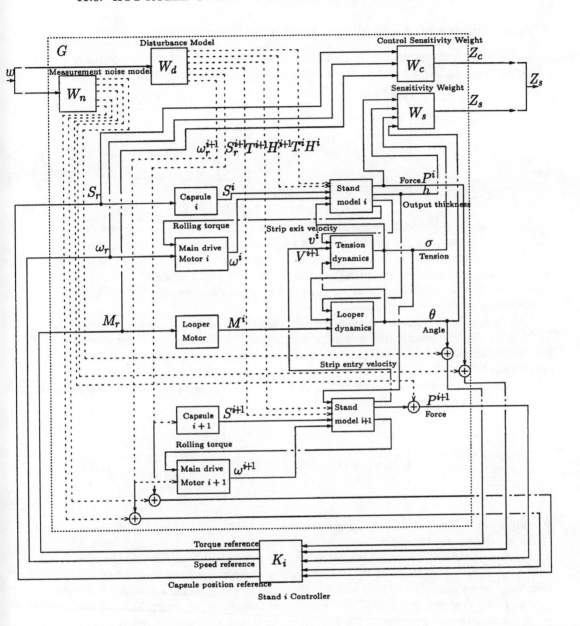

**Fig. 11.23** : Standard Model Representation for Design

## 11.6.6   Plant and Controller Frequency Responses

Consider the plant model for the dynamics of one mill stand and the downstream interstand region. The transfer function model relationship has the partitioned form:

$$
\left[ \begin{array}{c} \theta^i \\ P^i \\ \hline \sigma^i \\ h^i \\ T^i_{in} \\ V^i \end{array} \right] = \left[ \begin{array}{cc} P^i_{11} & P^i_{12} \\ P^i_{21} & P^i_{22} \end{array} \right] \left[ \begin{array}{c} \omega^i_r \\ S^i_r \\ M^i_r \\ \hline \sigma^{i-1} \\ H^i \\ T^i_{en} \\ V^{i+1} \end{array} \right] \tag{11.31}
$$

The inputs to the model are partitioned into the actuator references and the disturbances. The outputs are partitioned into local outputs of stand $i$ and interstand $i$ and the outputs which are either fed-forward or fed-back to other stand/interstand models. The strip thickness $h_i$, interstand temperature $T^i_{in}$ and tension $\sigma^i$ of unit $i$ is fed-forward to the downstream unit $i+1$, while the trip input velocity $V^i$ is fed-back to the immediate upstream unit $i-1$.

The frequency responses, for the standard system model, are shown in Figures 11.24 to 11.27. The plots for each stand are illustrated in these figures. The results for stand 1 are plotted with dashes, stand 2 with dash-dot, stand 7 with dots and the rest with solid lines.

Recall that the controller can be partitioned such that $K^i_1$ and $K^i_2$ are the feedback and feedforward parts, respectively.

$$
\left[ \begin{array}{c} \omega^i_r \\ S^i_r \\ M^i_r \end{array} \right] = \left[ \begin{array}{cc} K^i_{fb} & K^i_{ff} \end{array} \right] \left[ \begin{array}{c} \theta^i \\ P^i \\ P^{i+1} \\ \hline \omega^{i+1}_r \\ S^{i+1}_r \end{array} \right] \tag{11.32}
$$

Figures 11.28 and 11.29 show the gains and phase shifts of the individual elements of the controllers for the scaled plant for each stand. Controller 1 is shown with a dashed line, controller 2 with a dash-dot, controller 6 with a dotted line and the rest with solid lines. The phase of the transfer-function from the force to the capsule reference has a phase shift of 180° which is the same as the conventional AGC.

Figure 11.30 shows the closed-loop sensitivities and the inverse magnitude of the weighting function which weights the sensitivity for Stand 6. Each sensitivity represents the individual transfer functions from each disturbance to the output variable. The sensitivities which are close to the inverse weights represent the specifications which are the limiting factors in the design. For all six units the tension sensitivity, and in particular the closed-loop gains from $T^{i+1}_{en}$ to $\sigma^i$, is the closest to the inverse weighting which indicates that reducing this gain was the limiting factor on the performance. From the open-loop frequency responses, the looper motor torque was the least effective actuator in influencing the variables that are to be controlled. The control sensitivity function for the torque reveals that the controller uses this actuator the least, since the sensitivities lie far below the inverse weighting function.

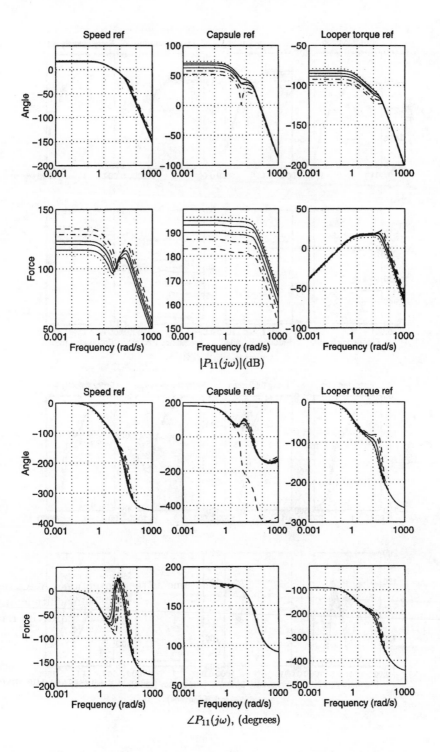

**Fig. 11.24** : Standard System Model Element (1,1) Frequency Responses

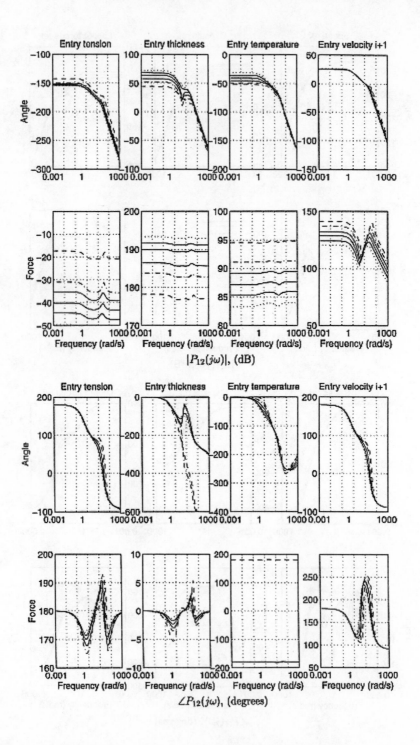

**Fig. 11.25** : Standard System Model Element (1,2) Frequency Responses

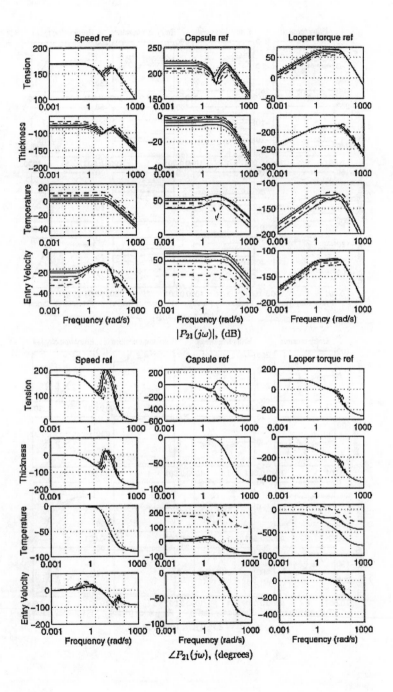

**Fig. 11.26** : Standard System Model Element (2,1) Frequency Responses

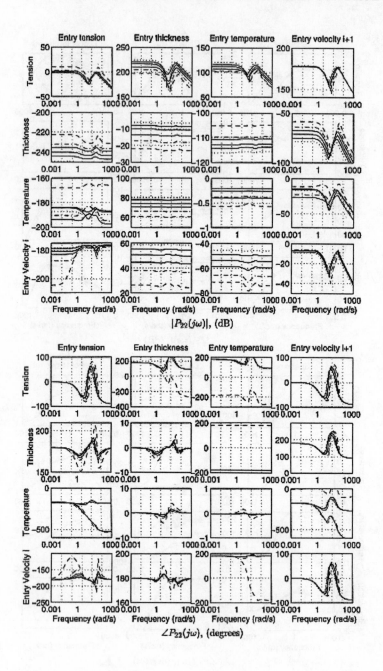

**Fig. 11.27** : Standard System Model Element (2,2) Frequency Responses

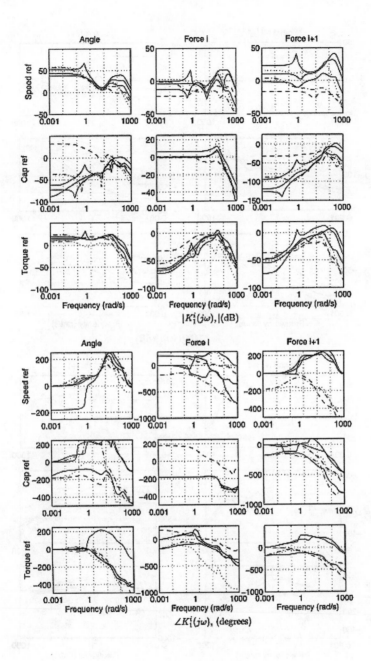

**Fig. 11.28** : Gains and Phases of the Partitioned Controllers Gain Elements $(K_{fb}^i)$

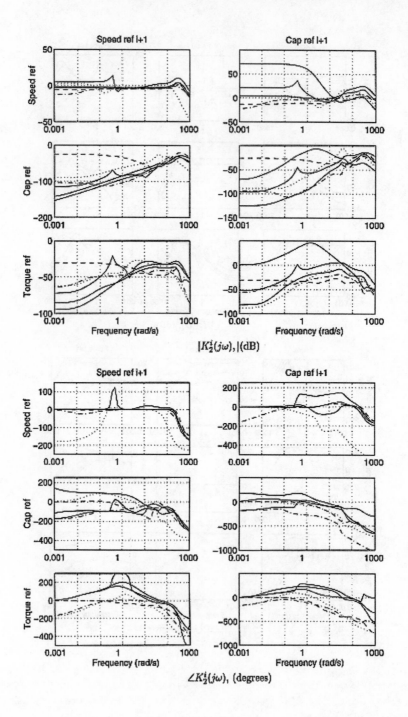

**Fig. 11.29** : Gains and Phases of the Partitioned Controllers Gain Elements $(K_{ff}^i)$

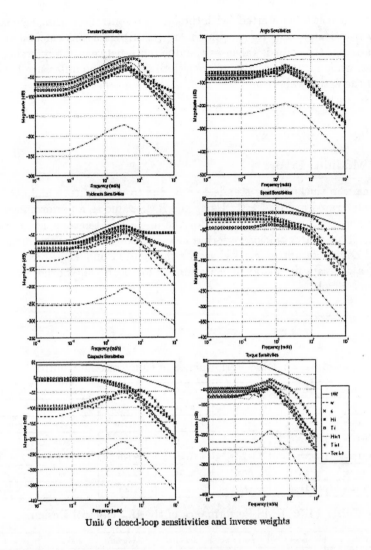

Unit 6 closed-loop sensitivities and inverse weights

**Fig. 11.30** : Closed Loop Sensitivities and Inverse Magnitude
of the Weighting for Stand 6

## 11.6.7   Hot Strip Mill Simulation Results

The objective in this section is to assess the performance and robustness of the multivariable controllers, using a nonlinear simulation (Hearns and Grimble, 1997). The mill will be run from threading to de-threading. The controllers are tested over the full range of the mill modulus by running a simulation with respectively: 80% of the nominal mill modulus for each stand ($0.8\ M_m$); the nominal mill modulus for each stand ($M_m$) and 120% of the nominal mill modulus for each stand ($1.2\ M_m$). The multivariable controllers are turned on sequentially starting with stand 1 at 14 seconds and finishing with stand 6 at 24 seconds. The conventional control is suspended and the multivariable controller started (with the last two controllers using safe startup to prevent any transients). The root mean square (RMS) error was calculated. For the regulated variables the error is taken as the difference between the reference and the actual values, while for the actuator references the error is the difference between the average and the actual values.

### -20% Mill Modulus Error

The changes in thickness for stands 1 to 6, due to temperature and other disturbances, are illustrated in Fig. 11.31. Using the RMS value as a performance measure the $H_\infty$ control law has a smaller error for the angle and tension. For most stands the $H_\infty$ control has better thickness performance, although the improvement over the conventional control is not as marked as for the tension and angle. For the actuator references the capsule energy is very similar for both the conventional and the multivariable controls.

**Table 11.2 : RMS Performance of Conventional Control**

|  | Stand 1 | Stand 2 | Stand 3 | Stand 4 | Stand 5 | Stand 6 | Stand 7 |
|---|---|---|---|---|---|---|---|
| $\theta$ | 1.50 | 0.543 | 1.06 | 0.586 | 0.874 | 2.215 | - |
| $h$ | 0.000374 | $7.837 \times 10^{-5}$ | $2.928 \times 10^{-5}$ | $3.335 \times 10^{-5}$ | $4.899 \times 10^{-5}$ | $3.388 \times 10^{-5}$ | $2.690 \times 10^{-5}$ |
| $\sigma$ | $1.426 \times 10^{-6}$ | $1.192 \times 10^{-6}$ | $1.956 \times 10^{-6}$ | $1.618 \times 10^{-6}$ | $1.145 \times 10^{-6}$ | $1.1345 \times 10^{-6}$ | - |
| $S$ | 0.000129 | 0.000108 | $4.611 \times 10^{-5}$ | $8.621 \times 10^{-5}$ | $4.792 \times 10^{-5}$ | $4.411 \times 10^{-5}$ | -0.0001061 |
| $\omega$ | 0.149 | 0.258 | 0.271 | 0.146 | 0.172 | 0.268 | - |
| $M$ | 2113.9 | 490.5 | 566.1 | 234.1 | 274.7 | 286.1 | - |

**Table 11.3: RMS Performance of the $H_\infty$ Control Design**

|  | Stand 1 | Stand 2 | Stand 3 | Stand 4 | Stand 5 | Stand 6 | Stand 7 |
|---|---|---|---|---|---|---|---|
| $\theta$ | 1.69 | 0.528 | 0.240 | 0.157 | 0.454 | 0.643 | - |
| $h$ | 0.000386 | $5.802 \times 10^{-5}$ | $1.422 \times 10^{-5}$ | $3.809 \times 10^{-5}$ | $3.402 \times 10^{-5}$ | $2.042 \times 10^{-5}$ | $3.285 \times 10^{-5}$ |
| $\sigma$ | $1.196 \times 10^{-6}$ | 732695 | $1.453 \times 10^{-6}$ | 298766 | 956574 | 465515 | - |
| $S$ | 0.000159 | 0.0001308 | $5.734 \times 10^{-5}$ | $8.820 \times 10^{-5}$ | $7.019 \times 10^{-5}$ | $5.687 \times 10^{-5}$ | $8.816 \times 10^{-5}$ |
| $\omega$ | 0.199 | 0.289 | 0.334 | 0.194 | 0.09899 | 0.238 | - |
| $M$ | 480.9 | 172.6 | 85.7 | 40.7 | 50.6 | 116.5 | - |

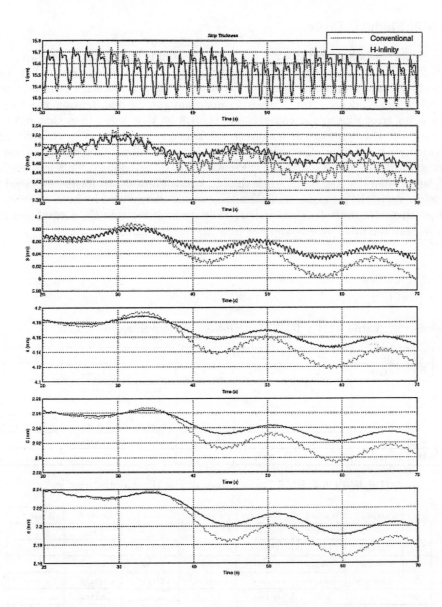

**Fig. 11.31** : Strip thickness Variations for 6 Stands and a Mismatch

Between Actual and Design Mill Modulus (0.8 $M_m$)

## No Mill Modulus Error

The variations in the strip thickness for stands one to six are shown in Fig. 11.32 for this case. The angle and tension error are less for the $H_\infty$ control but stands 3 and 6 do have a constant steady-state error. This arises when the multivariable control is switched on, since it tries to maintain the current tension value and if there is an error due to the conventional control at that time, then that error will be maintained, since there is no direct tension measurement. The strip thickness performance is substantially better with the $H_\infty$ controller. This can be explained, since to ensure robust stability the conventional AGC is designed for a mill modulus 25% larger than the true value, while the $H_\infty$ control is designed for the expected mill modulus. As in the previous simulation the capsule energy used by both conventional and multivariable controllers are similar, while the multivariable $H_\infty$ control uses more energy from the drive motors and less from the looper motors.

**Table 11.4 :** RMS Performance of Conventional Control

|          | Stand 1 | Stand 2 | Stand 3 | Stand 4 | Stand 5 | Stand 6 | Stand 7 |
|----------|---------|---------|---------|---------|---------|---------|---------|
| $\theta$ | 1.61656 | 0.546912 | 0.714776 | 0.493145 | 0.61665 | 1.11096 | - |
| $h$ | 0.00041 | $3.952 \times 10^{-5}$ | $1.669 \times 10^{-5}$ | $1.199 \times 10^{-5}$ | $3.684 \times 10^{-5}$ | $2.369 \times 10^{-5}$ | $1.049 \times 10^{-6}$ |
| $\sigma$ | $1.510 \times 10^{-6}$ | $1.265 \times 10^{-6}$ | $1.842 \times 10^{-6}$ | $1.616 \times 10^{-6}$ | $1.127 \times 10^{-6}$ | $1.153 \times 10^{-6}$ | - |
| $S$ | 0.000133 | 0.00011 | $5.0097 \times 10^{-5}$ | 0.000109 | $6.44 \times 10^{-5}$ | $5.187 \times 10^{-5}$ | $8.358 \times 10^{-5}$ |
| $\omega$ | 0.152 | 0.272 | 0.271 | 0.07287 | 0.105 | 0.162 | - |
| $M$ | 2265.2 | 495.1 | 375.2 | 203.9 | 177.2 | 183.1 | - |

**Table 11.5 :** RMS Performance of the $H_\infty$ Control Design

|          | Stand 2 | Stand 2 | Stand 3 | Stand 4 | Stand 5 | Stand 6 | Stand 7 |
|----------|---------|---------|---------|---------|---------|---------|---------|
| $\theta$ | 1.875 | 0.567 | 0.259 | 0.145 | 0.413 | 0.175 | - |
| $h$ | 0.000425 | $1.96 \times 10^{-5}$ | $3.693 \times 10^{-6}$ | $1.853 \times 10^{-5}$ | $2.029 \times 10^{-5}$ | $6.142 \times 10^{-6}$ | $1.326 \times 10^{-6}$ |
| $\sigma$ | $1.396 \times 10^{-6}$ | 758105 | $1.634 \times 10^{-6}$ | 396969 | 921806 | 490524 | - |
| $S$ | 0.0001649 | 0.0001335 | $6.011 \times 10^{-5}$ | $9.908 \times 10^{-5}$ | $9.211 \times 10^{-5}$ | $6.860 \times 10^{-5}$ | $5.195 \times 10^{-5}$ |
| $\omega$ | 0.1972 | 0.297 | 0.3301 | 0.1414 | 0.0791 | 0.2574 | - |
| $M$ | 532.6 | 185.2 | 95.0 | 43.4 | 48.7 | 36.9 | - |

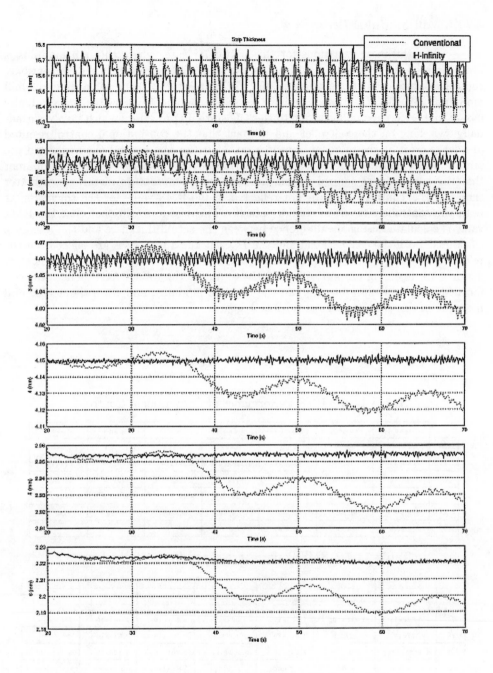

**Fig. 11.32** : Strip Thickness Variations for 6 Stands Given
Modelled Mill Modulus Equals Design Value ($M_m$)

**+20% Mill Modulus Error**

The thickness variations for this particular case, where the mill modulus is overestimated is shown in Fig. 11.33. At the start of the simulation the conventional control exhibits oscillations which is due to the mill setup being for a different mill modulus. For the $H_\infty$ control design the angle at stand 5, at switch on, gradually moves to its reference. This is due to the safe start-up procedure which involves gradually switching on the controller and the fact that the conventional control resulted in a large error at the moment of switch on. For the $H_\infty$ control the angle and the tension performance are better. From the strip thickness time-responses it is apparent that the conventional control rejects the skid-chill temperature disturbances better than the $H_\infty$ control. This is not reflected in the RMS value for the thickness, as both the controllers result in similar values. This is due to the fact that the RMS value is calculated using the difference between the actual thickness and the reference thickness. Both the conventional and the $H_\infty$ thickness responses are offset from the reference value due to errors in the mill setup. The conventional control has the better disturbance rejecting properties for thickness, due to the robust AGC being designed for a mill modulus 25% bigger than the nominal modulus, and since the actual mill modulus in this case happens to be 20% bigger.

**Table 11.6** : RMS Performance of Conventional Control

|          | Stand 1 | Stand 2 | Stand 3 | Stand 4 | Stand 5 | Stand 6 | Stand 7 |
|----------|---------|---------|---------|---------|---------|---------|---------|
| $\theta$ | 1.702 | 0.603 | 0.849 | 0.820 | 0.574 | 1.1784 | - |
| $h$ | 0.000436 | $1.355 \times 10^{-5}$ | $1.224 \times 10^{-5}$ | $1.214 \times 10^{-5}$ | $2.278 \times 10^{-5}$ | $1.856 \times 10^{-5}$ | $9.924 \times 10^{-6}$ |
| $\sigma$ | $1.587 \times 10^{-6}$ | $1.439 \times 10^{-6}$ | $2.217 \times 10^{-6}$ | $2.147 \times 10^{-6}$ | $1.543 \times 10^{-6}$ | $3.240 \times 10^{-6}$ | - |
| $S$ | 0.0001368 | 0.000111 | $5.373 \times 10^{-5}$ | 0.000119 | $8.502 \times 10^{-5}$ | $6.406 \times 10^{-5}$ | $6.569 \times 10^{-5}$ |
| $\omega$ | 0.155 | 0.282 | 0.269 | 0.0286 | 0.0403 | 0.123 | - |
| $M$ | 2382.6 | 542.6 | 485.6 | 311.4 | 127.5 | 119.6 | - |

**Table 11.7** : RMS Performance of the $H_\infty$ Control Design

|          | Stand 2 | Stand 2 | Stand 3 | Stand 4 | Stand 5 | Stand 6 | Stand 7 |
|----------|---------|---------|---------|---------|---------|---------|---------|
| $\theta$ | 2.012 | 0.606 | 0.286 | 0.164 | 0.613 | 0.192 | |
| $h$ | 0.000452 | $1.419 \times 10^{-5}$ | $1.182 \times 10^{-6}$ | $1.094 \times 10^{-5}$ | $1.207 \times 10^{-5}$ | $1.263 \times 10^{-6}$ | $1.237 \times 10^{-6}$ |
| $\sigma$ | $1.521 \times 10^{-6}$ | 827710 | $1.825 \times 10^{-6}$ | 699147 | $1.596 \times 10^{-6}$ | $2.571 \times 10^{-6}$ | |
| $S$ | 0.000169283 | 0.000135491 | $6.267 \times 10^{-5}$ | 0.000108118 | 0.00118011 | $8.429 \times 10^{-5}$ | $2.394 \times 10^{-5}$ |
| $\omega$ | 0.198809 | 0.305434 | 0.33005 | 0.117578 | 0.132402 | 0.367308 | |
| $M$ | 571.1 | 187.4 | 108.0 | 60.1 | 124.7 | 97.7 | |

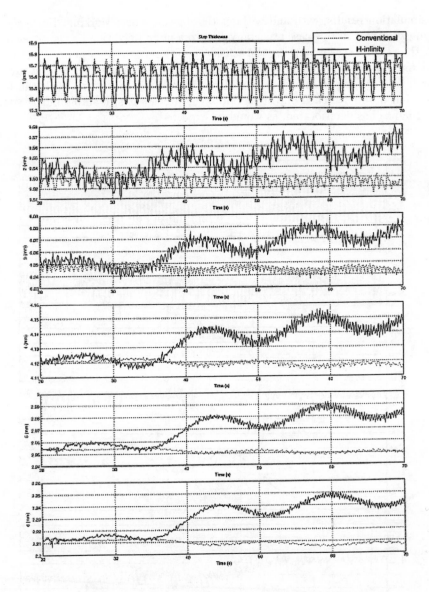

**Fig. 11.33 : Strip Thickness Variations for 6 Stands and a Mismatch
Between Actual and Design Mill Modulus (1.2 M$_m$)**

## 11.6.8 Reliable Tension Control

The theory of reliable control introduced in Chapter 1.3 (Veillette, 1992; Leland et
al. 1994; Medanic, 1993 and Hearns et al. 1998) can be applied to design a controller
which uses a tension measurement but can tolerate the loss of that measurement.
The design will be the same as in the previous multivariable design, but with the
addition of the input and output weightings for the fault model. Figure 11.34 shows

the simulation results, with and without the tension sensor working. Since the fault output weighting in this case was equivalent to a large weighting on the tension, then with the tension sensor working the tension performance is very good. The price for good tension performance is that the looper angle no longer has zero steady-state error but is drifting, although it is still within reasonable limits. With the tension sensor failed and switched out of the loop, the process is still stable but the tension performance has deteriorated, as would be expected.

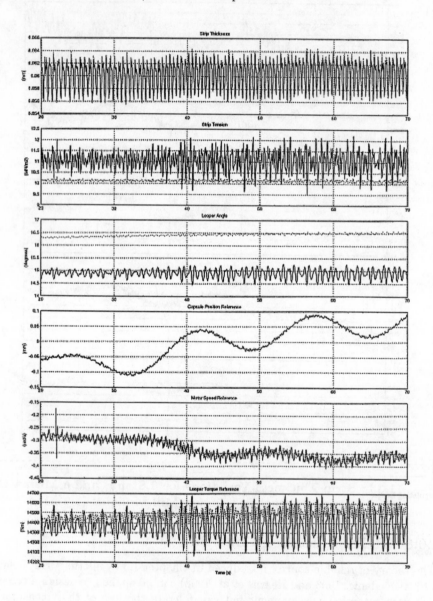

**Fig. 11.34** : Reliable Control Simulation Results With and Without
Tension Sensor Loss

# 11.7  Coordinated Control

The coordinated control system design must meet two main objectives:

1. To improve product quality by ensuring that the error between the desired and actual quality variables (gauge, shape and profile), at the exit of the last stand, is minimised.

2. To ensure that the actuator and process constraints are not violated.

Existing systems are based on an upper level control which take in the gauge error at stand seven and changes the set points to the regulating loops using PI controllers. The constraints are (at best) handled locally using some anti-wind up scheme. The problem with this approach is that it corrects the gauge but it can produce a bad shape and profile, since the constraints are only handled locally.

The supervisory or co-ordinated control system sits on top of the regulating loop setpoints, as illustrated in Fig. 11.35. The main function of the (lower) regulating loops is to quickly regulate against the effect of disturbances. The co-ordinated control system utilises measurements and then adjusts the regulating loop setpoints to remove the slower, almost steady-state, tracking errors. The supervisory level control structure can be based on simplified models of the regulating loops and is relatively slow.

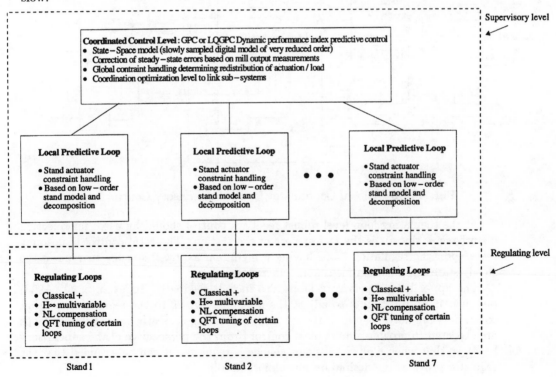

**Fig. 11.35** : Coordinated and Regulating Loop Control Structures
for a Hot Strip Mill

Predictive Control can be employed at the upper level to take into account the effect of the stand interactions and to develop a dynamic set up scheme to make the trade-offs between the quality parameters. Predictive control has particularly advantages for this supervisory level (Morari and Zafiriou, 1989; Ordys and Grimble, 1996). Predictive control allows constrained optimisation to be performed on, for example, the redistribution of load across stands, where it is important to remain within given mechanical and metallurgy limits. Moreover, predictive control algorithms can be made numerically efficient for the large systems involved. This is indicated in Fig. 11.36, which illustrates the global co-ordination, or supervisory control, architecture (Findeisen, 1980).

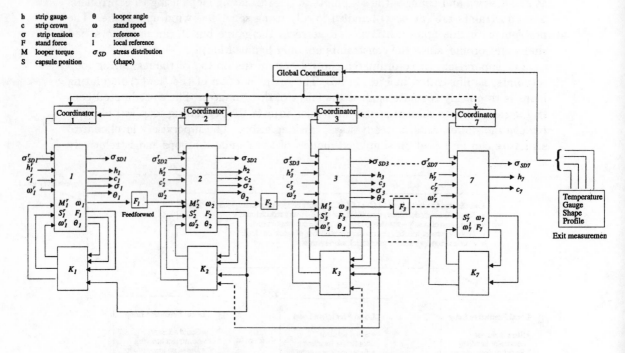

| h | strip gauge | $\theta$ | looper angle |
|---|---|---|---|
| c | strip crown | $\omega$ | stand speed |
| $\sigma$ | strip tension | r | reference |
| F | stand force | l | local reference |
| M | looper torque | $\sigma_{SD}$ | stress distribution |
| S | capsule position | | (shape) |

Fig. 11.36 : Global Coordinator for the Supervisory Control System

At the regulating loop level robust feedback control should be used, for rejecting disturbances, since it is the fastest part of the system. The total control philosophy relies upon the regulating loops having a high gain at low-frequency to remove the steady-state errors (when measurements are available).

The upper level supervisory loops also include feedback (Toguyeni et al. 1993) but their main function is to adjust the set-point signals to the regulators, so that constraints are not violated. If the system is operating correctly the regulating loops are expected to provide almost ideal control (from the perspective of the supervisory level), so that when a given set-point change is requested it can be assumed that the regulator provides the desired output almost ideally.

The difference between the supervisor and regulator occurs when the regulator is demanding actuator changes, which violate the actuator bending constraints. In this case the upper level supervisory control should reallocate the distribution of actuator power across the mill. However, once the upper-level control has decided what thickness changes to make, it is up to the regulators to provide the desired response. For example, on the last stand the regulator has feedback from a thickness measurement and if the upper level supervisory control decides a constraint is being violated it will give the last stand a new thickness reference. This thickness reference can then be regulated using closed-loop feedback on the last stand and it will be the regulator for this stand that is responsible for achieving the desired thickness reduction.

This strategy enables the upper level control to be based on a very low order approximate model of the combined plant and regulator loops. The supervisory coordinated control involves a model for the total hot mill which is so large a system that a reduced order approximation is imperative, if the predictive algorithm is to be of reasonable complexity. An important role for the coordinated control level is to ensure the static distribution of load is appropriate.

Measurements are not always available for regulator feedback, as in the case of the profile and flatness control. The current instruments to measure these variables are slow, so that any intermediate feedback loops (between the regulator-level force trims on the bending forces and the supervisory setpoints) would not be faster than the supervisory control itself. In these cases it cannot be assumed that the profile and flatness regulators can deliver a zero steady-state error.

**Robustness**

It is well known that different variants of predictive control are not robust particularly in the output constrained mode. Hence, a more robust formulation, based on LQGPC, can be employed to ensure robustness of the upper level control loop. In this case the time delays in the system can be modelled and included in a Kalman filter based prediction model. The advantage of this architecture is that the error in the constraint model can be estimated and incorporated into the future calculation of set points. Wang et al. (1995) considered the robust control design of decentralized control structures.

**Numerical Complexity**

Due to the complexity of the process and the speed of the machinery (10 msec at the regulating loop level), a *centralised* LQGPC is neither feasible nor desirable. The *decentralised* solution (Lopez and Athans, 1993) is optimal in the sense that it generates the same inputs as the centralised solution. The solution has two levels. The first level comprises the LQGPC controller, based on the model of a signal stand. The second level contains the coordinator, which is required to incorporate the effect of interactions between the stands (Fig. 11.37). It has the advantage of significantly reducing the computational burden of LQGPC. For this approach to work over the full operating range of hot rolling mill, the linear model should be adapted to compensate for possible modelling errors.

**Conflicting Control Requirements**

Because of the complexity of the control system and the interactions between the local controllers, it is likely that these will compete to meet their own local objectives. For example, an unreasonable demand on tension may lead to strip necking or speed mismatch between the stands which will lead to cobbling. To avoid these problems it is necessary to make a trade-off between different control objectives at the design stage. Furthermore, on-line procedures must be used to resolve these conflicts by manipulating the controller set points or readjusting the local controllers. The main trade-offs should be carried out between the gauge, tension, profile and shape and temperature control systems. One possible approach to solving this problem is to include output constraints on the supervisory or top level controllers.

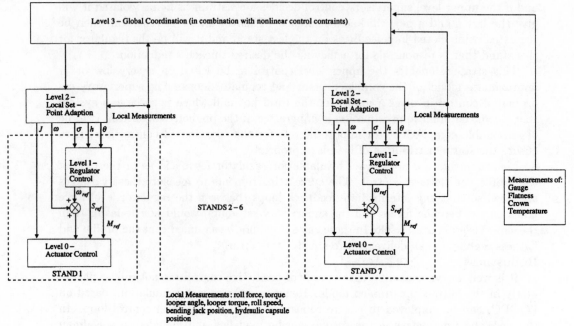

**Fig. 11.37** : Global Control Employing a Decentralised Control Architecture

**Summary of Operation of the Upper-level Controller:**

1. The temperature, gauge, profile and shape errors (global variables) are measured at the last stand. These errors are minimised by a feedback loop, which operates on these errors and modifies the set points of the regulators for each stand. This is an upper-level controller.

2. The upper-level controller is based on predictive control and is formulated in state-space. It comprises a Kalman filter to predict the states and a QP optimiser to generate the regulator set-points.

3. The upper level controller is designed using the closed-loop model of the regulator loops. This model will contain only the low frequency dynamics of the system and has relatively low order.

4. The upper level controller compensates for the global steady state error. This error cannot be minimised using integral action at the regulator level if no measurements are available.

5. The weighting functions will be used to compensate for interactions and trade-offs between the quality parameters, i.e. thickness, profile, shape and finished temperature. These can indirectly influence the way in which the constraints are satisfied (in a linear system fashion).

6. The optimisation is performed subject to the constraints on the actuator inputs at the regulator level (lower level inequality constraints) and the mass flow control (equality constraints).

7. The Kalman filter operates on the local stand measurement (speed, tension, force, looper angle) as well as the global measurements (shape, thickness, profile, temperature at Stand 7). The filter is decentralised to ensure a low computational load.

8. The optimiser generates the set points to the lower-level feedback loops subject to the process constraints. The actuation will therefore be redistributed, based on the error in the global measurements. The optimisation process can also be decentralised for each stand.

9. The stability robustness at high frequencies is achieved by robust regulator design. The system stability will be ensured by using an integrated design approach for the upper and lower level controllers.

## 11.8 Concluding Remarks

Several companies are currently in the process of implementing and assessing the $H_\infty$ steel mill designs. The main benefits over classical techniques is the natural way the multivariable nature of the problem is treated. The general design strategy for hot mills applies in a similar manner to cold mills even though there are not loopers and the roll force models (Grimble, 1978) are very different. The strategy may be summarised as follows:

(i) The effects of modelling uncertainty can be reduced by using robust control design techniques at the regulator level.

(ii) The sequential nature of the process should be exploited and feedforward control utilised, to cope with interactions.

(iii) Inferential control design principles can be employed when measurements are not available.

(iv) The control system for a stand zone should be as autonomous as possible, both for design and ease of tuning and commissioning (Lopez and Athans, 1993).

(v) Simplficiations can be achieved through the application of a multivariable controllers for each stand/ interstand unit to control strip thickness, strip tension and looper angle, without measurement of strip thickness or tension.

(vi) Controllers can be designed to provide robust stability for changes of the mill modulus, while preserving the nominal thickness performance.

(vii) The LQG optimal time-delay controllers for X-ray feedback thickness control will provide improved stochastic properties.

(viii) Reliable control design can be employed to accommodate failure of a strip tension sensor and enhance security of operation.

(x) Gain scheduled controllers should be used to reduce the effect of modelling errors, where the changes are too large for a robust controller.

(xi) Mechanical constraints should be accommodated by using model-based predictive control at the supervisory-level

(xii) Co-ordinated control may be used to resolve conflicting control design requirements, employing model-based predictive control, in combination with performance monitoring indices.

A number of results from the hot mill studies have much wider applicability and may be detailed as follows:

- Inferential control, where direct measurements are not available, provides good performance, given reasonable models.

- A decentralised control structure with feedforward and overlapping measurements can provide almost the full benefits of multivariable controllers, with the least complexity.

- A combination of a multivariable controller with a safe startup procedure will prevent transients at switch on and enable controller performance to be assessed and tuned (Stoustrup and Niemann, 1997).

- Upper level predictive control for supervisory and co-ordination functions, can be very effective for total system control.

# 11.9   References

Anbe, Y., Sekiguchi, K. and Imanari, H., 1996, Tension control of a hot strip mill finisher, *IFAC 13th Triennial World Congress*, San Francisco, 439-444.

Bryant, G.F., Ed, 1973, *Automation of trandem mills*, The Iron and Steel Institute, London.

Clark, M., Versteeg, H. and Konijn, W., 1996, The joint development of new Davy-Hoogovens high-performance loopers, *AISE Spring Convention*, Ohio.

Duysters, S., Govers, J. and van der Weiden, A.J.J., 1994, Process interactions in a hot strip mill; possibilities for multivariable control, *3rd Conference on Control Applications*, Glasgow, pp. 1545-1550.

Findeisen, W., Bailey, F.N., Bryds, M., Malinowski, K.,Tatjewiski, P. and Wozniak, A., 1980, *Control and coordination in hierarchical systems,* John Wiley and Sons Ltd, Chichester, UK.

Fukushima, K., Tsuji, Y., Ueno, S., Anbe, Y., Sekiguchi, K. and Seki, Y., 1998, Looper optimal multivariable control for hot strip finishing mill, *Transactions ISIJ*, Vol. **28**, pp. 463-469.

Ginzburg, V.B., 1989, *Steel rolling technology*, Marcel Dekker, New York.

Grimble, M., 1975, Shape control for rolled strip, *Inst. Mech. Engineers*, pp. 91-93.

Grimble, M., 1976a, Tension controls in strip processing lines, *Metals Technology*, pp. 445-453.

Grimble, M., 1976b, A roll-force model for tinplate rolling, *GEC Journal of Science and Technology*, Vol. *43*, No. *1*, pp. 3-12.

Grimble, M., 1978, Solution of the nonlinear functional equations representing the roll gap relationships in a cold mill, *J. of Optimization Theory and Applications*, Vol. **26**, No. 3, pp. 427-451.

Grimble, M., Fuller, M.A. and Bryant, G.F., 1978, A non-circular arc roll force model for cold rolling, *International Journal for Numerical Methods in Engrg.* Vol. *123*, pp. 643-663.

Grimble, M. and Fotakis, J., 1982, The design of strip shape control systems for Sendzimir mills, *IEEE Trans. on Auto. Control*, Vol. AC-**27**, No. 3, pp. 656-665.

Grimble, M.J. and Hearns, G., 1999, Advanced control for hot strip mills, *Semi-Plenary European Control Conference*, Karlsruhe, contributed chapter, Advances in Control (Editor : P.M. Frank), Springer-Verlag.

Grimble, M.J. and Hearns, G., 1998, LQG controllers for unstable systems with transport delays : Thickness Control in Rolling Mills, *37th IEEE Conference on Decision and Control*, Tampa, Florida, pp. 3150-3155.

Gunawardene, G.W.D.M., Grimble, M.J. and Thomson, A., 1981, Static model for Sendzimir cold rolling mill, *Metals Technology*, pp. 274-283.

Hamada, K., Ueki, S., Shitomi, M., Doi, K., Ishikawa, K. and Okuda, T., 1985, Finishing mill tension control system in the Mizushima Hot strip mill, Kawasaki Steel Technical Report, No. 11, pp. 35-43.

Harakawa, T. and Furuta, K., 1995, Decentralised control of multivariable looper system in hot strip finishing mill control, *4th IEEE Conference on Control Applications*, New York, pp. 856-861.

Hayashi, Y., Tanimoto, S., Saito, M., Kataoka, T., Sasao, H. and Yabuuchi, H., 1987, A new tension control method for hot strip finishing mill, *IFAC 10th Triennial World Congress*, pp. 114-119.

Hearns, G. and Grimble, M.J., 1998a, Fault tolerant strip tension control, Proceedings of the *17th American Control Conference*, Philadelphia.

Hearns, G. and Grimble, M.J., 1998b, Robust hot strip mill thickness control, *Proceedings of the 9th IFAC Symposium on Automation in Mining, Mineral and Metal Processing*, Cologne, Germany, pp. 115-120.

Hearns, G., and Grimble, M.J., 1997, Multivariable control of a hot strip mill, *Proceedings of the 16th American Control Conference*, Albuquerque, pp. 3775-3779.

Hearns, G., Katebi, M.R. and Grimble, M.J., 1996a, Robust control of a hot strip mill looper, *Proceedings of the 13th IFAC World Congress*, San Francisco, pp. 445-450.

Hearns, G., van der Molen, G.M. and Grimble, M.J., 1996b, Hot strip mill mass flow control, *IEE Conference Control 96*, University of Exeter.

Hearns, G., Grimble, M.J. and Johnson, M.A., 1998, Integrated fault monitoring and reliable control, *Control 98*, Swansea, pp. 1175-1179.

Holton, L.J. and Forsythe, J.A., 1991, Hot strip mill looper control - theoretical and practical considerations, *AISE seminar*, Dearborn MI.

Imanari, H., Morimatsu, Y., Sekiguchi, K., Ezure, H., Matuoka, R., Tokuda, A. and Otobe, H., 1995, Looper $H_\infty$ control for hot strip mills, *IEEE, IAS*, Florida pp. 2133-2139.

Johnson, M.A., Hearns, G. and Lee, T., 1999, *The hot strip rolling mill looper : a control case study*, contributed chapter, Mechatronic Systems, Techniques and Applications (Editor : C.T. Leondes), Gordon and Breach Science Publishers, New Jersey, USA.

Kawaguchi, T. and Ueyama, T., 1988, Optimal multivariable control of loopers at hot finishing mill, Steel Industry II, Control Systems, Gordon and Breach, pp. 104-111.

Kotera, Y. and Watanabe, F., 1981, Multivariable control of hot strip mill looper, *IFAC 8th Triennial World Congress,* Kyoto, Japan, pp. XVIII-1 XVIII-5.

Leland, P.M., Medanic, J.V. and Perkins, W.R., 1994, Reliable $H_\infty$ norm bounding controllers with redundant control elements, *Conference on Decision and Control,* pp. 1536-1541.

Lopez, J.E. and Athans, M., 1993, On synthesizing partially decentralized controllers, *American Control Conference,* pp. 2386-2390.

Lundstrom, P., Sigurd, S. and Wang, Z., 1991, Performance weight selection for $H_\infty$ and control methods, *Transactions Institute of Measurement and Control*, Vol. **13**, No. 5, pp. 241-252.

McNeilly, G., van der Molen, G.M., Katebi, M.R. and Grimble, M.J., 1996, Control of thickness profile whilst maintaining flatness in hot strip tandem rolling *IFAC World Congress*, San Francisco.

Medanic, J.V., 1993, Design of reliable controllers using redundant control elements, *American Control Conference*, pp. 3130-3134.

Morari, M. and Zafiriou, E., 1989, *Robust process control*, Prentice Hall.

Nakagawa, S., Miura, H., Fukushima, S. and Amasaki, J., 1990, Gauge control system for hot strip finishing mill, *29th Conference on Decision and Control,* Honolulu, pp. 1573-1578.

Okada, M., Iwasaki, Y., Murayama, K., Urano, A., Kawano, A. and Shiomi, H., 1996, Optimal control systems for hot strip finishing mill, *IFAC 13th Triennial World Congress*, San Francisco.

Ordys, A.W. and Grimble, M.J., 1996, A multivariable dynamic performance predictive control with application to power generation plants, *IFAC World Congress*, San Francisco.

Randall, A., 1994, Hot strip mill tension control multivariable techniques, Conference on Systems Engineering, University of Coventry.

Reeve, P.J., MacAlister, A.F. and Bilkhu, T.S., 1999, Control, automation and the hot rolling of steel, *Phil. Trans Royal Society London*, **357**, pp. 1549-1571.

Saito, M., Tomimoto, S., Hayashi, Y., Hirokawa, T., Yabuuchi, K. and Miyai, Y., 1985, Development of new tension control method for hot strip fishing mill, Nippon Kokan technical report, Vol. **107**, pp. 12-20.

Seki, Y., Sekiguchi, K. and Anbe, Y., 1991, Optimal multivariable looper control for hot strip finishing mill, *IEEE Transactions on Industry Applications*, Vol. **27**, No. 1, pp. 124-130.

Shioya, M., Yoshitani, N. and Ueyama, T., 1993, Development of high-response looper control system based on multivariable control theory, Nippon Steel Technical Report, No. 57, pp. 57-61.

Skogestad, S. and Postlethwaite, I., 1996, *Multivariable feedback control*, John Wiley & Sons Ltd, Chichester UK.

Stoustrup, J. and Niemann, H., 1997, Starting up unstable multivariable controllers safely, Department of Control Engineering, Aalborg University, Denmark, Doc. No. R-1997-4171.

Toguyeni, A.K.A., Craye, E. and Gentina, J.C., 1993, An approach for the placement of sensors for on line diagnostic purposes, *IFAC 12th Triennial World Congress*, Sydney, pp. 5-10.

Tsuji, Y., 1987, High quality control in hot strip mill, *IFAC 10th Triennial World Congress*, pp. 120-125.

Veillette, R.J., Medanic, J.V. and Perkins, W.R., 1992, Design of reliable control systems, *IEEE Transactions on Automatic Control*, Vol. **37**, No. 3, pp. 290-304.

Wang, Y., Xie, L. and de Souza, C.E., 1995, Robust decentralised control of interconnected uncertain linear systems, *IEEE Conference on Decision and Control*, pp. 2653-2658.

Zhou, K., 1995, A comparative study of $H_\infty$ controller reduction methods, *American Control Conference*, Seattle, pp. 41015-4019.

Zhou, K., Doyle, J.C. and Glover, K., 1996, *Robust and optimal control*, Prentice Hall.

# 12

# Marine Control Systems

## 12.1 Introduction

Both $H_2$ and $H_\infty$ design procedures are valuable for the control of marine control systems. The former technique is appropriate when stochastic behaviour dominates the requirements and the latter approach is useful when robustness problems are central to the design problem. Frequency respones are very important in marine systems, since the main disturbance is due to the waves. Most of the results presented are therefore obtained using the polynomial system models in Chapters 2 and 5. The actual applications were considered using continuous-time models and the results are therefore presented in this form, although the discrete-time computations would be similar.

The fin roll stabilisation control problem is considered first. In this case the design specification concerns the so called *roll reduction ratio*. This function reveals the amount by which the ship motion, due to waves, will be attenuated, or amplified. In control design terms this function is better known as the sensitivity function which represents the gain between the wave disturbance inputs and the roll angle of the vessel. The minimisation of the roll of the vessel, in terms of the $H_2$ criterion, is equivalent to the minimisation of the variance of the roll angle, for the given sea state.

The second design problem involves the $H_\infty$ robust control of ship positioning systems. The main interest in this case lies in the reduction of commissioning time. In many cases the models used for design are not very representative of the actual ship equations. The result is that LQG Kalman filtering solutions can take a long time to tune, which is expensive for both the control systems supplier and the ship operator, and it does not provide confidence in the product. The $H_\infty$ design approach is therefore a candidate design philosophy. In state-equation form this has the advantage that the basic structure is identical to that for an LQG solution, which is currently the most popular advanced ship positioning control system.

Long periods for tuning and commissioning vessels, and the need for re-tuning on occasions, often occurs because of poor model information on which the designs are based. This certainly applies in the design of most commercial vessels. Over the last decade the $H_\infty$ robust design techniques have been developed in the aerospace industry for systems where the models are poorly defined. The method has the advantage that is can easily be automated within the latest design packages. The marine control system designer can readily use this rather complex mathematical solution with the aid of packages such as Matlab, MatrixX, Program CC, Mathematica, etc.

## 12.2   Fin Roll Stabilisation Control Systems

The fins on a vessel are used to stabilise its rolling motions. These were initially applied on ocean liners where passenger comfort was a priority and on military vessels where stable gun platforms were needed. Most new vessels are now fitted with some form of roll stabilisation system and fins remain a simple and popular technique (Wong et al. 1990).

The fins are not required to correct for the static heel of a vessel but are used to stabilise the rolling motion due to waves. The feedback control problem is a little unusual, since the plant has zero gain at zero frequency. Moreover, the bias from the roll rate sensor suggests that the controller should not include integral action, so that the loop gain must be zero at zero frequency. The structure of a typical fin roll stabilisation regulation system is shown in Fig. 12.1.

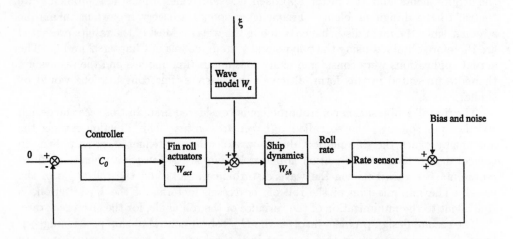

Fig. 12.1 : Regulating Loop for the Fin Roll Stabilisation
Control System

## 12.2.1   System Description

The definition of the system model is as important as the specification of the cost function weightings in the design. In fact, rather more freedom is available in the specification of the system description than might be thought. In particular, the wave disturbance model can be considered a design model with tuning variables, simply because it cannot be defined with certainty. A typical ship model is a lightly damped second-order system with the frequency response shown in Fig. 12.2.

**Ship Model :** The ship and actuator dynamics were defined as:

$$W_{sh} = 0.36s/(s^2 + 0.12s + 0.36)$$

**Actuator Model :**

$$W_{act} = 0.17/(1 + 0.374s)$$

**Plant Model :**

$$W = W_{sh}W_{act}$$

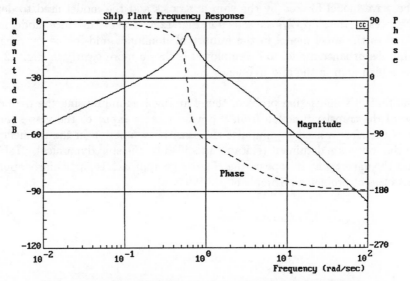

**Fig. 12.2** : Ship Transfer Function Frequency Response

### Disturbance Model

The wave disturbance is often generated by an approximation to a standard sea spectrum, such as the Pierson Moscowitz spectrum (Pierson and Marks, 1952). However, the vessel motions, due to this wave disturbance depend upon the dynamics of the vessel. This model is dependent upon the direction of the waves, since clearly the effect of the waves is very different if the angle changes from a head sea to a beam sea (see Fig. 12.3). This directionality will be neglected for the present and it will be assumed that the wave disturbance forces act on the vessel at the same point that the actuators affect the vessel motions. This is indicated in the block diagram of Fig. 12.1.

### Wave Model

The wave disturbance model was defined as :

$$W_{wa} = 0.5345s/(s^3 + 0.5919s^2 + 0.4624s + 0.1161)$$

The actual disturbance model to be used for design will include the wave model feeding the ship model, together with a white noise component. This helps the numerical conditioning of the problem and also approximates the unknown disturbance

components outside the wave frequency range.  The wave model is therefore defined
to satisfy :

$$W_{d1}W_{d1}^* = W_{sh}W_{wa}W_{wa}^*W_{sh}^* + 0.5$$

where $W_{d1}$ is a minimum phase and stable model.

The wave model filtered by the ship dynamics and the model used to define the
disturbances on the ship $W_{d1}$ are shown in Fig. 12.4. The disturbance model is clearly
the same as the wave model in the important dominant mid-frequency range.  The
resulting disturbance model to be employed is shown more clearly in Fig. 12.5. This
reveals a high gain in the mid-frequency range.

The fin roll stabilisation problem therefore involves minimising the frequency response of the transfer function from the wave motion input to the vessel roll angle.
There is therefore a need to minimise the sensitivity function in the frequency range
where the waves are dominant (suitably modified by the ship dynamics).  Taking into
account the uncertainty it appears the $H_\infty$ design approach is particularly appropriate
for this class of problems (Grimble et al. 1993a,b).

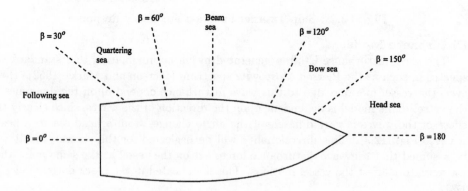

**Fig. 12.3** : Angles Between Ships Heading and the Direction of the Waves
(Encounter Angles)

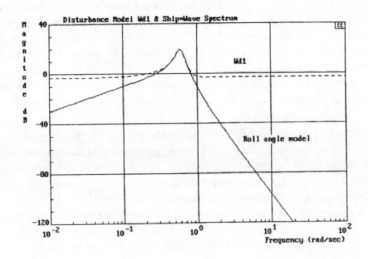

**Fig. 12.4** : Disturbance Model Frequency Response $W_{d1}$ and the Ship and Wave Model Frequency Responses

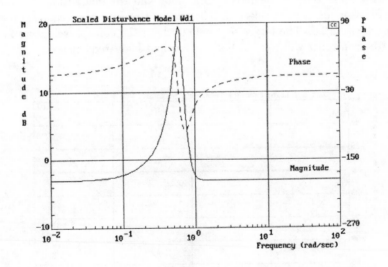

**Fig. 12.5** :  Total Disturbance Model $W_{d1}$ Frequency Response

## 12.2.2   Selection of $H_\infty$ Cost Function Weightings

The weighting selection in this roll stabilisation problem is unusual, since the characteristics required for the system do not follow the usual rules. The controller must have zero gain at zero frequency so that the sensor bias is not amplified. It must also reduce the sensitivity function at low frequencies, which suggests a high gain. Unfortunately, the high gain will amplify the roll sensor bias unless the controller frequency response is shaped very carefully. The main objective of the control system is to reduce the sensitivity function peaks. The sensitivity function represents the so-called *roll reduction ratio* that should be as small as possible. The peak at low frequencies is particularly important and can be difficult to reduce, while maintaining reasonable reduction at higher frequencies.

The sensitivity function weighting is normally chosen to have a high gain at low frequencies so that the controller has approximate integral action. In the present problem, it is the peaks in the sensitivity function response which are so important. The cost weighting on the sensitivity function or error spectrum will therefore be chosen to be a constant. If the control sensitivity or control weighting function were then defined as zero, the optimal controller would provide a sensitivity function which is the inverse of the disturbance spectrum. This is the best that can be achieved if there is no limit on the actuator power. This limiting case is not of course a very practical solution.

The control weighting frequency response $R_{c0}$ must be selected so that the gain of the controller at both low and high frequencies is low. There must therefore be a heavy weighting on the control sensitivity at both low and high frequencies. This is indicated in Fig. 12.6. To reduce the parameters to be chosen, a repeated time constant term is employed. Thus, only the time constant value and the magnitude of the control weighting function needs to be selected. These weighting function frequency responses are shown in Fig. 12.6. The larger the magnitude of the control weighting function, the faster the controller will roll off, both at low and high frequencies.

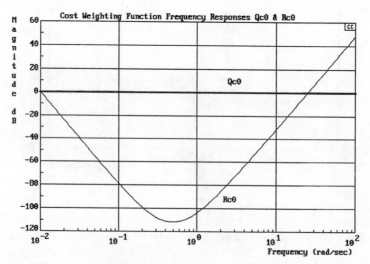

Fig. 12.6 : Frequency Response of the Cost Function Weightings

The weightings shown in Fig. 12.6 are defined below

**Sensitivity and Error Weighting**

$$Q_{c0} = H_{q0}^* H_{q0} , \qquad \text{where } H_{q0} = 1 \tag{12.1}$$

**Control Sensitivity Weightings**

$$R_{c0} = H_{r0}^* H_{r0} , \quad \text{where} \quad H_{r0} = 0.0001 \frac{(1 + s/0.5)^4}{s^2} \tag{12.2}$$

The higher the order of the control sensitivity weighting, the faster will be the fall off in gain of the controller at both low and high frequencies.

**Computed Controller**

The controller transfer function which minimises the $H_\infty$ criterion was obtained, after cancelling near common factors, from the polynomial results of Chapter 5 as:

$$C_0 = \frac{\begin{array}{c} 1394(s + 0.0001413)(s + 0.389849) \\ \times(s^2 + 0.533878s + 0.52713)(s + 2.673797) \end{array}}{\begin{array}{c} (s + 0.052793)(s + 0.30962)(s^2 + 0.2851947s + 0.3757) \\ \times(s^2 + 7.55925s + 21.79)(s + 20.3196) \end{array}} \tag{12.3}$$

The minimal cost was computed as: $\lambda_0^2 = 0.89617^2$ .

## 12.2.3    Frequency Responses Results

The frequency response of the open loop transfer function and the controller frequency response are shown in Fig. 12.7. Observe that the controller has the desired low gain at low frequencies and at high frequencies. Also note the unusual form of the open loop transfer function, which has the desired low gain at low frequencies.

One of the main objectives of the design is to reduce the peaks in the sensitivity function or roll reduction ratio. This must provide a large attenuation at the dominant wave frequency and must also limit the amount of amplification at both low and high frequencies. The frequency response shown in Fig. 12.8 clearly includes the desired frequency response shape. The peaks are reasonably low and the attenuation is slightly greater than is necessary.

The control sensitivity function is shown in Fig. 12.9. This is the transfer function which will affect the actuator responses and the gain must therefore be limited.

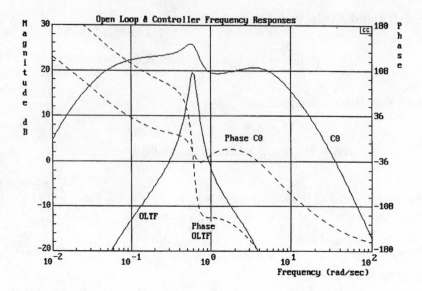

Fig. **12.7** :  Open Loop Transfer Function and Controller

Frequency Responses

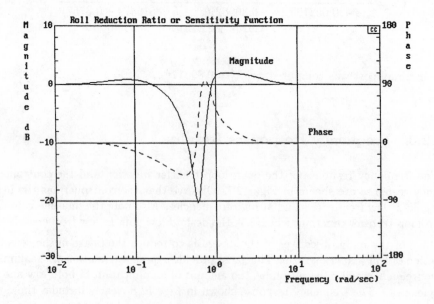

Fig. **12.8** : Roll Reduction Ratio or Sensitivity Function

Frequency Response

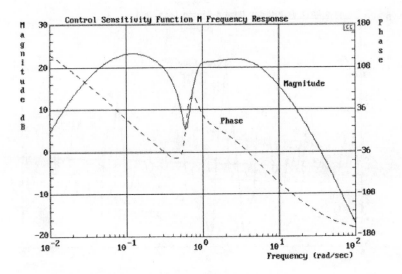

**Fig. 12.9** : Control Sensitivity Function Frequency Response

## 12.2.4 Time Response Results

The roll rate response of the vessel is shown in Fig. 12.10 with and without the feedback loop closed. Clearly there is significant attenuation when the loop is closed. Integrating the roll rate response gives the roll angle variations as shown in Fig. 12.11. Closed loop control clearly provides the necessary attenuation of the wave motion. Note that Figs 12.10 and 12.11 show the disturbances as modelled, but in Fig. 12.12 only the wave disturbance component is included and not the white noise part of the disturbance model. This figure is also plotted over a much longer time scale and the results reveal a considerable attenuation of the ship motions due to the waves, once the feedback loop is closed.

The main problem in closing the feedback loop is the roll amplification indicated in Fig. 12.8 at low frequencies. It is normally easier to introduce gain at higher frequencies and it is the low freqency peak which is difficult to reduce. This is partly because of low frequency sensor noise which requires the controller to have a low gain at low frequencies. The $H_2$ design clearly provides good roll attenuation, as illustrated in the time response results in Fig. 12.12.

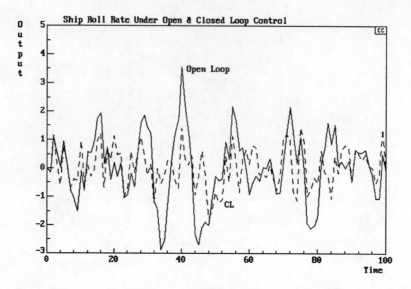

**Fig. 12.10** : Variations of the Ship Roll Rate due to Wave Motion under Open and Closed Loop Control

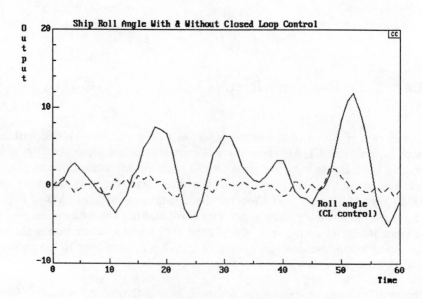

**Fig. 12.11** : Variations of Ship Roll Angle due to the Waves With and Without Closed Loop Control

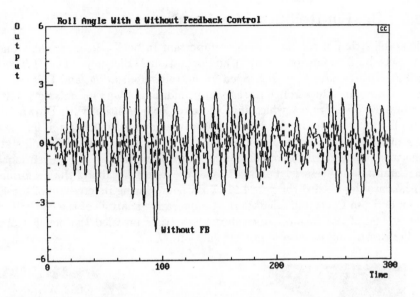

**Fig. 12.12** : Variations of the Ship Roll Angle With and Without
Feedback Control over an Extended Time Period

### Reduced control weighting

If the weighting on the control sensitivity function is reduced the consequence is
larger actuator movements and an even better roll reduction ratio. This is illustrated
in Fig. 12.13, which reveals a large attenuation in mid-frequencies and smaller peak
magnitudes.

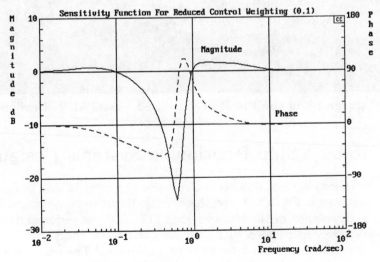

**Fig. 12.13** : Sensitivity Function or Roll Reduction Ratio when
Control Weighting is Reduced by a Factor of 10

### 12.2.5    $H_\infty$ Fin Roll Stabilisation Progress

The stabilisation of rolling motion is important in both passenger ships and military vessels, both to improve comfort and operational efficiency. The $H_\infty$ computer control algorithms have been developed for fin roll stabilisation, and these offer more reliable performance in a range of sea state conditions. Very promising results have been obtained on a project which involved close collaboration with Brown Brothers (now Rolls-Royce), Edinburgh. This company was founded in 1871 and has become a leading manufacturer of motion control equipment for ships. Active fin roll stabilisers are the most efficient means of controlling ship rolling motion on conventional commercial and military vessels. Full sea trials were used to evaluate the performance of the $H_\infty$ controller aboard the vessel M.V. Barfleur, during its crossing of the English Channel between Poole and Cherbourg. Comparative analysis of the controller's performance, and of the data obtained during sea trials revealed the benefits of robust controller design and its ease of implementation. [1]

**Photograph of the Fin Roll Stabilised Vessel M.V. Barfleur**

## 12.3    Robust Ship Positioning Systems Design

There are various types of thruster configuration employed in ship positioning systems, as illustrated in Fig. 12.14, and these enable thrust to be generated in a given direction, as determined by the control system. The first generation of dynamic ship positioning systems relied upon PID controllers and notch filters. The second generation involved LQG/Kalman filtering control solutions. The next generation will

---

[1]The photograph of the vessel MV Barfleur was kindly provided by Dr. Reza Katebi who supervised the $H_\infty$ fin roll stabilisation system sea trials.

be aimed at reducing commissioning and tuning time and require improvements in robustness, as well as more flexibility for a wider range of applications.

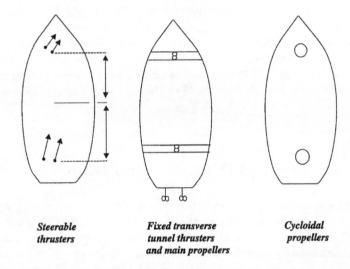

**Steerable
thrusters**

**Fixed transverse
tunnel thrusters
and main propellers**

**Cycloidal
propellers**

**Fig. 12.14**: Possible Bow and Stern Thruster Configurations

The LQG design of the feedback loop in ship positioning systems has been considered by Balchen et al. (1976) and by Grimble et al. (1978, 1979a,b, 1980). The $H_\infty$ design approach offers a step forward in the technology of dynamic ship positioning systems which is equally as important as the introduction of the Kalman filtering optimal control schemes by Jens Balchan and co-workers in 1976 and Grimble et al. in 1979. The advantages were so obvious that ship operators demanded this capability and the consequence was that the first company offering this Kalman filtering system was able to make considerable market gains. The multivariable frequency domain design methods, such as the characteristic locus method (Fotakis et al. 1982), did not attract the same industrial interest.

The $H_\infty$ robust control design technique was developed for uncertain systems and its main objective was to provide a design which is more robust than can be obtained with LQG/Kalman filtering methods or other techniques. The fundamental problem with Kalman filtering methods is that there is no formal procedure, in the usual problem description, in which modelling errors in the plant can be taken into account. In many cases, as will be explained below, information is available on the frequency ranges where the ship models are poor. However, there is no direct method of incorporating this information in Kalman filtering/optimal control schemes. The consequence is that many hours can be spent on commissioning. Moreover, it is likely that unpredictable performance will occur in unusual sea state conditions, or when unusual ship loading conditions arise.

The robust control design methods of which the $H_\infty$ technique is the best established and simplest, enable the modelling errors in systems to be taken into account in a relatively straightforward manner. The controllers developed are usually found to be cautious, in the frequency ranges where the modelling errors are the largest.

The $H_\infty$ method enables nominal models to be used for design and some bounds to be given to the maximum possible modelling errors. Although uncertainty bounds are not of course exact, by allowing for uncertainty better results are obtained than if models are assumed to be exactly known. This might be considered as taking out an insurance policy, since although it does not guarantee performance in the presence of ship modelling errors, there is nevertheless a much better chance the design will be successful on real ship applications. The first $H_\infty$ ship positioning solution (Katebi et al. 1997) was developed in a cooperative project involving: The University of Strathclyde, Nautronix (San Diego) and Industrial Systems and Control Ltd (Glasgow).

## 12.3.1   $H_\infty$ Design of Dynamic Ship Positioning Systems

As noted above, there are cases where the modelling error is very well known and understood. For example, consider the case where a vessel like a mine hunter is to be controlled. In this case the 60th order non-linear differential equations are known which very adequately describe the real vessel. However, a low order controller is normally based on a ship model of about 9th or 15th order. Conventional optimal DP systems are based upon this type of low order model. To check the design it is usual to simulate the full nonlinear system and inspect the performance of the controller based on the reduced order design.

With $H_\infty$ design the error between the full realistic simulation and the low order design model can be computed. This information can be used in choosing the optimal control cost function. It is then possible to give guarantees on stability and robust performance. These guarantees only apply if the uncertainty does not exceed the limits specified. However, it is clearly better to take uncertainty into account and it is then likely that the controller will be more robust and provide more reliable performance (Doyle and Stein, 1981; Doyle et al. 1990).

It is not practical to assume the uncertainty is excessively large, because even if the control system provides guaranteed performance in such a case the performance will be rather poor. As usual there is therefore a trade-off in the assumption of the size of the uncertainty likely to be encountered. An advantage of the $H_\infty$ robust design approach is that the trade-offs needed are very easy to achieve. Recalculation of the controller is a trivial task once the basic model and weighting have been chosen. The best compromise to be made between measurement noise rejection, disturbance rejection, performance and stability robustness is relatively easy to achieve. Indeed one of the advantages of this technique lies in the ability to parameterise the cost function so that the design process is very straightforward and re-design can be achieved very quickly.

The stochastic properties of the system regarding noise and disturbance rejection should not be compromised by this approach. Moreover, it is possible to set up the $H_\infty$ design problem in a state-space framework, so that engineers only see it as an extension of the familiar LQG optimal control Kalman filtering method. This simplifies training requirements for engineers experienced in the use of the LQ design methods.

## 12.3.2  Advantages of $H_\infty$ Design in Ship Positioning

The $H_\infty$ design method marks a significant stage in the development of marine control systems, since it is the first time errors in modelling the plant and disturbances can be taken into account rigorously. The polynomial approach to $H_\infty$ design is particularly appropriate for *adaptive* dynamic positioning (DP) systems, since polynomial based algorithms can easily be made adaptive. The $H_\infty$ self-tuning control approach is is more difficult to implement using standard (computationally intensive) state-space based algorithms.    The advantages of dealing with uncertainty in ship and wave models are clear:

- more consistent performance over wider operating conditions

- shorter commissioning periods

- improved station keeping, particularly in unusual sea state conditions

- a simple structure and hence more reliable software implementation and system integrity

- less sensitive to faults and sensor or actuator degradation

- possibility of using lower specification hardware but meeting the same performance.

The arguments for developing an $H_\infty$ dynamic ship positioning system are as follows :

 (i) Existing systems are designed assuming adequate models are available for ship, wind and waves.

 (ii) In practice (i) is seldom ever true and the consequence is that either poor control or long tuning periods must be accepted.

 (iii) The solution is to allow for modelling errors and to then obtain a more realistic design.

 (iv) $H_\infty$ optimal control provides a simple method of achieving a robust controller.

Some of the problems with existing Kalman filtering schemes can be easily be explained. That is, a Kalman filter is calculated assuming the model of the system is known exactly. The result can be poor wave filtering and current force estimation. To obtain reliable control, the control system must be made less sensitive to wind and wave force changes.

Thruster saturation arises mainly because of poor wave filtering action. This is particularly important when thruster power is limited, as for example in military vessels like the Single Role Minehunters built for the Royal Navy by Vosper Thornycroft. The $H_\infty$ designs can be made less sensitive to degradation in sensors and actuators. It is also possible to optimise the use of thrusters which may enable smaller thrusters to be used, or to be utilised more effectively.

The two photographs of DP vessels were kindly provided by Nautronix Ltd based in Aberdeen and San Diego. The first photograph is the vessel Peregrine I that has a dual redundant DP system. The second photograph is of the vessel Alloha close to an oil production platform.

**Photograph of the Vessel Peregrine I**

**Photograph of the Vessel Alloha and an Oil Production Platform**

### 12.3.3 Polynomial Based $H_\infty$ Ship Positioning Design

Combined feedback and feedforward control of ship positioning systems is now considered. Feedforward control is normally introduced using classical control ideas and it is not normally treated as part of the optimal solution. The following scalar design will illustrate the benefits of including feedforward action in the total $H_\infty$ optimal solution.

**Ship dynamics:**

$$W = \frac{B_0}{A_0} = \frac{0.16}{s(s + 0.0546)(s + 1.55)} \tag{12.4}$$

This model represents the transfer-function between actuator input and position output and has a $1/1.55$ denotes the thruster time-constant. It represents one axes of motion and is a much simplified model of a 3 axes model provided by Alstom Drives and Controls Ltd., Rugby.

**Wave model:**

$$W_h = \frac{C_h}{A_h} = \frac{3.52s^2}{(s + 0.086385)((s + 0.09043)^2 + 0.4487074^2)(s + 8.28776)}$$

**Input disturbance model:**

$$W_{d1} = \frac{0.01(1 + 0.1s)^2}{s^2(s + 0.0546)}$$

This represents an integrator disturbance model feeding into the ship dynamics . In fact the disturbance forces enter the ship equations at the same point as the thruster force model. The magnitude of this disturbance is chosen so that it dominates at low frequencies but $W_n$ dominates at high frequencies.

**Measurable disturbance model:**

$$W_{d2} = \frac{C_{d2}}{A_{d2}} = \frac{0.316228[(s + 0.6287)^2 + 0.6287^2]}{s^2}$$

The measurable disturbance model corresponds to the wind forces in this problem. This model was chosen to illustrate the benefits of feedforward control and would normally represent a Davenport spectrum, (1978).

**Wind spectrum corruption:**

$$W_s = A_s^{-1}B_s = (1 + 10s)$$

**Measurable disturbance path dynamics:**

$$W_{di} = A_{di}^{-1}B_{di} = 1/(s + 1)$$

**Reference model:**

$$W_r = A_e^{-1}E_r = 0.1/s$$

The reference model is of comparable magnitude as the disturbance models in the mid frequency range.

**Ideal response model:**

$$W_i = A_i^{-1}B_i = 1/(1+5s)$$

**Robustness Weightings**

The $H_2$ polynomial algorithm can be used to obtain a solution to the $H_\infty$ design problem, using the embedding result. To utilise the results in Chapter 5.2 compute the weightings: $W_{\sigma i}$ which satisfy $\Sigma_i = W_{\sigma i}W_{\sigma i}^*$ for $i = \{0,1,2\}$. These enable the desired $H_\infty$ solutions to be obtained for the three controllers:

**Feedback:**

$$W_{\sigma 0} = \frac{0.00941(s+0.133415)}{(s^2+0.324376s+0.05508)}$$

**Tracking:**

$$W_{\sigma 1} = \frac{2.6244(s+0.053764)}{(s^2+0.19378s+0.040729)}$$

**Feedforward:**

$$W_{\sigma 2} = \frac{0.062238(s+5.511885)}{(s^2+4.314248s+10.8)}$$

**H$_\infty$ Cost Function Weighting Definitions**

**Feedback Weightings:**

$$H_{qco} = \frac{1+200s}{s} \qquad and \qquad H_{rco} = 0.5(1+100s)^2$$

**Tracking Weightings:**

$$H_{qc1} = \frac{1+s/50}{s} \qquad and \qquad H_{rc1} = 10(1+s/10)^2$$

**Feedforward Weightings:**

$$H_{qc2} = \frac{1+1000s}{s} \qquad and \qquad H_{rc2} = 10^{-4}(1+1000s)$$

The weighting $Q_{ci}, R_{ci}$ and $G_{ci}$ are defined from the generalised weightings in (5.6), using the above definitions of $H_{qci}$ and $H_{rci}$

**Feedback Controller:**

$$C_0 = \frac{54857(s+0.005)[(s+0.12034)^2+0.056358^2](s+1.55)}{s(s+1.79816\times10^{-4})(s+2.295745)(s+8.69389)(s+11.1668)(s+98.979)}$$

**Tracking Controller:**

$$C_1 = \frac{\begin{array}{c}181.635(s+0.00496)(s+0.042286)(s+0.11974)(s^2+0.52838s+0.072063)\\ \times(s+1.540707)(s+1.55)(s^2+19.9487s+99.593)(s-202.67)\end{array}}{\begin{array}{c}(s+1.79816\times10^{-4})(s+0.053764)(s+0.2)(s+0.25444)(s+1.54389)\\ \times(s+2.295745)(s+8.69389)(s^2+20s+100)(s+11.16685)(s+98.979)\end{array}}$$

**Feedforward Controller:**

$$C_2 = \frac{\begin{array}{c}62.2224(s^2-2.352x10^{-4}s+5.49928\times10^{-7})\\ \times(s+0.052133)(s+1.55)(s^2+1.39759s+0.76194)\\ \times(s+1.84517)(s^2+19.334s+96.087)(s+99.969)\end{array}}{\begin{array}{c}(s+1.79816\times10^{-4})(s+0.1)(s^2+1.2574s+0.79057)\\ \times(s+1)(s+2.29574)(s^2+5.917967s+14.8147)\\ \times(s+5.51188)(s+8.69389)(s+11.1668)(s+98.979)\end{array}}$$

In practice some attempt would be made to reduce the order of the tracking and feedforward controllers. These have a high order, since they effectively cancel the dynamics of the feedback loop in their effort to introduce more desirable tracking and feedforward dynamics.

## 12.3.4 Frequency Responses

The frequency responses for the measurable and unmeasurable disturbance models and the reference model are shown in Fig. 12.15. The reference model is a simple integrator which is the stochastic equivalent of the step function input. Both the measurable and unmeasurable disturbances have similar responses at low frequencies but the measurable disturbance is assumed to have a high frequency component. This is indicated by the lead term that is introduced in the model $W_{d2}$, shown in Fig. 12.15.

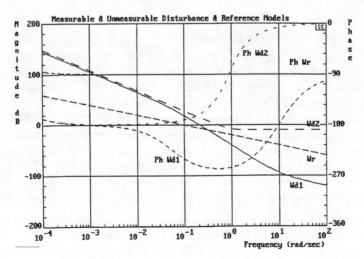

**Fig. 12.15** : Tracking Reference and Measurable and Unmeasurable Disturbance
Model Frequency Responses

It is assumed that the measurable disturbance signal affects the ship motions in a different manner to the measured feedforward control input. The two models that represent the differences in the path dynamics are shown in Fig. 12.16. The transfer $W_s$ represents the measurement system dynamics and $W_{di}$ the ship dynamics affecting wind gusts. The ship transfer function response is also shown in Fig. 12.16.

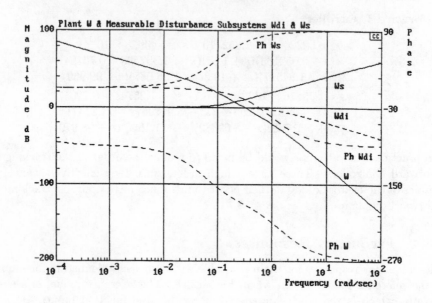

**Fig. 12.16** : Ship Model W and Measurable Disturbance Subsystem Models $W_{di}$ and $W_s$

The feedback controller cost weighting terms $Q_{co}$ and $R_{co}$ are shown in Fig. 12.17. Notice that the crossover frequency between the two weighting functions occurs at approximately 0.2 rads/sec. This is therefore likely to be close to the unity gain crossover frequency for the resulting system (Grimble, 1999). This figure also shows the weighting functions for the tracking and feedforward controllers. In all cases the control weighting function includes a lead term to ensure the controllers roll off at high frequencies and the error weighting function includes integral action to ensure good zero frequency behaviour. In fact, the feedforward controller weighting term $R_{c2}$ was chosen to be small compared to the error weighting term $Q_{c2}$. This will provide a comparison in results with the case to be considered later. This explains why the feedforward control time responses are so good relative to the case without feedforward action.

The feedback controller $C_0$ frequency response is shown in Fig. 12.18. This figure also includes a plot of the wave motion model. The dominant wave frequencies are in the same frequency region as the maximum controller attenuation. This is of course needed for good wave filtering action.

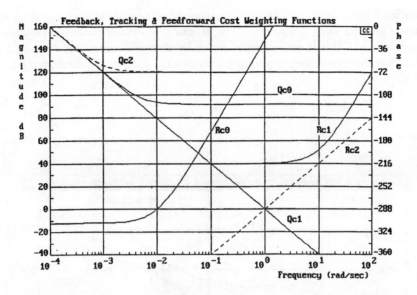

**Fig. 12.17**: Frequency Responses of Cost Weighting Functions for
Feedback, Tracking and Feedforward Controllers

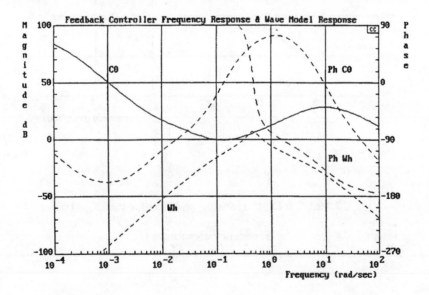

**Fig. 12.18**: Frequency Response of Feedback Controller
and Wave Model

The frequency responses of the 2 DOF tracking controller $C_1$ and the feedforward
controller $C_2$ are shown in Fig. 12.19. The frequency responses of both of these
controllers are formed from the differences of $C_{01}$ and $C_{02}$ and the feedback controller
response $C_0$ , respectively. However, observe that the feedforward controller does not
have too high a gain at high frequencies, which is important for measurement noise

attenuation (Seraji, 1987). The tracking controller acts on the set point signal and does not have the same limitation. The tracking controller includes a high gain at low frequency but this is unnecessary in the feedforward controller, since in fact the feedback loop includes integral action which ensures good zero-frequency disturbance rejection.

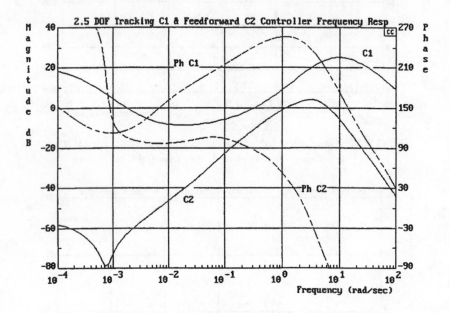

**Fig. 12.19** : 2.5 DOF Tracking and Feedforward Controller

Frequency Responses

The compensators $C_{01}$ for tracking and $C_{02}$ for feedforward control, shown in Fig. 12.20, are only used as devices in the calculation of the respective tracking $C_1$ and feedforward $C_2$ controllers. In fact, they do have a physical significance which is related to the equivalent open loop controllers. Since one component in these compensator terms is due to the feedback controller, the form of the responses at low frequencies is not surprising. It is only in the mid-frequency range where the frequency responses are shaped to improve either tracking (via 2 DOF) or feedforward action.

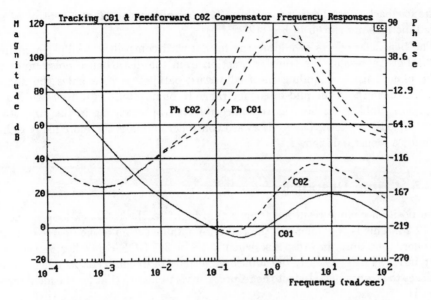

**Fig. 12.20** : Tracking $C_{01}$ and Feedforward $C_{02}$ Compensator
Frequency Responses

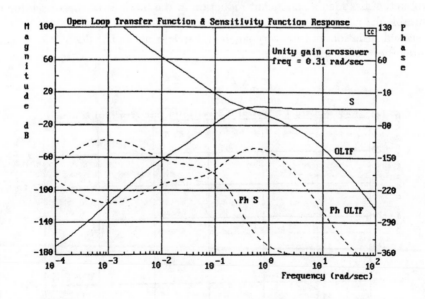

**Fig. 12.21** : Open Loop Transfer Function and Sensitivity Function
Frequency Responses

The open-loop frequency response with unity-gain crossover frequency of 0.3 rads/sec is shown in Fig. 12.21. Note that the weightings $Q_{c0}$ and $R_{c0}$ were chosen to achieve a crossover at about 0.2 rads/sec and this has clearly been achieved. The frequency response of the sensitivity function is also shown in Fig. 12.21. The overshoot achieved by the feedback controller is acceptable. The complementary $T$ and control $M$ sensitivity functions are shown in Fig. 12.22.

The beneficial effects of adding the feedforward controller are illustrated in Fig. 12.23. This represents the transfer function from the disturbance noise input point to the plant output. Adding the feedforward parallel path, enables the dominant disturbance effects in the mid-frequency range to be negated. In practice such a good reduction is unlikely to be achieved since the control weighting in this first stage of the example is very low. The feedforward controller action is therefore likely to be too active for a realistic design.

## 12.3.5    Time Responses

The position change resulting from a step input to the feedback loop is shown in Fig. 12.24. This is equivalent to the response obtained from a one degree of freedom controller. The unit step response when a 2.5 or 3.5 DOF controller is used is also shown in Fig. 12.24. Clearly the additional degrees of freedom enable much improved responses to be obtained. In fact the response is very close to that for the ideal response model $W_i$.

The time responses for the motion of the vessel due to both measurable (wind) and unmeasurable (current) disturbances are shown in Fig. 12.25. If feedforward control action is added a significant reduction in the motions of the vessel due to the measurable disturbance (wind) is obtained as shown in Fig. 12.26. This is of course an idealised situation and for very small $R_{c2}$ weighting, but it does demonstrate the advantages of feedforward control.

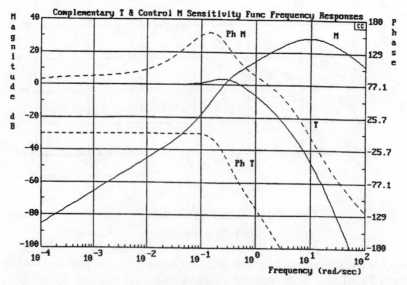

Fig. 12.22 : Complementary and Control Sensitivity Function
Frequency Responses

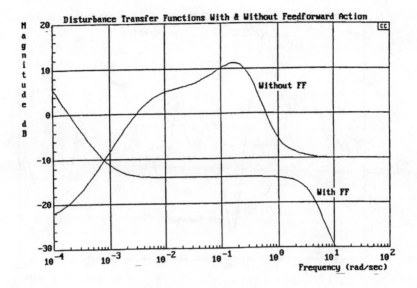

**Fig. 12.23** : Measurable Disturbance Closed-loop Responses
With and Without Feedforward Action

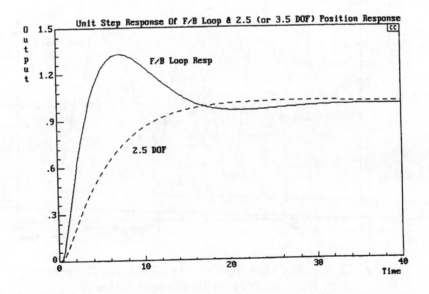

**Fig. 12.24** : Unit Step Response of Feedback Loop and 2.5 DOF Position
Change Responses

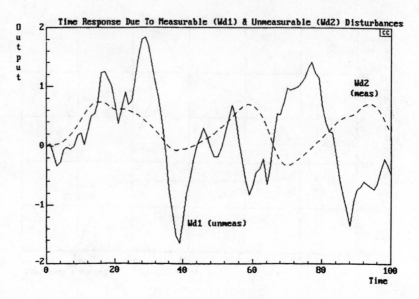

**Fig. 12.25** : Ship Motions due to Measurable and Unmeasurable Disturbances

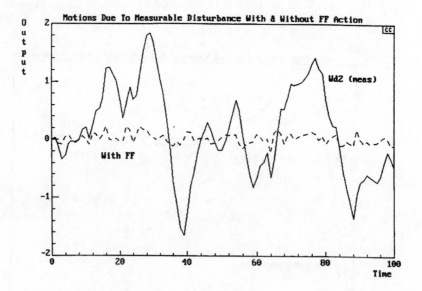

**Fig. 12.26** : Ship Motions due to the Measurable Disturbance
With and Without Feedforward Action

The motions of the vessel due to the waves when the feedback loop is open and when the feedback loop is closed, are shown in Fig. 12.27. One of the objectives of the control system is to introduce effective wave filtering action. This is to ensure the thrusters do not respond significantly to the wave motions. This figure indicates that closing the feedback loop does not change significantly the motions of the vessel due to the waves and this suggests good wave filtering action has been achieved. Further evidence of this is clear from the feedback controller frequency response, shown in Fig. 12.18.

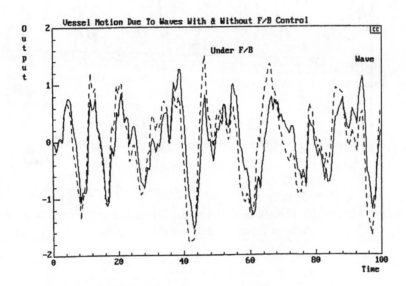

**Fig. 12.27** : Ship Motions due to Waves With and Without
Feedback Loop Closed

## 12.3.6   Feedforward Controller Tuning

Feedforward control design was considered in some detail in Chapter 2.   If the knowledge of the feedforward disturbance model is very uncertain it is unrealistic to introduce a controller with too higher gain. One of the advantages of the actual control problem formulation presented is that separate costing of the feedforward and feedback controllers is possible. By increasing the wind feedforward control weighting $R_{c2}$ by a factor of 10,000 the results shown in Fig. 12.28 are obtained. Clearly the disturbance rejection is not as good but it is still significant.

These results are also confirmed from the frequency domain responses shown in Fig. 12.29 for the measurable disturbance signal path with and without forward action. The feedforward control clearly improves disturbance rejection but not as much as in the previous case.   The cost weightings for this situation are shown in Fig. 12.30. The crossover frequency in this case is at a more reasonable frequency point.   For convenience the frequency responses of the other weightings are also shown.

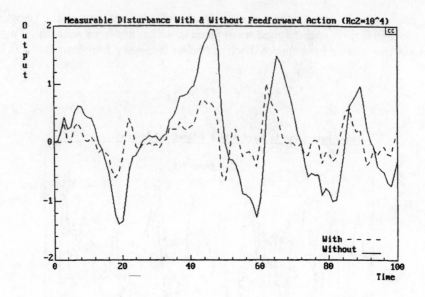

**Fig.12.28 :** Measurable disturbance With and Without Feedforward Action with Increased Feedforward Control Weighting

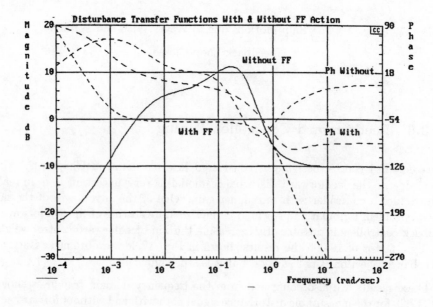

**Fig. 12.29 :** Disturbance Frequency Responses in Closed-loop With and Without Feedforward Action

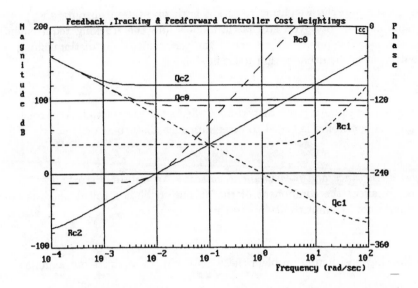

**Fig. 12.30** : Cost Weightings Functions for Feedback, Tracking and Feedforward Controllers (Increased Control Weighting)

## Summary of Models and Weightings

*Ship model* : Usually second order/axis with one integrator for position. Sway and yaw coupled and surge decoupled.

*Wave Model* : Second or fourth order/axis with $s$ or $s^3$ respectively in the numerator to be approximation to Pierson Moskowitz spectrum.

*Current force* : Normally an integrator for each axis, driven by white noise.

*Second order wave force* : This may be omitted leaving the current force model to represent both of these low frequency effects.

*Wind model* : This is often only introduced after the main feedback system design is complete and involves a feedforward element.

*Error weighting* : An integrator with a lead term introduced a decade before the desired bandwidth point is usual.

*Control weighting* : The rate of increase of the lead term should be chosen according to the desired roll-off rate for the controller. The lead term should come in before the desired bandwidth point for the system and the two weightings should cross-over at this point.

# 12.4 Multivariable State-space Ship Positioning

The design of a Dynamic Positioning (DP) system involves a compromise between robustness and stochastic requirements. The $H_\infty$ robustness of the solution will be addressed through the weighting selection both on system outputs and possibly on exogenous inputs. However, stochastic models and the tracking requirements form the easiest starting point. First recall the terminology for distinguishing between different types of disturbance and noise model:

*Input disturbance* : An input disturbance affects the output to be controlled directly and the controller is tuned to attenuate such disturbances.

*Output disturbance and measurement noise* : An output disturbance does not directly affect the component of the output to be controlled and the controller should not attempt to attenuate this disturbance.

In the ship positioning problem these disturbances separate out as follows:

*Input disturbances* : Wind, second-order wave forces, current forces.

*Output Disturbances* : First-order wave forces.

The output disturbances enter the equations in the same way as the coloured measurement noise, and the feedback controller is required to behave in the same way for both types of signal.

From Newton's equations of motion all forces on the ship enter at the same point including the thruster forces. However, for the present approach the wave model forces will not be added at this point and instead the wave motion signal will be moved to the output. The transfer functions will be arranged to be the same but by this device it is possible to separate out the position responses due to the different types of disturbance. The wave motion must not be offset by control action and it is therefore an output disturbance in the above terms. In fact, this modelling approach is probably the most common in the DP industry. That is, the low and high frequency motion signals are separated. The motion errors in the cost-function can then be defined in terms of the output $\{y_\ell(t)\}$ due to the low frequency motions, caused by the input disturbances. The control signals must not respond to the motion $\{y_h(t)\}$, due to the higher frequency output disturbances or wave motion.

The ship positioning problem will be considered using algorithms which require the system to be in standard system model form. The state-space structure shown in Fig. 12.31 can be employed to model the system. The correspondence between Figs.12.31 and 12.32 should be clear. The wind feedforward control is normally added using a separate classical control design but it can be included in the $H_\infty$ solution, as in the previous example.

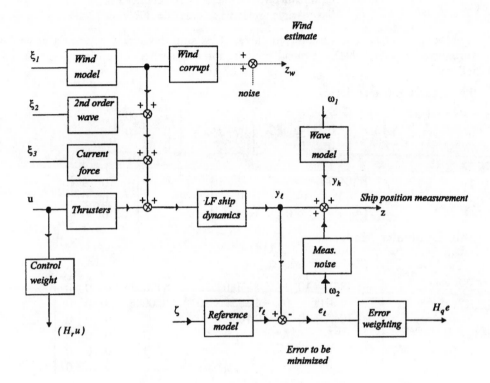

**Fig. 12.31** : Block Diagram of Ship Positioning Plant Model

The dynamic ship positioning system is one of the largest users of power on a vessel. This suggests very close links exist between the DP and power production systems which might be exploited for improved power management. There is therefore a need to integrate the diesel engine power management systems with the DP system in future designs. However, this aspect of the problem will not be considered further here.

## 12.4.1 Dynamic Ship Positioning Multivariable Models

The models to be used in a DP design (English and Wise, 1976) are obtained for a given nonlinear operating point. The operating conditions for the state-space system of interest are defined below:

$$
\begin{aligned}
\text{Surge Position} &= 1 \text{ metre} \\
\text{Sway Position} &= 1 \text{ metre} \\
\text{Yaw Direction} &= 60^\circ \\
\text{Average Wind Velocity} &= 10 \text{ m/s} \\
\text{Wind Angle (relative to earth)} &= 30^\circ \\
\text{Average Current Velocity} &= 0 \text{ m/s} \\
\text{Current Angle (relative to earth)} &= 30^\circ
\end{aligned}
$$

Significant Wave Height = 10 meter
Wave Angle (relative to earth)= 30°

Models must include thruster characteristics, ship dynamics, wind and wave disturbances (Fossen, 1994). Typical models, for a linearised operating point, are as follows:

**Thruster Characteristics**

$$
A_{th} = \begin{bmatrix} -2 & 0 & 0 \\ 0 & -2 & 0 \\ 0 & 0 & -2 \end{bmatrix} , \quad B_{th} = \begin{bmatrix} 1 & 0 & 0 \\ 0 & 1 & 0 \\ 0 & 0 & 1 \end{bmatrix} ,
$$

$$
C_{th} = \begin{bmatrix} 0.4572 & 0 & 0 \\ 0 & 0.2468 & 0 \\ 0 & 0 & 0.0401 \end{bmatrix} , \quad D_{th} = \begin{bmatrix} 0 & 0 & 0 \\ 0 & 0 & 0 \\ 0 & 0 & 0 \end{bmatrix}
$$

**Ship Dynamics Model**

$$
A_{lf} = \begin{bmatrix} -1.4361 \times 10^{-5} & -0.0357 & -0.0326 & 0 & 0 & 0 \\ 0.0106 & -4.8851 \times 10^{-4} & 0.0068 & 0 & 0 & 0 \\ -2.7047 \times 10^{-7} & -9.7632 \times 10^{-7} & 0 & 0 & 0 & 0 \\ 1 & 0 & 0 & 0 & 0 & 0 \\ 0 & 1 & 0 & 0 & 0 & 0 \\ 0 & 0 & 1 & 0 & 0 & 0 \end{bmatrix}
$$

$$
B_{lf} = \begin{bmatrix} 1 & 0 & 0 \\ 0 & 1 & 0 \\ 0 & 0 & 1 \\ 0 & 0 & 0 \\ 0 & 0 & 0 \\ 0 & 0 & 0 \end{bmatrix} \quad \text{and} \quad C_{th} = \begin{bmatrix} 0 & 0 & 0 & 1 & 0 & 0 \\ 0 & 0 & 0 & 0 & 1 & 0 \\ 0 & 0 & 0 & 0 & 0 & 1 \end{bmatrix}
$$

**Wind Disturbance Model**

$$
A_{win} = \begin{bmatrix} -0.1021 & -0.0015 & 0 & 0 & 0 & 0 \\ 1.0000 & 0 & 0 & 0 & 0 & 0 \\ 0 & 0 & -0.1021 & -0.0015 & 0 & 0 \\ 1 & 0 & 1.0000 & 0 & 0 & 0 \\ 0 & 0 & 0 & 0 & -0.1021 & -0.0015 \\ 0 & 0 & 1 & 0 & 1.0000 & 0 \end{bmatrix}
$$

$$
B_{win} = \begin{bmatrix} 1 & 0 & 0 \\ 0 & 0 & 0 \\ 0 & 1 & 0 \\ 0 & 0 & 0 \\ 0 & 0 & 1 \\ 0 & 0 & 0 \end{bmatrix} \quad \text{and} \quad C_{win} = \begin{bmatrix} -0.0071 & 0.0000 & 0 & 0 & 0 & 0 \\ 0 & 0 & 0.0111 & 0.0000 & 0 & 0 \\ 0 & 0 & 0 & 0 & 0.0007 & 0.0000 \end{bmatrix}
$$

**Wave Disturbance Model**

$$A_{wav} = \begin{bmatrix} -0.0159 & -0.0061 & 0 & 0 & 0 & 0 \\ 1.0000 & 0 & 0 & 0 & 0 & 0 \\ 0 & 0 & -0.0163 & -0.0061 & 0 & 0 \\ 1 & 0 & 1.0000 & 0 & 0 & 0 \\ 0 & 1 & 0 & 0 & -0.0050 & -0.0038 \\ 0 & 0 & 1 & 0 & 1.0000 & 0 \end{bmatrix}$$

$$B_{wav} = \begin{bmatrix} 1 & 0 & 0 \\ 0 & 0 & 0 \\ 0 & 1 & 0 \\ 0 & 0 & 0 \\ 0 & 0 & 1 \\ 0 & 0 & 0 \end{bmatrix} \quad \text{and} \quad C_{wav} = \begin{bmatrix} 0.1749 & 0 & 0 & 0 & 0 & 0 \\ 0 & 0 & 0.1476 & 0 & 0 & 0 \\ 0 & 0 & 0 & 0 & 0.0052 & 0 \end{bmatrix}$$

The wind and current disturbances affect the system at the same point as the thruster input and the wave disturbance at the ship output. The white noises used to drive the system disturbances have unity-variance and can be offset by a constant signal. All of the through terms (D terms) in the state equation models are taken to be null.

## 12.4.2 The Ship Positioning Criterion

Recall that the expression for the tracking error involves the sensitivity function and the expression for the control signal involves the control sensitivity function. The weighted signals therefore enable the $H_\infty$ cost problem to be established in traditional sensitivity minimisation terms. The system may be given the simple state-space block diagram structure in Fig. 12.32. This system may easily be rewritten in the standard system form shown in Fig. 12.33, so that a standard commercial toolbox can be used to compute the optimal controller, based on the well known results by Doyle et al. (1989). Integral action may be introduced in much the same way as for LQG state-space problems (Grimble 1979).

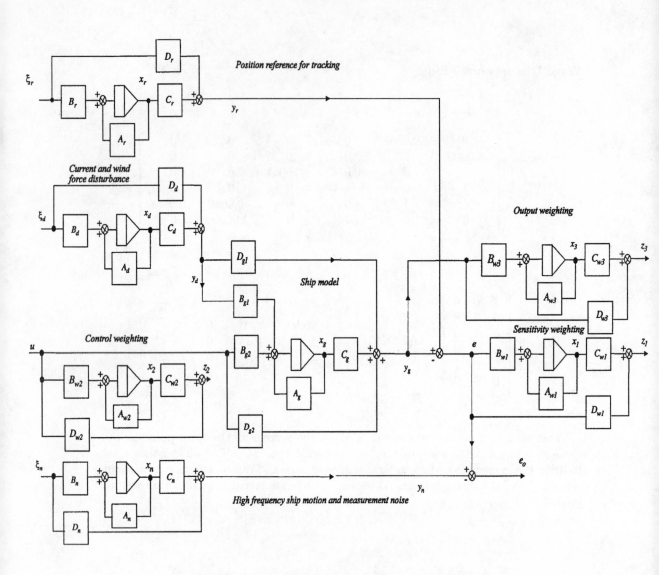

**Fig. 12.32** : Dynamic Ship Positioning Disturbance, Plant, Noise, Reference and Weighting Models

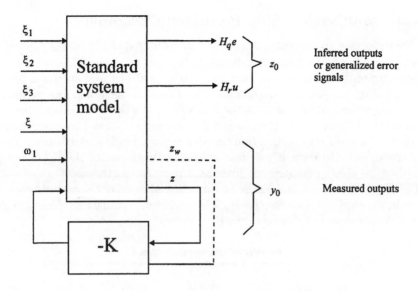

**Fig. 12.33** : Standard System Description DP Problem

It may be desirable to pass the standard system description inputs ($w_o$) through input weighting functions but since these can be absorbed into the disturbance models they are not shown separately. The $H_\infty$ cost-function being minimised, if the preceding system model is employed, is as follows:

$$J = \left\| \begin{array}{c} H_q G_\ell \\ H_r G_u \end{array} \right\|_\infty = \left\| \begin{array}{c} \text{Weighted low frequency error transfer} \\ \text{Weighted control energy transfer} \end{array} \right\|_\infty \qquad (12.5)$$

This cost function requires that the low frequency tracking errors are minimised and the control energy is limited in some sense. The criterion does not at this stage allow for modelling errors. The sources of modelling errors include:

(a) *Wave model uncertainty* : This does not affect the stability of the system and can normally be accommodated by switching the wave model.

(b) *Ship dynamics uncertainty* : This can be represented by additive or multiplicative modelling errors which can be interpreted as either sensitivity or control sensitivity function constraints, respectively.

(c) *Current or low frequency force models uncertainty* : This type of disturbance has uncertainty which can be allowed for through the use of integral action in the controller.

The first stage of a design procedure therefore involves the minimisation of the above cost index, which takes into account most of the above objectives. The second stage involves a modification of the criterion, particularly if the uncertainty under (b) above dominates.

### 12.4.3   Multivarible Ship Positioning Example

There follows an example, taken from an industrial study, of the type of design steps and results obtained for a multivariable $H_\infty$ controller (Katebi et al. 1997). The performance of the controller was analysed given frequency responses and both linear and non-linear time-domain simulations. The robust stablity was first verified, using the non-linear simulation, without the presence of the sea wave disturbances. The sea model disturbances were then included to examine the low frequency disturbance rejection and the thruster modulation that determines robust performance. The ship's surge speed, sway speed and yaw rate are defined in terms of the ship's body fixed axis, while the ship's positions are given relative to the earth axis.

The Fig.12.34 shows that range of linearised ship model uncertainty for a 40% change in the sway and yaw directions. The maximum, middle and minimum singular value plots are shown at different operating points.

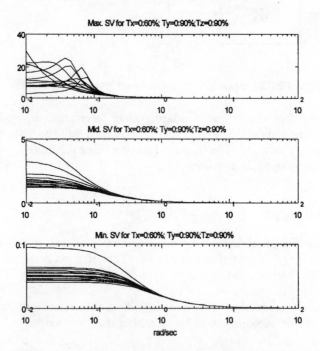

**Fig. 12.34** : Maximum, Middle and Minimum Singular Value Plots
for Changes in the Thruster Forces

### Robustness Test

As decribed in Chapter 1.2.2 if the plant has an additive uncertainty model a conservative condition for robust stability is given by

$$\sigma_{\max}(\Delta(e^{-j\omega})) < 1/\sigma_{\max}(M(e^{-j\omega}))$$

where $\sigma_{\max}(\Delta(e^{-j\omega})) < 1$ and $M$ denotes the control sensitivity function, over all frequencies $\omega$. The maximum, middle and minimum singular value plots for the control sensitivity function are shown in Fig. 12.35. The maximum uncertainty allowed in the plant model is inversely proportional to the magnitude of the maximum singular value which is 10% in the mid frequency range and large at both low and high frequencies.

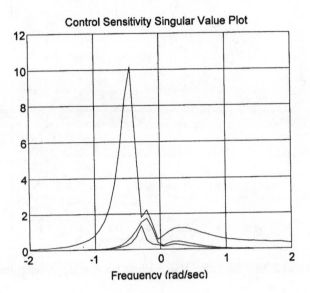

**Fig. 12.35** : Maximum, Middle and Minimum Singular Values for Control
Sensitivity Function

## Controller and Open Loop Singular Values

The singular values for the $H_\infty$ optimal controller are shown in Fig. 12.36.  A suitable controller response for the DP problem should have the following features:

- Sufficient gains at low frequencies ($<0.4$ rad/sec) to compensate the current and wind forces.

- Notch type frequency response about the wave centre frequencies to reduce thruster modulation and saturation.

- High cut-off rate at high frequencies to filter out the measurement noise.

- A trade-off must be made beween the low frequency disturbance rejection and the wave filtering performance.

The singular values for the open loop frequency response are shown in Fig. 12.37. These enable the open loop low frequency gain and the cut-off rate to be determined.

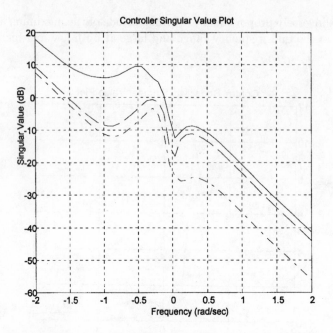

**Fig. 12.36**:  Singular Value Plots for the $H_\infty$ Controller

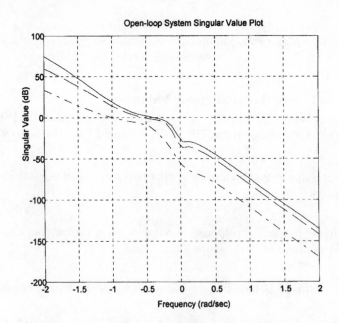

**Fig. 12.37** : Open-loop System Singular Value Plots

**Sensitivity Function Singular Values**

The singular values for the sensitivity function are shown in Fig. 12.38.  These determine robustness and disturbance rejection properties, as discussed in Chapter 1.

The function should be small at low frequencies and will tend to unity gain at high frequencies.

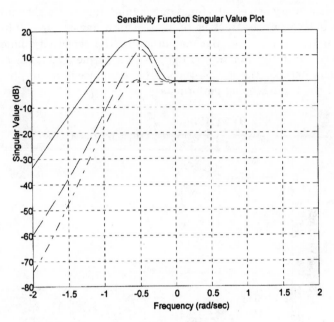

**Fig. 12.38** : Sensitivity Function Singular value Frequency Responses

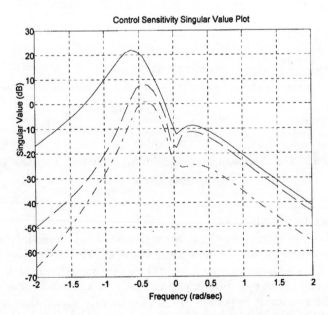

**Fig. 12.39** : Control Sensitivity Function Singular Value Frequency Responses

The control sensitivity function responses are shown in Fig. 12.39. These have a similar notch filter characteristic for wave filtering, that is present in the controller singular value responses. This is necessary to reduce thruster modulation at the wave frequencies.

The complementary sensitivity function singular values are shown in Fig. 12.40. This function determines the tracking properties for a single degree of freedom design. Recall that in the scalar case, the magnitude of the sensitivity function plus the complementary sensitivity function is unity for all frequencies. Thus, a trade-off is always necessary between the station-keeping (disturbance rejection) properties of the controller and the track-keeping performance.

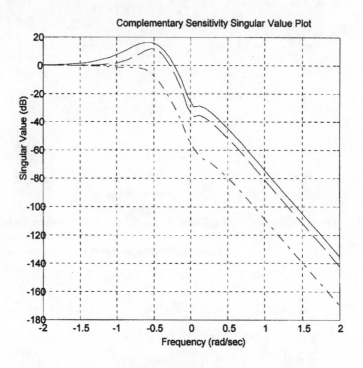

Fig. 12.40 : Complementary Sensitivity Function Singular Value
Frequency Responses

**Linear Simulation Responses**

It is beneficial to investigate the linear response of the closed-loop system before evaluating the controller on the non-linear system. The linear step responses for this example, for a setpoint change in Northing, Easting and Heading, are shown in Fig. 12.41. The linear step responses can easily be modified using the weighting functions. The non-linear responses will not necessarily match the linear responses, unless the non-linear model is at an operating point where the non-linear dynamics were linearised, and the magnitude of the step change is within the linear range. The set point-tracking can be improved signficiantly, without affecting the disturbance rejection and wave filtering properties, by using a two-degrees of freedom controller.

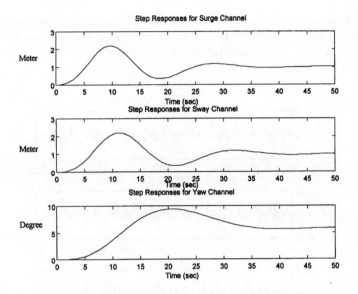

**Fig. 12.41** : Step Reponses of the Surge, Sway and Yaw Channels

## Nonlinear Simulation Responses

The wave filtering action of the controller is investigated by introducing the wave motion as the main high frequency disturbance. The effect of the wave motion was added at the output of the non-linear simulation. It is assumed that the wave forces have an equivalent effect of ±10 m, in the surge and sway directions and ±5 degrees deviation in heading. A birds eye view of the ship's position is shown in Fig. 12.42.

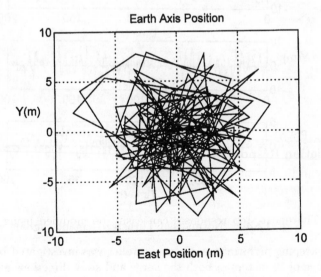

**Fig. 12.42:** The Ships Position due to Wave Disturbances

The positions as a function of time are shown in Fig. 12.43. The thruster demand is shown in Fig. 12.44. This figure reveals that the demand is well below the saturation limits of the thrusters. Note that the thruster demands in the X, Y and Z directions were normalised to unity.

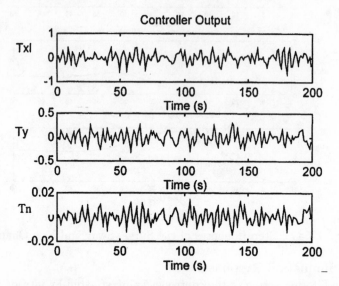

**Fig. 12.43** : The Ship Position in Surge, Sway and Yaw due to Wave Disturbances

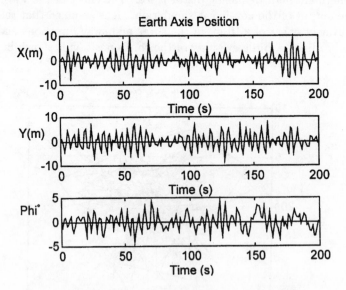

**Fig. 12.44** : Normalised Thruster Demand Signals

The track-keeping performance of the controller was investigated by demanding a set point change of 10 meters in both the surge and sway directions and a 10 degrees change in the heading. The position of the ship is shown in Fig. 12.45. The ship positions as a function of time are shown in Fig. 12.46.

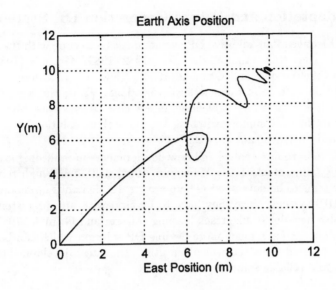

**Fig. 12.45**: Ship's Position Variations Due to a Reference Change

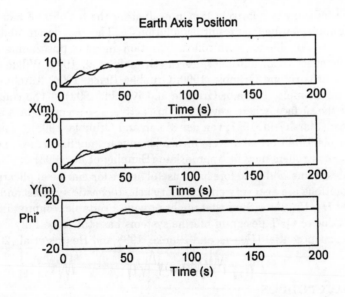

**Fig. 12.46** : Time Responses of Ship's Position Variations Due to a Reference
Signal Change

Although the controller was not designed for set-point tracking, the simulation results show that the tracking error is small and the controller reacts quickly to a reasonable set-point change. Note that the step change was smoothed by a prefilter of time constant 20 seconds, to simulate the described position reference change.

### 12.4.4   Adaptation and Future Generation DP Systems

The main objective of an adaptive DP system must be to cope with the wave model variations due to sea state Fung et al. (1981, 1982, 1983), Grimble (1981), Grimble and Fairbairn (1989), Moir and Grimble (1984). However, $H_\infty$ design attempts to provide a solution with robustness to such variations. Thus, the need for adaption is reduced which is a significant advantage in low cost DP systems. Should adaption be required some form of simple switching between robust solutions is possible. The switching procedure can be automated.

New advances in robust control and new developments in fault diagnosis and monitoring systems (Ding et al. 1991; Mangoubi, 1995) will enable high reliability ship positioning systems to be developed. There are greater demands on modern Dynamic Positioning (DP) systems since they have to be used with anchor systems, mooring lines and involve new duties like track keeping (Messer and Grimble, 1993). There is therefore requirement for a much more flexible DP solution which can be specialised to each of the different roles and which can utilise the latest developments to provide a more robust and reliable solution.

## 12.5   Concluding Remarks

The marine industry is relatively fast in evaluating the benefits of new design approaches and implementing the resulting solutions. The $H_\infty$ design philosophy has been shown to be valuable for both roll stabilisation and ship positioning systems. It may be equally effective for autopilot design (Grimble et al. 1994; Wilkie et al. 1987; Byrne, 1989; Fairbairn and Grimble, 1990; Grimble, 1993a, 1994; Katebi et al. 1985, 1986; Song and Grimble, 1989a,b; Grimble and Katebi, 1999). The control of submarines, near the surface, where wave forces are active, also seems to be a natural $H_\infty$ design problem (Marshfield, 1991; van der Molan and Grimble, 1993). The control of marine power plant and prime movers involve significant nonlinearities and a form of scheduled $H_\infty$ controllers may be appropriate (Banning et al. 1993).

The $H_\infty$ design method therefore has a useful future for marine applications. However, LQG solutions are also very effective for such stochastic systems and for the future combined $H_2/H_\infty$ algorithms that might provide a suitable compromise (Grimble, 1990). The value of QFT design in marine systems has still to be fully explored but it has significant potential (Hearns and Blanke, 1998 and Hearns et al. 2000).

## 12.6   References

Balchen, J.G., Jenssen, N.A. and Saelid, S., 1976, Dynamic positioning using Kalman filtering and optimal control theory, *Automation in Offshore Oil Field Operation*, 183-188.

Banning, R., Grimble, M.J. and Johnson, M.A., 1993, Modelling and advanced control of a marine diesel engine propulsion system, *European Control Conference*, Groningen, June.

Byrne, J.C., 1989, *Polynomial system control design with marine applications*, PhD thesis, Industrial Control Unit, Dept. of Electronic and Electrical Eng., University of Strathclyde, Glasgow.

Davenport, A.G. et al., 1978, Wind Structure and Wind Climate in *Safety Structures under Dynamic Load*ing, Tapir, Trondheim.

Ding, X. and Frank, P.M., 1991, Frequency domain approach and threshold selector for robust model-based fault detection and isolation, *IFAC Fault Detection, Supervision and Safety for Technical Processes*, Baden Baden, Germany, pp. 271-276.

Doyle, J.C. and Stein, G., 1981, Multivariable Feedback Design: Concepts for a Classical/Modern Synthesis, *IEEE Transactions on Automatic Control*, Vol.AC-**26**.

Doyle, J.C., Stein, G., Banda, S.S. and Yeh, H.H., 1990, *Lecture Notes for the Workshop on $H_\infty$ and $\mu$ Methods for Robust Control*, San Diego, California, American Control Conference.

Doyle, J.C., Glover, K., Khargonekar, P.P. and Francis, B.A., State-space solutions to standard $H_2$ and $H_\infty$ control problems, IEEE Trans. On Auto. Contr. 1989, Vol. 34, No.8, pp 831-846.

English, J.W. and Wise, D.A., 1976, Hydrodynamic aspects of dynamic positioning, *Transactions*, Vol. **92**, No. 3, North East Coast, Inst. Of Engineers and Shipbuilders, pp. 53-72.

Fairbairn, N.A. and Grimble, M.J., 1990, $H_\infty$ marine autopilot design for course-keeping and course changing, *Ship Control Symposium*, Washington.

Fossen T.I., 1994, *Guidance and Control of Ocean Vehicles*, John Wiley Sons, Chichester.

Fotakis, J., Grimble, M.J. and Kouvaritakis, B., 1982, A comparison of characteristic locus and optimal designs of dynamic ship positioning systems, *IEEE Trans. On Auto. Contr.*, Vol. AC-**27**, No. 6, pp. 1143-1157.

Fung, P.T.K. and Grimble, M.J., 1983, Dynamic ship positioning using a self-tuning Kalman filter, *IEEE Trans. on Automatic Control*, Vol. AC-**28**, No. 3.

Fung, P.T.K. and Grimble, M.J., 1981, *Self-tuning control of ship positioning systems*, IEE Control Engineering Series 15, Self-tuning and adaptive control : Theory and Applications, 1st Edition, Peter Peregrinus.

Fung, P.T.K., Chen, Y.L. and Grimble, M.J., 1982, Dynamic ship positioning control systems design including non-linear thrusters and dynamics, *NATO Advanced Institute on Nonlinear Stochastic Problems*, Algarve, Portugal.

Grimble, M.J., 1978, Relationship between Kalman and notch filters used in dynamic ship positioning systems, *Electronics letters*, Vol. **14**, No. 13.

Grimble, M.J., 1979, Design of optimal stochastic regulating systems including integral action, *Proc., IEE,* Vol. **126**, No. 9.

Grimble, M.J., 1980, Use of Kalman filtering techniques in dynamic ship-positioning systems, *IEE Proceedings*, Vol. **127**, Pt. D. No. 3.

Grimble, M.J., 1981, Adaptive Kalman filter for control of systems with unknown disturbances, *IEE Proc.*, Vol, **128**, Pt. D, No. 6.

Grimble, M.J., 1990, Design procedure for combined $H_\infty$ and LQG cost problem, *Int. J. Systems*, Vol. **21**, No. 1, pp. 93-127.

Grimble, M.J., 1994, *Robust Industrial Control,* Prentice Hall, Hemel Hempstead

Grimble, M.J., 1999, Polynomial solution of the $3\frac{1}{2}$DOF $H_2/H_\infty$ feedforward control problem *IEE Proc. Control Theory Appl.* Vol. **146**, No. 6, pp. 649-560.

Grimble, M.J., Patton, R.J. and Wise, D.A., 1979a, The design of dynamic ship positioning control systems using extended Kalman filtering techniques, *IEEE Oceans '79 Conference*, San Diego, California.

Grimble, M.J., Patton, R.J. and Wise, D.A., 1979b, The design of dynamic ship positioning control systems using stochastic optimal control theory, *Optimal Control Applications & Methods*, Vol. **1**, pp. 167-202, (1980).

Grimble, M.J. Patton, R.J. and Wise, D.A., 1980, Use of Kalman of filtering techniques in dynamic ship positioning systems, *IEE Proc.* Vol. **127**, Pt.D., No. 3, pp. 93-102.

Grimble, M.J., Katebi, M.R. and Wilkie, J., 1984, Ship steering control vehicles modelling and control design, *Ship Control Symposium* at University of Bath.

Grimble, M.J., Katebi, M.R. and Byrne, J., 1985, LQG self-tuning control with applications to ship steering systems, *IEE Meeting Balliol College*, Oxford.

Grimble, M.J. and Johnson, M.A., 1988, *Optimal multivariable control and estimation theory and applications : Vol I and II*, John Wiley and Sons, Chichester.

Grimble, M.J. and Fairbairn, N.A.,1989, The F-iteration approach to $H_\infty$ control, *IFAC Conf.* Glasgow, Adaptive Control and Signal Processing.

Grimble, M.J., Zhang, Y. and Katebi, M.R., 1993a, $H_\infty$-based ship autopilot design, *Ship Control Symposium*, Ottawa, Canada.

Grimble, M.J., Katebi, M.R. and Zhang, Y., 1993b, $H_\infty$-based ship fin-rudder roll stabilisation design, *Ship Control Symposium*, Ottawa, Canada.

Grimble, M.J. and Katebi, M.R., 1999, Robust ship autopilot control system design, *12th Ship Control System Symposium*, The Hague, The Netherlands.

Hearns, G. and Blanke, M., 1998, Quantitative analysis and design of a rudder roll damping controller, *4th IFAC Conference on Control Applications in Marine Systems*, Fukuoka, Japan, pp. 115-120.

Hearns, G., Katebi, R. and Grimble, M.J., 2000, Robust fin roll stabiliser controller design, *5th IFAC MCMC 2000 Conference*, Aalborg, Denmark.

Katebi, M.R., Grimble, M.J. and Byrne, J., 1985, LQG Adaptive Autopilot Design, *7th IFAC/IFORS Symposium,* York, U.K

Katebi, M.R., Grimble, M.J. and Wilkie, J., 1986, Ship steering control problem and performance criterion for energy minimisation, *American Control Conference,* Seattle, Washington.

Katebi, M.R. and Grimble, M.J., 1999, Design of dynamic ship positioning using extended $H_\infty$ filtering, *12th Ship Control System Symposium*, The Hague, The Netherlands.

Katebi, M.R., Grimble, M.J. and Zhang, Y., 1997, $H_\infty$ robust control deisgn for dynamic ship positioning, *IEE Proc. Control Theory and Applications*, Vol. **144**, No. 2.

Mangoubi, R., 1995, *Robust estimation and failure detection for linear systems*, PhD thesis, MIT Drapers Laboratory.

Marshfield, W.B., 1991, Submarine periscope-depth keeping using an $H_\infty$ controller together with sea-noise reduction notch filters, *Trans Inst. M.C.* Vol. **13**, No.5., pp 233-240.

Messer, A.C. and Grimble, M.J., 1993, Introduction to robust ship track-keeping control design, *Trans Inst MC* Vol. **15**, No. 3.

Moir, T. and Grimble, M.J., 1984, Optimal self-tuning, prediction and smoothing for discrete multivariable processes, *IEEE Trans. on Automatic Control*, Vol. AC-**29**, No. 2.

Pierson, W.J., and Marks, W., 1952, The power spectrum analysis of ocean wave records. *Trans. American Geophysical Union*, Vol. **33**, No. 6, pp. 834-844.

Seraji, H., 1987, Design of feedforward controllers for multivariable plants, *Int. J. Control*, Vol. **46**, No. 5, pp. 1633-1651.

Song, Q. and Grimble, M.J., 1989a, Application of sector theory to the robust design of an LQG autopilot, *IFAC Workshop on Expert Systems and Signal Processing in Marine Automation,* 1989.

Song, Q. and Grimble, M.J., 1989b, Application of self-tuning techniques to ship autopilot design - a survey, *IMAS 89*, Applications of New Technology in Shipping.

van der Molen, G.M. and Grimble, M.J., 1993, $H_\infty$ submarine depth and pitch control, *32nd IEEE Conference on Decision and Control*, San Antonio, Texas, Dec. 15 - 17.

Wilkie, J., Byrne, J., Grimble, M.J. and Katebi, M.R., 1987, State space LQG self-tuning autopilot, *American Control Conference*.

Wong, D., Johnson, M.A., Grimble, M.J., Clarke, M., Parrott, E.J. and Katebi, M.R., 1990, Optimal fin roll stabilisation control system design, *Ship Control Symposium,* Washington.

# 13

# Aero-engine and Flight Control Design

## 13.1  Introduction

Aircraft flight and engine control systems must adhere to stringent safety and performance requirements over a wide operational envelope. In the case of military aircraft the open loop flight dynamics will be unstable for high manoeuvrability. The control design must therefore provide high performance in the presence of difficult aircraft dynamics. Robustness to uncertainties in the aerodynamic/engine models and wind turbulence rejection capabilities are also highly desirable controller characteristics (Glover and McFarlane, 1989).

Traditionally control law design has been based on classical techniques. However, improved computing facilities allow more advanced control solutions to be exploited, such as the *Loop Transfer Recovery* design method (Athans et al. 1986, Pfeil et al. 1986, Song et al. 1993). The $H_\infty$ robust control design approach has been developed specifically for systems which are subject to uncertainty and disturbances and it provides guaranteed closed loop stability properties. The inevitable trade-offs between command tracking, actuator activity and turbulence/measurement noise rejection can be performed using $H_\infty$ cost weighting functions. When these weightings are parameterised these can be easier to tune than PID controller and notch filter coefficients.

Two applications of $H_\infty$ robust multivariable control systems design for the aero space industry are considered in the following. Control of aeroengine gas turbines is first discussed, followed by flight control systems design with application to advanced fighter aircraft. In both cases the results presented were obtained using state-space models and a standard system description. The state-space methods are particularly favoured in the North American aerospace industry.

## 13.2  Aero-engine Gas Turbine Control

Before considering the aero-engine control problem the basic system components and models will be introduced. A good simulation of the nonlinear characteristics

of the aero-engine is essential to be able to assess the performance of the controllers
designed. As with most industrial control design the most difficult part is often the
physical system modelling and the development of a suitable simulation tool (St.John
Olcayto, et al. 1996). The chapter therefore begins with a detailed description of the
aero-engine dynamics and behaviour (Blakelock, 1991). The notation for this problem
is introduced in Table 13.1.

**Table 13.1** : Gas Turbine Modelling Notation

| | |
|---|---|
| AJ– Nozzle area | IGV - Input guide vane |
| IGV - Inlet guide vane | IGVPOS - Input guide vane position |
| WF - Fuel flow | SM - Surge margin |
| HP - High pressure | MN - Mach number |
| HPC - High pressure compressor | TBT - Turbine blade temperature |
| NH - High pressure compressor speed | WRE - Re-heat fuel flow |
| LP - Low pressure | BOV - Blow off valve |
| LPC - Low pressure compressor | PR - Pressure ratio |
| LPT - Low pressure turbine | WPR - Working pressure ratio |
| HPT - High pressure turbine | SPR - Surge pressure ratio |
| NL - Low pressure compressor speed | POFT - Power offtake |
| PRCT2 - Bypass pressure ratio | $\Delta$PT1 - Turbulence disturbance signal |

The aero-engine gas turbines control problem is dominated by robustness requirements, multivariable interaction and non-linear effects (Binns and North, 1991). Increasing demands on engines, for increased reliability, fuel efficiency, low emissions
and noise has resulted in the use of more measurements and actuators. Thus, not
only must the performance and robustness of the system be improved but the multivariable nature of the control problem is also becoming more difficult. A possible
approach is to use an adaptive controller (El-Sheikh. et al. 1994) or a neural network
identification algorithm (Song et al. 1994; Chen and Khalil, 1992). However, it is
difficult to certify such systems and guarantee sufficient reliability.

## 13.2.1  Introduction to Gas Turbine Control

A gas turbine engine, illustrated in Fig. 13.1, takes in air from the atmosphere, after
which it is compressed and heated as the fuel is burned. The hot gases expelled from
the engine provide a reaction force giving a thrust in the opposite direction. In a basic
engine air is drawn through the front of the engine and the low pressure compressor
blades inside the engine then compress the air. After the air has been compressed it
is channelled into the combustion chambers. Fuel is also injected into the combustion
chamber containing the compressed air. The hot gases resulting from the fuel and air
mixture are forced through the engine and this turns the turbine which in turn drives
the compressors expelling gases from the back of the engine. The momentum forcing
the aircraft forwards is equal and opposite to the momentum of the gases ejected from
the engine or engines. After the low pressure compressor the air is also diverted into
the bypass stream and it rejoins the air flow down stream of the LPT. The total airflow
passes through the reheat section and is exhausted through the propelling nozzle.

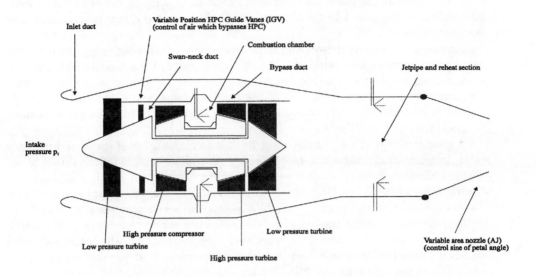

**Fig. 13.1** : Schematic Diagram of an Aero-engine

There are various modes of operation to be considered including regulation at an operating point, transient manoeuvres, surge and flame-out (Moellenhoff and Rao, 1990). The former is normally dealt with by closed-loop control while surge, flame-out and some transient manoeuvres are dealt with by open-loop control. The main objective of the control system is to ensure an adequate response to pilot thrust demands. Since thrust measurement on an aircraft is difficult the high pressure spool speed (NH) is often used to provide an indirect measure of thrust and is one of the outputs to be controlled. The range and rate of change of engine temperature must also be controlled to minimise thermal fatigue. Other limits include acceleration and deceleration limits, actuator limits, and limits to protect structural integrity.

The aims of the control system may be summarised as:

- Control of engine thrust

- Regulation of compressor surge margins

- Limitation of parameters to safe operating ranges

- Maintaining consistency of thrust response over engine life and between engines

- Regulation of thrust trajectory during transient demands.

**Need for improved aero-engine control**

Modern aero-engines include variable geometry features such as variable nozzle petal angles and inlet guide vanes (Fig. 13.1). Engines are therefore very complex and the multivariable aspects of the problem are more demanding, requiring the use of more effective multivariable design procedures. The control problem is difficult, since engine efficiency should be improved both in terms of fuel usage and compressor efficiency, while at the same time instability caused by surge in the compressor must be avoided. Disturbances in the system include those due to power off-take and inlet flow distortion.

Increased numbers of sensors and improved actuator technology offer the potential for improved performance but also add to the complexity of the design. There are, for example, improved pyrometric techniques for temperature sensing which can provide significant improvements in the life of the turbine blades and enable more accurate control to be achieved. Variables such as turbine blade and jetpipe temperature must be monitored very effectively.

A good multivariable controller must be able to regulate spool speed and pressure ratio. In traditional designs the spool speed is controlled to regulate the thrust of the engine while the auxiliary outputs such as temperature are only monitored. However, if variables go outside working limits the multivariable controllers are switched to a different mode until the auxilary variables re-enter an acceptable working range.

In traditional systems the fuel flow is continuously varied to achieve good control. Inlet guide vanes to the engine are scheduled open loop against the rotational spool speeds. Similarly, the nozzle area is scheduled to operate at pre-defined positions, according to the operating condition for the engine.

**Need for robustness**

There are several causes of uncertainty including:

1.  *Variations between units* : Aero-engines are constructed to a high degree of precision but there are tolerances which allow differences between units.

2.  *Difficulty in dynamic modelling* : Aero-engines are very complex systems and the models and simulations employed involve approximations leading to uncertainty in the descriptions used for control design.

3.  *Intake geometry variation* : It is normally assumed for modelling purposes that the airflow into an engine is parallel to its axis and undistorted which may not be the case.

4.  *Change in operating points* : If the input air is turbulent, or the aircraft operates at different angles of incidence, the pressure distribution across the compressor can vary with time, resulting in significant model variations.

## 13.2.2   The Engine and Gas Turbine Control Problem

The Rolls-Royce Spey engine to be considered is an axial, twin spool, reheated turbofan developed to power the Royal Air Force's Phantom II aircraft. The term twin spool refers to the fact that there are two compressors and two turbines in this engine. The engine includes variable geometry components, which are also likely to be included in future engines.

The objective is to control the power output of the gas turbine while keeping within the boundaries of safe working conditions. Fast and accurate thrust modulation in response to pilot power lever demands is required. The gas turbine has two shafts, one of which drives the low pressure compressor (LPC), or fan, from the power supplied by the low pressure turbine (LPT) and the second shaft drives the high pressure compressor (HPC) from the power output of the high pressure turbine (HPT). The high pressure flow is combusted and expands through the HP and LP turbines. Part of the flow from the LPC bypasses the high pressure area and forms a jet of cool air surrounding the hot jet from the combustion chamber. At the nozzle exit the flow from the LPT expands to atmospheric pressure producing the desired thrust.

The LPC therefore has two purposes. It compresses and channels the bulk of the airflow to the HPC where it is compressed further. It also compresses and directs part of the flow into the bypass section. The flow to the HPC is known as the inner airflow or core airflow. The air flow which is bypassed is termed the outer airflow or bypass airflow.

Modern gas turbine engines include the possibility of controlling extra inputs and measuring several extra variables. So called variable geometry engines enable the nozzle area (AJ) to be changed and inlet guide vanes to be moved. These inputs and the fuel flow are the manipulated variables. Outputs which can be measured include high (NH) and low (NL) pressure spool speeds, the fan exit Mach number (MN) and the turbine blade temperature (TBT).

Some of the main inputs include the main fuel flow (WF), the re-heat fuel flow (WRE), nozzle petal angle or normalised nozzle area (AJ), the high pressure compressor inlet guide vane angle (IGV) and the blow off valve (BOV) actuator position. The BOV allows the air to pass from the core to the bypass duct and can be scheduled against speed. This action causes a disturbance input to the feedback system.

As a starting point for single-input single-output (SISO) design the input fuel flow and high pressure spool speed output can be considered. The SISO design problem is therefore to control NH so that it tracks the pilot's demands for thrust control. Note that the pressure ratio (PR) is defined as the ratio of the HP compressors outlet pressure to the engine inlet pressure. It is difficult to measure engine thrust directly and hence the product of engine pressure ratio and air flow rate is used as a thrust indicator.

### 13.2.3 Major Engine Components

There follows a brief summary of the main components in an aero-engine. This also includes details of the operating limits, engine inputs and outputs, and the types of control.

*Air inlet duct* : Air is channelled into the engine via an inlet duct.

*Low pressure compressor:* After the inlet duct the air flow is directed into the LPC by input guide vanes. After compression in the LPC the airflow is directed partly into the engine core and partly into the bypass duct which surrounds the engine core. Before the air in the inner section reaches the HPC it flows through the swan-neck duct (Fig. 13.1).

*High pressure compressor* : The compressed air enters the HPC via input guide vanes (IGV's). These ensure the air has the correct angle of incidence with respect to the HPC rotor blades.

*Blow-off valve*: This is placed between the LPC and the HPC and acts as a surge protection device. It allows air to pass from the core to the bypass duct. The BOV is normally scheduled against the HP shaft speed (NH).

*Combustion chambers* : The compressed air is fed into can-type combustion chambers or combustors. The controller must ensure the temperature of the hot gasses is limited to minimise thermal stresses in the HPT blades.

*High pressure turbine*: The high pressure turbine extracts work from the hot gas to power the HPC.

*Low pressure turbine* : The LPC is driven by the low pressure turbine.

*Jetpipe and reheat section* : A large part of the hot gas expansion takes place in the jetpipe and reheat section. The flow from the bypass section is introduced into the hot core flow at this point. The flow rate of the after-burner fuel (WRE) controls the additional thrust for supersonic speeds. This heats the hot gasses exiting from the engine and increases the energy in the gas and hence the engine thrust.

*Propelling nozzle* : This nozzle accelerates the hot gasses to increase the kinetic energy and thrust. By varying the area of the nozzle the position of the working line relative to the surge line, for the engine, can be controlled.

## Engine operating limits

A number of parameters must be controlled to maintain the engine in a safe operating condition:

(i) Maximum rotational speed - to limit rotational stresses.

(ii) Maximum compression pressure - to restrict casing stresses.

(iii) Entry temperature to turbine - for material strength and creep.

The reliability and life of the engine is determined by the compressor variables for surge protection and the rates of acceleration and deceleration.

## Aero-engine plant inputs

Possible input variables to the aeroengine include:

1. *Main Fuel Flow* : This provides control of the volumetric flow (WF) of fuel into the combustion chambers measured in litres/sec.

2. *HPC Input Guide Vanes* : The input guide vanes for the HPC are movable surfaces positioned around the inner section of the duct to provide a smooth airflow into the HPC (after compression in the LPC).

3. *Nozzle Petal Angles* : Hinged plates attached to the perimeter of the engine outlet have angles which can be changed to vary the engine nozzle area (AJ).

4. *Blow off* : The BOV position is used as a surge protection device. This prevents excessive pressure ratios during engine acceleration.

5. *Re-heat Fuel Flow* : The volumetric flow of fuel (WRE) into the re-heat section can be controlled but this will not be considered further.

## Aero-engine plant outputs

Many more variable can be monitored than are actually required for control purposes. Some of the more important variables include:

1. *Low Pressure Spool Speed :* The LP Spool Speed (NL) is required for working line control.

2. *High Pressure Spool Speed :* The HP Spool Speed (NH) is needed to give an indication of the thrust produced.

3. *Combustion Exit Temperature :* Thermal stress in the HPT must be kept to a minimum and temperature limits must not be exceeded. The temperature of the hot gasses at the exit of the combustion chamber can be measured by a pyrometer and must be controlled.

4. *Dynamic Pressure in By-pass Duct :* The by-pass Mach number is a function of the ratio of the by-pass dynamic to static pressures and can give an indication of the proximity to surge conditions.

**Controls**

The most effective controls are the main fuel flow (WF) and the nozzle area (AJ). The fuel flow input WF moves the equilibrium point up and down the working line. The nozzle area moves the working line closer to or further from the surge line. The input actuators are driven by hydraulic/hydromechanical systems, from the digital control unit (DCU).

**Engine fuel system motor dynamics :**

$$\frac{k_1}{1+0.013s}$$

**Nozzle actuator electrically controlled hydraulic jack dynamics :**

$$\frac{k_2}{1+0.03s}$$

**Inlet guide vane electrically controlled jack dynamics :**

$$\frac{k_3}{1+0.03s}$$

The fuel flow channel includes a pure transport delay which is more than one sample period in the discrete control implementation. The time delay represents the delay which occurs between the injection of the liquid fuel into the combustion chamber and the vapourisation and air mixing which must occur before it burns.

## 13.2.4 Surge Margin and Working Lines

Surge in an engine is associated with a sudden drop in delivery pressure and represents an aerodynamic pulsation. A reduction in fuel flow and pressure ratio results when aerodynamic stalling of some of the blading occurs. The change from the non-stalled to the stalled condition causes a change in the operating point. The flow pattern within the engine may break down if the outlet pressure becomes too large. The engine may then enter a condition where the flow surges backwards and forwards through the compressor, due to the pressure gradient. This high frequency phenomenon is the most serious behaviour and is known as surge.

The performance of the aeroengine control system is often described in terms of the compressor characteristics shown in Fig. 13.2. This shows the low pressure compressor (LPC) bypass characteristics. The horizontal axis represents the LPC airflow and the vertical axis denotes the pressure ratio (PR) across the LPC. During normal operation the engine operates in the region to the right of the *surge line*. The compressor map in Fig. 13.2 indicates the lines of constant spool speed for varying pressure ratios. An important control objective is to maintain an appropriate pressure ratio across the compressor. The operating point of the aeroengine, on the compressor map, is determined by the spool speed and the compressor pressure ratio.

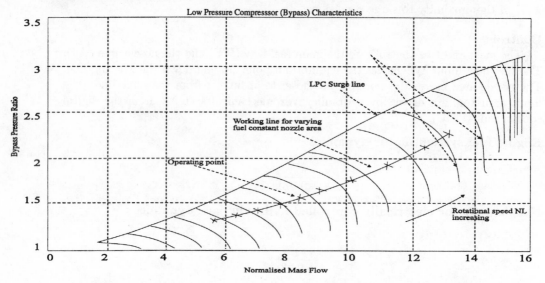

**Fig. 13.2** : Gas Turbine Low Pressure Compressor Characteristic

It is well known that axial compressors become more efficient the closer they operate to the surge line for the engine. However, this line also represents a limit on the safe operation of the engine. The surge line represents the boundary over which reverse flow through the compressor occurs, with potentially disastrous results. The surge line should not therefore be crossed even transiently. It is impossible to operate on the surge line and hence a *working line* some distance away from the surge line is defined. The control system must maintain this distance or surge margin (SM) from the working line. It is common practice to specify a safe limit, which is some distance from the surge line under steady-state conditions. This safety margin is termed the surge margin and relates the working pressure ratio to the actual surge pressure ratio. The avoidance of surge must be a primary design objective.

**Surge margin**

Points on the working line are described by the working pressure ratio (WPR). That is, the pressure ratio sustained by the compressor for a particular rotational speed. The surge pressure ratio (SPR) is the maximum pressure ratio that can be sustained for a particular rotational speed and flow. The surge margin is defined as:

$$SM = \frac{SPR - WPR}{WPR - 1} \tag{13.1}$$

The steady-state surge margin must be around 0.2 but the minimum value during transients can reduce to say 0.07.

Recent aero-engines operate at higher pressure ratios and hence lower specific fuel consumptions than their predecessors. This means that such compressors operate closer to the surge line, reducing safety margins. Entering the surge regime is not only dangerous but can also cause serious damage to the engine. It is therefore important to track the compressor working line to avoid this condition.

**Modes of working line operation**

There are 3 basic modes of operation including:

(i) Working line control in steady-state operating conditions

(ii) Working line control during acceleration

(iii) Working line control during deceleration

**Working line control**

Working line control aims to maintain efficient, surge free operation of the gas turbine. This is achieved by controlling the engine operating pressure along a pre-defined working line. The working line can be stated in terms of a relationship between engine output variables, such as low pressure compressor shaft speed (NL) and pressure ratio (PR) (or NL and fan exit Mach number MN). The desired relationship must be maintained closely both in the steady-state and in transient operation. Thus, a major objective of the control action is to provide good tracking of the LPC working line which is the most critical margin. The existing open-loop working line control systems require hydromechanical systems with high accuracy components and specifications but these might be relaxed if closed-loop working line control can be employed.

## 13.2.5 Classical Gas Turbine Control and Design

In classical design PI controllers are often employed and scheduled with the operating point (Wilkie and St.John Olcayto, 1992). A controller for the full range of nonlinear operation (Garg, 1997) can be provided using controller parameter switching and interpolation between parameters. The use of robust $H_\infty$ controllers, scheduled with operating points, should reduce the need for controller switching and interpolation. Improved disturbance rejection can also be achieved, at the expense of requiring more realistic models for the disturbances. In the aero-engine example there are two major disturbance sources due to a power off-take and due to inlet turbulence and distortion. Heat soakage in the engine is also sometimes treated as an additional disturbance. Sudden increases in the power demand by ancillary equipment, such as air conditioning units, can be modelled as step disturbances. The effects of power demands, due to aircraft flight control surface changes, can be modelled by a coloured noise signal.

A nonlinear model of the engine and actuators is required on which the robust designs can be tested. The problem considered below is for a Rolls-Royce Spey engine which is a twin spool bypass engine that includes reheat and a variable nozzle area. The engine and subsystems are represented by a 24 state nonlinear model. The operating region or working line employed was used to find the linear models which are used for control design. The working line should provide an adequate surge margin which is normally a constant but can be a variable. The surge margin is not directly measurable and is a function of the pressure ratio (PR).

Large reference changes on the thrust demand can cause the LP compressor spool speed (NL) and the hot duct entry temperature to be exceeded. Clearly higher thrust levels result in greater fuel flows and higher temperatures and speeds. A method of handling such constraints, by the engine reference management system, is therefore required.

## 13.2.6  Gas Turbine Control Design Study

The aim of the gas turbine control design study which follows was to investigate advanced control strategies for aeroengines. The full range of control of engine thrust, from idle to maximum engine speed, was considered, while satisfying the constraints of compressor stability margin and minimising the effects of disturbances (due to air bleed and power off-take). The controller was also required to regulate against the most important of the structural limits, while at the same time maintaining the stall margins for each compressor (St.John Olcayto et al. 1994).

The engine was therefore to be controlled, according to the pilot's demand, while ensuring the fan and the core working lines were maintained close to prespecified values. The limits to protect structural integrity relate to the turbine blade temperature, high pressure and low pressure shaft speeds. For the small change step responses, to avoid the acceleration and deceleration limits, the output response was required to attain approximately 90% of the demanded change within 0.5 seconds and the overshoot and output variations were restricted to a band of 10% of the demanded value. The interaction in the system was also limited so that a reference change in one channel would result in only a small change in other signal outputs.

The rejection of disturbances had to be adequate so that step disturbance effects would be removed within about 1 second. About 50% of the maximum amplitude of the response due to the disturbances was required to be off-set within 0.5 seconds and at least 80% removed within 1 second.

There are various sources of the disturbances:

(i) Air flow through the engine at non-zero angles of incidence.

(ii) Cross wind disturbance effects.

(iii) Power off-take from the HP spool.

Inlet distortion disturbances can be considered to be 0-2% of the surge margin. Step disturbances arise from the power demands of equipment like air conditioning units and aircraft flight surface actuators. Heat soakage can account for a disturbance equivalent to about 5% of NH.

The two-input, two-output, multivariable problem of interest may now be summarised. The high pressure compressor spool speed NH needed to track the pilot's demands for thrust control. To maintain an adequate surge margin, PRCT2 was required to follow a reference demand calculated as a function of the required NH. This was achieved through control of the fuel flow WF, nozzle area AJ and input guide vane positions IGVPOS. Auxiliary variables such as the temperature and acceleration were also monitored. In the case of excessive rises, in for example the temperature, the primary control objective was changed until the temperature was brought back within an acceptable range.

In the present design study, three input variables were manipulated including WF, AJ, IGVPOS. In the classical control approach the only fully continuous input was the fuel flow WF. The inlet guide vanes were open loop scheduled against rotational spool speed and the nozzle area was scheduled to operate at pre-defined positions for specified operating conditions. The IGV's do not have a large effect at the chosen design points.

### Linearised state-space model

The continuous-time two-input, two-output, linear state-space model to be used for $H_\infty$ design can be evaluated from the non-linear engine model, at a number of working line operating points, as:

$$\dot{x}_p = A_p x_p + B_p u + Ld \tag{13.2}$$

$$y_p = C_p x_p + D_p u \tag{13.3}$$

where for the example $x_p \in R^{15}$, $u_p \in R^2$, $d \in R^2$, so that $A_p \in R^{15\times15}$, $B_p \in R^{15\times2}$, and $L \in R^{15\times2}$, and $y_p \in R^2$, hence $C_p \in R^{2\times15}$ and $D_p \in R^{2\times2}$. The model parameters are functions of operating pressure, engine speed and are also affected by manoeuvres, turbulence and inlet temperature. The inputs, outputs and disturbance model signals for this study were selected as follows:

**Inputs:**

$$u = \left[\begin{array}{c} WF \\ AJ \end{array}\right] = \left[\begin{array}{c} \text{Fuel flow} \\ \text{Nozzle area} \end{array}\right]$$

**Outputs:**

$$y_p = \left[\begin{array}{c} NH \\ PRCT2 \end{array}\right] = \left[\begin{array}{c} \text{Percentage high pressure spool speed} \\ \text{Bypass pressure ratio} \end{array}\right]$$

**Disturbances:**

$$d = \left[\begin{array}{c} POFT \\ \Delta PT1 \end{array}\right] = \left[\begin{array}{c} \text{Power off take} \\ \text{Turbulence} \end{array}\right]$$

The inlet guide vanes (IGV) represent a further control input but this is not included since it is open-loop scheduled. The disturbance POFT has a spectrum equivalent to a pulse of duration 1/8 second and a peak of 200. The disturbance $\Delta PT1$ can be assumed to be a white noise signal (Gaussian) with zero mean and unity variance.

**Design objectives**

(i) The closed loop bandwidth should be in the region of 10 radians per second or slightly greater in both loops.

(ii) A small overshoot on NH is allowable but zero overshoot on PR is desirable.

(iii) A zero steady-state error to step inputs in both channels is ideal, although a 1% steady state error on NH would be acceptable.

With the system shown in Fig. 13.3, the only reference change due to the pilot is $NH_{dem}$ and the pressure ratio reference change or demand $PR_{dem}$ may then be calculated by the command generator. These values are determined according to the compressor spool speeds.

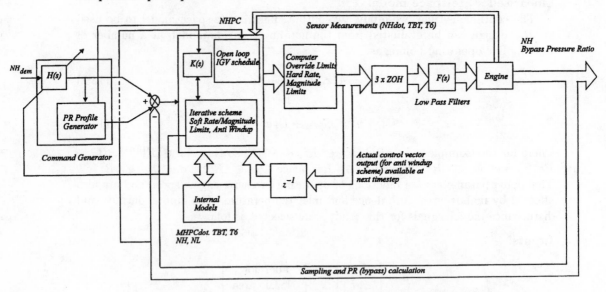

**Fig. 13.3** : Total System Architecture Including Outer Working Line Control Logic

**Example 13.1** : *Gas Turbine Control Models*

The gas-turbine model will be described in the form obtained from the basic physical equations, namely continuous-time form.

**Disturbance models:**

$$W_{d1} = 5000/(1000s + 1) \quad , \quad W_{d2} = 100/(1000s + 1)$$

**Reference signal model:**

$$W_{r1} = 10^{-4}(s + 10000)/(10s + 1) \quad , \quad W_{r2} = 10^{-4}(s + 10000)/(10s + 1)$$

**Measurement noise model:**

$$W_{n1} = 10^{-2} \quad and \quad W_{n2} = 10^{-2}$$

**Continuous-time plant model (Case NL 35):**

$$G_{11}(s) = \frac{\begin{array}{c}1.706126 \times 10^7 s^{10} + 4.562066 \times 10^9 s^9 + 3.910072 \times 10^{11} s^8 - 1.133707 \times 10^{18} s^7 \\ -1.208705 \times 10^{15} s^6 + 9.50472 \times 10^{15} s^5 + 4.781899 \times 10^{18} s^4 + 1.402759 \times 10^{20} s^3 \\ +1.057519 \times 10^{21} s^2 + 2.755543 \times 10^{21} s + 1.868649 \times 10^{21}\end{array}}{\begin{array}{c}s^{14} + 672.4391 s^{13} + 206302.8 s^{12} + 3.78709 \times 10^7 s^{11} + 4.599174 \times 10^9 s^{10} \\ +3.864981 \times 10^{11} s^9 + 2.285654 \times 10^{13} s^8 + 9.479886 \times 10^{14} s^7 + 2.6939087 \times 10^{16} s^6 \\ +4.994838 \times 10^{17} s^5 + 5.540925 \times 10^{18} s^4 + 3.210053 \times 10^{19} s^3 + 8.929646 \times 10^{18} s^2 \\ +1.03552 \times 10^{20} s + 4.131334 \times 10^{19}\end{array}}$$

$$G_{22}(s) = \frac{\begin{array}{c}139326.7 s^{11} + 1.56719 \times 10^7 s^{10} - 1.250063 \times 10^9 s^9 - 1.62149 \times 10^{11} s^8 \\ +1.30117 \times 10^{13} s^7 + 2.518501 \times 10^{15} s^6 + 1.35811 \times 10^{17} s^5 + 3.009805 \times 10^{18} s^4 \\ +2.0573 \times 10^{19} s^2 + 5.281216 \times 10^{19} s^2 + 4.904767 \times 10^{19} s + 1.370827 \times 10^{19}\end{array}}{\begin{array}{c}s^{14} + 672.4391 s^{13} + 206302.8 s^{12} + 3.78709 \times 10^7 s^{11} + 4.599174 \times 10^9 s^{10} \\ +3.864981 \times 10^{11} s^9 + 2.285654 \times 10^{13} s^8 + 9.479886 \times 10^{14} s^7 + 2.6939087 \times 10^{16} s^6 \\ +4.994838 \times 10^{17} s^5 + 5.540925 \times 10^{18} s^4 + 3.210053 \times 10^{19} s^3 + 8.929646 \times 10^{19} s^2 \\ +1.03552 \times 10^{20} s + 4.131334 \times 10^{19}\end{array}}$$

$$G_{21}(s) = \frac{\begin{array}{c}85256.54 s^{11} + 2.414119 \times 10^7 s^{10} + 2.301551 \times 10^9 s^9 + 2.441153 \times 10^{10} s^8 \\ -6.689519 \times 10^{12} s^7 + -5.518416 \times 10^{13} s^6 + 2.485491 \times 10^{16} s^5 + 1.075327 \times 10^{18} s^4 \\ +1.55702 \times 10^{19} s^3 + 7.740803 \times 10^{19} s^2 + 1.338387 \times 10^{20} s + 7.164967 \times 10^{19}\end{array}}{\begin{array}{c}s^{14} + 672.4391 s^{13} + 206302.8 s^{12} + 3.78709 \times 10^7 s^{11} + 4.599174 \times 10^9 s^{10} \\ +3.864981 \times 10^{11} s^9 + 2.285654 \times 10^{13} s^8 + 9.479886 \times 10^{14} s^7 + 2.6939087 \times 10^{16} s^6 \\ +4.994838 \times 10^{17} s^5 + 5.540925 \times 10^{18} s^4 + 3.210053 \times 10^{19} s^3 + 8.929646 \times 10^{19} s^2 \\ +1.03552 \times 10^{20} s + 4.131334 \times 10^{19}\end{array}}$$

$$G_{12}(s) = \frac{\begin{array}{c}-2992956 s^{10} - 3.370146 \times 10^8 s^9 + 2.681872 \times 10^{10} s^8 + 3.487037 \times 10^{12} s^7 \\ -2.79148 \times 10^4 s^6 - 5.41411 \times 10^{16} s^5 - 2.923352 \times 10^{18} s^4 \\ -6.4904272 \times 10^{19} s^3 - 4.4453334 \times 10^{20} s^2 - 1.083405 \times 10^{21} s - 7.136006 \times 10^{20}\end{array}}{\begin{array}{c}s^{14} + 672.4391 s^{13} + 206302.8 s^{12} + 3.78709 \times 10^7 s^{11} + 4.599174 \times 10^9 s^{10} \\ +3.864981 \times 10^{11} s^9 + 2.285654 \times 10^{13} s^8 + 9.479886 \times 10^{14} s^7 + 2.6939087 \times 10^{16} s^6 \\ +4.994838 \times 10^{17} s^5 + 5.540925 \times 10^{18} s^4 + 3.210053 \times 10^{19} s^3 + 8.929646 \times 10^{19} s^2 \\ +1.03552 \times 10^{20} s + 4.131334 \times 10^{19}\end{array}}$$

**H∞ gas turbine cost function error weightings** (Grimble, 1993):

$$H_{q1} = (s + 0.04)/(s + 10^{-5}) \quad , \quad H_{q2} = (s + 0.1)/(s + 10^{-5}) \qquad (13.4)$$

The error weighting includes near integral action.

**H∞ gas turbine cost function control weighting:**

$$H_{r1} = 10^3(2s + 1)/(s + 10^4) \quad , \quad H_{r2} = 10^3(s + 0.5)/(s + 10^4) \qquad (13.5)$$

The control weighting represents a lead-lag transfer function, which is made proper by adding a high frequency lag term.

### Gas turbine linear design results

The frequency response of the plant model is shown in Fig. 13.4. Note that the behaviour of the 2-input 2-output gas turbine model is dominantly low frequency. The disturbance model for the system also has a low-pass characteristic, as shown in Fig. 13.5. The time responses of the open-loop plant model are well damped but rather slow. This is indicated in Fig. 13.6, for a high speed case NL70, where the 2 outputs are shown for open loop step inputs on the two control channels, respectively.

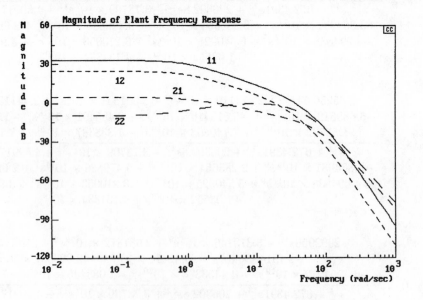

**Fig.13.4**: Magnitude of the Gas Turbine Plant Frequency
Response for Each Channel

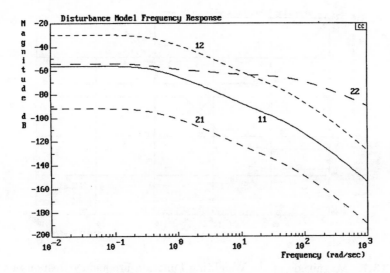

**Fig. 13.5** : Magnitude of the Frequency Response of the Disturbance
Model for Each Channel

The weighting function frequency responses, shown in Fig. 13.7, have the traditional forms considered previously. That is, there is integral action at low frequencies
on the error weighting terms and the control weighting functions include high pass
characteristics. In fact the control weighting functions are of course improper. In
the continuous-time state space realisation a high frequency lag term is introduced to
ensure these terms are strictly proper.

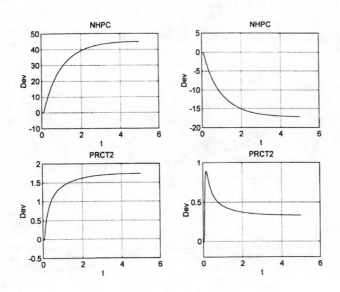

**Fig. 13.6** : Open Loop Step Responses for the Case NL70

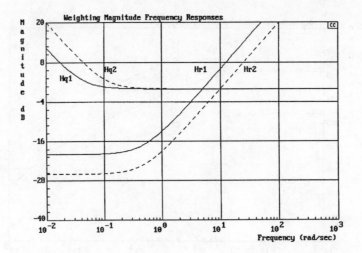

**Fig. 13.7** : Magnitude of the Weighting Function Frequency Responses for the Error and Control Signals on the Two Channels

The $H_\infty$ mixed sensitivity controller frequency response is shown in Fig. 13.8 for the different elements of the 2 by 2 transfer-function matrix. Observe that the controller has a traditional form, including integral action at low frequency and roll-off at high frequencies. The gains in the different controller channels are very different but this is indicative of the poor plant scaling. It is a simple matter to scale the plant inputs and outputs to improve these characteristics. In fact the controller has 24 state variables but these may easily be model reduced with little loss of performance. If the number of states is reduced to 17 there is little discernible difference compared with the controller responses in Fig. 13.8.

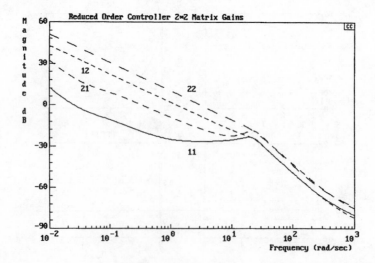

**Fig. 13.8** : Magnitude of the Frequency Responses for the Full Order Controller Gains

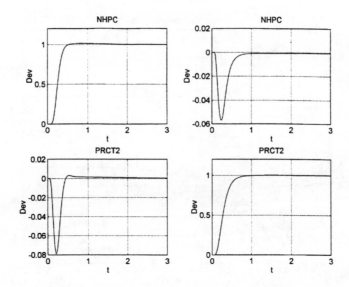

**Fig. 13.9** :Unit Step Responses for the Closed Loop System for Reference
Changes, Case NL70

The closed loop performance of the resulting system is illustrated in Fig. 13.9.
This reveals the system outputs for step reference changes in the two channels, respectively. The speed of response, degree of interaction and well damped behaviour
are all appropriate for this case. If the reduced order controller is employed, the
characteristics shown in Figs. 13.10 to 13.13 are obtained, which are also suitable.
The output responses and control actions are both shown in these figures.

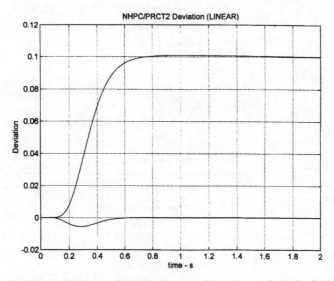

**Fig. 13.10** : Outputs Due to a Step Reference Change on Speed of Magnitude 0.1
Using the Reduced Order Controller

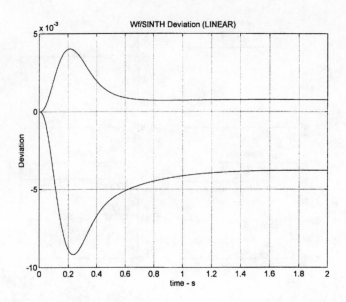

**Fig. 13.11** : Control Signal Responses for Step Reference Change of
Magnitude 0.1 on the Speed Signal

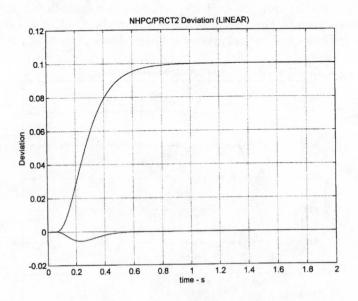

**Fig. 13.12** : Outputs Due to a Step Reference Change in Pressure Ratio of
Magnitude 0.1 Using Reduced Order Controller

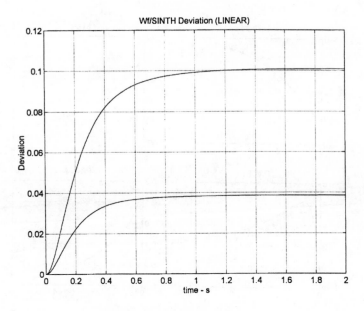

**Fig. 13.13** : Control Signal Responses for Step Reference Change
of Magnitude 0.1 on the Pressure Ratio

## Case NL35

This case corresponds to a low speed condition and it is not therefore surprising that the plant step responses, shown in Fig. 13.14, are rather slow. Their characteristics are also very different to those for the high speed condition. However, the same weightings give reasonable results for this case. The controller has the characteristics shown in Fig. 13.15. The closed-loop step responses are shown in Fig. 13.16 and these are also acceptable.

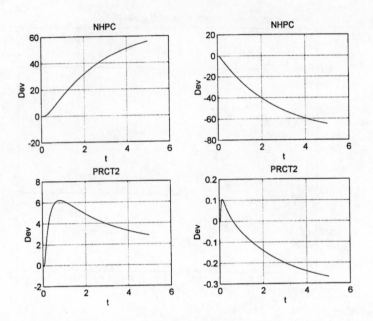

**Fig. 13.14** : Unit Step Responses for the Plant Model for Inputs on the Two
Control Channels, Respectively.

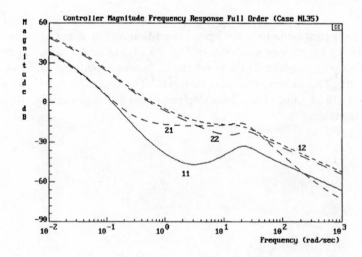

**Fig. 13.15** : Magnitude of the Controller Frequency Responses for Each Channel
(Case NL35)

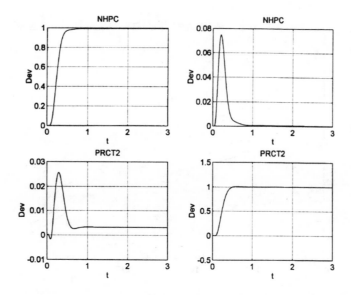

**Fig. 13.16** : Unit Step Responses for the Closed Loop System for Step Changes in the Reference Signals on the Two Channels

## Case NL33

The lowest speed condition for which designs were completed corresponds with the very different plant responses shown in Fig. 13.17. The $H_\infty$ controller for this case (Fig. 13.18) must also have quite a different response compared with the previous speed condition. However, using the same weightings as before the responses shown in Fig. 13.19, for the closed loop system, are again reasonable but quite slow.

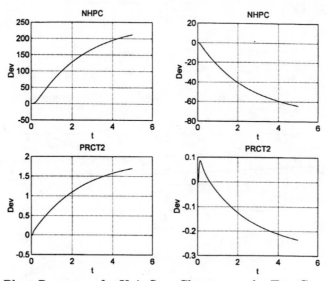

**Fig. 13.17**: Plant Responses for Unit Step Changes on the Two Control Channels, Respectively (Case NL33)

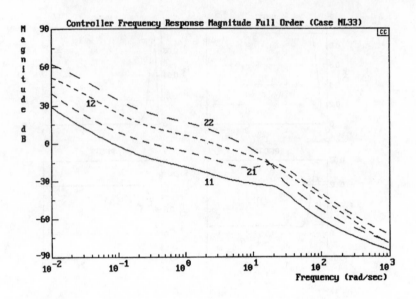

**Fig. 13.18** : Magnitude of the Controller Frequency Responses
for Case NL33

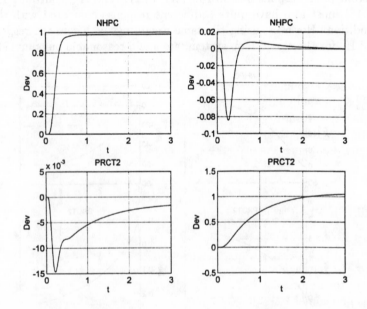

**Fig. 13.19** : Output Responses of Closed Loop System for Unit
Step Reference Changes on Two Channels

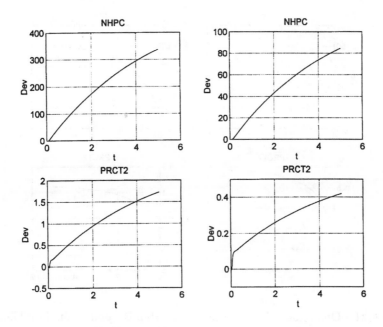

**Fig. 13.20** : Open Loop Plant Responses to Step Responses on Two Channels
(Case NL 28)

## Nonlinear Simulation Results

The nonlinear simulation of the aero-engine was based upon a thermodynamic model. The thermodynamic quantities were interpreted from data tables compiled from engine test results provided by Lucas Aerospace and the Defence Research Agency at Pyestock. Many of the variables are functions of the rotational speed of the shafts, and since the engine has two spools there are two rotational speeds of interest.

## Variability in Plant Step Responses

Figures 13.20, 13.21 and 13.22 illustrate the variation in the plant dynamics from low speed to the high speed range. Operating points NL 28, NL 60 and NL 76 are considered. In the high speed range (NL 76) the step responses are reasonably fast, particularly to the pressure ratio. Note the poor responses represent those due to a step input in channel 1 and due to a step input in channel 2, respectively. The speed and power step responses, for the low speed case NL28, shown in Figure 13.20, are much slower and the interaction effects (particularly to speed) are much larger. The wide variability in the responses is matched by the large change in the DC gains for different operating points.

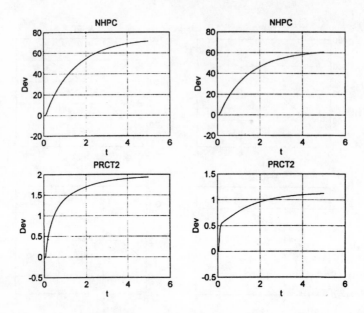

**Fig. 13.21** : Open Loop Plant Responses to Step Responses on Two Channels
(Case NL 60)

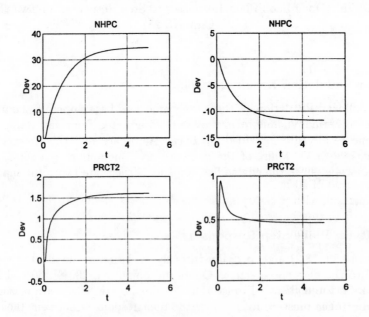

**Fig. 13.22** : Open Loop Plant Responses to Step Responses on Two Channels
(Case NL 76)

**Plant Model DC Gains**
**Case NL 28:**

$$W_{dc} = \left[ \begin{array}{cc} 520.\ 219 & 131.112 \\ 2.597 & 0.6002 \end{array} \right]$$

**Case NL 60:**

$$W_{dc} = \left[ \begin{array}{cc} 73.972 & 62.457 \\ 1.9706 & 1.146 \end{array} \right]$$

**Case NL 76:**

$$W_{dc} = \left[ \begin{array}{cc} 34.722 & -11.8276 \\ 1.602 & 0.4621 \end{array} \right]$$

There is clearly a very large change in the DC gains of the plant with speed. This suggests that either the controller gains will have to be scheduled, or that a single robust controller with modest performance will have to be accepted.

Summary of control system requirements:

- Control of the engine input-output pressure ratio.

- Control of high pressure compressor spool speed.

- Control of low pressure compressor surge margin.

- Limit on the low pressure compressor spool speed and certain working temperatures.

### Designs based upon step response testing

Step response testing on the nonlinear simulation provided crude models to enable one set of low order designs to proceed. Several $H_\infty$ controllers were designed for different operating points, and one controller was found to be suitable across the range, although much improved performance could be achieved by scheduling. A sampling period of 30 milliseconds was used and the controllers were discreted using Tustin's method, with pre-warping at 10 rads/sec (St.John-Olcayto et al. 1996). The scheduling algorithm involved parameterising the poles and zeros of the discrete time controllers, as functions of the spool speed. It provided a continuous scheduling capability. The PRCT2 reference signal was generated as a function of the NH reference demand.

### Implementational issues

The controller must include anti-windup protection. If the controller is implemented in state-feedback form it will involve a filter (or observer) and control gain structure (Rao et al. 1988). Major nonlinearities can then be taken into account in the spirit of Extended Kalman filtering. When the controller is switched between operating points there must be a bumpless transfer which requires appropriate controller initialisation. The controller coefficients can be scheduled as functions of speed NH.

## Step response models

The following transfer-function models were obtained from step responses on the nonlinear simulation which was used to tune the controllers (prior to engine tests). These models were used for robust control design, where a low order model was desirable. The models had the general matrix form:

$$g = \begin{pmatrix} g_{11} & g_{12} \\ g_{21} & g_{22} \end{pmatrix}$$

### Case : NL28

$$g_{11}(s) = \frac{190}{(s+0.8403)} \qquad g_{21}(s) = \frac{(-0.2346s + 1.5484)}{(s+1.0526)}$$

$$g_{12}(s) = \frac{-4.11}{(s+0.8)} \qquad g_{22}(s) = \frac{(0.068s + 0.0041)}{(s+1.4085)}$$

### Case : NL33

$$g_{11}(s) = \frac{194.4482}{(s+0.8403)} \qquad g_{21}(s) = \frac{(-0.2042s + 2.8387)}{(s+1.1236)}$$

$$g_{12}(s) = \frac{-5.0899}{(s+0.8)} \qquad g_{22}(s) = \frac{(0.0817s + 0.0198)}{(s+1.2048)}$$

### Case : NL37

$$g_{11}(s) = \frac{67.7407}{(s+0.6757)} \qquad g_{21}(s) = \frac{3.4838}{(s+1.6949)}$$

$$g_{12}(s) = \frac{-5.205}{(s+0.7299)} \qquad g_{22}(s) = \frac{(0.1021s + 0.0209)}{(s+1.7857)}$$

### Case : NL48

$$g_{11}(s) = \frac{115.8689}{(s+0.8)} \qquad g_{21}(s) = \frac{6.1934}{(s+2.0833)}$$

$$g_{12}(s) = \frac{-6.1923}{(s+0.8)} \qquad g_{22}(s) = \frac{1.497s}{(s+1.8868)}$$

### Case : NL64

$$g_{11}(s) = \frac{63.7408}{(s+0.9901)} \qquad g_{21}(s) = \frac{5.08}{(s+2.381)}$$

$$g_{12}(s) = \frac{-6.4258}{(s+0.9901)} \qquad g_{22}(s) = \frac{(0.3642s + 0.1107)}{(s+2.1277)}$$

**Case : NL76**

$$g_{11}(s) = \frac{54.8377}{(s + 1.6949)} \qquad g_{21}(s) = \frac{5.4193}{(s + 2.7778)}$$

$$g_{12}(s) = \frac{-9.5086}{(s + 1.6949)} \qquad g_{22}(s) = \frac{(0.6894s + 0.4293)}{(s + 3.3333)}$$

Note that the discrete controller was not implemented as one expression:

$$C_0(z^{-1}) = \frac{(b_0 + b_1 z^{-1} + b_2 z^{-2} + \cdots + b_m z^{-m})}{(1 + a_1 z^{-1} + a_2 z^{-2} + \cdots + a_n z^{-n})}$$

but as first and second-order filters in parallel form:

$$C_0(z^{-1}) = \frac{K_1}{(1 - z^{-1})} F_1(z^{-1}) + F_2(z^{-1}) + F_3(z^{-1}) + \cdots$$

where

$$F_k(z^{-1}) = \frac{b_{0k} + b_{1k} z^{-1} + b_{2k} z^{-2}}{1 + a_{1k} z^{-1} + a_{2k} z^{-2}}$$

Integral windup protection can be placed around the integrator in the above expression. The numerical precision of controller implementation can be improved using delta-transforms in place of the z-transform.

### Engine test results

The results of the step rest results on the Rolls-Royce aeroengine are illustrated in Figs. 13.23 and 13.24. The responses are relatively slow but this is for the case of one controller which can be used across the whole speed range.

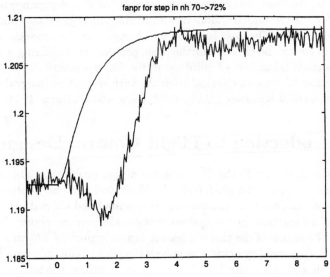

**Fig. 13.23** Engine Test Results for $N_h$ Against Time

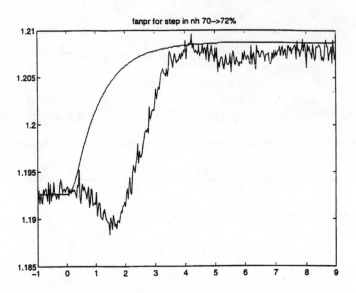

**Fig. 13.24** : Engine Test Results for Fan Pressure Against Time

### 13.2.7    Remarks on the Gas Turbine Control Problem

The design of an $H_\infty$ controller using the Glover-McFarlane (1989) coprime factori-sation approach for the gas turbine aero-engine considered was described by Postleth-waite et al. (1995). Test results were presented for the Rolls Royce turbofan engine and multi-mode control logic was developed to preserve structural integrity by limiting engine parameters to specified safe limits.

Classical frequency domain multivariable design methods can be used to design gas turbine controls. However, one of the advantages of the $H_\infty$ design approach is that it allows explicitly for the multivariable nature of the problem (Watts and Garg, 1995; Frederick et al. 1996). The cost function weightings and disturbances and reference models must, of course, be chosen but the basic problem of stabilising a multivariable, interactive system is taken care of quite naturally. Implementational problems for $H_\infty$ designs are similar to those of classical controls, with respect to integral wind-up, rate limiting and numerical accuracy (Garg, 1993; Garg and Mattern, 1994).

## 13.3    Introduction to Flight Control Design

Attention may now turn to the flight control design problem.    The primary flying controls for an aircraft are the pitch control, roll control and yaw control mechanisms. These affect motions about the transverse, longitudinal and normal axes, respectively, but because of interaction in the system motion also occurs about the other axes. Thrust control by means of the throttle leavers is also important but this is determined by the engine management system. The yoke is the pilots main input mechanism which is used for pitch and roll control. When the yoke is pulled towards or pushed away from the pilot the elevator is moved accordingly.

The classical approach to flight control systems design normally involves using proportional plus integral feedback of various aircraft motion variables on a loop by loop basis. Modern combat aircraft are designed to be aerodynamically unstable, since this provides performance benefits but this gives rise to a more difficult control problem. There is also a trend towards the use of higher levels of system integration, particularly with respect to flight and power plant control systems. There is therefore a need for truly multivariable solutions, which are robust and have high integrity.

The design of a flight control system (Breslin, 1996) can be separated into two main components: (i) Design of the low frequency rigid body mode control system and (ii) Design of the high frequency flexible body mode control system.

The main functions of the rigid body control system are as follows:

- Stabilisation of the unstable airframe.

- Manoeuvre demand control to allow carefree handling, limiting the angle of attack and normal acceleration.

- Provision of good handling qualities to allow the pilot to carry out the operational role with minimum effort.

The purpose of the flexible body design is to reduce transmission of structural resonances to aircraft motion sensors. The rigid body control loop design must aim to satisfy the pilot's instinctive requirements. The notation and symbols utilised for the flight control study are summarised in Table 13.2.

**Table 13.2** : Aircraft Flight Control Notation

| | |
|---|---|
| ASTOVL | Advanced Short Take Off and Vertical Landing |
| FCS | Flight Control System |
| PCB | Plenum Chamber Burning |
| $\theta_F$ | Front nozzle angle (radians) |
| $\theta_R$ | Rear nozzle angle (radians) |
| X | Axial (forward) thrust |
| Y | Side force |
| Z | Normal thrust |
| V | Velocity in direction of flight (metres/second) |
| $\alpha$ | Angle of attack (radians) |
| $\theta$ | Pitch attitude (radians) |
| q | Pitch rate (radians/second) |
| $\delta$ | Flap angle (radians) |
| m | Pitching moment |
| $n_z$ | Normal acceleration (metres/second$^2$) |
| w | Normal velocity (metres/second) |
| h | Altitude or Height (metres) |
| $\dot{h}$ | Height rate (metres/second) |
| q | Pitch rate (radians/second) |
| $a_x$ | Longitudinal acceleration (metres/second$^2$) |

### 13.3.1   Dynamic Modes of Aircraft

As the performance of the aircraft increases, dynamic modes which were previously controllable became uncontrollable. If a dynamic mode has a period which is greater than about 10 seconds it will be within the pilot's bandwidth and controllable by the pilot. If the dynamic mode has a period of less than about 4 seconds, then it will not be controllable by the pilot and additional damping must be introduced by feedback. The important dynamic modes include, for the longitudinal plane, the *short period*, and *phugoid modes*, and in the lateral plane the *dutch roll*, and the *spiral* modes. In a high performance aircraft a *Stability Augmentation System* (SAS) is required to augment the damping of nodes which are outside the pilot's bandwidth. The short period modes must normally be damped by adding a SAS system, which simply involves a feedback loop designed to increase the damping of particular modes. An *Automatic Flight Control System*, (AFCS) not only controls troublesome modes but also provides the pilot with a particular type of response, to satisfy handling quality requirements. A model following capability may easily be included through the use of an ideal response model.

Structural coupling or aero-servo elasticity is the interaction between the aircraft dynamics and the flight control system. The motion sensors for the aircraft, can pick up the structural vibrations of the aircraft, which are fed back to the actuators by the control system. Oscillatory actuator demands are then generated, which can amplify the structural vibrations. The transmission of the worst structural resonances to the flight control system must be suppressed by appropriate design of the control loops. In classical design this is accomplished by using deep notch filters centred at the resonant frequencies. Unfortunately, these filters introduce considerable phase lag into the system, which has to be offset to some extent by including phase advance in the rigid body controllers.

A typical fighter aircraft has the first bending mode at about 7 Hz, that is 44 radians/second. The open-loop unstable mode is usually about 6 radians/second. The stability margins to be achieved are $\pm$ 6 dB with 45° phase margin for the nominal loop design and $\pm$ 4.5 dB with 45° for the uncertain (perturbed) loop.

### 13.3.2   Vectored Thrust Aircraft

In a jump jet some of the thrust from the engine is directed downwards to enable take off and landing or a slow hovering motion to occur. The BAe Harrier jump jet is probably the best known vertical take off and landing aircraft. In this aircraft four air nozzles can be swivelled to any angle between vertical and horizontal, at a rate of up to 100 degrees per second. One of the delicate manoeuvres involves the move from hovering to wingborne flight. In this case the pilot swivels the nozzles to direct the gases at a desired angle. As the aircraft moves forwards, the wings begin to produce lift. When the forward sp

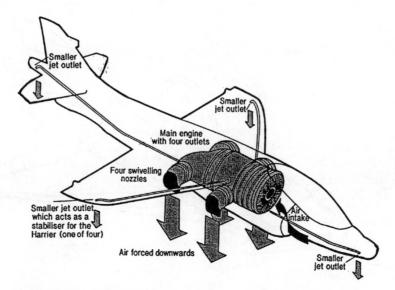

**Fig. 13.25** : ASTOVL Aircraft Engine Thrust Configuration

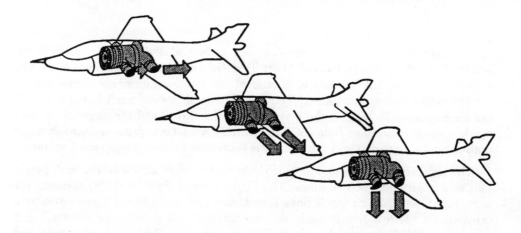

**Fig. 13.26** : Engine Thrust Vectors in Transition From Hover to Wingborne Flight

## Multivariable control aspects

Multivariable design techniques are needed for vectored thrust aircraft, particularly because of the high levels of dynamic coupling. There are also large changes in the aerodynamic characteristics which arise over relatively modest speed range changes. There is clearly a difficult transition to be managed between conventional flight and powered-lift and it entails a complex schedule of control actions. The aircraft model that will now be considered is for a generic canard-delta type of configuration, illustrated in Fig. 13.27.

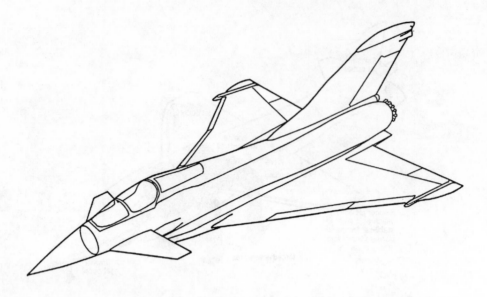

**Fig. 13.27** : Generic Canard-Delta Aircraft Configuration

The main benefits of a $H_\infty$ controller, over a classical design, are apparent in the multivariable case. The Advanced Short Take Off and Vertical Landing (ASTOVL) aircraft of interest has two sets of nozzles at the front and the rear of the aircraft through which thrust is obtained. The front nozzle angles are fixed but the aircraft can be controlled by varying the magnitude of the thrust and the angle of the rear nozzle. In the transition from jet born to fully wing born flight no aero-dynamic control can be employed and the aircraft is supported by the engine thrust vectors.

The rigid aircraft dynamics include those for the pitch axis together with height rate and longitudinal acceleration. The Flight Control System (FCS) converts the three input demands for pitch rate, height rate and longitudinal acceleration into demands on the rear nozzle angle and the forward and rear engine thrusts. The thrust from the forward and rear nozzles, together with the rear nozzle angle, can be transformed using a 3×3 force transformation matrix into total forces along the longitudinal and perpendicular axis and the pitching moment. This combination of forces provides the inputs to the rigid aircraft model. The block diagram of the system is illustrated in Fig. 13.28.

### Axis transformation matrix

The axis transformation matrix is a 3×3 matrix which transforms pilot demands for pitch attitude, height rate and longitudinal acceleration into demands on rear nozzle angle, forward and rear thrusts. It effectively performs the transformation from the earth axis to the engine axis.

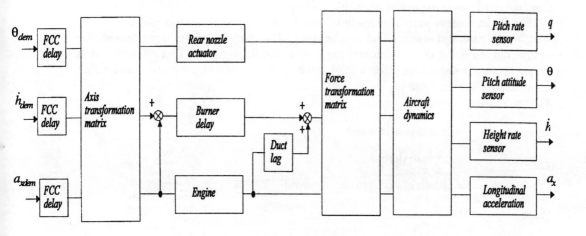

**Fig. 13.28** : ASTOVL Aircraft and Engine Model

### Force transformation matrix

The force transformation matrix which resolves the forward and rear thrusts into an axial force, a normal force and a pitching moment, has the form:

$$
\begin{bmatrix} X_{force} \\ Z_{force} \\ M \end{bmatrix} = F \begin{bmatrix} \theta_R \\ T_F \\ T_R \end{bmatrix}
\tag{13.6}
$$

where the forces and moments are as illustrated in the aircraft diagram of Fig. 13.29.

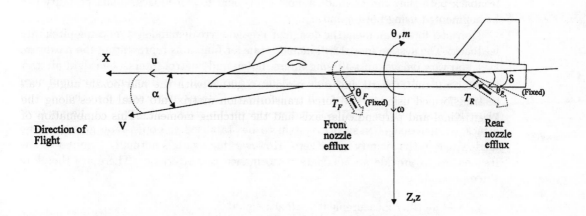

**Fig. 13.29** : ASTOVL Aircraft Thrust Actuator Arrangements
(Forces, X, Y, Z ; displacements, x,y,z)

**Control and measurement signals**

The flight control system considered has three input reference demands of pitch rate, height rate and longitudinal acceleration. The control system must provide the control demands for the rear nozzle angle and the forward and rear thrusts. Measurements are obtained of pitch angle, pitch attitude, height rate and longitudinal acceleration. To summarise :

*Reference signal demands* :

- pitch attitude demand

- height rate demand

- longitudinal acceleration demand.

*Plant outputs*:

- height rate

- pitch attitude

- pitch rate

- longitudinal acceleration.

## 13.3.3   Problems In Flight Control Systems Design

The principle role of the flight control system is the stabilisation of the unstable airframe dynamics. To stabilise the system stability derivatives must be augmented, particularly $M_\omega$ which represents the change in pitching moment with vertical air speed. This can be achieved using proportional (angle of attack) feedback. There exists a proportional relationship between $q$ (pitch rate), $n_z$ (normal acceleration) and $\alpha$ (angle of attack). If the proportionality constants for $q$ and $n_z$ are included in the feedback path they can be made almost equivalent to $\alpha$ feedback. Thus, stability can be augmented using both $q$ and $n_z$.

Consider for the moment the design of a manoeuvre demand system using pitch rate feedback. The system model will include transfer functions representing the hardware lags, software delays, anti-aliasing filters, notch and control filters. The rigid aircraft dynamics of interest are unstable and have one small low frequency non-minimum phase zero. This zero is a function of the angle of attack and starts off in the left-half plane for small angles of attack and moves to the right half-plane, as the angle of attack is increased. For such a system to be stabilised the open-loop gain must be made small in the vicinity of the zero. However, high gain is normally required at low frequencies to provide an adequate performance characteristic. There are therefore three possibilities:

1. Add another measurement feedback signal.

2. Relax the requirement for closed-loop stability.

3. Expolit the multivariable structure.

The problem with the former case is that pilots like the use of a pitch rate demand system. The second option, at first sight, appears to be dangerous but is sometimes employed in practice. A very small unstable mode in the vicinity of the non-minimum phase zero is allowed, with the knowledge that the pilot can always correct for such a small unstable mode. This rather unlikely situation allows much better transient responses to be obtained from the system, whether it be designed by classical or more advanced methods. If, for example, the open-loop gain is defined to be high at low frequencies this will result in an unstable closed loop system but the unstable mode will be at very low frequencies, where the pilot, or autopilot, can easily correct for any drift. As noted, the stability specifications for the different motion control loops, based on military specifications, are $\pm$ 6 dB with 45 degrees phase margin.

**Aircraft Model Dynamics and the Design Problem**

The main difficulty in the FCS design is the near pole-zero cancellations which occur between non minimum phase zeros and open loop unstable plant poles which are close to the $j\omega$ axis. Inspection of the transfer functions for each of the paths in the 4×3 transfer function matrix for the system reveals the presence of these small pole-zero terms. The most difficult loop to control is the longitudinal acceleration which has a small gain at low frequency and a small Non-Minimum Phase (NMP) zero. Attempting to push up the gain in this loop at lower frequencies results in poor step responses and sensitivity because of the presence of the NMP zero. This is therefore an unusual problem from the viewpoint that the multivariable nature of the system can actually assist the designer by giving additional degrees of freedom to mitigate the difficulties due to the NMP zero.

As discussed in Breslin et al. (1996), there are several limitations to achieving good step responses for systems which have small non-minimum phase zeros. The longitudinal acceleration loop is particularly difficult to stabilise because of the near pole-zero cancellation of the zero which is very small and non-minimum phase. A possible design approach is to remove these pole zero terms, which are very close and complete a design based on the resulting system. When the controller is used in the actual system the closed-loop will have a very slow unstable mode, close to the non-minimum phase zero. However, much improved step responses can be obtained by this procedure. The loop is of course closed by the pilot and the slow stable drift is corrected through this outer (human) mechanism. In the problem considered here this procedure was unnecessary because, as explained, the multivariable nature of the system mitigated the problem.

## 13.3.4 The Aircraft Model

The aircraft model of interest represents the dynamics of the generic, canard-delta configuration similar to the aircraft shown in Fig. 13.27. The control surfaces employed include:

1. Full span trailing edge flaperons.

2. Foreplanes (or canards).

3. Rudder.

4. Leading edge droop.

The trailing edge flaperons are used symmetrically for primary pitch control and are also used asymmetrically for roll motion control. The foreplanes are used for pitch control in combination with the symmetric flaperon actions. The *foreplanes* also enable a drag compensation function to be introduced by appropriate scheduling with the flaperons. This provides a degree of performance optimisation. The rudder provides directional stabilisation and control functions.

The aircraft model of interest has been deliberately destabilised in pitch by locating the centre of gravity aft of the centre of pressure. For the present purposes it will be considered that the pitch of the aircraft can be controlled using the flaps alone and the canards will be assumed fixed. A linear model of the aircraft will be used which represents the decoupled longitudinal dynamics. The model for the system is multivariable with reference signal inputs corresponding to pitch attitude, height rate and longitudinal acceleration. The system has four outputs which can be measured including height rate, pitch rate, pitch attitude and longitudinal acceleration, as illustrated in Fig. 13.30.

The aircraft model describes a complete pitch axis system with flap and foreplane angles fixed. In hover, the aircraft is controlled using vectored thrust produced from the engine nozzles at the front and the rear of the aircraft. The thrust is split with one part being directed through the rear nozzle at a variable angle, $\theta_R$ to the horizontal, producing a thrust $T_R$, and the remainder through the front nozzle included at a fixed angle to the horizontal, $\theta_F$ producing a thrust $T_F$. The forward thrust can be augmented with plenum chamber burning (PCB) when required. The aircraft is controlled by varying the rear nozzle and the magnitudes of the thrusts. The pilot demands for pitch attitude, height rate and longitudinal acceleration are converted into demands on rear nozzle angle, forward and rear thrusts, through the 3×3 force transformation matrix shown in Fig. 13.28. The thrusts produced by the engine are then resolved via the force transformation matrix into a normal force, an axial force and a pitching moment which are the inputs to the aircraft. In this way the actuator channel is effectively decoupled.

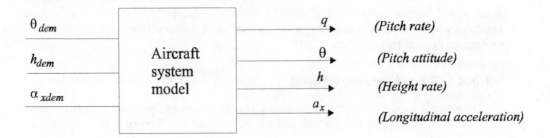

**Fig. 13.30** : Three Input and Four Output Aircraft Model

The model represents the linearised, rigid-body dynamics of the canard-delta configured aircraft. The model is linearised for straight and level flight at 100 ft, Mach

No. 0.151 and 6 degrees angle-of-attack. The dynamics at this operating condition are unstable. The full order model is 30th order and is open-loop unstable. It includes the flight control computer delays and the sensor hardware. There are three pilot demands for pitch attitude ($\theta$), height rate ($\dot{h}$) and longitudinal acceleration ($a_x$). There are four outputs available for feedback purposes including pitch rate ($q$), pitch attitude ($\theta$), height rate ($\dot{h}$) and longitudinal acceleration ($a_x$). Here $q$ is in radians/second, $\theta$ is in degrees and the height rate $\dot{h}$ is in metres/second. The plant model is only 4th order and is represented by the following continuous-time state equation matrices.

**Aircraft 4×3 Plant Model State Matrices**

$$
A = \begin{bmatrix} -0.0017 & 0.0143 & -5.3257 & -9.7565 \\ -0.0721 & -0.3393 & 49.5146 & -1.0097 \\ -0.0008 & 0.0138 & -0.2032 & 0.0009 \\ 0 & 0 & 1.0000 & 0 \end{bmatrix}, \quad
B = \begin{bmatrix} 0.2086 & -0.0005 & -0.0271 \\ -0.0005 & 0.2046 & 0.0139 \\ -0.0047 & 0.0023 & 0.1226 \\ 0 & 0 & 0 \end{bmatrix}
$$

$$
C = \begin{bmatrix} 0 & 0 & 57.2958 & 0 \\ 0 & 0 & 0 & 57.2958 \\ 0.1045 & -0.9945 & 0.1375 & 51.3791 \\ -0.0002 & 0.0045 & 0 & 0 \end{bmatrix}, \quad
D = \begin{bmatrix} 0 & 0 & 0 \\ 0 & 0 & 0 \\ 0 & 0 & 0 \\ 0.0212 & 0 & 0 \end{bmatrix}
$$

**Parametric uncertainty**

Parametric uncertainty arises in the flight control problems in three areas:

1. The engine and actuator models.

2. The force transformation matrix.

3. The aero-dynamic derivatives.

There are also unmodelled dynamics which can be included through the unstructured uncertainty models. In this particular case a multiplicative uncertainty model for both the actuators and sensors may be employed.

**The engine and actuator model**

The engine and actuator models are shown in Fig. 13.31. As noted, the aircraft is controlled by varying the angle of the rear nozzle actuator and the dynamics of the rear nozzle actuator may be described by a second-order lag function. The thrust demand is converted into an engine speed demand before it is passed through a second-order dynamic representation of the engine model. The plenum chamber burning and the duct lag are represented by first-order delay functions. The forward and rear thrusts are resolved into an actual force, a normal force, and a pitching moment which act as the inputs to the ridged aircraft dynamic model.

Uncertainties in the first order lags representing the plenum chamber burning, duct lag and engine time delay are expressed as tolerances on the time constants of each element. The uncertainty in the rear nozzle actuator and engine dynamics are represented by tolerances on their natural frequencies and damping ratios.

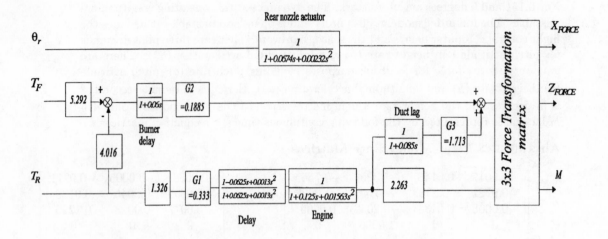

Fig. **13.31** : Block Diagram of Engine and Actutor Models

**Disturbance model**

As with previous work it is convenient to consider the disturbances from a stochastic point of view, using white noise into appropriate colouring filters to generate the disturbance spectrum. In the absence of good disturbance model information it was decided to derive the disturbance by inserting white noise (scaled by a gain matrix) into an integrator for each channel of the inputs. It therefore provides a pragmatic starting point for disturbance modelling which recognises that steady-state or low frequency disturbances must be offset and has the virtue that it does not increase significantly the number of states in the system. This latter point was important in this design problem, where the system was of relatively high order. The commercial packages often ran into computational problems, partly due to the near pole-zero cancellations which occurred. The disturbance model had the form:

$$W_d = diag\{0.01, \quad 0.01, \quad 0.01\}/s$$

Similarly, the reference model:

$$W_r = diag\{10^{-4}, \quad 1, \quad 1, \quad 0.1\}$$

**Combined system model**

The total system for the aircraft model, disturbances and cost-function weightings is shown in Fig. 13.32. The state space model shown in this Figure represents a more detailed view of the system shown in the block diagram form in Fig. 13.30. The error, control and output weightings have the outputs $z_1$, $z_2$ and $z_3$, respectively. The system can be put in standard system model form, as discussed in Chapter 9 and the $H_\infty$ controller can then be calculated using commerical software.

The $H_\infty$ controller for this system and dynamic cost-function weightings and disturbance model is 42nd order. This can be reduced significantly with little loss of performance and robustness. The uncertainty in the aircraft model which is taken into account, stems from uncertainty in the stability and control derivatives of the equations of motion.

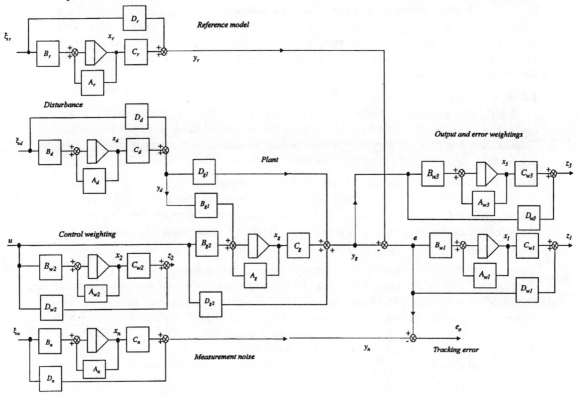

Fig. 13.32 : **Disturbance, Plant, Noise, Reference and Weighting Models**

## 13.3.5 Flight Control Loop Design

The design procedure followed here was truly multivariable but the classical approach used depends upon successive loop closures (Breslin et al. 1996). It is of interest to consider each of the individual loop control requirements and problems. The system is 4×3 and the loop formed from output 1 to input 1 is treated as a regulating rather than a tracking problem. That is, the output is not required to follow a reference signal.

In classical control design the open loop unstable aircraft model is normally first stabilised by closing the pitch rate feedback loop. Integral action in the pitch rate loop is not necessary since the pitch rate feedback is only used for stabilisation purposes.

### Loop 1 : Output 1 / Input 1 : Height Rate Control

The reference for the output 1 should really be set to zero. However, to satisfy the rank condition on the standard system model matrix $D_{21}$, either the reference or the disturbance models must have some white noise component on each system output. This is necessary for the Doyle, Glover, Kharnegor and Francis (1989) state-space $H_\infty$

solution to hold.  Since the disturbance model was only added to each of the 3 control inputs it did not provide such a term, and hence the reference model was assumed to have a small white noise component on all the outputs (being particularly small on the output one).  This particular output was not to be controlled in this study and hence the error weighting term on output one was also assumed to be small.  By this means the importance of the loop 1 controller was downgraded and its role was simply for stability enhancement.

### Loop 2 : Output 2 / Input 1 : Pitch Angle Control

A good step response performance for this loop is required and fortunately it is not one of the loops which is the tightest coupled for the model used. Like the remaining loops a fast overdamped response with a small steady-state error is required.  The interactions between loops must also be minimised.

### Loop 3 : Output 3 / Input 2 : Pitch Rate Output

The control of pitch rate is a more difficult problem and this is tightly coupled to the longitudinal acceleration loop. The high order and the non-minimum phase terms complicate this design but it was possible to achieve a good solution, satisfying the step-response requirements, with a reasonable response time.

### Loop 4 : Output 4 / Input 3 : Longitudinal Acceleration

The final loop is the most difficult to control adequately. There were problems in choosing the weightings, to obtain reasonable transient responses and to achieve the desired steady-state values.  It has already been noted that it is difficult to increase the gain at low frequencies when the plant contains a small NMP zero.  However, the coupling between loops was used to advantage in this problem. By choosing the output weightings for outputs 3 and 4 in an interactive manner (tuning by alternating between the loops) good results were obtained with only a small steady-state error. The size of the steady state error is not of major importance in this problem.

### $H_\infty$ design and weighting selection

The general guidelines normally used in $H_\infty$ design were applied in this problem. The error (or sensitivity) weighting was chosen to have high gain at low frequency and the control weightings were chosen to have high gains at high frequencies. These provided small sensitivities at low frequencies and adequate roll-off for the controller at high-frequencies, respectively. The fine tuning of the weightings required an iterative design procedure (Grimble, 1993).

**Error weighting:**

$$H_q = diag\{H_{q1},\ H_{q2},\ H_{q3},\ H_{q4}\} \tag{13.7}$$

where

$$H_{q1} = 0.01(s+1)/(s+10^{-4})\quad,\quad H_{q2} = 10(s+1)/(s+10^{-6})$$

$$H_{q3} = 10(s+2)/(s+10^{-4})\quad,\quad H_{q4} = 20(s+1)/(s+0.01)$$

**Control weighting:**

$$H_r = diag\{H_{r1},\ H_{r2},\ H_{r3}\}$$

where

$$H_{r1} = 0.05(s^2 + 2s + 1)/(10^{-6}s^2 + 2 \times 10^{-3}s + 1)$$

$$H_{r2} = 0.5(10^{-4}s^2 + 0.003s + 0.0225)/(10^{-6}s^2 + 2 \times 10^{-3}s + 1)$$

$$H_{r3} = (0.0002s^2 + 0.004s + 0.02)/(10^{-6}s^2 + 2 \times 10^{-3}s + 1)$$

Observe that the error weightings mainly involve approximate integrators with the exception of the 4th output which represents acceleration and where steady state error need not be zero. The lead term on the error weighting is to ensure the error is weighted in the mid-frequency region as well as at low frequencies. The control weighting functions introduce mid-frequency and high frequency control costing to ensure the controller rolls off at high frequencies. The high frequency pole terms in these weighting functions were in fact unnecessary, but were introduced since in the state-space description the system description must be proper.

### $H_\infty$ Flight Control Problem Results

The dynamic responses of the closed-loop system to the four outputs are illustrated in Figures 13.33 to 13.36. The time response plots represent the following signals : Pitch rate = 1, pitch angle = 2, height rate = 3,  acceleration = 4.  The pitch rate response in Fig. 13.33 is rather poor but this loop is not to be controlled. However, it was useful to test the response of the loop to a step function, even though this is an unrealistic reference or disturbance change.

The pitch angle time response shown in Fig. 13.34 is very much more important and in this case there is a good well damped response with relatively modest interaction to the other loops. A unit-step reference change in pitch angle was made and the steady-state tracking error clearly goes to zero. The height rate is also important and this is shown in Fig. 13.35. Once again low interaction occurs and good steady-state tracking.

The response to the reference demand in longitudinal acceleration is shown in Fig. 13.36. There is a small steady-state error which is not so important in this loop and relatively small interaction occurs once again.

A more detailed analysis of the design results is provided in the thesis by Breslin (1996). The work by Breslin also considers the addition of different types of uncertainty to the system description and compares the results with the $\mu$-synthesis problem results. The thesis by Steen Toffner-Clausen (1995) also considers this problem and provides a realistic design procedure. However, the order of the $\mu$-synthesis controller (60th order) is far too high, which is of course a common problem with this approach. The $H_\infty$ controllers, when reduced, are of a much more realistic order (Typically 14th). It is rather incongruous that although $H_\infty$ design has always been promoted as a robust design philosophy, in fact the results are often used to achieve good performance requirements. In this particular problem the decoupled time responses of the aircraft are equally as important as robustness.

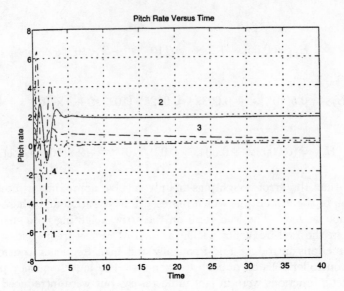

**Fig. 13.33** : Pitch Rate Time Response (Pitch rate = 1,

pitch angle = 2, height rate = 3, acceleration = 4)

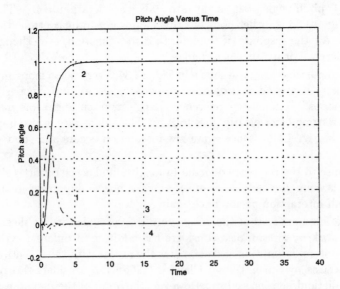

**Fig.13.34** : Pitch Angle Time Response to a Unit Step change in

Pitch Angle Demand (Pitch rate = 1, pitch angle = 2,

height rate = 3, acceleration = 4)

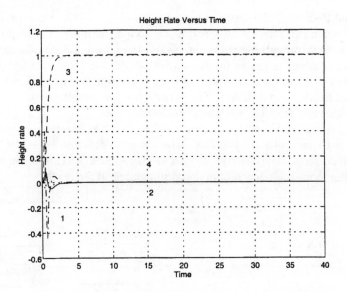

**Fig. 13.35** : Height Rate Time Response to a Unit Step Change in Height Rate
Demand
(Pitch rate = 1, pitch angle = 2, height rate = 3, acceleration = 4)

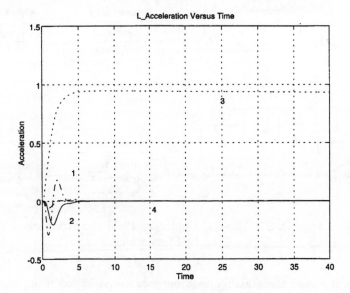

**Fig. 13.36** : Longitudinal Acceleration Time Response to a Unit Step Reference
Demand
(Pitch rate = 1, pitch angle = 2, height rate = 3, acceleration = 4)

**Classical Design**

A classical controller can be designed for this model using successive loop closures
and pitch rate feedback to stabilise the unstable dynamics. The resulting controller

has a diagonal P+I controller structure, as shown in Fig. 13.37. A precompensator can also be used which can attempt to decouple the loops. Although the step responses for this system achieve acceptable transient results they do not decouple the height rate and longitudinal acceleration as much as a true multivariable controller can achieve.

In a classical flight control system PI controllers are employed with pitch rate feedback used for the initial design. The other loops for the pitch attitude, height rate and longitudinal acceleration, can then be closed.

The nonlinear model used for design was linearised at Mach number 0.151, angle of attack $\alpha = 6^\circ$ and altitude 100 feet. The classical controller was designed by the method of successive loop closures. Pitch rate feedback was used for stabilisation. The remaining loops included proportional plus integral feedback.

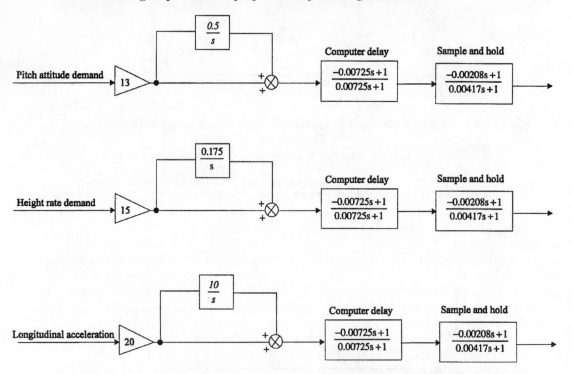

**Fig. 13.37**: Classical Controller PI Model Structure
Used for Simulation

### Stability

In classical design the stability requirements are specified using bounds on the open-loop Nichols chart which give $\pm$ 6dB and $45^\circ$ phase margin. Handling qualities for an unstable aircraft are rather different from those of a conventional aircraft. The handling qualities criterion can be specified via a Nichols plot template for the stick to pitch attitude response.

### Handling Qualities

The main design objective for this example was to generate good step responses to improve the handling qualities of the aircraft. Since time-domain transient responses often reflect the frequency domain sensitivity function responses this also implies, but does not guarantee, that reasonable frequency responses will be obtained. The frequency responses of the sensitivity functions were of course also checked by plotting their singular values.

## Summary of Classical Design Procedure

(a) Use rate feedback $q$ to stabilise the unstable modes and achieve good damping and stability margins.

(b) Design the pitch attitude $\theta$ feedback loop to obtain a deadbeat attitude response with a settling time of approximately 2 seconds, ensuring the stability margins remain satisfactory.

(c) Design the height rate feedback loop to obtain a deadbeat response with a settling time of about 2 seconds and attempt to obtain decoupled responses.

(d) Close the acceleration $a_x$ feedback loop and obtain a deadbeat response with a settling time of about 2 seconds that is decoupled.

(e) Assess the stability margins and the time responses for the feedback system.

### Classical Design Results

The unit step responses of the 3×3 classically designed feedback loop are shown in Fig. 13.38. These represent an optimised design and indicate good time responses but with rather poor interaction in some cases. The height rate response includes a negative step because of the sign notation used.

The classical solution was reasonably robust to changes in the system model. The relatively low order of the classical design appears to be an advantage in this situation, since the gain and phase shifts in the frequency range, around the unity gain crossover frequency, are not so large as in $H_\infty$ design. However, achieving good designs by classical methods is very difficult and time consuming. If the plant model changes there is also a lengthy redesign exercise. An advantage of the $H_\infty$ approach is that the $H_\infty$ design procedure can be formalised, so that the engineer can follow a *cook book* procedure to complete a redesign (in the case of changes).

### $\mu$ Analysis in Flight Control System Design

To use $\mu$ analysis the uncertainties in the system model are represented by individual models and collected together into one diagonal matrix whose magnitude has been normalised to be less than unity. Singular value robustness tests are based on the small gain theorem which refers to stable transfer functions in a closed-loop system. The closed-loop will remain stable if the magnitude of the product of the two transfer functions is less than unity. Results indicate that the classical and $H_\infty$ designs are insensitive to parametric uncertainties except at the rear nozzle actuator. The $H_\infty$ design was found to be more robust to multiplicative uncertainty at the plant output.

### Typical Civil Aircraft Utilities

There are many other systems engineering problems in civil aircraft, in addition, to the flight control system, that involve some aspect of control design including: Fuel systems, Environmental control systems, Electrical systems (*generation - control - distribution*), Hydraulic systems, Secondary flight control system, De-icing system,

Passenger systems (*galley - lighting- entertainment*), Engine bleed air system, Freight system, Emergency systems, Handling systems (*door actuation*).

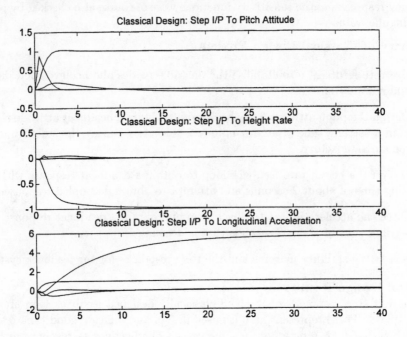

**Fig. 13.38** : Unit Step Responses of a Classical Control Design with Steps into Pitch Attitude, Height Rate and Longitudinal Acceleration

## 13.4   Concluding Remarks

The gas turbine and the flight control design studies have some similar common elements. In both cases the classical designs gave reasonable results, but the benefits of the $H_\infty$ design philosophy was that it coped with the multivariable nature of the problem more directly. Low order classically designed controllers do have some advantages for implementation and better robustness than might be expected. However, the chances of providing an automated design procedure, are much greater for the $H_\infty$ approach. This enables $H_\infty$ designs to proceed quickly, and to some extent the skill level needed is less. This benefit arises because questions of interaction and stability are accommodated naturally in the $H_\infty$ approach.

In the classical design method it was more difficult to achieve decoupled responses and adequate stability margins. In the flight control system design example, very good transient responses were obtained by the $H_\infty$ approach. In fact the classical designs were achieved by engineers having a lot of previous experience. However, to some extent the improvement in the transient characteristics of the $H_\infty$ solutions is illusory, since robustness requirements would probably demand some reduction in the gains employed, so that unexpected uncertainties could be accommodated.

Experience suggests the QFT design approach in Chapter 9 is also an excellent candidate for aerospace applications. The original work by Horowitz (1973), that was expanded on by other reseachers, such as Houpis and Rasmussen (1999), was directed at flight control problems (Breslin and Grimble, 1997; McLaren et al. 1998). There is also the exciting possibility of using QFT for *fault tolerant control*, in for example remote pilotless vehicles (Wu et al. 1999). The QFT design method (Horowitz, 1973) has also been used successfully in other applications such as an optical disk drive control. These are employed in advanced fighter aircraft to store terrain following information (Hearns and Grimble, 1999). However, QFT is most valuable for single-input single-output robust control designs. It is not so appropriate for providing an all encompassing integrated flight control and engine management system In fact other frequency domain multivariable design methods, such as *Individual Channel Design* may be more relevant for such applications (Akbar et al. 1993). The $H_\infty$ approach is suitable for such large integrated systems and was therefore preferred for the above applications.

# 13.5 References

Akbar, M.A., Leithead, W.E. and O'Reilly, J., 1993, Design of robust controllers for a fighter aircraft using individual channel design, *Proc. of 32nd CDC Conference*, San Antonio, Texas, pp. 430-435.

Athans, M., Kapasiouris, P., Kappos E. and Spang, H.A., 1986, Linear-quadratic Guassian with loop transfer recovery methodology for the F-100 Engine, *Journal of Guidance, Control and Dynamics*, Vol. **9**, No. 1.

Binns, J. and North, M., 1991, The design of multivariable controller for gas turbine, *Proceedings of Control 91*, Edinburgh, UK.

Blakelock, J.H., 1991, *Automatic control of aircraft and missiles*, Second Edition, John Wiley and Sons, Chichester.

Breslin, S.G., 1996, *On aircraft flight control*, PhD Thesis, University of Strathclyde, Glasgow.

Breslin, S.G. and Grimble, M.J., 1997, Longitudinal control of an advanced combat aircraft using quantitative feedback theory, *American Control Conference*, New Mexico, pp. 1-6.

Breslin, S.G., Toffner-Clausen, S., Grimble, M.J. and Andersen, P., 1996, Robust control of an ill-conditioned aircraft, *IFAC World Congress*, San Francisco.

Chen, F. and Khalil, H.K., 1992, Adaptive control of non-linear systems using neural networks, *Int. J. Control.* Vol. **55**, No. 6, pp. 1299-1317.

Doyle, J.C., Glover, K., Khargonekar P.P. and Francis, B.A., 1989, State-space solutions to standard $H_2$ and $H_\infty$ control problems, *IEEE Trans on Automatic Control*, **34**, 8, pp. 831-846.

El-Sheikh, G., Grimble, M.J. and Johnson, M.A., 1994, On the performance of $GH_\infty$ self-tuning for aeroengine control, *IEE Control 94 Conference*, Warwick University.

Frederick, D., Garg, S. and Adibhatla, S., 1996, Turbo fan engine control design using robust multivariable control technologies, 32nd Joint Propulsion Conference, Lake Buenavista, Florida.

Garg, S. and Mattern, D., 1994, Application of an integrated methodology for propulsion and airframe control design to a STOVL aircraft, *Guidance, Navigation and Control Conference*, American Institute of Aeronautics and Astronautics, Scottdale, Arizona.

Garg, S., 1993, Robust integrated flight/propulsion control design, for a STOVL aircraft using $H_\infty$ control design techniques, *Automatica*, Vol. **29**, No. 1, pp. 129-145.

Garg, S., 1997, A simplified scheme for scheduling multivariable controllers, *IEEE control systems magazine*, pp. 24-30.

Glover, K. and McFarlane, D., 1989, Robust stabilization of normalized coprime factor plant descriptions with $H_\infty$ bounded uncertainty, *IEEE Trans. AC*, Vol. **34**, No. 8, pp. 821-830.

Grimble, M.J., 1993, Weighting selection and robustness of $H_\infty$ designs, *IEEE Mediterranean Symposium on New Directions in Control Theory and Applications*.

Hearns, G. and Grimble, M.J., 1999, Limits of performance of an optical disk driver controller, *Proceedings of the 18th American Control Conference*, San Diego.

Horowitz, I.M., 1973, Optimum loop transfer function in single-loop minimum phase feedback systems, *Int. J. Control*, Vol. **22**, pp. 97-113.

Houpis, C.H. and Rasmussen, S.J., 1999, *Quantitative feedback theory, Fundamentals and applications*, Marcel Dekker, New York.

McLaren, I., Grimble, M.J. and Breslin, S.G., 1998, QFT flight control system design using Mathematica, *Control 98*, University of Wales, Swansea.

Moellenhoff, D.E. and Rao, S.V., 1990, Design of robust controllers for gas turbine engines, *The Gas Turbine and Aeroengine Congress and Exposition*, Brussels, June 11-14.

Pfeil, M., Athans, M. and Spang, H.A. 1986, Multivariable control of GE T700 engine using the LQG/LTR design methodology, *American Control Conference*, Seattle, Washington.

Postlethwaite, I., Samar, R., Choi, B-W. and Gu, D.W., 1995, A digital multi-mode $H_\infty$ controller for the SPEY turbo fan engine, *Proceedings of the 3rd European Control Conference*, Rome, Italy, pp 3881-3886.

Rao, S.V., Moellenhoff, D. and Jaeger, J.A., 1988, Linear state variable dynamic model and estimator design for Allison T406 gas turbine engine, *ASME Gas Turbine and Aeroengine Congress*, Amsterdam, The Netherlands.

Song, Q., Wilkie., J. and Grimble, M.J., 1994, An integrated robust/neural controller with gas turbine applications, *IEEE Conference on Control Applications*, Glasgow.

Song, Q., Wilkie, J. and Grimble, M.J., 1993, Robust controller for gas turbines based upon LQG/LTR design with self-tuning features, *Journal of Dynamic Systems, Measurement and Control*, Vol. **115**.

St. John-Olcayto, E.,Wilkie, J. and Grimble, M.J., 1996, Computer control design and simulation of a gas turbine aeroengine, *Proceedings of Conference on Computers in Reciprocating Engines and Gas Turbines*, IMA., London.

St.John-Olcayto, E., Wilkie, J. and Grimble, M.J., 1994, Aeroengine thrust modulation and surge margin regulation using robust, multivariable control, *IEE Symposium, Multivariable Methods for Flight Control Applications*.

Toffner-Clausen, S., 1995, *System identification and robust control, A synergistic approach*,PhD Thesis, Electical and Electronic Eng., Dept. of Control Eng., Aalborg, University Denmark

Wilkie, J., and St.John Olcayto, E., 1992, Comparison of modern and classical techniques for aeroengine control *I MechE Symposium*, Controls for Engines.

Watts, S.R. and Garg, S., 1995, A comparison of multivariable control design techniques for a turbo fan engine control, *40th Gas Turbine and Aeroengine Congress and Exposition*, American Society of Mechanical Engineers, Huston, Texas.

Wu., S., Grimble, M.J. and Wei, W., 1999, QFT based robust/fault tolterant flight controller design for a remote pilotless vehicle, *IEEE Conference on Control Applications*, Kona, Hawaii.

# Index